School Mathematics Textbooks in China

Comparative Study and Beyond

East China Normal University Scientific Reports
Subseries on Data Science and Engineering

ISSN: 2382-5715

This book series reports valuable research results and progress in scientific and related areas. Mainly contributed by the distinguished professors of the East China Normal University, it will cover a number of research areas in pure mathematics, financial mathematics, applied physics, computer science, environmental science, geography, estuarine and coastal science, education information technology, etc.

Published

Vol. 13 *Load Balance for Distributed Real-time Computing Systems*
by Junhua Fang (Soochow University, China),
Rong Zhang (East China Normal University, China) and
Aoying Zhou (East China Normal University, China)

Vol. 12 *Biological Language Model: Theory and Application*
by Qiwen Dong (East China Normal University, China),
Xiaoyang Jing (Yunnan University, China),
Aoying Zhou (East China Normal University, China) and
Xiuzhen Hu (Inner Mongolia University of Technology, China)

Vol. 11 *Probabilistic Approaches for Social Media Analysis: Data, Community and Influence*
by Kun Yue (Yunnan University, China), Jin Li (Yunnan University, China), Hao Wu (Yunnan University, China), Weiyi Liu (Yunnan University, China) and Zidu Yin (Yunnan University, China)

More information on this series can also be found at https://www.worldscientific.com/series/ecnusr

(Continued at end of book)

East China Normal University Scientific Reports | Vol. 2

School Mathematics Textbooks in China

Comparative Studies and Beyond

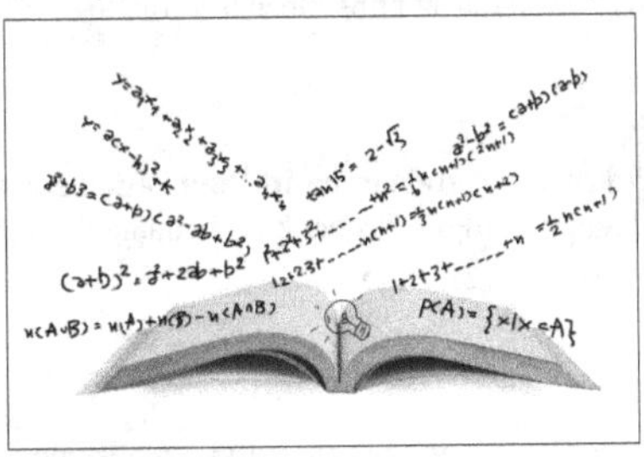

Editor

Jianpan Wang

East China Normal University, China

Co-editors

Lianghuo Fan

Binyan Xu

East China Normal University, China

World Scientific

NEW JERSEY • LONDON • SINGAPORE • BEIJING • SHANGHAI • HONG KONG • TAIPEI • CHENNAI • TOKYO

Published by

World Scientific Publishing Co. Pte. Ltd.

5 Toh Tuck Link, Singapore 596224

USA office: 27 Warren Street, Suite 401-402, Hackensack, NJ 07601

UK office: 57 Shelton Street, Covent Garden, London WC2H 9HE

Library of Congress Cataloging-in-Publication Data

Names: Wang, Jianpan, 1949– editor.

Title: School mathematics textbooks in China : comparative studies and beyond / editor, Jianpan Wang, East China Normal University, China.

Description: New Jersey : World Scientific, [2021] | Series: East china normal university scientific reports, 2382-5715 ; vol 2 | Includes bibliographical references.

Identifiers: LCCN 2020051742 | ISBN 9789814713931 (hardcover) | ISBN 9789814713948 (paperback) | ISBN 9789814713955 (ebook for institutions) | ISBN 9789814713962 (ebook for individuals)

Subjects: LCSH: Mathematics--Study and teaching--China. | Mathematics--China--Textbooks.

Classification: LCC QA14.C6 S36 2021 | DDC 510.71/051--dc23

LC record available at https://lccn.loc.gov/2020051742

British Library Cataloguing-in-Publication Data

A catalogue record for this book is available from the British Library.

For any available supplementary material, please visit https://www.worldscientific.com/worldscibooks/10.1142/9753#t=suppl

Desk Editor: Liu Yumeng

https://doi.org/10.1142/9789814713955_fmatter

Contents

Chapter 15

https://doi.org/10.1142/9789814713955_fmatter

Preface

The articles in this collection mainly come from a national key educational research project — *Comparative Study of High School Mathematics Textbooks of Certain Countries* (hereinafter referred to as Comparison Project), which is a key educational project of 2010 under the Eleventh Five-Year Plan of the National Social Science Foundation. Also, a few articles on mathematics textbook studies from relevant domestic teams and research projects are also selected.

1

The Ministry of Education of China issued a *High School Mathematics Curriculum Standards* (hereinafter referred to as 2003 Standards) in April 2003, and based on the standards, different publishers published several sets of textbooks afterwards. In the following years, different areas of China successively moved onto the experimental and implementation phase guided by the 2003 Standards. Based on the textbooks, teachers made great efforts in their teaching practices to implement the ideal goal of the high school mathematics curriculum, and the relevant departments of the Ministry of Education paid close attention to the curriculum implementation progress. In order to summarize the experiences and to drop lessons, as well as to prepare for the modification of the 2003 Standards, the Ministry of Education convened researchers to study the textbooks by project bidding and entrusted professional teams to carry out large-scaled investigation. The Comparison Project, led by Jianpan Wang from East China Normal University (ECNU), is one of the main research projects, which was divided into ten subprojects, carrying out international comparative studies on

— the organizations and presentation patterns of core contents,
— the development of basic concepts and skills,
— the characteristics of problem-solving and arrangement of examples and exercises,
— the difficulty level and difficulty characteristic,
— the history of mathematics and mathematical culture in textbooks, and
— the application of ICT in textbooks.

We chose the United States, the United Kingdom, France, Germany, Russia, Japan and Singapore, along with China itself as the "certain countries". The research findings were successively published in periodicals at home and abroad, putting forward cutting-edge opinions and suggestions for the revision of Chinese curriculum standards and textbooks.

In 2012, entrusted by the Department of Basic Education of the Ministry of Education, a national survey team, also led by Jianpan Wang, conducted a large-scale investigation on the implementation of the 2003 Standards. The survey was mainly done by the methods of stratified sampling, in-depth investigation and interviews. The sampling covered seven provincial administrative areas with a reasonable geographical distribution and ethnic consideration. The interviewees included the members of the 2003 Standards developing group, the textbook compilers, mathematicians, mathematics teaching researchers, school leaders, school mathematics teachers, and high school students. The research was carried out with focus on the following aspects:

— objectives and value orientation of the curriculum,
— rationality of the curriculum framework,
— relevancy, difficulty and cohesion of the curriculum contents,
— implementation of the 2003 Standards, and
— readability and guidance of the texts of the 2003 Standards.

Textbook is the main path from curriculum standards to the classrooms, thus, interviewees put forward constructive opinions on how to reflect flexibility and diversity in compilation of textbooks in the context "one set of standards, multiple versions of textbooks". In view of the extensive use of "extracurricular teaching materials" and its relationship with textbooks, a profound view is also put forward. The rich, diverse and enlightening findings have a positive impact on the revision of the curriculum standards and related textbooks.

Internationally, the Second International Mathematics Study (SIMS, 1981–1982) has studied and discussed the influence of the participating states' mathematics textbooks on the learners' mathematical achievements. Later, in the Third International Mathematics and Science Study (TIMSS, 1995, 1999) and the Trends in International Mathematics and Science

Study (TIMSS, 2003, 2006, 2009), studies over all the participant states' mathematics textbooks as important background factors have been carried out, and a series of research frameworks, methods and tools have been correspondingly put forward. The researches by Chinese scholars at home on mathematics curriculum and textbooks started relatively late. It included introduction of theoretical studies in this direction from abroad such as translation of the *Curriculum Development in Mathematics* by Geoffrey Howson et al., published by Shanghai Education Publishing House in 1992, and independent research by Chinese researchers such as *Introduction to Middle School Mathematics Curriculum* edited by Ersheng Ding, published by Shanghai Education Publishing House in 1994. It is followed by certain preliminary international comparative studies, for example, *Mathematical Educational Comparison and Study* edited by Changping Chen from ECNU, published by ECNU Press in 1995. Since the turn of the century, a number of young students have been engaged in educational studies in mathematics with textbooks as the research object, and then completed their master theses or doctoral dissertations. Therefore, soon after the Comparison Project was initiated, the research team organized an international conference titled *International Conference on School Mathematics Textbooks* in October 2011 at ECNU. The invited speakers of the conference include Zalman Usiskin from the University of Chicago, Keith Jones from the University of Southampton, Frederick Leung from the University of Hong Kong, and Naiqing Song from the Southwest University. Professor Geoffrey Howson from University of Southampton, who had accepted the invitation but couldn't come due to health reason, videoed his report in advance and had an online conversation with participants. Howson reported the context and driving forces of curriculum textbook development and analyzed the trend of curriculum and textbook development in the future and possible study challenges with sharp but insightful views and elegant and humorous expressions. This international conference became a new booster for international textbook study. Thanks to the effort of Lianghuo Fan, then a professor from the University of Southampton, the international mathematics educational journal ZDM (Zentralblatt für Didaktik der Mathematik, now ZDM Mathematics Education) published a special issue on textbook research in October 2013, which presented the achievements of textbook researches including those by Chinese researchers, becoming an important archive for textbook research. As the continuation of *International Conference on School Mathematics Textbooks* at ECNU in 2011, a new series of international conference titled *International Conference on Mathematics Textbook Research and Development* has been designed and initiated by Lianghuo Fan. So far, three conferences were held successively at Southampton, UK (July 2014), Rio de Janeiro, Brasil (May 2017), and Paderborn, Germany (September 2019). Scholars from ECNU took active roles in these conferences by joining the program committees, giving plenary lectures, etc.

Thus, it can be seen that in the last two decades, textbook research has been a hot spot and highlight in international mathematics education research. In a sense, the mathematics education team of ECNU is an important initiator and promoter of the research upsurge.

An array of achievements was harvested during the research of the Comparison Project, with the ECNU team as the core in the research group. Meanwhile, a number of domestic counterparts dedicated to textbook research also obtained outstanding achievements. We believe it is necessary to invite colleagues within and outside the research group, within and outside ECNU, to write papers concerning school textbook research to display Chinese mathematics textbook research achievements and its characteristics in the international arena. Coincidentally, the Journal of East China Normal University and World Scientific intend to jointly publish a book series *East China Normal University Scientific Reports*. We took this opportunity and this platform to compile this book with textbook research as the theme.

2

This book consists of fifteen chapters, containing research achievements of three aspects:

— international comparison of high school mathematics textbooks,
— mathematics textbook analysis at the stage of compulsory education, and
— relevant special research of mathematics textbooks.

As most of the achievements collected in this book come from the Comparison Project, the papers in this book are mainly based on high school textbook research, nine chapters in total. Chapter 1 is the study of core knowledge selection in high school mathematics textbook design; the author is a senior editor of People's Education Press. The study is based on detailed data and rich experience, illustrating how to design textbooks with core mathematics knowledge as the focus. The chapter revolves around the questions such as what the selection standards of core contents are, what the core contents are and how core contents should be organized. The chapter puts forward relevant conclusions and points out that the selection of the core contents of high school mathematics should consider the demands of students' development, mathematics development and social development; the core knowledge of the high school mathematics textbook design should give full play to the self-growing ability of core concepts to establish a consistent and logically coherent study program for students; the textbook structure centering on core concepts should be designed to keep systematic and logically coherent.

Chapter 2 to Chapter 5 have mathematics content design as research objects and comparative study as the method, focusing on algebra, geometry, probability and statistics, and calculus, respectively. The chapters pay attention to the organization or presentation methods and the arrangement system of these contents. Through comparison, the characteristics and reasons of content arrangement in Chinese mathematics textbooks are highlighted. Chapter 2 compares the contents concerning complex numbers in chosen textbooks of China, Japan and Singapore. First, the research sets up an analytical framework on the basis of abundant documents, and then analyzes the characteristics of selection and layout of contents concerning complex numbers in the textbooks of different countries in detail, illustrating the features of content presentation of different textbooks. For instance, by establishing conceptual graphs of complex numbers in textbooks, it is found that the contents of complex numbers in Japanese mathematics books (JPB) not only covers complex numbers themselves, but also include the applications of complex numbers, while the latter does not appear in the other textbooks. Singaporean books (SGB) do not even talk about the connection of complex numbers and real numbers, which may cause problems for the overall grasp of the number systems for students.

Chapter 3 chooses high school mathematics textbooks from China and Germany and carries out a comparative study on the organization and presentation pattern of their contents of geometry. The methods used are mainly content analysis and descriptive statistics. It turns out that the contents of geometry in the LS edition of German textbooks are relatively narrow, but they are profound, well organized, emphasizing the applications in presentation, attentive to the cultivation to the comprehensive application capability of students. The geometry contents of Chinese PEP textbooks are relatively wide and more systematic, laying more emphasis on the integrity of mathematics knowledge and the cultivation of spatial imagination with intuitive perception of knowledge presentation.

Chapter 4 studies statistics and probability contents. The authors select six sets of textbooks from China, Japan, Singapore, the United States and the United Kingdom, and analyze respective characteristics in knowledge selection, task design and presentation patterns with binomial distribution as a sample. The study shows that Chinese textbooks do not lay due emphasis on binomial distribution, and they have only a small amount of study tasks with monotonous activity type. But Chinese textbooks pay attention to the design of knowledge leading-in. American textbooks, when it comes to the design of such statistics contents, lay emphasis on the connection between these contents and other knowledge with diversified activity types, making it convenient for students to learn from various perspectives through colorful activities. Singapore textbooks are unique due to their solving numerical problems and mathematics contents through a graphic calculator, and for giving detailed guidance for the operation of students.

Chapter 5 compares calculus content in high school textbooks from China, Japan, Singapore, Germany, France and Russia. Six sets of textbooks are selected from the six countries, aiming at probing into the positioning of teaching goals of calculus in high school curricula, selection of calculus contents and the balance between the scientific and instructive nature of a calculus course. The study reveals that teaching calculus is not targeted at making preparation for advanced courses in universities, but at understanding its thought and methods of variable, movement, infinity, limit, and dialectic for students. Those six sets of textbooks are highly consistent in the content range of calculus, namely, the main concepts and methods of derivatives, indefinite integral and definite integral of functions with one variable based on the notion of limit, together with their applications. Particularly, the Russian and Japanese textbooks not only emphasize the rigor of calculus, but allow appropriate "reasoning" to replace strict "proof" as well, and correct intuition is also allowed to be taken as truth.

Chapter 6 to Chapter 9 are devoted to certain hot topics in textbook research. For example, Chapter 6 targets on mathematical inquiry. By establishing an analytical framework, the chapter explores, by quantitative encoding and qualitative interpretation, how textbooks from different countries present the inquiry contents. The study finds that the German textbooks GR-EDM focuses on creating rich real situations and designing activities of peer cooperation, the Chinese textbooks CN-PEP highlight designing open-end questions and describing questions mainly with interrogative sentences, the Japanese textbooks JP-SYP prepares rich reading materials for students, the English textbooks UK-SMP pay attention on creating team cooperation situations, the US textbooks US-PHM make use of columns with inquiry material column and pay attention to the application and development of knowledge in the inquiries, the French textbooks FR-EDB put high priority on the exploration of information technology for students or the inquiries on mathematics with the aid of information technology, and, finally, the Singaporean textbooks SG-PPE, influenced by the Singaporean way of problem-solving, greatly highlight cultivating students' ability to predict and explain when designing the inquiry contents.

Chapter 7 focuses on the comparison of the difficulties of examples in high school mathematics textbooks. A comparison model with five factors (background, cognition, operation, inference and comprehensive knowledge) and four cognitive levels (calculating-operating memory level, concept-conceptive memory level, understanding-reasoning level and analyzing-exploratory understanding level) is designed, and five sets of textbooks from five countries (China, US, France, Russia and Australia) are compared and analyzed, revealing the difficulty level and characteristics of examples in different textbooks.

Chapter 8 is a comparative study of the example and exercises designs in high school mathematics textbooks. A comparison framework and related coding system are established and used to compare and analyze the example and exercise designs in Chinese, US and German mathematics textbooks, particularly the contents of functions. The study finds that all the three sets of textbooks have suitable arrangement of examples in the sense of the correspondence of examples with knowledge involved. Talking about exercises, the three sets of textbooks show great diversity. US ranks first in its numbers of exercises, Germany ranks second and China the last. On the hierarchical design of exercises, the American textbooks place emphasis on consolidation by imitation and divergence, the German textbooks counterpart place emphasis on imitation, highlighting variance and expansibility, while the exercise design in the Chinese textbooks attaches more importance to the comprehensiveness and complexity.

Chapter 9 compares and analyzes the features in employing information technology in high school mathematics textbooks. In this research, textbooks from China, Singapore, US, Australia and Japan were chosen as samples, and a four-dimensional (technical tools, presentation modes, content links, and application forms) analytical framework was constructed. By encoding and statistical analysis of different textbooks, research has it that textbooks of the five countries show their own characteristics in the application of information technology. For instance, with the exception of the Chinese and the US textbooks, the technical tools used by the other three countries are all monotonous. On the manner of presenting technology, all the textbooks show similarities: most of the technologies are integrated with the mathematical contents, only a small proportion presented as special introductions or narrations.

In the last two decades, textbook research in China is not just limited within the range of high schools. For instance, the research team of Beijing Normal University is engaged in the study of secondary school mathematics textbooks. We include as Chapter 10 a study on the international comparative study on exercises in secondary mathematics textbooks. The authors collected 10 sets of currently used textbooks from China, Australia, US, UK, France, Germany, Korea, Singapore, Japan and Russia and carried out, with both quantitative and qualitative approaches, a comparative study on the total number of exercises, types, openness of conclusions and difficulty degree from five major themes including "number and arithmetic", "equation", "triangle", "spatial figures" and "statistics". Study finds that in all ten sets of textbooks exercise show their regional characteristics and different emphases. The hierarchy and the relationship with real life and other subjects is taken into consideration in the design of exercises.

The following two chapters are devoted to the research of primary school mathematics textbook. Chapter 11 analyzes and investigates the influence of the usage of new mathematics textbooks under the background of mathematics curriculum reform to the mathematics performances of pupils. The researchers did a survey in 2011 by testing 437 Grade Six pupils from 8 schools in 3 provinces. The test papers designed with reference to the mathematics papers of the "2010 National Survey of Academic Level in Japan". Research shows that, with the use of new edition of mathematics textbooks for primary schools, the mathematics performance of pupils in 2011 remarkably outperforms that in a similar survey conducted in 2008.

Chapter 12 is a contribution of a research team in Southwest University. The team investigates the compiling features of primary school mathematics textbooks with an analytical framework with five factors including writing philosophy, content selection, content organization, content presentation, and column setting. Six sets of textbooks from US, Japan, Australia, Singapore and UK together with eights sets of the Chinese textbooks are chosen to do comparison. Different compiling philosophy and different compiling features are summarized, and suggestion for the further development of Chinese primary school mathematics textbooks are proposed.

Chapter 13, written by scholars from Northeast Normal University, discusses the specific embodiment of basic mathematical thoughts in textbooks. Based on the related literature, the authors adopt a combined method of content analysis and critical thinking, and analyzes the possible presentation forms of basic mathematical thoughts in textbooks with certain typical contents of Chinese school mathematics textbooks as examples. The study finds that the basic mathematical thoughts should be an indispensable part of mathematics textbooks, and introduces three ways to present its basic thoughts in mathematics textbooks: First, setting a special chapter or section to systemically and explicitly elaborate the basic thoughts; second, presenting basic thoughts by combining the contents of mathematical concepts, formulae and theorems with the knowledge and skills; and third, adopting interposed narration and sporadic penetration in explicit presentation of knowledge and skills.

The book also includes two related topics — history of mathematics and mathematics textbooks, and mathematics contest and mathematics textbooks. Chapter 14 introduces mathematics history in American mathematics textbooks in the late 19th and early 20th century. The author uses a content analysis method to explore the early American textbooks via analyzing the presentation forms of mathematics history in textbooks and the way of its application. Its presentation forms can be divided into explicit and implicit ways, and its application methods can be divided into five types: embellishment, add-on, replication, compliance and restructure. The study of the history of mathematics among the 27 kinds of textbooks shows that the use of the history of mathematics in mathematics textbook is closely related to the research status of the history of mathematics, the editor's literacy and value of history of mathematics and the scientific humanism ideological trend at that time.

Chinese schools, students and their parents are very concerned about extracurricular teaching materials, and materials related to mathematics contests in particular. They are regarded as an important supplement to textbooks with which teachers train the mathematical logic and skills of students. Chapter 15 discusses how to supplement and improve classroom teaching with mathematics contests. In this chapter, the authors discuss the organization situations, training systems and learning materials of mathematics contests, reflect on the effect of mathematics contests on the cultivation of students' mathematical abilities and list the problems in contest-oriented education. Though Chapter 15 does not probe into mathematics textbooks in a direct way, the special position of extracurricular teaching materials in China's classroom teaching is revealed.

3

The publication of this book is due to the joint efforts of colleagues. First of all, we are very grateful to all authors, especially those who are not members of the research team of the Comparison Project. We chief editors cannot make something out of nothing, so we are grateful for their hard work and perseverance. We are also very grateful to Yan Zhu, associate professor of the Faculty of Education in ECNU (FE-ECNU), and Yingkang Wu, associate professor of the School of Mathematical Sciences in ECNU (SMS-ECNU). They could have become authors. However, they were willing to undertake a lot of miscellaneous work to collect and sort out manuscripts and remind the authors to turn over or revise drafts in a friendly and patient manner. Our thanks would also go to Mr. Shanping Wang from the editorial office of Journal of East China Normal University, who not only did a lot of organization and liaison work, but also directly participated in the collation and revision of Chinese and English Manuscripts. We are also grateful to all the reviewers, who gave professional reviews and revision suggestions of the papers. In the process of translation, Yan Zhu, Yingkang Wu, as well as Associate Professor Guangtian Zhu from FE-ECNU, and Associate Professor Jun Li and Dr. Xiaoli Lu from SMS-ECNU have all done meticulous proof-reading on the tentative translated version, and made corrections, revisions or suggestions for revisions; in addition, faculty of the Mathematical and Educational team in the University of Southampton in the UK and their graduate students, including Manahel Alafaleq, Mailizar Mailizar, Yi Wang, Kun Xiang, and Sharon McGaughey who is at the Institute of Education, University College London, also made great efforts to improve the English text. Our thanks go to all of them. We would also be grateful to related editors in World Scientific Publishing and manager Rebecca Fu from its division in China for their great support and patience in the whole process of editing and compilation of this book. Finally, we are grateful to Weian Zheng, chief editor of Journal of East China Normal University

(Natural Science Edition). It would be impossible to publish this book without his innovation, recommendation and concern.

As the second volume of the East China Normal University Scientific Reports, this book should have come out earlier. Because the chief editors and the editing team were deeply involved in the bidding and organizing affairs of the 14th International Congress on Mathematical Education, the publication has been postponed several times. During this period, the 2017 version of the Chinese *High School Mathematics Curriculum Standards* has been released, and several versions of high school mathematics textbooks revised according to the new curriculum standards will be put into use soon. However, the traditions and basic characteristics of Chinese mathematics textbooks would not change substantially due to a revision of curriculum standards. Hopefully, the research done in this collection will still be a good reference for the current Chinese mathematics education and its future trend. We look forward to your recognition of this statement and apologize for our belated publication of this book.

Jianpan Wang (editor)
Lianghuo Fan and Binyan Xu (co-editors)
Professors of East China Normal University

February 2020, Shanghai

https://doi.org/10.1142/9789814713955_0001

Chapter 1

Core Content Selection and Textbook Design for High School Mathematics

Jianyue Zhang*, Dan Zhou, Rong Wang and Lili Song

Division of School Mathematics, People's Education Press
Beijing 100081, China
**zhangjy@pep.com.cn*

Curriculum content is an integral part of mathematics courses; it is the basic elaboration of the manifestation for humanities underpinned by mathematics. Therefore, curriculum content selection and organization are deemed as essential in the preparation of mathematics curriculum. The curriculum content of mathematics is based on its curriculum goals, the core part of which is selected from the basic mathematical system, organized and designed according to some logical order to form a knowledge and experience system. Therefore, the determination and organization of mathematical core content are given top priority in the preparation of mathematics curriculum. Currently, curriculum content has aroused widespread concern in international mathematics education and study. This chapter explores three basic questions with regard to the core mathematical content in high schools: (1) what is the benchmark for the selection of the core content, (2) what constitutes core content, and (3) how to organize core content.

1. Literature Review

Currently, the study of "core content" typically focuses on "core concepts", what are also often mentioned as "big ideas", "key concepts" and so on.

1.1 What are "core concepts"

R. Charles (2005) held it that "a big idea is a statement of an idea that is central to the learning of mathematics, one that links numerous mathematical understandings into a coherent whole" (AAMT 2009, p. 10). NCTM (2006) pointed out that "big ideas are important mathematical topics that provide focus of the mathematics experience for all students at each grade level. They are related ideas, concepts, skills and procedures from the foundation of understanding, permanent learning and successes at higher mathematics. They are indispensable elements in developing problem solving, reasoning, and critical thinking skills, which are important to all mathematics learning". By contrast, NCTM (2009) took mathematical reasoning and sense making as the focus of high school mathematics study on account of the following reasons: (1) It helps students cope with future challenges; (2) It is an inherent component of mathematical ability, enabling students to find the global coherence of mathematics and guiding them to establish connections between new concepts and the existing knowledge; (3) It is an effective way to learn mathematics. As can be seen, although reasoning and sense making are not any specific foci as knowledge point, its definition is consistent with that of "big ideas" in NCTM (2006). In 2010, the NGA Center and CCSSO jointly issued the *Common Core Learning Standards for Mathematics* (NGA and CCSSO, 2010), which absorbed the view of NCTM, but shifted "toward greater focus and coherence". As is pointed out in the introduction of this document: "what and how students are taught should reflect not only the topics that fall within a certain academic discipline, but also the key ideas that determine how knowledge is organized and generated within that discipline."

According to the AAMT's *Discussion Paper: School Mathematics for the 21st Century* (AAMT, 2009), "Big Ideas ... are key to connecting other aspects of mathematics, both between and within the Mathematical Concepts and Mathematical Actions." They are overarching ideas that are neither "concepts" nor "actions". Most of the Big Ideas pervade a number of conceptual areas of mathematics, and provide connections between them. For example, the idea of symmetry infiltrates in all four areas of mathematics.

In Chinese *Mathematics Curriculum Standards of Compulsory Education (2011 edition)* (MOE of PRC, 2012), ten core concepts of mathematics were put forward, and their specific meanings are explained.

Meanwhile, the connotations of such core concepts were given authoritative interpretation by the experts of the development group. For example, Xiang Huang (2012) indicated that the core concepts tend to be the nucleus or focus points for a type of curriculum content which is helpful for researchers to understand the nature of the course content and grasp its clues as well as seize the key to teaching; Shangzhi Wang and Fengjuan Hu (2012) also mentioned that the core concepts could focus on curriculum content, embody the essence and ideas of mathematics, and reflect the targeted objectives of the curriculum; Dan Zhang and Yongxiao Bai (2012) deemed that the core concepts do not refer to the specific content itself, instead, they refer to the reflected basic thought and methods of thinking as well as the understanding, perception, consciousness and capability in mathematical learning. Core concepts reflect the kernel of a type of curriculum content; they are not only the objectives of mathematics learning, but also the key to mathematics teaching.

As has been pointed out in one influential project named "theory and practice of mathematics core concepts, thoughts and methods, structural system, and teaching design for middle school", whether a mathematics concept can be considered as the "core" requires some "conditions". Such conditions include: (1) It is important and occupies the primary status in mathematics itself; (2) It has an important, indispensable and fundamental position in students' mathematical cognitive structure; (3) It belongs to a "natural link" in the logical chain of mathematical concepts, and it is appropriate for the development for students' thinking and also acceptable to students (Project Group, 2008).

Obviously, there has been no consensus reached as to how to define the "core concept" (perhaps it won't be reached in future either), but as can be seen from the previous studies, there are some shared characters in "core concepts": they are fundamental, interconnected; a notion, an idea, a method with widespread use and so on.

1.2 Selection of core contents

The three curriculum standards developed between 2006 and 2010 in USA and their interpretations have provided lists of core concepts. For example, the Pennsylvania State Board of Education adopted the Core Standards in July 2010. Since that time, the decision was made to craft a set of *PA Core Standards* in (English Language Arts and) Mathematics[a] (PDE, 2014); in NGA and CCSSO (2010), contents of various curriculum domains are listed and introduced in detail, the curriculum focal points of which are noted in italics, e.g., the focal points for algebra include seeing structure in expressions (interpreting the structure of expressions, writing expressions in equivalent forms to solve problems), arithmetic with polynomials and rational functions (performing arithmetic operations on polynomials, understanding the relationship between zeros and factors of polynomials, using polynomial identity to solve problems, rewrite rational functions), creating equations (creating equations that describe numbers or relationships), and reasoning with equations and inequalities (understanding solving equations as a process of reasoning and explaining the reasoning, solving equations and inequalities on one variable, solving systems of equations, representing and solving equations and inequalities graphically). Big ideas presented by AAMT (2009) include dimension, symmetry, transformation, algorithm, patterns, equivalence, and representation. *Mathematics Curriculum Standards on Compulsory Education (2011 edition)* of China (MOE of PRC, 2012) also presented 10 core concepts, including number sense, symbol sense, spatial concept, geometric intuition, consciousness of data analysis, operational ability, reasoning ability, modeling thought, consciousness of application, and spirit of innovation.

More often than not, researchers have their own core concepts in minds. For example, R. Charles (2005) presented 21 big ideas: numbers, the base-10 numeration system, equivalence, comparison, operation meanings & relationships, properties, basic facts & algorithms, estimations, patterns, variables, proportionalities, relations & functions, equations & inequalities, shapes & solids, orientation & locations, transformations, measurement, data collection, data representation, data distribution, and chance.

In his APEC Mathematics Education Conference report, Z. Usiskin (2010) offered nine "learning progressions" in mathematics contents for Grades 7 to 12, which suggests the "big ideas" in school mathematical

---

[a]See https://www.education.pa.gov/Teachers%20-%20Administrators/Curriculum/Math/Pages/default.aspx.

curriculum — "from whole number to rational number to real number, and then to complex number and vector"; "from numerical expression to algebraic expression, and then to function as a relationship and then to function as an object"; "from properties of individual figures to general properties of all figures in a particular class"; "from inductive arguments to deductive ones and then to deduction within a mathematical system"; "from uses of numbers to uses of operations to modelling with functions"; "from estimation of a single measurement to statistics for sets of numbers, and from descriptive statistics to inferential statistics"; "from the idea of same size and shape (same shape) to a general definition of congruence (and similarity) applying to all figures, to conditions for the congruence (and similarity) of simple geometric figures, to the application to all figures and graphs and the Graph Transformation Theorems"; "from scientific calculators to graphing calculators to computer algebra systems"; "from a view of mathematics as a set of memorized facts to seeing mathematics as interrelated ideas accessible through a variety of means".

As can be seen, the selection of core concepts differs in different countries and by different scholars, with some based on a macro picture and some micro oriented. The main reason for such differences lies in the different perceptions of the mathematical contents and their educational value. However, comparative studies on the core concept selection and organization over the curriculum in different countries are rarely found.

1.3 Organization and representation of core contents

As for the organization approach, Yingzhi Zhang and Enshan Liu (2010) pointed out that "the selection of curriculum contents should center on the core concepts of the subjects, while the concrete facts should serve as the foreshadowing to help students develop deeper understanding". Their paper cites the way McCathy and Morris used to sort out core concepts: "imagining that there is a large umbrella under which a smaller one is set, and an even smaller one is again placed under the second umbrella", which illustrates that a superordinate key concept is usually separated into a series of more concrete concepts. Chao Liu and Zhijun Wang (2010) made similar views from the perspective of the learning of core concepts: the final goal of learning the core concepts is to form a conceptual system. To understand a mathematical concept, students have to construct gradually a conceptual network around the concept. The more nodes in the network and the more channels between nodes, the deeper understanding of the concept will be. In the previous study of this project, we put forward a concrete proposal on how to build a structural system of textbooks over the core concepts (Rong Wang *et al.*, 2013) — two perspectives could serve as the breaking points: one is to clarify the "concept map" centering on core concepts, including the longitudinal main line of development and the cross-sectional linkage nodes, thus to form a reticular system with explicit main lines and smooth connections. The other is to grasp a consistent conception of mathematics education, so that the textbooks provide appropriate learning method to fulfill corresponding aims for mathematics education, either from content determination, material selection, column setting representation methods and others.

Some researchers believe that the nature of the educational process, according to the "Law of Recapitulation", is a repetition of human culture, so historical background of a concept's development shall be given due attention in the organization and representation of a core concept. For example, based on the exploration of the origin and early forms of trigonometric function, Zhenxuan Chen (2009) analyzed the changing process of the definitions of trigonometric functions under the influence of the ideas of functions and coordinates, and revealed the central position of the concept of trigonometric functions in trigonometry, then accordingly he put forward suggestions for the writing of textbooks on trigonometric functions. Another example is found in Suijun Jia (2008), where the main power backing the historical evolution period of functions and its early development was analyzed, upon which the author proposed that the content of functions in textbooks should be designed from the rich background in reality and the concrete functions to present the nature of function via abstract generalization, and the function shall be taught from various perspectives.

Through the analysis of existing textbooks, we can find two approaches to organizing and presenting the content around the core concepts of mathematics — one is that curriculum content should be organized around the core concepts for a particular topic, the other is that curriculum content should be organized according to the relevant core concepts of various topics. The former can be called the "linear" approach, while the latter the "hybrid" approach. For example, the UCSMP textbook takes "function" as the core and all the algebra knowledge is organized into a radialized network, some sub-concepts of function and their relationships with other mathematical domain/topics serve as the carriers to introduce the relationships of various topics; meanwhile, operations are designed to run through the whole algebra domain, and algebraic properties, which are extended from the rules and properties of arithmetic operations, are used to govern symbolic operations and operations of equations, inequalities and functions (Jianyue Zhang *et al.*, 2013). In the curriculum of *Core-plus Mathematics Project* (1999), a "standard-based" mathematics educational reform project in USA, each grade is designed to involve contents from various mathematics domains, including algebra, geometry, trigonometric functions, statistics and probability, and discrete mathematics, and the concrete contents of the same domain are often presented around one common topic, similar methods or important themes, while the content from different domains are linked via important mathematical concepts such as symmetry, recursion, function, data analysis, graph approximation etc., seeking their common grounds from the perspective of mathematical thinking and behavior (Bikai Nie *et al.*, 2010).

In summary, there have been some studies over the connotation of core mathematical contents, selection of core contents, organization and representation of teaching materials around the core contents conducted by mathematics education researchers at home and abroad, but on the whole, research in this area needs to be further strengthened.

Although different environments, national identities and the level of social development will result in differences in students' development, from an anthropological perspective, human cognitive development follows a universal pattern. The core contents of mathematics are determined by the inherent objective rules in mathematics, therefore, it would be a significant and desirable research theme to design a mathematics curriculum in line with the cognitive development level of the public and the requirement of social development based on the shared core contents of school mathematics that are confirmed through international comparative studies on the mathematics curriculum standards and representative textbooks. Below we will give the core content set through the international comparison of high school mathematics textbooks based on the established selection criteria of core mathematical contents for high schools, and then study the organization and representation of textbooks on core contents.

2. Core Content Selection Criteria

The evolving history of mathematics curriculum shows that, although there are certain objective criteria for core and non-core mathematical concepts, methods, or ideas from the perspective of mathematical science itself (at least there is consensus within the mathematical community), there are other constraints as to whether certain core content can be included in the mathematics curriculum. Next we will discuss the selection criteria for core contents in high school mathematics from macro to micro.

2.1 Basic characteristics of core contents

Macroscopically, the selection of core contents of high school mathematics reflects the needs for the developments of mathematics and students. For example, the NCTM (2006) proposes that a curriculum focal point had to pass the following three rigorous tests:

- Is it mathematically important, both for further study and for use in applications in and outside of school?
- Does it "fit" with what is known about learning mathematics?
- Does it connect logically with the mathematics in earlier and later grade levels?

China traditionally selected contents (Curriculum and Teaching Materials Research Institute, 2001) that are broadly applicable in modern social life, production, and science and technology; necessary for further study; fundamental in theories, methods and thoughts; and acceptable by students.

The *National Mathematics Curriculum Standard for High Schools (Experimental)* of China (MOE of PRC, 2003) proposes the principles of content selection:

- The contents in the compulsory courses should be selected to satisfy people's basic demands of mathematics, and to prepare necessary mathematical basis for students' further learning.
- The contents in the optional courses should be selected to meet students' interests and the need of their future development, to lay the foundation for further study, and thus to obtain higher mathematics accomplishment.

In our research (Project Groups, 2012), we also put forward three criteria for content selection. They are:

- Fundamental — basic foundation in the domains of algebra, geometry, statistics and probability, etc., needed by people in this information society;
- Developmental — essential for further learning, self-growing, easy to develop new ideas and new methods in new situations;
- Feasible — commensurate with students' thinking level, can be learnt and grasped by the students.

From the viewpoint of structural system of mathematics curriculum, contents are similar to cells which constitute the basic units of a living organism, whereas the nuclei are the control centers of the cells, play an important role in the metabolism, growing, and differentiation of cells, and are the main part to keep genetic materials. While the core contents are the basic units of the curriculum structure and function with core concepts, being the control center of core contents, they play an important role in the occurrence and development of mathematics, and are the main growing points of mathematics.

Mathematics is gradually expanded as science progresses, and its content is huge. For example, years ago calculus was only taught at university level, but now the preliminary knowledge of calculus appears in high school curriculum. High school mathematics must be something that high school students can learn within a three-year time frame. So the core contents of high school mathematics should be commensurate with the development of students' thinking level and be the essence of mathematics, which not only reflects the nature and the basic ideology of mathematics, but also is easy to understand and to learn, and is very useful in exploring various problems and their mathematical rules in nature.

The above discussions are made from a macro point of view. Now we will discuss the restricting factors against the selection of core contents of high school mathematics from an intermediate point of view.

2.2 Main factors influencing the core content selection

Regardless of the type of courses, society, students and subjects are the main influencing factors to the content selection (Zhexun Liao and Huisheng Tian, 2003). The selection of core contents of high school mathematics should also be considered from these three aspects (Jianyue Zhang, 2011).

2.2.1 Social factors

School mathematics education should prepare students for life in the information society. General requirements of citizens in the information society are the objective basis for the content selection. The change in information is extensive. New knowledge, new tools and new methods of mathematics study and communication are emerging and developing nonstop. Relevant materials show that, the sales of lap-tops in China in 2002 was just about 600,000, but that number exploded to hundreds of millions in 2012. Various statistics that were confidential in the past, such as the data of production, consumption, investment, and demographics, etc., are now released and spread widely through mass media on television, internet, etc. The rapid development and change of society has also enhanced people's demands, understandings and applications of mathematics more than ever. There is an increasing need for the support of mathematics and modern technology in daily life. Many areas of expertises also have a growing need for mathematical thinking and problem-solving skills. Moreover, not only mathematicians, scientists, statisticians, economists and engineers, but also finance and insurance practitioners need to be well prepared in mathematics through education.

In order to make students adapt to the needs of social development, and be ready for realizing an ideal life in society, the selection of the mathematics contents should be based on the social realities and developmental needs. History suggests that mathematics topics always change with the social and historical conditions. In the face of new demands for social development, people will consciously consider "what knowledge of mathematics is most valuable".

2.2.2 Patterns of students' psychological development

There are a lot of important contents in mathematics, but students may not be able to accept them all. Their thinking levels will have a restriction on the depth and breadth of mathematics courses. For example, the failure of New Mathematics Movement is due to the reformers ignoring the intelligence of students, making them learn difficult contents too early. Meanwhile, the selection of contents should also promote students' development. Ideal course content should be close to the scope of students' thinking, which can be learnt by those students who have made an effort. Students' interests, hobbies and developments should be taken into consideration as the fundamental of the selection of mathematics contents.

2.2.3 The characteristics of mathematics and its development trends

Mathematics core knowledge is the fundamental element of mathematics course and the basic factor restricting the choice of contents. This kind of restriction effect is reflected in the scope and structure of content, and its speed of update. For example, the mathematics course in the past is mainly about algebra and geometry. With the development of modern mathematics, there is an urgent need to update the contents, such as incorporating the basic knowledge of vectors, statistics, probability, calculus, etc. and refining the contents of plane geometry and the operations of algebraic expressions by deleting some unnecessary contents. The updating cycle is also getting shorter. The structure of the mathematics courses is changing accordingly — from a single course in the ancient time to branched courses for quite a long period, then to the coexistence of both currently.

Next, we will study the selection criteria of the contents of mathematics course from the micro point of view.

2.3 Detailed criteria of the core content selection

By summarizing the existing studies such as R. Tyler (1949), G. Hawson *et al.* (1982), Liangfang Shi (1996), Yongchun Zhang (1996), A. Ornstein and F. Hunkins (1997), Xiaoda Zhang (2000), Deyong Hao (2000), Zhexun Liao and Huisheng Tian (2003), we come to the following criteria of selecting core contents for high school mathematics courses.

2.3.1 Fundamental

"Fundamental" is mainly reflected as meeting the demands of the students and the needs of society. First, to meet students' basic needs of mathematics, that is, the selected contents should be what the students are interested in, and closely related to their practical needs; Second, to accomplish the objectives of mathematics curriculum, that is, to improve students' understanding of basic concepts, and to develop their mathematical ability and rational spirit, so as to provide necessary mathematics basis for the society and future development.

2.3.2 Reliable

"Reliable" means the core contents must be generally acknowledged and play a definite role in students' development. Contents that are being researched cannot be selected for high school mathematics course. Besides, the contents should also not be determined by course designers' personal preference or their subject background. Otherwise, the course will lose its authority due to the arbitrariness of content selection. For example, the contents of "trisection of an angle and expansion of a number field" and "operation research and graph theory" selected in Chinese *National Mathematics Curriculum Standard for High Schools (Experimental)* (MOE of PRC, 2003) are arbitrary, because it is not clear whether those contents will play an expected role in students' development.

2.3.3 Developmental

"Developmental" refers to the fact that the core contents should be in tune with time, with the development of mathematics, and should also be updated regularly. When selecting the core contents, we should take into consideration the application of mathematics in various areas of society, especially how the integration of information technology and mathematics have brought changes to the teaching of mathematics. We should include those contents that are needed currently, and discard outdated contents. In order to meet the needs of the information age, data processing, statistics and probability should be regarded as core contents. Those skillful algebraic operations and trigonometric transformations could be completed by computers/calculators.

2.3.4 Adaptable

"Adaptable" is the consideration from students' ages and cognitive levels. If it is adaptable, it will commensurate with students' thinking levels and arouse their learning interests. Meanwhile, it will increase students' mathematics learning experiences and widen the scopes of their interests. In other words, the core content should play a role in transforming the student's immediate development area into the current development level. It is necessary to consider whether students' existing interests have sustained individual value and social value.

2.3.5 Practical

First, the core contents should play a continuing role in the subsequent mathematics learning, because the main criterion is "to learn in order to practise". The knowledge that can be used to solve problems in life, production, technology and society should be selected. Derivatives and differential equations, for example, should be included in the scope of core contents of school mathematics because they are useful in solving practical problems. The main function of mathematics teaching is to educate, and therefore the core contents should be able to stimulate students' potential, cultivate their mathematics abilities, help them to gain confidence, etc. In the current climate of mathematics teaching in China, too much emphasis is put on formal training, thus it is important to emphasize the practicability of core contents.

2.3.6 Feasible

Feasibility refers to the objective conditions for implementation. For example, if there are no enough qualified teachers, even the most important knowledge cannot be taught. Likewise, even if the integration of information technology and the mathematics course can effectively promote students' understanding of mathematics, the corresponding advantages cannot be realized if there are no appropriate hardware devices, or teachers are not skillful with the devices. This means that the selection of the core contents should take educational investment into consideration.

The above criteria provide guidance for the selection of core contents for high school mathematics. Next, we will make a comparison of textbooks from selected countries, so as to suggest a set of contents to be selected.

3. An International Comparison of Content Selection

3.1 Research method

We did text analysis and comparative analysis on mathematics curriculum materials from USA, UK, Germany, France, Japan, Russia, and China, including interpretative documents of the curriculum standards (e.g., AAMT 2009, NCTM 2009, NGA and CCSSO 2010, Guantie Deng and Zhiqiang Gao 2009, Fenping Wang 2009, Chen Yu 2012, Wenfang Zhu 2009) and textbooks (see Tab. 1-1). We first draw concept maps to identify core contents of the chosen textbooks, then list the core contents and form a union. In this way we finally draw a table to visually show the differences and similarities in content selections among these textbooks.

The concept map is drawn in the following steps:

Step 1: List the knowledge points[b]. Select the basic knowledge points from algebra, geometry, statistics and probability, and calculus as the nodes of concept map.

---

[b]Knowledge point is a commonly used term in school curricula, teaching practice and examinations in China, covering concepts, skills and method of thinking, etc. It can be regarded as a unit to describe knowledge, and to transfer, grasp and measure knowledge understanding. Basic knowledge points are the key nodes identified

Step 2: Define the level of knowledge points. Define the level of a knowledge point according to its generality and inclusivity, and the knowledge points are layered hierarchically from the superordinate to the subordinate. Each knowledge point can only occur once on the concept map.

Step 3: Build the hierarchical connection. Connect related points with a line, and this connection can be established within the same module or cross connected among different modules. The connection type is usually unidirectional to indicate the relation of upper-lower subordination of concepts within the same module or correlation among concepts, and a linking phrase is used on the connection line when it is necessary to explain the relation of concepts.

It is stipulated in this study that domain names (such as algebra) are the root concept and belong to the zero layer, and the concept directly connected to the root concept is the first layer concept. Because the first layer concept is of the highest generality and inclusivity, it's taken as the core knowledge set of the domain. The concept in direct connection with core knowledge set is treated as the sub-concept generated from the core knowledge set. The vertical development context of core knowledge set in high school mathematics is presented by way of a concept map in the four domains, namely, algebra, geometry, statistics and probability, and calculus.

3.2 Textbooks from seven countries for the comparison

The core contents in high school mathematics can be obtained by comparing mathematical knowledge in chosen textbooks from chosen countries. Tab. 1-1 is the basic information of the textbooks used for our comparative study.

---

from the knowledge point network given by a curriculum standard. A knowledge point being basic depends on the situation it appears, regardless of its importance in other situations. For example, the description of concept "measurement" involves "iterative approximation", "upper and lower bound", and "limit". However, in current situation the latter three notions are used only in the description of the notion "measurement", so they cannot be regarded as basic knowledge points.

Tab. 1-1. Sample Textbooks

Country	Code	Textbooks	Publisher	Grades	Edition(s)
USA	US-PHM	Algebra 1, Algebra 2, Geometry	Prentice Hall	9-12	2009
UK	UK-SMP	Core 1-4 for AQA, Statistics 1-2, Mechanics 1-2, Further Pure Mathematics 1-4	Cambridge University Press	9-12	2010
Germany	DE-LSM	Lambacher Schweizer Mathematik für Gymnasien 10-12	Ernst Klett Verlag	10-12	2008, 2009, 2010
France	FR-EDB FR-HLT	Math 1[re]s, Maths 2e Déclic Maths Terminal S enseignement obligatoire	Editions Belin Hachette Livre	10-12	2009 2004
Japan	JP-SKS	Math I, Math II, Math III, Math A, Math B, Math C	Suken Shuppan	10-12	2007
Russia	RU-MGU	Алгебра и начала анализа, класс10, класс 11; Геометрия, классы10—11	Просвещение Издательство	10-12	2006
China	CN-PEP	Version A, Compulsory courses Math 1-Math 5, Optional courses Opt. 2-1, Opt. 2-2, & Opt. 2-3	People's Education Press	10-12	2007

The basic foundation of the selection of the above textbooks is that they are written under the background of curriculum reform in their countries and can preferably reflect the spirit of their curriculum reforms. For instance, Prentice Hall, the publisher of US-PHM, is one of the well-known educational publishers in the world and mainly publishes school textbooks, accounting for a large proportion of school textbooks market in USA. The core textbook series UK-SMP published by Cambridge University Press fully covers the mathematics contents in the A level curriculum, recommended as the textbook for the GCE test. With a history traced back to year 1897, Ernst Klett Verlag is the pioneer of German publishing industry and one of the most important publishers in Europe, whose publications Lambacher Schweizer-10, 11, 12 (GE-LBS) are widely used textbooks in the country. Among a dozen series of high school mathematics textbooks in France, Math 1[re]s and Math 2e from Editions Belin (FR-EDB) and Décuc Maths Terminal S Enseignement Obligatoire from Hachette Education (FR-HLT; we also use FR-EDB/HLT for all the 3 volumes together) are the most widely used for Senior 1, Senior 2 and Senior 3 students respectively, and they are written according to 2009 national syllabus Nouveau Programme 2009 (Published on August 10, 2008). Suken Shuppan Publishing Company is a relatively authoritative textbook publisher in Japan, and the series of textbooks JP-SKS selected here started to be used from 2003, though we included the 2007 edition in this study. For Russia, we chose the currently used typical high school mathematics textbooks RU-MGU published by Russian Education Press (Просвещение Издательство) in the series MGU-School (Moscow

University School Textbook). Finally, Version A of high school mathematics textbooks CN-PEP published by People's Education Press are used in 70% of high schools in China currently.

3.3 A list of high school mathematical knowledge

Due to the limited space, we present directly the basic knowledge points included in the above textbooks after comparison. The concept maps of basic knowledge refer to relevant research reports of the sub-project "*Comparison of Core Contents*" of the research project *Comparative Study of High School Mathematics Textbooks of Certain Countries.*

3.3.1 Algebra

Complex number: concept of complex number; algebraic, trigonometrical and exponential forms of complex numbers; modulus and argument of a complex number; geometric meaning of complex numbers; operations of complex numbers.

Number theory: properties of integral numbers; indeterminate equations; congruence; factorization and Euclidean division of integers; the binary system; greatest common factor and least common multiple; the Gauss theorem and the Bezout theorem; properties and ordering of numbers.

Polynomial theory: factorization; multiplication and division; factorization theory; remainder theorem; simplification of rational expressions.

Equation and inequality: solving equations; solving inequalities; relationship among equations, inequalities and functions; relationship among equations, inequalities and plane area; mean inequality; equations of higher degrees; proofs with inequalities.

Function: concept of function; representations of functions; graphs of functions; properties of functions; operations of functions; elementary functions (polynomial functions, exponential functions, logarithmic functions, trigonometric functions, rational functions, etc.).

Trigonometry: trigonometric ratio of acute angle; definition of unit circle; angle identities; angle sum and angle difference identities; double-angle identities; half-angle identities; universal substitution formulas; sine theorem; cosine theorem.

Number sequence: concept and representation of number sequence; sum of the first n terms; arithmetic sequence; geometric sequence; mathematical induction.

Matrix: concept and representation of a matrix; matrix operations; matrices and the system of linear equations; matrices and transformations.

Set: set of numbers; relations between sets; operations of sets.

Common logic: proposition; converse and/or negative propositions; necessary and sufficient conditions; proof by contradiction; reasoning and proof.

Algorithm: basic meaning of algorithm; program chart; basic algorithmic sentences; algorithm cases (binary system, Euclid division and its algorithm).

3.3.2 Geometry

Plane geometric figures: parallel lines and perpendicular lines; relations between figures; circles and related properties; sector; affine theorem and similarity transformations; congruent triangles; similar triangles; properties of triangles; geometric construction.

Positional relation of points, straight lines and planes in space: deductive reasoning of axiomatic systems, geometric reasoning, equations of lines and planes in space; space line; parallelism and perpendicularity of planes; handling problems with space vectors.

Solid geometry: Euler formula; polyhedron; solid of revolution; cylinder; cone; prism; pyramid; platform body; sphere; plane sections of a geometric body; drawing of solid figures; parallel projection; projection area.

Vector: definition and representation of vectors; addition, subtraction, and scalar multiplication of vectors; inner product; exterior product; the vector method; linear dependence and independence of vectors; the vector decomposition theorem.

Coordinate geometry: equation of a straight line; equation of a sphere; equation of a conic section; equation of a sphere and a spherical surface; algebraic representations of a cone and a cylinder; parametric equations; Cartesian coordinate and polar coordinates.

Graphic transformation: translation; rotation; reflection; magnification; simple transformation and composite transformation.

Measurement: distance, area and volume; degree of an angle.

3.3.3 Probability and statistics

Counting: permutation and combination; combinatorial numbers; the binomial theorem.

Data analysis: data collection; sampling; data processing and presentation; common statistics; variable correlation; cluster analysis; regression analysis.

Statistics evaluation: test of independence; hypothesis testing; type I ("reject when true") error and type II ("accept when false") error; quality of evaluation.

Probability of event: classical probability model; geometric probability model; independent tests; independent events; exclusive events; complementary event; conditional probability; probability formulas.

Probability distribution: random variables and their expectation; variance; discrete probability distribution; continuous probability distribution.

3.3.4 Calculus

Limit: concept of limit; properties of limit, operations of limits, limit of a function, continuity of a function, limit of a number sequence, the idea of approximation.

Derivative: concept of derivative, the tangent line of a curve and the physical meaning of derivative, differentiation and derivation; derivatives of elementary functions; the second derivative and application of derived functions.

Integral: concept of definite integral; application of definite integrals (calculation of the area enclosed by a curve); indefinite integral; properties of integral; integration by parts and by displacement; the fundamental theorem of calculus; generalized infinite integral; applications of integrals.

Differential equation: first-order and second-order linear ordinary differential equations; applications of differential equations.

3.4 Selection of core contents in textbooks from various countries

The textbooks from various countries have their own selection of core contents. Tab. 1-2 lists the basic information of the different selections:

Tab. 1-2. A set of core contents in the textbooks of seven countries

Domain	Set of core knowledge points	US-PHM	UK-SMP	DE-LSM	FR-EDB/HLT	JP-SKS	RU-MGU	CN-PEP
Algebra	Complex number	√	√		√	√		√
	Number theory	√			√	√		
	Polynomial theory		√	√		√		
	Equation & inequality	√	√	√	√	√	√	√
	Function	√	√	√	√	√	√	√
	Trigonometry	√	√	√	√	√	√	√
	Number sequence		√		√	√		√
	Matrix	√	√					
	Set				√	√		√
	Common logic				√	√		√
	Algorithm					√		√
Geometry	Plane geometric figures	√	√	√	√	√		
	Relation of points, lines and planes in space			√	√	√	√	√
	Geometric solid	√		√	√	√	√	√
	Vector	√	√	√	√	√	√	√
	Coordinate geometry	√	√	√	√	√	√	√
	Graphic transformation	√	√	√	√	√	√	√
	Measurement	√	√	√	√		√	√
Probability and statistics	Counting	√		√	√	√	√	√
	Data analysis	√	√		√	√	√	√
	Statistics evaluation	√	√	√		√		√
	Probability of event	√	√	√	√	√	√	√
	Probability distribution	√	√	√	√	√		√
Calculus	Limit	√	√	√	√	√	√	√
	Derivative		√	√	√	√	√	√
	Integral		√	√	√	√	√	√
	Differential equation			√	√			

Note: contents such as "set", "common logic", "algorithm" are not described in a separate chapter in US-PHM, UK-SMP, DE-LSM, FR-EDB/HLT and RU-MGU; instead, they are found in the relevant contents.

3.5 Conclusions

(1) The core contents of high school mathematics are still dominated by traditional mathematics contents; however, each country handles them with modern educational ideas and methods. This is because "the development of individual's cognitive structure repeats the process of human knowledge" and only these contents fit the development level of high school students' thinking. Meanwhile, processing with a modern mathematics viewpoint, for instance, handling geometry contents with the viewpoint of transformations or connecting trigonometry, complex numbers and geometry by means of vectors, can make the mathematics textbooks meet the needs of the development of mathematics and society.

(2) In the case of algebra contents, the textbooks of various countries take functions as the core. Types of specific functions include polynomial functions, rational functions, exponential functions, logarithmic functions and trigonometric functions, etc. The importance of equations and inequalities corresponding to these functions varies among different textbooks. Only a few countries such as China set up separate chapters for algorithms, reasoning and proof, commonly used logical terminologies, etc.

(3) The core contents of geometry differ greatly among the countries, both in contents themselves and in their appearances. There is a tendency to solve geometric problems with algebraic approaches. Comparatively speaking, the contents and appearances of solid geometry in Chinese textbooks are relatively "neutral". On one hand, the positional relations, such as the parallelism and perpendicularity of straight lines and planes, are dealt with the purely geometric approach, and, on the other hand, quite a lot of geometric problems, such as measurement, positional relation, are solved with vector calculations. Various countries include vectors in their core contents with the main knowledge points basically identical. A small number of countries also include the exterior product of vectors.

(4) Statistics and probability have been emphasized in every country. However, greater differences exist in the selection of core contents.

Comparatively speaking, the relatively comprehensive contents have been included in CN-PEP and the main difference lies in the selection of materials. CN-PEP needs to include more materials that have close relations with reality.

(5) Calculus also differs greatly among countries. In USA, it is in advanced university placement courses (AP courses, not included in Tab. 1-2) with complete contents with great attention to application instead of the rigorousness of theory. Textbooks FR-EDB/HLT, RU-MGU and JP-SKS include limit theory. The contents of derivatives differ among countries with some countries introducing second derivative and its application, which are not included in CN-PEP. For the integral part, textbooks from many countries explain the properties of integrals, integration methods such as integration by parts and integration by displacement, as well as an introduction to indefinite integrals, which are not mentioned in the CN-PEP either. Textbooks US-PHM, UK-SMP, DE-LSM and FR-EDB/HLT also include differential equations, which, again, are not in the CN-PEP.

4. Structural Design and Organization of Core-Knowledge-Based Textbooks

4.1 Different types of textbook structure

Mathematics textbook structures may have different designs which reflect different educational philosophies and curriculum views (Jianyue Zhang, 2011).

4.1.1 Knowledge-centered textbook structure

Such textbook design emphasizes the systematicness of knowledge and the rigorousness of logic. "General concepts" such as algebra, geometry, statistics and probability, and calculus run through the textbooks and form the basic frame. The knowledge system is constructed by way of the intrinsic logical order of mathematics, and the study of previous contents forms the foundation of follow-up study, or the subsequent issues are the special cases of concepts mastered at the beginning. In other words, the textbook is taken as a systematic structure that widens and deepens knowledge continuously by means of a basic and general idea. The theoretical basis of such textbooks is what was proposed by Brunner that "the foundations of any subject may be taught to anybody at any age in some form" (Bruner, 1960). The failure of the New Mathematics Movement indicates that this practice is too idealistic. In fact, textbooks should have both "general concepts" and "sensitive materials" (problem situations, intuitive materials, etc.), application materials (various forms of exercises). As textbooks are written for students, and, in considering the rigorousness and systematicness of mathematics, students' interests, feelings and their mathematics learning rules should also be taken into account.

4.1.2 Experience-centered textbook structure

Such textbooks emphasize taking the students' daily life experiences as the teaching materials and focus on the connections between mathematics and students' real life. John Dewey is an active advocator of such textbooks and he explicitly proposed to use the students' individual experiences as the main line of a textbook. He thought that textbooks should include both the experimentally logical aspect and the psychological aspect, where the former is about textbooks themselves and the latter is about the relationship between textbooks and students. Only promoting the psychologization of textbook design can realize the organic connection between logic and psychology. Experimental textbooks are classified into three categories, one for children, experimental organization with society and human as the center; one for adults, organization with more objective knowledge; and the third for subject experts, with more scientific and logic knowledge. Such kind of philosophy has certain reference meanings to the mathematics textbook design. In fact, when the mathematics textbooks of various countries all endeavor to design the textbook structural system with core contents as the mainline, active exploration is also made in the aspects of rendering the textbook contents based on students' learning psychology by selecting learning materials from students' life experiences and connecting knowledge with the students' surrounding world.

4.1.3 Human-centered textbook structure

Such textbook design is based on the principle of developing students' individual talents and adapting to students' diversity. It is the embodiment of humanitarian ideas in textbook design. It emphasizes respecting students' personal integrity and requires a focus on students' overall life experiences. It gives full play to the synergistic effect of intelligence and non-intelligence factors, linking the textbook with students' life experiences and the textbook contents with students' physical and mental growth processes. Moreover, it treats students as the participator of textbook design for the actual realization of psychologization of textbooks. The reference signification of such philosophy to mathematics textbook design lies in that we should emphasize students' individual differences, stress the role of non-intelligence factors in the organization of teaching contents. However, we should also overcome the defect that it overlooks the systematicness and integrity of mathematics knowledge.

4.2 Textbook structural design with core mathematical concepts being the nodes

Core contents should be considered as "knowledge clusters" with kernels, while the core concepts are exactly the kernels. Therefore, core contents are knowledge hierarchies constituted with core concepts and the derived sub-concepts. There are two approaches that one can start with to construct the structural system of the textbooks around the key concepts: one is to clarify the "concept map" centering on core concepts, including the longitudinal main lines of development and the horizontal linkage nodes, to form a reticular system with explicit main lines and smooth connecting channels, which is critical; the other is to grasp a consistent philosophy of mathematics education, thus to take the corresponding aims of education and to provide appropriate learning methods by content determination, material selection, column setting, representation methods, etc. in the textbook design. For example, to the content "function", all the countries surveyed attach great importance to the great "growing" capacity of the function concept, and, in this way, build a radial structural system centering on the core concept "function". This system is featured by its compact and omni-directional relations of concepts, being a typical carrier for students to understand mathematics and apply mathematics to solve problems.

As mentioned in the above, a core concept is placed at the center of a conceptual system in mathematics with a super self-growing capacity, and it can generate a "concept cluster". Therefore, in core concept identification and representation, we shall not only consider the occurrence and formation of the concept, but also take a macro view to investigate how to form the system and nodes of the concept network with such a concept as the center. Below we will take *Advanced Algebra* (UCSMP, 2010; not listed in Tab. 1-1), a textbook from the UCSMP (University of Chicago School Mathematics Project) series as an example to discuss how to construct a concept map and to identify the kernel "function" from the map.

4.3 A concept map of algebra domain in UCSMP's Advanced Algebra

Fig. 1-1 is a concept map of algebra domain in UCSMP's *Advanced Algebra*. As we can see from Fig. 1-1, all the core contents in the algebra domain constitutes a criss-crossing net, and there are four obvious nodes directly under the domain node "algebra" — "algebraic symbols", "functions", "mathematical models" and "change in various contexts", among which the node "functions" is placed at the central position of the whole net. Four sub-concepts are generated from "functions", which are further linked to other concepts and their sub-concepts as well as other domains via various chains.

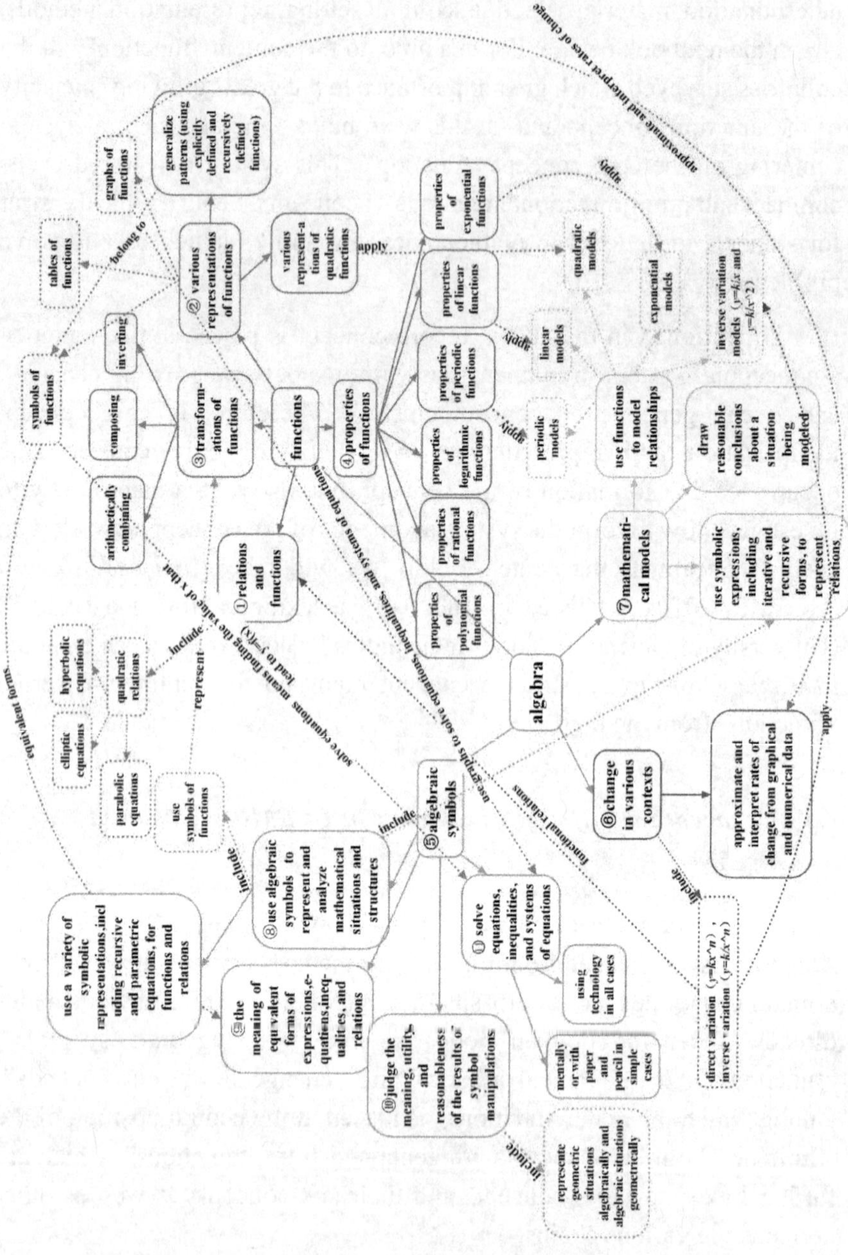

Fig. 1-1. Core concept map of algebra in UCSMP's *Advanced Algebra*

Notes: (1) Some nodes in Fig. 1-1 are attached with serial numbers ①, ②, ③, … for convenience of reference.

(2) The sub-concept "relationship and function" (node ①) indicates that in UCSMP textbooks, function is deemed as a special relation, whereas the discussion over quadratic curves and their classification in "quadratic relations" thereof reflects the relationship between algebra and geometry.

(3) The node ②, "various representations of functions" indicates that there are mainly three forms to represent functions — symbols, tables and graphs. The inductive modes employed in the UCSMP textbook, either explicitly defined, such as $y = ax^2 + bx + c$ or recursively defined, such as $b(n,m) = \begin{cases} 26\,335, & n = 0; \\ b(n-1,m) - m, & n > 0 \end{cases}$ fall into the category of symbolic representations of functions. The symbolic representations of functions have the same forms with algebraic expressions and equations, which indicates the relation between "functions" and "algebraic symbols" (node ⑤). This connection regards the solving of equations as a process of "evaluating the corresponding argument according to a known function value", and makes it possible to solve part of the equations, system of equations and inequalities, etc. with the help of function graphs. The graphic representation of functions can also be used to understand the concepts of approximation and rate of changes. It provides means to analyze the variation relations. Here "change in various contexts" (node ⑥) consist of two types: direct variation, such as in $y = kx^n$, y varies directly with the change of x^n, and inverse variation, such as in $y = k/x^n$, y varies in a reversed way with the change of x^n. They are all functional relations, in which the inverse variation relationship serves as the base to construct the inverse variation models. Among the "various representations of functions" (node ②), different representations of quadratic functions are also given additional attention. The quadratic function knowledge involved is applied to construct quadratic models, thus reflecting the relationship of "functions" and "mathematical models" (node ⑦).

(4) The "transformations of functions" (node ③) mainly include arithmetically combining, composing and inverting. The use of symbols enables the formal representations of such transformations. For example, the four arithmetic operations of functions can be represented as $f \pm g, f \cdot g$

and f/g, thus reflecting the relationship between functions and "algebraic symbols" (node ⑤).

(5) The "properties of functions" (node ④) include the properties of polynomial functions, rational functions, logarithmic functions, periodic functions, linear functions and exponential functions, among which the knowledge of linear functions, periodic functions and exponential functions provide the foundation to construct linear models and exponential models, thus reflecting the relationship between "functions" and "mathematical models" (node ⑦).

Moreover, the conceptual development in algebra domain of the UCSMP textbooks also reflects two main lines. One is "algebraic expression" → "function as a relationship" → "function as an object", which is incorporated in "use algebraic symbols to represent and analyze mathematical situations and structures" (node ⑧), "relationship and function" (node ①), "transformations of functions" (node ③) and other concepts. The other line is that operations run through the whole algebra domain. The rules and properties of arithmetic operations also serve as the foundation for symbolic operations and operations of equations, inequalities and functions. Such design embodies the practice of "use algebraic symbols to represent and analyze mathematical situations and structures" (node ⑧), "the meaning of equivalent forms of expressions, equations, inequalities, and relations" (node ⑨), "judge the meaning, utility, and reasonableness of the results of symbol manipulations" (node ⑩), "solve equations, inequalities and systems of equations" (node ⑪), and "transformations of functions" (node ③). Then the sub-concepts of "algebraic symbols" (node ⑤) build a connection between algebra and geometry: in "judge the meaning, utility, and reasonableness of the results of symbol manipulations" (node ⑩), the idea of "represent geometric situations algebraically and algebraic situations geometrically; using connections in solving problems" is exhibited, and examples like using an area model to solve the equation $x^2 + 10x = 144$ are included (Jianyue Zhang *et al.*, 2013).

In brief, when constructing the knowledge system of algebra domain, UCSMP's *Advanced Algebra* takes "function" as the kernel, and all the knowledge in the domain is organized into a radial network, some sub-concepts of function and their relationship with other mathematical domains/topics serve as the carriers to develop the relationship between domains/topics; in the meanwhile, operations run through the whole domain, and algebraic properties of operations, which grow up from the rules and properties of arithmetic operations, are used to govern symbolic operations and operations of equations, inequalities and functions. Such a knowledge system gives full play to the powerful organizing function of core concepts. With the self-growing ability and potential linking function of core concepts, the mathematics knowledge is arranged into an integrative and closely-connected network structure — all knowledge points are connected with a clear logical relation, and relations of different knowledge from various domains are explicit as well. Thus the structure is very useful for students to understand the development context of knowledge and to conduct self-directed learning.

5. Core-Knowledge-Based Textbook Design — The Case of Vectors

In all the high school mathematics textbooks, vectors are deemed as core contents. Herein we will compare some aspects on the design of the content of vectors in various textbooks.

5.1 Concrete contents and logic clues of vectors

For the sake of concision yet without sacrificing its representativeness, here we compare and analyze the textbooks from four countries: France (FR-EDB/HLT), Germany (DE-LSM), Japan (JP-SKS) and China (CN-PEP). In the following we exhibit the concrete contents and logic clues of vectors in the four series of textbooks.

FR-EDB/HLT: Concepts relevant to vectors → coordinates of vectors (coordinate formula of two points) → vector operations (definition, properties and coordinate representations of vector addition and scalar multiple) → directed angle and angle between vectors (studying trigonometry by means of vectors) → inner product of vectors (definition, properties and coordinate representation) → application of the inner product (equation of a line or circle, metric relations within a triangle such as the length of a median line and cosine theorem, trigonometric formulas) → concept and linear operations of space vectors → coplanar space vectors → coordinate representation of space vectors (orthonormal basis, the distance of two points) → space coordinate equations (equation of a plane parallel to coordinate axis, equation of a sphere with center at the origin, equation of a cylinder with a coordinate axis as its axis, equation of a cone with a coordinate axis as its axis) → the center of gravity problem on the plane or in the space (the weighted center of gravity for two points, three points and n points) → vectors and transformations (magnifications and translations) → distance from a point to a line → modulus of a space vector and inner product of space vectors → equation of a plane through a given point and with a given normal vector → equation of a space straight line (equation of the straight line through a given point and with a given normal vector, parametric equation of a straight line) → relationships of planes (discussing geometric relationship with algebraic equations).

DE-LSM: Concepts relevant to vectors → linear operations of vectors (addition, subtraction and inner multiple) → modulus of a vector and angle between vectors → inner and exterior products of vectors → studying equations of circles and spheres by means of vectors→ linear dependence and independence of vectors → application of vectors in coordinate geometry (equations of space straight-line and plane, positional relation, distance and angle).

JP-SKS: Directed line segment and vectors → operations of vectors (addition, subtraction and inner multiple) → zero vector and parallel vectors → the decomposition theorem of plane vectors → coordinates of vectors (coordinate representation and operations in coordinate forms) → inner product of vectors (definition, coordinate representation and properties of inner product, norm of a vector and angle between vectors) → coordinate formula for internal point of division and external point of division, coordinate formula for the barycenter of a triangle → application of vectors in geometry (collinearity of three points, relationship of inner product and perpendicularity, parametric equation of a straight line, equation of the straight line through a given point and along a given direction, equation of the straight line through a given point and with a given normal vector) → space vector (similar to the plane vector, including concepts, decomposition, coordinate representation, coordinate operations, inner product etc.) → application of space vectors (equation of plane parallel to a coordinate plane, equation of a sphere).

CN-PEP: Practical background of plane vectors → definitions → related concepts (zero vector, unit vectors, equivalent vectors and collinear vectors etc.) → linear operations and operational rules → fundamental theorems of plane vectors (orthogonal decomposition of plane vectors) and coordinate representation of plane vectors → inner product of plane vectors and its coordinate representation → application (in geometry and physics); Concept of a space vector and its representation → linear operations → the fundamental theorem of a space vector (orthogonal decomposition of space vectors) and coordinate representation of space vectors → inner product of space vectors and its coordinate representation → application of space vectors (to solve problems in solid geometry, mainly problems in directional vectors of straight lines and normal vector of planes, parallelism and perpendicularity, calculations of angles between two lines, a line and a plane or two planes).

As can be seen from the above, the main logic lines for vectors in the textbooks from the surveyed countries are consistent at large: the concept of a vector and its representation → linear operations of vectors and their coordinate representations → inner product (and exterior product) of vectors → applications. Moreover, all of them go from plane vectors to space vectors. But there is a huge difference in the selection of other concepts. A typical example is that DE-LSM contains exterior product, linear dependence and independence, etc. For the applications of vectors, there are even more striking differences. Comparatively, FR-EDB/HLT is comprehensive — besides the shared content such as processing distance, angles, parallelism and perpendicularity of straight lines and planes, etc., there are also other applications referred to in the surveyed textbooks, including formulating equations for straight lines, planes, circles, spheres, cylindrical and conical surfaces, and then discussing the relevant problems. However, CN-PEP does not involve such applications.

In our opinion, such differences are not surprising, as they are due to the differences in the circumstances of different countries, their varied educational philosophy, educational traditions as well as the differences in the writers' understanding of mathematics knowledge. Such "differences" and the resulted discrepancies are not only reflected in the construction of the concept network system, but also embodied in the focal points of such concepts and the designed learning methods for students, etc. Regardless of these differences, it would be the unanimous purpose of all the countries to systematically introduce the desirable vector knowledge and to pay attention to the logical soundness of such structural systems, so that teachers could deliver lectures smoothly and students could solve the questions encountered in geometry and physics.

5.2 Concerns in textbook design for vectors

5.2.1 A tool to solve geometric and physical problems

A vector has both magnitude and direction, and it has characteristics of both geometry and algebra. Thus, it is the common sense of textbooks from various countries to study geometric problems by using vectors as a tool:

Firstly, arrange relevant geometric content together with vectors and study them via vector approach. Textbooks from various countries prefer to study trigonometry and space geometric bodies with vectors. For instance, CN-PEP derives the angle sum (difference) identities, sine theorem, and cosine theorem through vectors. FR-EDB/HLT introduces the "unit circle" definition of trigonometric functions by means of vectors. DE-LSM uses inner product and exterior product of vectors to study the positional relation of space line-plane (distance, included angle), and deduce the volume formula of a pyramid by using exterior product (Fig. 1-2). The textbooks JP-SKS, DE-LSM and FR-EDB/HLT apply vectors into the study of the equations of straight lines, planes and spheres. FR-EDB/HLT also defines translations and magnifications in terms of vectors.

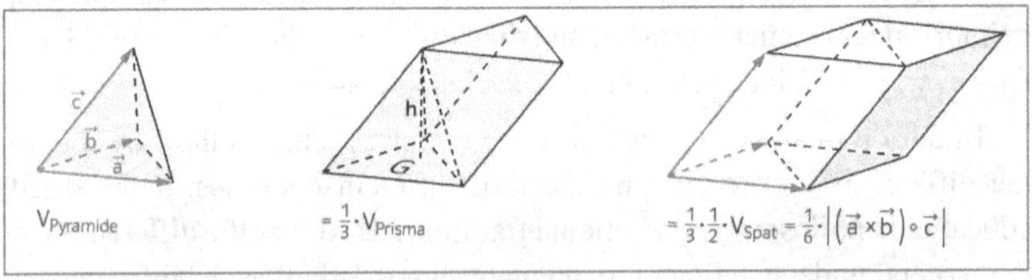

Fig. 1-2. A capture of page 112, DE-LSM, Mathematik für Gymnasien 11

Secondly, solve problems in related subjects by means of vectors. For example, textbooks from various countries introduce the vector concept from physical context such that displacement, forces, etc., connect vector addition, orthogonal decomposition of plane vectors, inner product of plane vectors, etc. with the corresponding physics problems, and trace the background of the triangular law and parallelogram law for vector addition to the composition of displacements and forces. DE-LSM studies vectors in physics as a theme in the chapter on vectors, including composition and decomposition of forces, work, centripetal force and electric field force, etc.

5.2.2 Establish the logically coherent learning process

In mathematics, different core contents may have the same structure system and similar logical routes, and the solutions of different problems may be greatly analogical. This is the subject characteristic of mathematics and it provides the possibility to construct a consistent and logically coherent learning process in textbook design. All surveyed textbooks have paid close attention to this point in the design of vector content. For instance, the exposition of space vectors in JP-SKS and CN-PEP is basically parallel to that of plane vectors, introducing space vector operations and coordinate representation by analogy with plane vector operations and coordinate representation, extending the decomposition theorem for plane vectors to that for space vectors, generalizing position vectors and internal and external points of division from plane context to space context, and applying space vectors in solid geometry in the spirit of the application of plane vectors in plane geometry, etc. This reflects the consistent logic in the aspects of knowledge, process and method. CN-PEP has also explicitly proposed the "three stages" of solving geometry problems with vector approach, namely, "convert shapes to vectors" → "apply vector operations" → "explain the results in terms of shapes". In order to enhance the importance of operations and operation rules, the textbooks point out that "with operations, vector has infinite power; without operations, it is only the road guide to show direction".

5.2.3 Enhance connection with reality

All the textbooks have paid attention to enhancing connections with reality and making students feel the necessity of learning vectors in order to stimulate students' strong zeal for learning. For example, embodying its persistent style, practical problems in DE-LSM appear in the introductory part of each section on vectors in order to tell students that "we need to learn the content in this section in order to solve this problem", leading the teaching and learning process to problem-solving. The problems come from a variety of circumstances, see Fig. 1-3.

Fig. 1-3. Vectors from real life circumstances: captures from DE-LSM, Mathematatik für Gymnasien 10 and 11

5.2.4 Moderately formalize logical expressions

In all the textbooks, content for vectors is consistent at large, and due attention is given to the requirement for moderately formalized expressions. For example, FR-EDB/HLT uses "DÉFINITION", "PROPERTY", "THEOREME" consistently to mark the corresponding items; they are listed one by one and numbered within a chapter, most of which are described by symbolic language to keep the text explicit and concise (Fig. 1-4):

DÉFINITION 6

Soit $\vec{u}$ et $\vec{v}$ deux vecteurs non nuls du plan.
Parmi toutes les mesures $x + 2k\pi$, avec $k \in \mathbb{Z}$, de l'angle orienté de vecteurs $(\vec{u}, \vec{v})$, il en existe une et une seule qui appartient à l'intervalle $]-\pi\ ;\ \pi]$; elle est appelée **la mesure principale** de l'angle orienté $(\vec{u}, \vec{v})$.

Exemple :
$(\overrightarrow{OA}, \overrightarrow{OB}) = (\overrightarrow{OA}, \overrightarrow{OB_1})$ et $(\overrightarrow{OA}, \overrightarrow{OB_1}) = -\frac{\pi}{2} + 2k\pi,\ k \in \mathbb{Z}$;
or : $-\frac{\pi}{2} \in\ ]-\pi\ ;\ \pi]$;
donc la mesure principale de $(\overrightarrow{OA}, \overrightarrow{OB})$ est $-\frac{\pi}{2}$.

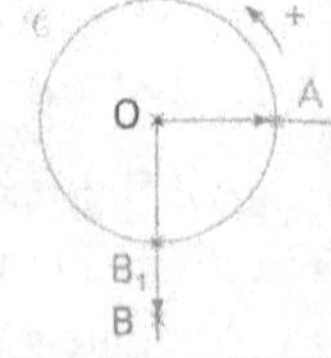

Fig. 1-4. A capture from page 244, FR-EDB, Math 1[re]s

Translation of Fig. 1-4

DEFINITION 6

Let $\vec{u}$ and $\vec{v}$ be two nonzero vectors in the plane. Among the measures $x + 2k\pi$, where $k \in \mathbb{Z}$, of the directed angle $(\vec{u}, \vec{v})$ of the vectors, there exists one and only one who belongs to the interval $]-\pi; \pi]$. This measure is called the **principal measure** of the directed angle $(\vec{u}, \vec{v})$.

Example:

$(\overrightarrow{OA}, \overrightarrow{OB}) = (\overrightarrow{OA}, \overrightarrow{OB_1})$, and $(\overrightarrow{OA}, \overrightarrow{OB_1}) = -\frac{\pi}{2} + 2k\pi,\ k \in \mathbb{Z}$;

but $-\frac{\pi}{2} \in\ ]-\pi; \pi\,]$, so the principal measure of $(\overrightarrow{OA}, \overrightarrow{OB})$ is $-\frac{\pi}{2}$.

In addition, in the presentation of solutions for examples, the French textbooks also mark the adopted theorems (properties) of the key steps, thus highlighting its logical reasoning (Fig. 1-5).

Énoncé 2

Démontrer en géométrie classique
Soit un parallélogramme ABCD. On appelle E, le milieu de [AB] et F celui de [CD]. La droite (AC) coupe la droite (DE) en I et (BF) en J.
a/ Montrer que le quadrilatère EBFD est un parallélogramme.
b/ Montrer que les points I et J partagent le segment [AC] en trois segments de même longueur.

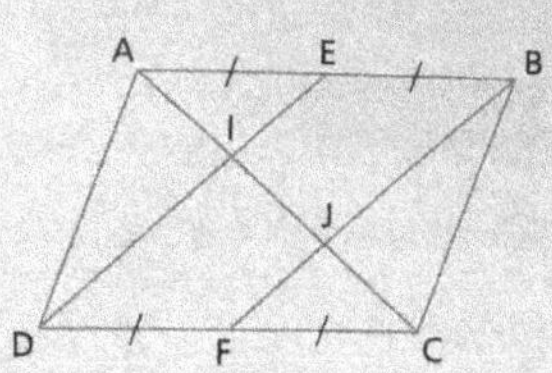

Solution 2

a/ On souhaite montrer, par exemple, que $\overrightarrow{EB} = \overrightarrow{DF}$.

$\overrightarrow{AE} = \overrightarrow{EB} = \frac{1}{2}\overrightarrow{AB}$, car E est le milieu de [AB]

et $\overrightarrow{DF} = \overrightarrow{FC} = \frac{1}{2}\overrightarrow{DC}$ car F est celui de [DC].

De plus $\overrightarrow{AB} = \overrightarrow{DC}$ (car ABCD est un parallélogramme),

donc $\frac{1}{2}\overrightarrow{AB} = \frac{1}{2}\overrightarrow{DC}$ donc $\overrightarrow{EB} = \overrightarrow{DF}$.

On en déduit que EBFD est un parallélogramme.

Méthode

Quelques propriétés utiles :
- ABCD parallélogramme ⇔ [AC] et [BD] ont même milieu ⇔ $\overrightarrow{AB} = \overrightarrow{DC}$;
- I milieu de [AB] ⇔ $\overrightarrow{AI} = \overrightarrow{IB} = \frac{1}{2}\overrightarrow{AB}$ ⇔ pour tout point M, on a $\overrightarrow{MA} + \overrightarrow{MB} = 2\,\overrightarrow{MI}$;
- (AB) // (CD) ⇔ $\overrightarrow{AB}$ et $\overrightarrow{CD}$ sont colinéaires ;
- A, B, C sont alignés $\overrightarrow{AB}$ et $\overrightarrow{AC}$ sont colinéaires.

Remarque : on pourrait penser que cette démonstration est longue : en effet, on a de façon relativement évidente le parallélisme de (EB) et de (DF) ainsi que l'égalité des longueurs de [EB] et [DF], mais cela ne suffit pas.

b/ D'après le théorème des milieux, puisque que (DE) et (BF) sont parallèles (car EBFD est un parallélogramme), et que E est le milieu de [AB], alors I est celui de [AJ]. De même, puisque F est le milieu de [DC], J est celui de [IC]. On en déduit que AI = IJ = JC.

Fig. 1-5. A capture from page 231, FR-EDB, Maths 2e

Translation of Fig. 1-5

Problem 2

A proof in classical geometry

Given a parallelogram ABCD. Call E the middle point of [AB] and F that of [CD]. Suppose (AC) intersects with (DE) at I, and with (BF) at J.
a/ Prove that the quadrilateral EBFD is a parallelogram.
b/ Prove that the points I and J cut the segment [AC] into three segments of equal length.

Solution 2:

a/ We want to show, for example, that $\overrightarrow{EB} = \overrightarrow{DF}$.
$\overrightarrow{AE} = \overrightarrow{EB} = \frac{1}{2}\overrightarrow{AB}$, for E is the midpoint of [AB];
and also $\overrightarrow{DF} = \overrightarrow{FC} = \frac{1}{2}\overrightarrow{DC}$, for F is the midpoint of [DC].
In addition, $\overrightarrow{AB} = \overrightarrow{DC}$ (since ABCD is a parallelogram), so $\frac{1}{2}\overrightarrow{AB} = \frac{1}{2}\overrightarrow{DC}$, and hence $\overrightarrow{EB} = \overrightarrow{DF}$.
Therefore, EBFD is a parallelogram.

Methods
Some useful properties:
- ABCD is a parallelogram ⇔[AC] and [BD] are bisectors to each other ⇔ $\overrightarrow{AB} = \overrightarrow{DC}$.
- I is the midpoint of [AB] ⇔ $\overrightarrow{AI} = \overrightarrow{IB} = \frac{1}{2}\overrightarrow{AB}$ ⇔ for any point M, $\overrightarrow{MA} + \overrightarrow{MB} = 2\overrightarrow{MI}$.
- (AB) // (CD) ⇔ $\overrightarrow{AB}$ and $\overrightarrow{CD}$ are collinear.
- A, B, C are on a straight line ⇔ $\overrightarrow{AB}$ and $\overrightarrow{AC}$ are collinear.

Remark. One might think that this proof is lengthy. In fact, it was relatively clear of the parallel of (EB) and (DF), as well as the equal lengths of [EB] and [DF]. But this is not enough.

b/ Since (DE) and (BF) are parallel (EBFD is a parallelogram) and E is the midpoint of AB, by the theorem of midpoints, I is the midpoint of [AJ]. Similarly, since F is the midpoint of [DC], J is that of [IC]. Therefore, [AI] = [IJ] = [JC].

We believe that traning of formalized logic expressions should be strengthened in high schools. This educational value of mathematics cannot be offered by other disciplines. Meanwhile, since the high school students' abstract thinking has been developed to certain degree, such requirements can be deemed as consistent with the students' cognitive level.

5.2.5 Strengthen the learning style of self-exploration

Textbooks could play a positive role in improving students' learning style, thus hierarchical learning activities shall be arranged in textbooks to guide the students towards self-exploring learning activities. In this regard, there are practices to be learned from all the textbooks of the surveyed countries. For example, each chapter of FR-EDB/HLT is divided into three parts: “activities” (*activités*), “essentials” (*l'essentiel*), and “exercises” (*exercices*). It separates activities and knowledge explanations. The “activities” part includes some primary inquiry, practices and cognitive activities, with focus on the understanding of the representation modes under various contexts and problem solving. The textbooks do not provide the analysis and answers of such activities, thus there is more free space for students to explore on their own. In addition, the solutions or answers of many activities could be multiple, so the diversity of learning styles are further illustrated. As is pointed out by some surveys (Jianyue Zhang, 2011) the “activities” part is paid the most attention and frequently used by French teachers.

Similarly, DE-LSM chooses a variety of examples with realistic backgrounds to guide students to engage in problem solving activities. The CN-PEP intersperses the questions in the texts with such columns as “Observation”, “Thinking” and “Exploration” to guide students to think and explore. For example, when introducing the fundamental theorem for plane vectors, the textbooks first guide students to construct vectors $3\boldsymbol{e}_1+2\boldsymbol{e}_2$ and $\boldsymbol{e}_1-2\boldsymbol{e}_2$ with two given vectors $\boldsymbol{e}_1$ and $\boldsymbol{e}_2$, and then think about whether any vectors in the plane can be expressed with the form $\lambda_1\boldsymbol{e}_1+\lambda_2\boldsymbol{e}_2$.

In addition, “task-based learning” is also designed for students to further study the concepts and realize the integrative reconciliation of knowledge through the approaches of practice and communication. This is also a good measure to strengthen initiative inquiring learning. For example, the column “Tasks: Vectors in Physics” in the DE-LSM introduces physical notions such as “composition of forces”, “decomposition of forces”, “work”, “centripetal force” and “electric field force”, closely related to mathematical themes like linear operations, inner product and exterior product of vectors. The column “Extracurricular Reading: Mathematicians and Philosophers Who Opened New Eras” tells stories of R. Descartes, J. Kepler, G. Galileo, L. Euler and others, enabling the infiltration of related mathematical history and mathematical culture to the learning of vectors. Another example is in the JP-SKS, where the columns "Inquiry" and "Topic Study" provide good platforms for self-learning. At the end of the chapter on vectors, topic studies "vectors and equation of a circle" and "equation of the plane through a given point and with a given normal vector" apply vectors to plane geometry and solid geometry respectively, widening students' scope for applications of vectors. Whereas the topic study after the section "inner product of vectors" links inner products to "doing work" in the real life, reflecting the tool role of vectors in physics. In addition, each of the above-mentioned topic studies is illustrated with vivid pictures to attract the students' interests in learning, as shown in Fig. 1-6.

Application of inner products of vectors	
Circles and vectors	
Develop-ment: equation of a plane	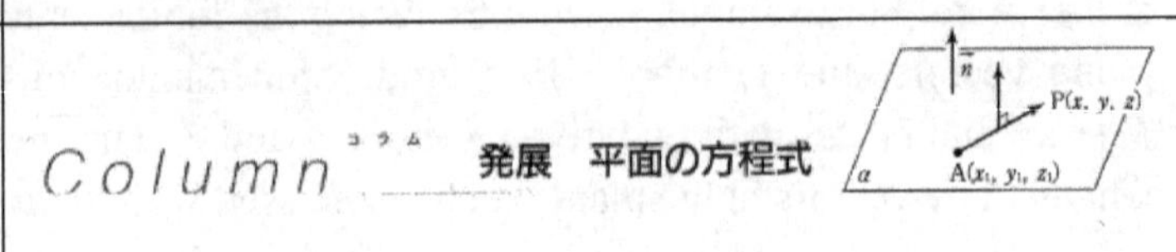

Fig. 1-6. Vivid pictures in the topic studies

6. Summary

This study has discussed three aspects on core contents in high school mathematics textbooks, namely, criteria of core content selection, core content collection, and core-knowledge-based textbook organization and presentation. The main conclusions are as follows:

(1) Core knowledge is the basic ingredient of the structure and function of high school mathematics curriculum, whereas core concepts are the "control centers" of core knowledge and the main growth points of knowledge.

(2) Selection of core contents should take into consideration the needs of students' further development, the inner structure of mathematics, and the requirements of social development. In determining the core contents, one should carefully consider from perspectives such as fundamentality, reliability, expansibility, adaptability, practicability and feasibility.

(3) Though the large parts of the core contents of high school mathematics are still the traditional mathematical knowledge, it should try to handle the modern ideas and methods. In terms of content update, this should take the spirit of "streamlining traditional contents, increasing and infiltrating modern contents" and treating the fundamental knowledge of calculus, statistics, probability, vector, matrix and logic algebra as the priority.

(4) The structure of mathematics textbooks may be constructed differently; however, the systematicness and logical consistency of mathematics should be given full attention. To construct a system around the core concepts there are two ways to start with: one is to clarify the "concept map" centering around core concepts, including the longitudinal main line of development and the horizontal linkage nodes, thus to form a reticular system with explicit main line and smooth connections; the other is to grasp a consistent conception of mathematics education, so that the textbooks in the content determination, material selection, column setting, presentation methods and other aspects can achieve the corresponding goal of mathematics education and provide an appropriate way of learning.

(5) The design of high school mathematics textbooks around the core knowledge should create enough room for the self-growth ability of core concepts and construct a consistent and logically coherent learning process for students. The design should pay attention to different educational functions of different core knowledge, enhance the connection with students' reality (reality of life, reality of mathematics), stimulate students' zeal for learning, and improve the effectiveness of learning. The textbook should strengthen the training of students' formal logic expressions, exercise students' logical thinking, and develop a mathematical way of thinking. Meanwhile, it should arrange hierarchical learning activities and guide students to carry out independent exploratory learning activities.

Our study did not use specific selection criteria of core contents to refine "core knowledge" from the knowledge points of textbooks from seven countries; we did not compare the degree of importance and organization of this knowledge in various textbooks either. Instead, we only made a preliminary illustration on the content structure (concept map) and modes of representation with examples. This is the main limitation of this study. We hope that we can come back to this topic and conduct more systematic study about the relevance of the three mentioned aspects in the future.

References

[1] AAMT (Australian Association of Mathematics Teachers) (2009). *Discussion paper: School mathematics for the 21st century*. Retrieved from http://www.aamt.edu.au.

[2] Bruner, J. S. (1960). *The Process of Education*. Cambridge, Massachusetts: Harvard University Press.

[3] Charles, R. I. (2005). Big ideas and understandings as the foundation for elementary and middle school mathematics. *Journal of Mathematics Education Leadership*, 7(3): 9–24.

[4] Chen, Zhenxuan (2009). The core position of trigonometric functions in high school mathematics [In Chinese]. *Shuxue Tongbao* (*Bulletin des Sciences Mathematics*), 48(6): 25–30.

[5] Curriculum and Teaching Material Institute (2001). *Collection of 20th Century Curriculum Standards and Syllabi for Chinese schools: Mathematics* [In Chinese]. Beijing: People's Education Press.

[6] Deng, Guantie and Gao, Zhiqiang (2009). Brief introduction of French mathematics curriculum standards [In Chinese]. *Shuxue Tongbao* (*Bulletin des Sciences Mathematics*), 48(1): 14–18.

[7] Hao, Deyong (2000). *Methodology of Curriculum Development* [In Chinese]. Beijing: Education Science Press.

[8] Howson, G., Keitel, C. and Kilpatrick, J. (1982). *Curriculum Development in Mathematics*. Cambridge: Cambridge University Press. Chinse version translated by Chen Yingshu. Beijing: People's Education Press, 1991.

[9] Huang, Xiang (2012). Ten core concepts in the curriculum standard of mathematics [In Chinese]. *Journal of Mathematics Education*, 21(4): 16–19.

[10] Jia, Suijun (2008). Evolution of the concept of functions and its inspiration for the teaching of high school mathematics [In Chinese]. *Curriculum, Teaching Material and Method*, 28(7): 51–54; 74.

[11] Liao, Zhexun, and Tian, Huisheng (2003). *New Curriculum Theory* [In Chinese]. Beijing: Education Science Press.

[12] Liu, Chao and Wang, Zhijun (2010). On the core mathematical concepts and teaching of them [In Chinese]. *Study of High School Mathematics*, (10): 1−4.

[13] MOE of PRC (Ministry of Education of the People's Republic of China) (2003). *Mathematics Curriculum Standard for High Schools (Experimental)* [In Chinese]. Beijing: People's Education Press.

[14] MOE of PRC (Ministry of Education of the People's Republic of China) (2012). *Mathematics Curriculum Standards for Compulsory Education (2011 Edition)* [In Chinese]. Beijing: Beijing Normal University Press.

[15] NCTM (National Council of Teachers of Mathematics) (2006). *Curriculum Focal Points for Prekindergarten through Grade 8 Mathematics: A Quest for Coherence*. Reston: NCTM. Retrieved from http://www.nctm.org.

[16] NCTM (National Council of Teachers of Mathematics) (2009). *Focus in High School Mathematics: Reasoning and Sense Making*. Reston: NCTM.

[17] NGA and CCSSO (National Governors Association and Council of Chief State School Officers) (2010a). *Designing High School Mathematics Courses Based on the Common Core State Standards*. Retrieved from http://www.corestandards.org/Math.

[18] NGA and CCSSO (National Governors Association and Council of Chief State School Officers) (2010b). *Common Core State Standards for Mathematics*. Retrieved from http://www.corestandards.org/Math.

[19] Nie, bikai, Zheng, tingyao, Sun, Wei, and Cai, Jinfa (2010). *Modern Mathematics Education Reforms in the USA* [In Chinese]. Beijing: People's Education Press.

[20] Ornstein, A. C. and Hunkins, F. P. (1997). *Curriculum: Foundations, Principles, and Issues* (3rd ed). Pearson Education Limited. Chinese version translated by Sen Ke, Nanjing: Jiangsu Education Press, 2002.

[21] PDE (Pennsylvania Department of Education) (2014). *PA Core-Mathematics, Grades PreK-12*. Pennsylvania Department of Education Official document, downloadable from https://www.pdesas.org/Page/Viewer/ViewPage/11.

[22] Pepin, B. and Haggarty, L. (2001). Mathematics textbooks and their use in English, French and German classrooms: a way to understand teaching and learning culture. *ZDM*, 33(5).

[23] Project Group (2008). An interim report on "Core concepts, thinking and method structure, and teaching design theory and practice" [In Chinese]. *Reviews of High School Mathematics Teaching*, (7): 1−4.

[24] Project Group (2012). Design of the research thought of "Study on the core contents in the primary and secondary school mathematics curricula and their teachings" [In Chinese]. *Reviews of High School Mathematics Teaching*, (7): 2−5.

[25] Shi, Liangfang (1996). *Curriculum theory: Foundations, Principles and Issues* [In Chinese]. Beijing: Education Science Press.

[26] Tyler, R. W. (1949). *Basic Principles of Curriculum and Instruction*. Chicago: University of Chicago Press. Chinese version translated by Liangfang Shi, Beijing: People's Education Press, 1994.

[27] UCSMP (University of Chicago School Mathematics Project) (2010). *Advanced Algebra*. Chicago: Prentice Hall.

[28] Usiskin, Z. (2010, March). Grades 7–12 learning progressions in mathematics content. *Paper presented at APEC Conference on replicating exemplary practices in mathematics education, Koh Samui, Thailand.* Chinese Translation by Lu Xiaoli, Mathematics Teaching, 2010(9).

[29] Wang, Fenping (2009). Introduction of Level A test of British AQA mathematics [In Chinese]. *School Mathematics (High School Edition)*, (7–8): 77–79; 81.

[30] Wang, Rong, Zhang, Jianyue, Song, Lili and Zhou, Dan (2013). International comparison of high school mathematics core concept textbooks' compilation — taking the example of function [In Chinese]. *Curriculum, Teaching Material and Method*, 33(6): 51–56.

[31] Wang, Shangzhi and Hu, Fengjuan. (2012). Understanding and grasping the core concepts in mathematics curriculum, I [In Chinese]. *Mathematics Education of Elementary School*, (7–8): 8–11.

[32] Yu, Chen (2012). Brief introduction on the working out of the learning guidelines in Japan [In Chinese]. *Curriculum, Teaching Material and Methodology*, 32(2): 113–116.

[33] Zhang, Dan and Bai, Yongxiao (2012). Core concepts in the new curriculum standard and their evolution [In Chinese]. *Elementary School Teaching: Mathematical Edition*, (6): 4–8.

[34] Zhang, Jianyue. (2011). *Curriculum Theory on the High School Mathematics* [In Chinese]. Beijing: Beijing Normal University Press.

[35] Zhang, Jianyue, Song, Lili, Wang, Rong and Zhou, Dan (2013). Core concept maps of the U. S. senior high school mathematics curriculum [In Chinese]. *Curriculum, Teaching Material and Method*, 33(11): 115–121.

[36] Zhang, Yingzhi and Liu, Enshan (2010). The significant role of key concepts in science teaching: The transforming from memorizing facts to understanding concepts [In Chinese]. *Journal of Educational Studies*, 6(1): 57–61.

[37] Zhang, Xiaoda (2000). *To Master Mathematics for the Public* [In Chinese]. Beijing: People's Education Press.

[38] Zhang, Yongchun (1996). *On Mathematics Curriculum* [In Chinese]. Nanning: Guangxi Education Press.

[39] Zhu, Wenfang (2009). Brief introduction of Russian national standard of mathematics education: high school part [In Chinese]. *Shuxue Tongbao* (*Bulletin des Sciences Mathematics*), 48(1): 18–23.

https://doi.org/10.1142/9789814713955_0002

Chapter 2

Comparison of Features in Content Coverage and Presentation of Complex Numbers in Textbooks of China, Japan and Singapore

Yuelan Chen* and Jun Xie

Department of Mathematics, East China Normal University
Shanghai 200241, China
**ylchen@math.ecnu.edu.cn*

This research made a detailed comparative study on the complex number contents in four series of textbooks from China (People's Education Press, Shanghai Education Press), Japan and Singapore, focusing on the following questions:

Question 1: what are the similarities and differences of the four series of textbooks in content selection and arrangement for teaching complex numbers?

Question 2: what are the similarities and differences of the four series of textbooks in the presentation of the content?

1. Introduction

1.1 Research background

The knowledge of number and algebra is the foundation of mathematics. It also plays an important role in our daily life and scientific work. Thus, each country attaches great importance to its teaching in curriculum reforms. Along with the development of the times, modern mathematics has also undergone significant change. For many countries, probably no major changes occurred in selecting main algebra topics for their new curriculum, but some changes such as selection of specific knowledge points and teaching approaches might have occurred. Such changes have evoked many echoes and controversies in many countries.

For example, according to the *1999 Syllabus for High School Mathematics* (hereinafter referred to as the *Japanese Syllabus*), the weekly learning time of students in school had been reduced since 2002 in the curriculum reform of Japan. This resulted in adjustments of mathematics contents from elementary school, middle school to high school. Teaching contents such as "complex numbers and complex plane" were removed from middle school mathematics and the curriculum arrangement was also adjusted accordingly. The new curriculum was more humane and gave students more selections. It paid attention to the cultivation of students' interest in mathematics; meanwhile it intensified mathematical thinking and applications (Yuelan Chen, 2006).

However, in the years of curriculum implementation, arguments from the society toward the curriculum have been growing. As a result, the Ministry of Education of Japan was forced to make partial revision to the *Japanese Syllabus* in 2005 and criticized the previously proposed "Tolerant Education", while the 6th revision of the *Japanese Syllabus* in 2008 brought order out of chaos by more complete actions to enrich the content of the mathematics and increase learning time (Yuelan Chen, 2010). Moreover, the new textbooks for middle school which were originally scheduled for 2012 had been exceptionally put in use in 2009 (Yuelan Chen, 2009).

These curriculum reforms in Japan rang an alarm to the curriculum reform in China. Next is a review of the curriculum reforms in recent years in China, related to the domain of number and algebra and also some opinions from experts and teachers.

Fenping Wang (2011) pointed out that compared with the textbooks used in the 20th century, the textbooks based on the National *High School Mathematics Curriculum Standards (Experimental)* (issued in 2003, hereinafter referred to as the *Curriculum Standards* for short) deleted some difficult contents such as trigonometrical form of complex numbers, rooting operation of complex numbers, solving quadratic equation within the scope of complex numbers, solving quadratic equation with complex coefficients, and the geometric meanings of operations of complex numbers.

As for the reform measures of reducing the difficulty by deleting contents in new curriculum, Boju Jiang (2005) argued that "... the new curriculum reform has significant deviation in direction, and the curriculum system starts all over again and has caused confusion to teaching in practice", and the overall level is "greatly lowered compared with proceeding syllabi and it leads to the wrong direction". He indicated that such reform by taking away contents could not achieve the original purpose of reducing the overweight burden of students; on the contrary, it would even aggravate the burden of students.

Many other scholars, e.g., Songmei Xiao (2007), have also done studies on this aspect. The problems involved are: What contents have been selected in current textbooks? To what extent have they been unfolded for teaching? How is the teaching implemented? To which level do we expect our students to reach after studying the textbooks? How do other countries schedule the teaching of these contents? This paper used "complex numbers" as the carrier to deeply investigate the above questions and tried to find out the compiling features of the content across four textbooks series from different countries, with the hope that it could provide some implications for the curriculum reform and textbook compilation in China.

1.2 Research questions

This research will make a detailed comparison of the treatment of the complex number contents in four series of textbooks from China (People's Education Press, Shanghai Education Press), Japan and Singapore, focusing on the following questions:

Question 1: what are the similarities and differences of the four series of textbooks in content selection and arrangement for teaching of complex numbers?

Question 2: what are the similarities and differences of the four series of textbooks in the presentation of the content?

2. Literature Review

The upper programmatic documents of textbook are the curriculum standards or syllabi, namely the expected curriculum. The documents serve as the principles of textbook compilation to facilitate an overall grasp of the philosophy of compilation, the content selection and arrangement, the knowledge points and requirements. Hence, this section mainly reviews the comparative studies on curriculum (especially on textbooks), either published in Chinese or in English. Following the curriculum comparative studies, algebra textbook research are examined.

2.1 Status quo of curriculum comparative studies

According to the accessible literature, the studies of secondary school mathematics curriculums cover two arenas, i.e., the comparison of curriculum standards (intended curriculum) and the comparison of textbook design (potentially implemented curriculum), which can be further divided into domestic comparison and international comparison.

After comparing the changes of text about the domain number and algebra between the contents of the current curriculum standards and that of the proceeding syllabi in China at the compulsory education stage, Qiping Kong and Songlin Hu (2002) argued that the new curriculum standard weakened the requirements on complicated computation and, at the same time, "attached great importance to students' experience of abstracting mathematical model out of actual background, the process of exploring relations of quantities and rules of changes, and to the development of students' number sense and symbolic sense", "strengthened practice and comprehensive application", and "emphasized the application of new technology".

Yuelan Chen (2010) compared the revision intentions, education goals, and the teaching contents of the 2009 and 1999 versions of Japan's Learning Guidelines for High School Mathematics. She found that the idea of the new version was to raise "survivability" and ensure students master the basic knowledge and basic skills necessary to "survive". The education goal of the new syllabus was developed around the concept of cultivating students' "survivability" and it emphasized the balance of moral, intellectual and physical developments, the cultivation of social responsibility, the comprehension of basic knowledge and basic skills, and mathematical literacy. The teaching domains of the new syllabus had changed back to six, including three core domains and three auxiliary ones.

There are many studies on curriculum across countries, such as the comparison of middle school mathematics curriculum standards of China and the United Kingdom by Yingfang Fu and Junhua Zhou (2007), the comparison of curriculums of China and Singapore by Jiao Liu (2009), and the comparison of mathematics curriculums and textbooks of United States, Japan, Macau, Hong Kong, Mainland China by Chunxia Qi (2008).

Concerning the comparison on textbooks, Weizhong Zhang and Fangqi Li (2009) compared the Singaporean New Mathematics Counts and textbooks of Zhejiang Education Press. Other studies include Guangwen Cheng (2009), Zhou Cheng (2008), Qinglin Li (2009), etc.

International researches such as the TIMSS studies and the ICMI studies play a leading role in this field. In the ICMI studies series, Kyungmee Park and Frederick Koon Shing Leung (2006) compared eighth-grade mathematics textbooks of five countries including China, Japan, South Korea, the United Kingdom and the United States.

In general, the research results of the comparative studies of mathematics curriculum show that the comparison of textbooks should be based on studies of curriculum standards (syllabi). Both domestic and international comparisons aim to examine the current situation, analyze the advantages and disadvantages, and finally make clear the direction of reform. Most of the studies on textbooks tend to discuss the arrangements of knowledge points, the presentations of textbooks, and the setup of examples and exercises, with differences mainly on the research methods, means and focuses. Grounded on the previous results, this chapter also studies the contents, presentations and expectations of mathematics textbooks in China, Japan and Singapore, starting with their curriculum standards. The following is a brief review of the current research situation of algebra textbooks at home and abroad from the three aspects.

2.2 Status quo of algebra textbook research

With the development of TIMSS and PISA, more and more researchers have conducted studies over the textbooks from various countries. In recent years, many domestic scholars have shown their interests in this research area and begun to study the overseas textbooks in order to provide opinions and suggestions to our domestic problems.

Scholars such as C. Y. Charalambousa, Lianghuo Fan and Yeping Li have studied the algebra textbooks from various countries. C. Y. Charalambousa *et al.* (2010) have suggested a two-dimensional framework and applied it to study the addition and subtraction of fractions in five series of textbooks from Cyprus, Ireland and Taiwan. The findings indicated that differences existed in the presentations of textbooks from the three regions. The textbooks from Taiwan provided more learning opportunity to students than the other two regions and they had high expectation to students in both cognitive requirement and challenging tasks' types.

Research Question 1 mainly focuses on the textbook content selection and arrangement. Currently various countries are proceeding with curriculum reforms. In order to keep school mathematics up to date, the key problem is which part of school mathematics could be considered as obsolete and should be cancelled and what new mathematical knowledge needs to be added; besides, the arrangement of the main contents may directly influence students' learning.

Fenping Wang (2011) compared the textbooks in China and the UK and found that the contents about complex numbers in the textbooks of Chinese high schools were much fewer than that of the UK. The former only included basic concepts of complex numbers and the four arithmetic operations. Thus "it is obvious that the reform of high school mathematics in China has a tendency to make the teaching of complex numbers too simple" which will result in a lack of integrity and systematicness of learning of complex numbers in high schools of China.

Yeping Li's (1999) research on textbooks of grade eight from Hong Kong, Mainland China, Singapore and the US showed that algebraic contents occupied the largest proportion in the US textbook, to which the greatest importance is attached. On the contrary, the other four U.S. textbooks give the least emphasis on it. The research also showed that, though the algebraic topics covered in the grade eight textbooks were not the same, the US algebra textbook has the broadest coverage of different algebra topics.

Other scholars dedicated to such study include Youcai Xue *et al.* (2006), Junbo Yang (2007), Xue Yao (2010), Weizhong Zhang and Fangqi Li (2009), etc.

The above researches show inconsistent suggestions to the depth and scope of contents should cover by textbooks in China. This drives the authors to the study of the status of currently used high school textbooks in China.

Question 2 focuses on the presentations of contents. Curriculum reform means not only content innovation but also transformation of teaching philosophy. Whether textbooks have embodied the transformation requires study on the ways of presenting knowledge contents.

Lianghuo Fan and Yan Zhu (2007) have compared the ways of presenting steps of problem solving by investigating the set-up of examples in textbooks from China, Singapore and the US. The research showed that in the aspect of general strategies textbooks of China gave the clearest problem solutions and their representation is the easiest one for the users to understand among the three textbook series. Concerning the specific problem-solving strategies, various countries scattered the distribution of heuristic methods to some degree. We question the necessity of teaching some heuristic methods.

Yeping Li *et al.* (2009), taking fraction division as the carrier, studied the content conceptualization and presentation in the textbooks of China, Japan and the US. They found that fraction division was constructed as an inverse computation of multiplication in China and Japan, while in the US it was introduced only as one kind of operations. The research explains the presentation features of textbooks from various countries in three aspects: i.e., types of examples, types of representations and example purposes. The types of examples include word problem and pure mathematics problem. The types of representations include pictorial, verbal and numerical explanations.

Related research also includes comparative study of algorithm contents in high school textbooks between China and Japan carried out by Ying Xin (2008), and a comparison done by Shuqing Yang (2007) on algebra contents in middle school textbooks published by Beijing Normal

University Press and in the textbooks published by Singapore Panpac Education Press.

Basing on the literature review, we found that examples and concept construction are the main components of studies on content presentation. This study will also focus on these two aspects from different angles and with different analysis indicators.

The above review has expanded our views and provided a general knowledge of ways and techniques for comparative study of mathematics textbooks. Generally speaking, the researchers mostly base themselves on the contents of textbooks and analyze from the various components (contents, organization, examples, exercises, etc.) presented in textbooks, in order to understand the interrelation of textbook, teaching and students achievements. By using their research methods as reference and combining the specific research questions, we designed a framework of this study which will be introduced in details in the next section.

3. Research Design

This section will show the objects of study, the research methods, and the research framework.

3.1 Objects of study

This study investigates the state of complex numbers in four series of high school mathematics textbooks of China, Japan and Singapore. The four series of textbooks are:

- **CN-PEP**: *Mathematics as per the National High School Curriculum Standard (experiment)* published by People's Education Press, Version A, China;
- **CN-SEP**: *Mathematics (trial version)* published by Shanghai Education Press, China;
- **JP-SKS**: *New Edition Mathematics* published by Sukan Shuppan Publishing Company, Japan; and

- **SG-PPE**: *New Express Mathematics* published by Panpac Education Pte Ltd, Singapore.

The main reason for selecting the mathematics textbooks from Singapore and Japan as the objects of the study is that both countries are in the East Asia region and deeply influenced by "Confucian culture", therefore there is a certain consistency in cultural tradition, social value orientation and other aspects. Though several rounds of curriculum reforms in Japan have reduced the teaching contents, the curriculum reforms in 2005 and 2009 have ended this tendency and added new teaching contents into the new curriculum (Chen Yuelan, 2010). To some degree, the detour experienced by Japan can bring about reflection of curriculum reform in China. All the four series of textbooks are under the background of curriculum reform, and, with exception of the CN-SEP textbooks, the other three series of textbooks are widely used in their own countries and of great influence. CN-PEP textbooks are very typical in China and may be considered as the representative of textbooks in China. Shanghai is one of the most economically developed cities in China, and was authorized by the central governor to take independent curriculum reform and to make independent test papers for college entrance examination. Thus it is significant to study the CN-SEP textbook written as per the *Shanghai Standards*. SG-PPE totally includes 6 volumes, among them, 4 volumes are written according to *Singapore 2007 Syllabus* (hereinafter referred to as *Singapore Syllabus*), the other 2 volumes are written according to A Level Test introduced in 2006 and implemented in 2007. With very distinctive features, this series of textbooks, are issued upon approval by Singapore Ministry of Education and include a large number of contents using graphic calculator to help students to learn. The textbooks by Sukan Shuppan are also widely used in Japan (Chen Yuelan, 2010).

Following the *Curriculum Standards*, CN-PEP version of textbooks includes 5 volumes for compulsory courses and 4 series for optional courses. Complex numbers appear in Optional Series 1 (for human-social science-orient students) and Optional Series 2 (for science-technology-orient students), or more precisely, in volumes Optional 1-2 and Optional 2-2. Though these two series are called optional, students who have the willing to take college entrance examinations are required to learn one of them, depending on their subject orientation.

Contents of complex numbers are compulsory in CN-SEP textbooks, appearing in the textbook for Grade 11 students.

In Singapore textbooks, complex numbers appear in H2 Mathematics. *Singapore Syllabus* points out that H2 Mathematics is designed to prepare students for a range of university courses, such as mathematics, science, engineering and related courses, and it includes themes like functions and graphs, sequences and series, vectors, complex numbers, permutations and combinations, and probability and statistics.

Japan Syllabus stipulates that complex numbers are in *Mathematics II*, following *Foundations of Mathematics* and *Mathematics I* and followed by *Mathematics III* and *Mathematics B*. Thus complex numbers are in the transitional position.

The selected textbooks are described in details in Tab. 2-1.

Tab. 2-1 Information of the four series of textbooks selected

	CN-PEP	CN-SEP	JP-SKS	SG-PPE
Textbook selected	Opt. 2-2	Grade 11, part 2	Math II	H2 Math, vol.1
Year of publication	2011	2008	2008	2008
Total chapters on complex numbers	3	3	6	5
Total pages on complex numbers	116	95	208	416

3.2 Research framework

This study mainly analyzes the text of complex number content in four series of high school textbooks from China, Japan and Singapore. The research methods are content analysis and comparison and the techniques used mainly include concept mapping and the multi-factor composite difficulty model (Jiansheng Bao, 2002). The following research framework is designed on the basis of the two-dimensional framework of C. Y. Charalambous *et al.* (2010). For the two research questions mentioned above, we carry on comparative study from horizontal and vertical perspectives (see Tab. 2-2). Question 1 mainly focuses on content selection and arrangement order. This can be explored from two dimensions — compiling arrangement and content coverage. Question 2 mainly concerns the presentation of contents. This can be explored from four dimensions — presentations of teaching contents for concepts and for principles, presentation of examples and of brief summaries. More detailed research framework is shown in Tab. 2-2.

Tab. 2-2 Research framework

Horizontal comparison	Content selection	Compiling arrangement	➤ Main contents and their arrangements; ➤ Knowledge points and their arrangements, described in linear form.
		Content coverage	➤ The logic relations of knowledge points, shown in the form of concept mapping to portray their scope and depth; ➤ Identifying and comparison of key concepts and principles.
Vertical comparison	Content presentation	Communicating with students	➤ Presentation of teaching content relating to key concepts, including the concept formation and concept comprehension; ➤ Presentation of teaching content relating to key principles, including the principle introduction and principle mastery, etc.; ➤ Presentation of examples, including explanation of problem solution, diversity of solutions, assistance means, etc.; ➤ Summary presentation, including its existence, forms, etc.
	Expectation for students	Requirement for students	➤ Types of problems, including types of question and types of answer ➤ Using the multi-factor composite difficulty model to illustrate the exercises and examples difficulty[a].

4. Results

Based on the research framework shown in the previous section, this section mainly focuses on the results of macroscopic and microscopic analysis on the four series of textbooks from three countries.

[a]Limited to the length of this chapter, the findings of using the composite difficulty model are not reported in this paper.

4.1 Macroscopic Analysis

Textbooks are generally written according to curriculum standards or syllabi. Thus, in order to compare the four series of textbooks, we need to study its guides, namely, the intended curriculum, i.e., the curriculum standards or syllabi.

Macroscopic analysis is mainly about the comparison of curriculum standards (syllabi) and the comparison of the content proportion of complex numbers in textbooks.

In terms of the curriculum principles and the overall goals of *Curriculum Standards* of China, *Shanghai Standards*, *Japan Syllabus* and *Singapore Syllabus*, the curriculum reforms in the three countries require students to grasp basic knowledge and basic skills of mathematics, develop a spirit of innovation and awareness of applying mathematics, and improve their ability of thinking. Except for the *Japan Syllabus*, the other three also require teachers to develop students' cooperation and communication ability in mathematics and be able to integrate information technology with mathematical curriculum. Generally speaking, the overall requirements of the curriculum standards or syllabi followed by in writing the four series of textbooks are basically the same, which results in certain consistence of East Asia countries in mathematics teaching requirements.

Though in *Curriculum Standards* of China the complex number content is in the optional series (Optional 2-2), it is compulsory for all science-technology-orient students. After studying this module, students should be able to understand the process of number system expansion and the necessity of introducing complex numbers. They should master basic knowledge of complex numbers, but tedious and over complicated calculation should be avoided. In *Shanghai Standards*, complex numbers are contained in the compulsory course. All students need to learn about the ideas of expansion from real numbers to complex numbers, master the concept of complex numbers and their basic operations in algebraic forms, and understand the consistence of operations of vectors and those of complex numbers. In *Japan Syllabus*, complex number content is optional and it requires students to be able to solve quadratic equations in the scope of complex numbers and to have some knowledge on equations of higher degree, emphasizing the application of complex numbers. *Singapore Syllabus* has listed the complex number contents to be mastered, including Cartesian form and polar form. Except for the *Japan Syllabus*, the other three standards (syllabi) have requirements on an understanding of the expansion of number systems.

For each series of textbooks, we calculate the proportion of number of pages of the complex number chapter to the total number of pages of the volume of the textbook containing the chapter, and, as a reference, we also calculate the average proportion of a chapter in that volume of textbook. The data are shown in Fig. 2-1, from which one can see that, in all four series of textbooks, the proportions allocated to complex numbers are smaller than the average proportions of chapters. However, the CN-PEP textbook shows the biggest gap which indicates that, to some extent, its teaching content would be the least.

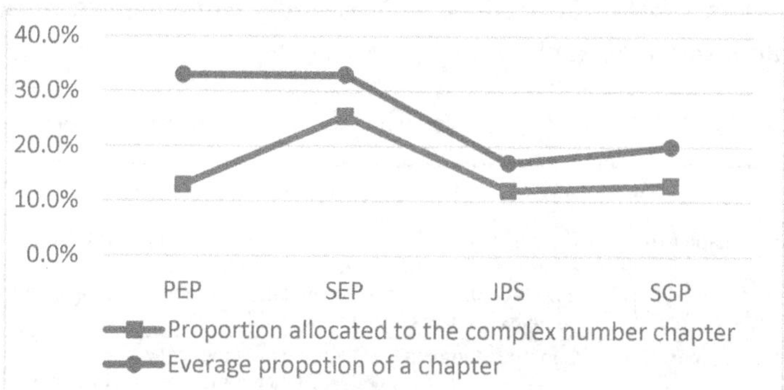

Fig. 2-1 Proportions allocated to the complex number chapters in four series of textbooks

4.2 Microscopic analysis

The microscopic analysis includes horizontal (content selection) and vertical (content presentation, expectation for the students) aspects.

4.2.1 Content selection

The data and analysis in this part is still concerning the first research question: what are the similarities and differences of the four series of textbooks in content selection and arrangement for teaching? Qualitative and quantitative analysis will be carried out from the two aspects of compiling arrangement and content coverage.

(1) Compiling arrangement

Compiling arrangement mainly refers to the teaching order of the main contents and key knowledge points relating to complex numbers as compiled.

The arrangement of complex number contents in the four series of textbooks is shown in Tab. 2-3. As the titles of the chapters shown, different series of textbooks focus differently. CN-PEP textbook lays more emphasis on the origin of complex numbers — the development of number systems, JP-SKS stresses the application of complex numbers — solving equations, while CN-SEP and SG-PPE textbooks emphasize some properties of complex numbers themselves. Judging from the titles, each series of textbook is basically consistent with the corresponding curriculum standards (syllabi).

Tab. 2-3 Chapters of complex numbers in four series of textbooks

	Chapter Order	**Chapter Title**	**Section titles in the chapter**
CN-PEP	Chapter III	Expansion of number system and introduction of complex numbers	➢ Expansion of number system and the concept of complex numbers ➢ Arithmetic operations of complex numbers in the algebraic form
CN-SEP	Chapter III	Complex numbers	➢ Complex number concept ➢ Coordinate representation of complex numbers ➢ Addition and subtraction of complex numbers ➢ Multiplication and division of complex numbers ➢ Square roots and cube roots of a complex number ➢ Quadratic equations with real coefficients
JP-SKS	Chapter II	Complex numbers and equations	➢ Complex numbers and solution of equations ➢ Equations of higher degree
SG-PPE	Chapter IV	Complex numbers	➢ Cartesian form of complex numbers ➢ Polar form of complex numbers ➢ Loci in Argand diagram

Let us continue to the bottom layer (subsection) subtitles to investigate the main contents and their arrangement in the four series of textbooks (Tab. 2-4).

Tab. 2-4 Main contents and their arrangement in the four series of textbooks

	Subtitle	Number of subtitles
CN-PEP	✧ Number system expansion and the concept of complex numbers ✧ Geometric meaning of complex numbers ✧ Addition and subtraction of complex numbers in algebraic forms and their geometric meaning ✧ Multiplication and division of complex numbers in algebraic form	4
CN-SEP	✧ Concept of complex number ✧ Equality of complex numbers ✧ Complex planes ✧ Vector representation of complex numbers ✧ Modulus of a complex number ✧ Addition of complex numbers ✧ Conjugate complex number ✧ Subtraction of complex numbers ✧ Distance between two points on complex plane ✧ Multiplication of complex numbers ✧ Powers of a complex number ✧ Division of complex numbers ✧ Moduli of products and quotients ✧ Square roots of a complex number ✧ Cube roots of a complex number ✧ Quadratic equations with real coefficients	16
JP-SKS	✧ Complex numbers ✧ Operations of complex number ✧ Square roots of a negative number ✧ Solutions of quadratic equation $x^2=k$ ✧ Rooting formula ✧ Discriminant ✧ Solution-coefficient Relation ✧ Determination of quadric expression ✧ Solving equations of higher degree (1) ✧ Solving equations of higher degree (2) ✧ Imaginary roots of equations of higher degree	10
SG-PPE	✧ Cartesian form of complex numbers ✧ Arithmetic operations of complex numbers ✧ Conjugate complex numbers ✧ Geometrical interpretation of kz ✧ Addition of complex numbers in Argand diagram ✧ Multiplying a complex number by i ✧ Modulus of a complex number ✧ Argument of a complex number ✧ Complex numbers in modulus-argument form ✧ Exponential form of complex numbers ✧ Conjugate complex number in polar form ✧ Multiplication and division in polar form ✧ De Movire theorem and its application ✧ Basic concepts in Argand diagram ✧ Standard loci — circle ✧ Standard loci — half line ✧ Standard loci — perpendicular bisector ✧ Loci defined by inequality	18

From Tab. 2-4, we can see that the numbers of themes in the complex number chapters of the four series of textbooks decreases by the order of SG-PPE, CN-SEP, JP-SKS and CN-PEP, with the numbers being 18, 16, 10, and 4, respectively, involving the concept of complex number, the relationship of complex numbers with geometry, operations of complex numbers, and applications of complex numbers, etc.

To be specific, in CN-PEP the chapter of complex numbers only involves the concept of complex numbers, the relationship of complex numbers with geometry, and the operations of complex numbers, which are arranged in the order of concept → relation → operation; CN-SEP includes concept of complex number, the relationship of complex numbers with geometry, operations of complex numbers, and application of complex numbers, in the order concept → relation → operation → application; JP-SKS concerns concept of complex number, operations of complex numbers, and applications of complex numbers, in the order concept → operation → application; SG-PPE covers the representations of complex numbers in various forms, operations and geometrical applications of complex numbers in different forms, and applications of complex numbers, arranged by the order algebraic form concept → algebraic form operations → geometrical form concept → geometrical form operations → geometrical applications.

As clearly shown from above table, CN-PEP lacks of the theme "applications of complex number". This will lessen students' overall understanding of complex numbers, and influence the integrity of complex numbers to some extent, and as a result, hinder the further learning of university mathematics. The comparison of CN-SEP with textbooks from Japan and Singapore shows considerable similarity. They all roughly arranged in the order of concept, operation and application, following the mathematical development of the contents. Thus it helps students to grasp the overall structure of complex numbers.

Particularly, SG-PPE shows an upward spiral from algebraic form to geometrical form and establishes a more complete mathematical structure of complex numbers. As seen from Tab. 2-4, SG-PPE firstly introduces the representation and operations of complex numbers and conjugate complex numbers from the perspective of algebra. Such introduction is rather basic and normal. On this basis, some operations are further explained from the perspective of geometrical transformations, enhancing students' understanding of mathematical content and form a transition from algebra. With all above as groundwork, new forms of complex numbers are introduced and the concept and operations of complex numbers are reinvestigated from a purely geometrical view, leading the instruction to the essence. Such a design allows students to get a first image of complex numbers from the form, then to go forward gradually to profound understandings through spiral learning, and finally to grasp the whole picture of complex numbers from the representation of essence. Though CN-PEP and CN-SEP textbooks also touch the geometrical interpretation, they only stay on the surface instead of getting deep into the core.

Then we study the arrangement of sections in the complex number chapters of the four series of textbooks and the specific knowledge points in the text of each section, see Fig. 2-2(1)–2-2(4).

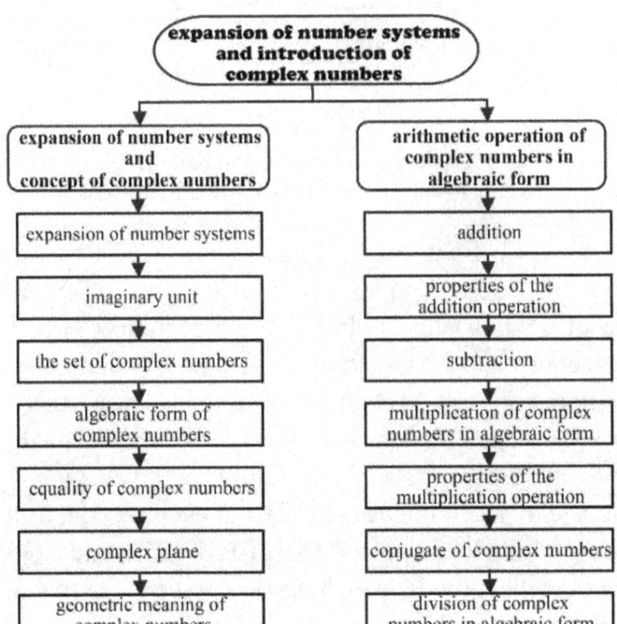

Fig. 2-2(1) The introduction order of the knowledge points relating to complex numbers in CN-PEP textbook

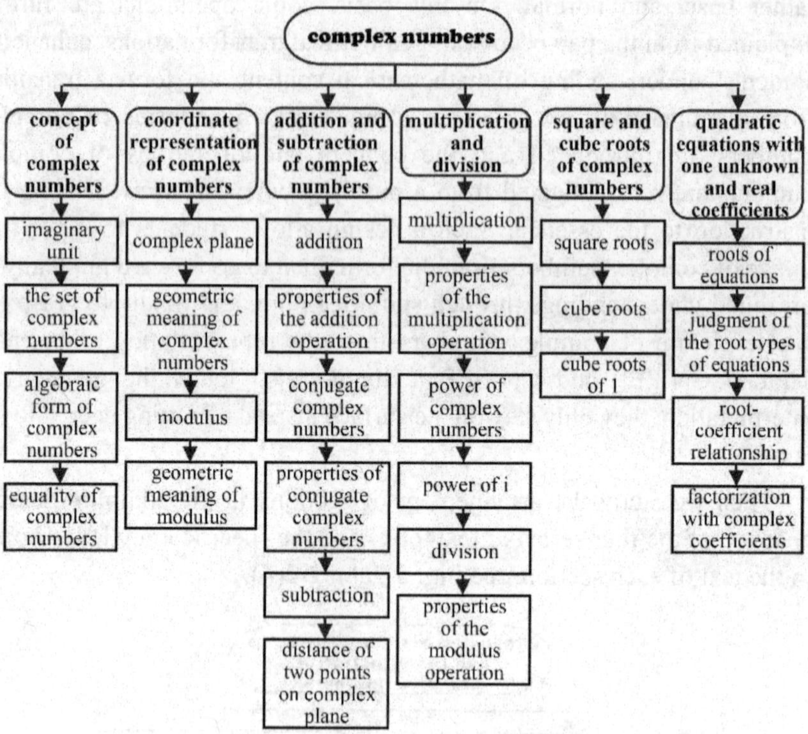

Fig. 2-2(2) The introduction order of the knowledge points relating to complex numbers in CN-SEP textbook

As can be seen from Fig. 2-2, there are 2 sections and **14** key knowledge points about complex numbers in the CN-PEP textbook with an average of 7 knowledge points in each section. CN-SEP textbook includes 6 sections and **27** key knowledge points with an average of 4.5 knowledge points in each section. JP-SKS textbook contains 2 sections and **23** key knowledge points with an average of 11.5 knowledge points in each section. SG-PPE consists of 3 sections and **40** key knowledge points with an average of 13.3 knowledge points in each section. Talking about the structure of knowledge points, CN-PEP, JP-SKS and SG-PPE adopt the traditional vertical mode, which has fewer sections with a long chain of knowledge points within a section; while CN-SEP textbook uses the American way of horizontal mode, which has more sections in a chapter with a shorter chain of knowledge points in a section. Generally speaking, the list of contents in a textbook only shows down to the section level, vertical structure is conductive to overall grasping by students while horizontal mode favors mastering the internal details. Thus we can argue that the arrangement of CN-PEP, JP-SKS and SG-PPE tends to focus on mastering the overall structure of mathematics while that of CN-SEP inclines to stress grasping the interior of mathematics.

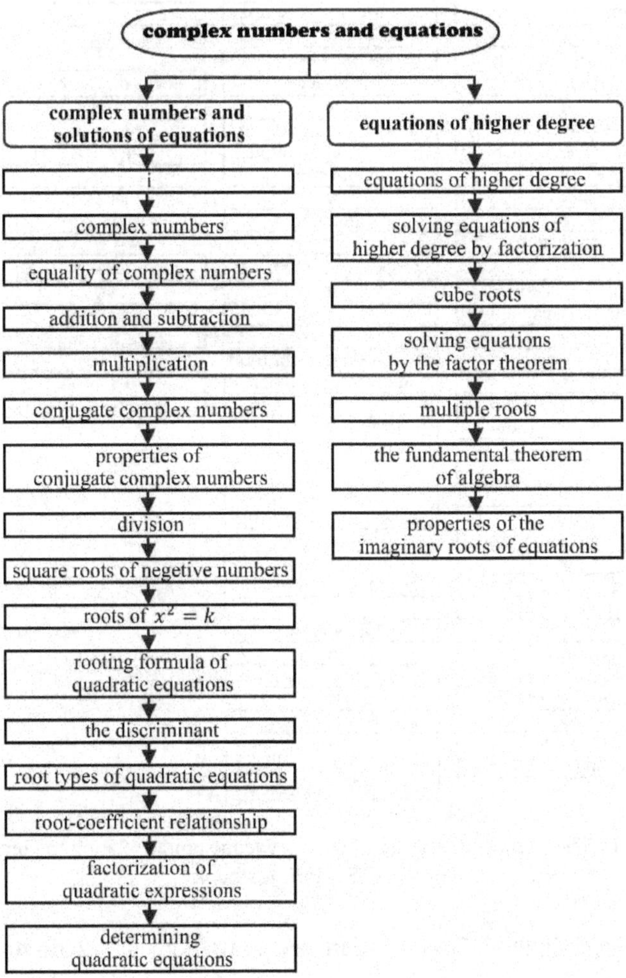

Fig. 2-2(3) The introduction order of the knowledge points relating to complex numbers in JP-SKS textbook

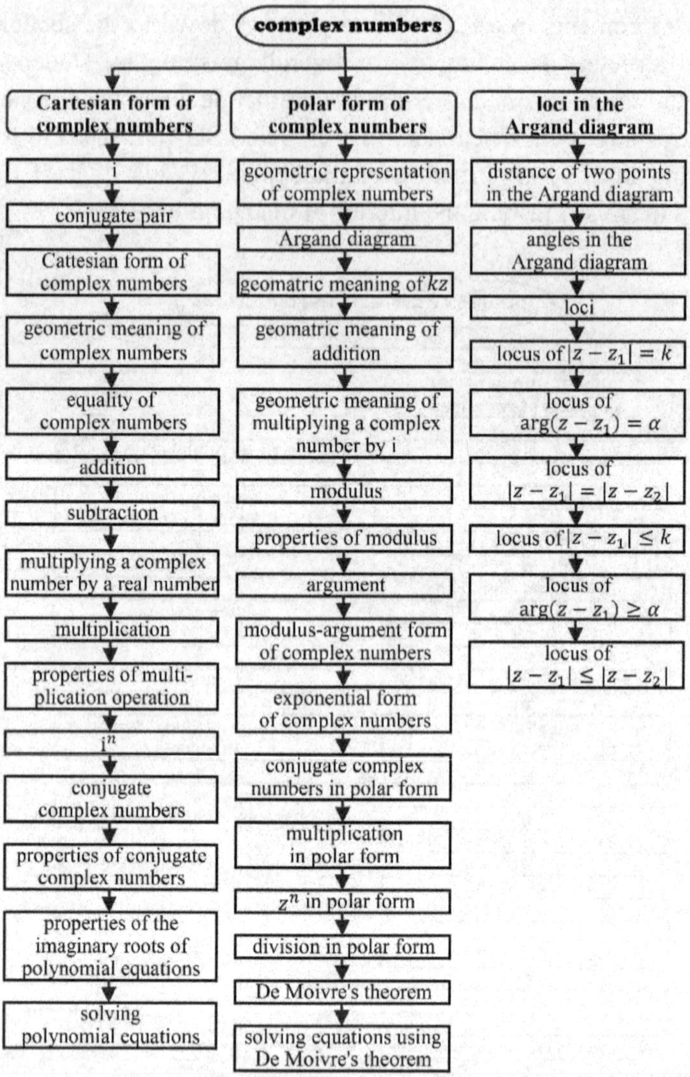

Fig. 2-2(4) The introduction order of the knowledge points relating to complex numbers in SG-PPE textbook

From Fig. 2-2(1)–2-2(4) we can see that there are 9 common key knowledge points in four series of textbooks, i.e., imaginary unit i, algebraic (Cartesian) form of complex numbers, equality of complex numbers, addition of complex numbers, subtraction of complex numbers, multiplication of complex numbers, division of complex numbers, and conjugate complex numbers. The four series of textbooks show basically the same arrangement sequence for these points. They firstly introduce imaginary unit i, then algebraic (Cartesian) form of complex numbers, followed by the complex number addition, subtraction, multiplication and division operations; however, the introduction of the conjugate complex numbers is located slightly different. In CN-PEP and JP-SKS the notion of conjugate complex numbers is located between multiplication and division. This arrangement is more directive and explicitly points out that the notion of conjugate complex numbers is a bridge from the multiplication to the division of complex numbers. This arrangement is based on the feature of the product of a pair of conjugate numbers. In CN-SEP conjugate complex numbers come after the addition of complex numbers, introducing the concept from its formal features followed by the study of its properties. This is more consistent with the mathematical logic of concept to property. In the four series of textbooks, conjugation is most frequently mentioned in SG-PPE: it firstly appears right after the establishment of the concept of complex numbers, then after the introduction of multiplication of complex numbers, and finally in the section of polar form. It starts from the appearance in an example, letting students know the term. When it comes on stage again, the notion of the conjugate number of an arbitrary complex number is introduced, the general features of conjugate numbers are exhibited, and their role in addition, subtraction and multiplication is studied in detail. Finally, the textbook enhances the understanding of conjugate complex numbers and explains it at geometry level. Actually, in addition to the concept of "conjugate complex numbers" in SG-PPE, "Argand diagram" also appears twice. This explains the spiral arrangement reflected in SG-PPE.

In summary, there are some differences in the compiling arrangement in the four series of textbooks. The amount of both main contents and knowledge points gradually decreases by the order from SG-PPE, CN-SEP, JP-SKS to CN-PEP. As for the arrangement of main contents and key knowledge points, SG-PPE follows the idea of spiral learning and the other three series of textbooks are in linear form. For section setup, CN-SEP textbook uses the horizontal mode, and the other three series are all in vertical mode.

(2) Content coverage

The findings of similarities and differences in logical order of the knowledge points and the presentation features of main concepts and principles will be reported in this section. We will analyze, by means of concept map, the depth and scope of the related knowledge points in four series of textbooks.

More specifically, we firstly list all knowledge points in relevant chapters in the four series of textbooks, including the knowledge points in the texts that have already listed in Fig. 2-2(1−4), knowledge points in reading material between sections, as well as certain “old” knowledge points related to complex numbers from earlier chapters of the textbooks. By “reading material between sections” we mean the “Read and Think” part in CN-PEP, “Reading Material” part in CN-SEP and the “Column” in JP-SKS. Secondly, we roughly divide these knowledge points into five categories according to their role in mathematics: the origin of complex numbers, connection among number fields, knowledge within complex numbers, relationship of complex numbers with geometry, and applications of complex numbers. Then all these knowledge points are connected by arrows as per the logical relation presented in textbook (knowledge points in text are connected with full line arrow and those in reading material with dotted line arrow). In this way, we draw four concept maps as shown in Fig. 2-3(1)−2-3(4) (limited to space, linking phrases were omitted in the maps).

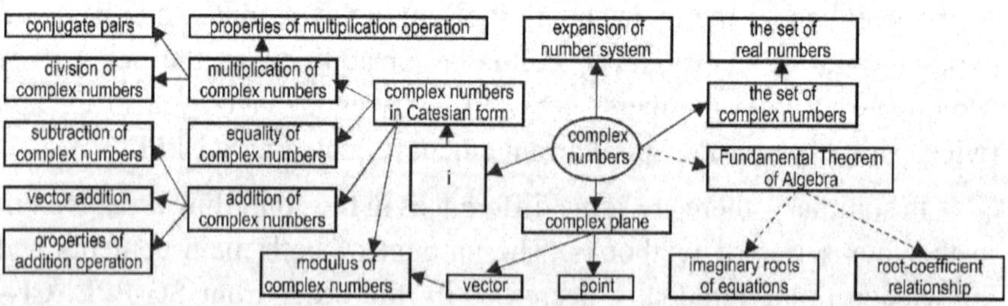

Fig. 2-3(1) Concept map of CN-PEP

It can be seen from Fig. 2-3(1) that the complex number concept map of CN-PEP covers all the five categories and includes totally 22 knowledge points. Most of the knowledge points confined themselves in the scope of complex numbers, and applications appear only in reading materials.

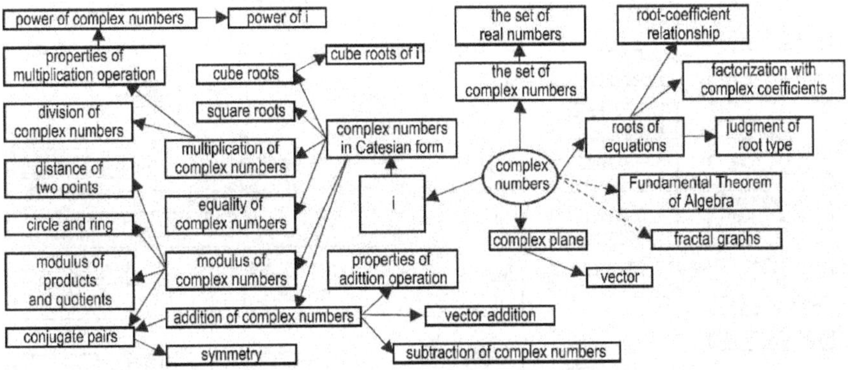

Fig. 2-3(2) Concept map of CN-SEP

Fig. 2-3(2) indicates that CN-SEP textbook does not include knowledge points in the category "origin of complex numbers". There are totally 32 knowledge points with most of them confined themselves in the scope of complex numbers, and applications are also designed in reading materials.

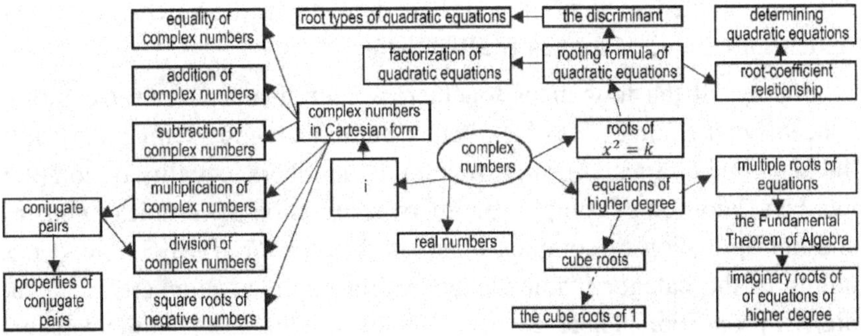

Fig. 2-3(3) Concept map of JP-SKS

Fig. 2-3(3) shows that JP-SKS does not include knowledge points in the categories "origin of complex numbers" and "relationship of complex numbers with geometry". There are totally 25 knowledge points. Most of them confined themselves in the scope of complex numbers, only 1 point, as a special case of cube roots, is given in the reading material.

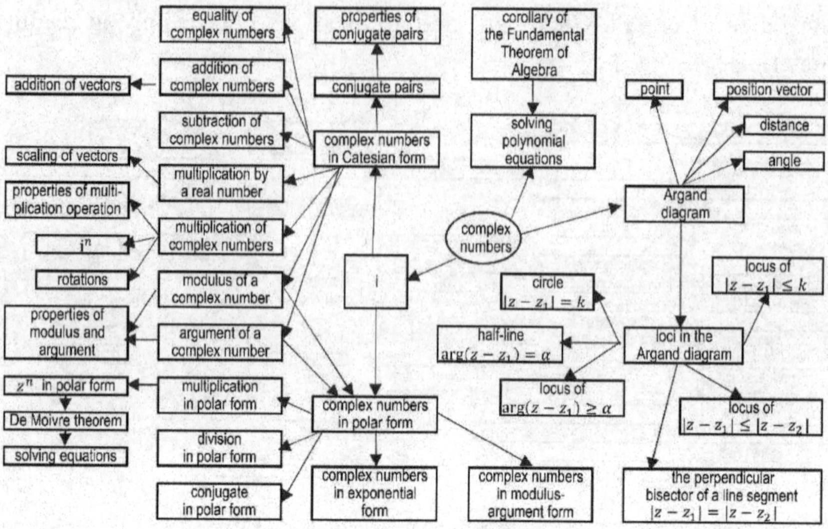

Fig. 2-3(4) Concept map of SG-PPE

It can be seen from Fig. 2-3(4) that SG-PPE does not include knowledge points in the categories "origin of complex numbers" and "connection among number fields". There are totally 41 knowledge points with most of them confined themselves in the scope of complex numbers. All the knowledge points are designed in the text, which means that students are required to master all of them.

Putting all the four maps together, we see that there are 7 common concepts in the four series of textbooks: complex numbers, imaginary unit, the algebraic (Cartesian) form of complex numbers, equality of complex numbers, addition of complex numbers, subtraction of complex numbers, and multiplication of complex numbers. Except for JP-SKS, knowledge points in the category "knowledge within complex numbers" take the biggest proportion. There are very few knowledge points in the category "application of complex numbers", and even if there are some, they exist outside of the main text, thus are often absent from teaching. On the contrary, most knowledge points in JP-SKS belong to the category "applications of complex numbers". Only CN-PEP textbook contains knowledge points of origin of complex numbers. These findings are consistent with previous analysis on the titles of complex number chapters.

Meanwhile, we notice that both the corresponding standard and syllabus for CN-SEP and SG-PPE have requirements on number system expansion but none of them has implemented accordingly. Moreover, there is a lack of links between complex numbers and real numbers in SG-PPE concept map. This may influence students' overall grasping of number systems.

The concept maps will be analyzed by means of structure/data analysis. To do this we need to define the terms and variables (see S. Hough *et al.* 2007).

Tab. 2-5 Terms in concept map analysis

Term	Definition
root	the first concept in the map
concept	concepts enclosed by circle or box in the map
link	line connection between two concepts
depth	length of the longest chain in the map
level	the number of links from the root to a given concept
width	the amount of concepts involved in a maximal level

Tab. 2-6 Variables in concept map analysis

Variable	Definition	Remark
number of concepts	total number of concepts in a concept map.	The number of concepts is used to evaluate the capacity of knowledge points in a textbook.
width	the amount of concepts involved in a maximal level in the map.	Width indicates the range of knowledge.
depth	length of the longest link in the map.	Depth reflects the profundity of knowledge.
HSS	width + depth.	HSS evaluates the complexity of a concept map

On basis of the above definitions, the data in the concept maps [Fig. 2-3(1)–(4)] about CN-PEP, CN-SEP, JP-SKS and SG-PPE are further analyzed. The results are summarized in Tab. 2-7 below.

Tab. 2-7 Statistics of concept maps in the four series of textbooks

	CN-PEP	CN-SEP	JP-SKS	SG-PPE
number of concepts	22	32	25	41
width	6	10	11	19
depth	4	6	5	6
HSS	10	16	16	25

As for the number of concepts, SG-PPE has 41 concepts and ranks first, followed by the CN-SEP and JP-SKS; while CN-PEP has the least concepts, only about half the number of SG-PPE.

As for the width of knowledge, SG-PPE is the most extensive (19), JP-SKS and CN-SEP come next, and CN-PEP ranks the last with less than one third of scope compared with PPE.

As for the depth of knowledge, CN-SEP and SG-PPE are the most profound (6), JP-SKS comes next, and CN-PEP again comes last.

In terms of HSS, SG-PPE is the most complicated, JP-SKS and CN-SEP textbook come next, and the CN-PEP textbook is the least complicated.

In summary, among the four series of textbooks SG-PPE towers over the other three series of textbooks in number of knowledge points, width and depth of knowledge as well as complexity of concept map. This shows that the textbook structure of complex number content in SG-PPE is the most complete among the four series of textbooks, which is consistent with the results obtained from previous analysis on compiling arrangement. By contrast, CN-PEP textbook ranks lowest in number of knowledge points, width and depth of knowledge as well as the complexity of concept map. This indicates that CN-PEP textbook is too simple in the complex number content and its content structure seems questionable. CN-SEP and JP-SKS are between CN-PEP and SG-PPE textbooks in every respect. There are more knowledge points in CN-SEP than in JP-SKS, mainly because JP-SKS ignores the relationship of complex numbers with geometry.

After specific research on the differences between CN-SEP and JP-SKS in width and depth, we can find that the two series of textbooks adopt different approaches to subtraction of complex numbers and division of complex numbers. The former only regulates addition and multiplication of complex numbers from the perspective of construction of an algebraic system, while subtraction and division are deductive from the view of inverse operations so that these two operations are no longer regulations. However, JP-SKS regulates the subtraction and division, putting four arithmetic operations on the same basis. Such difference also exists between CN-PEP and SG-PPE. The CN-PEP textbook treats subtraction and division operations as inverse operations, taking the same approach as

CN-SEP, while SG-PPE regards subtraction as one of the other operations and it even does not mention division operation generally. This reflects that the logical order of operations in the two series of textbooks from China is basically identical while that of JP-SKS is basically consistent with SG-PPE.

The completeness of knowledge system in SG-PPE makes it superior to the other three series of textbooks in width, depth, and complexity of concept map. On the contrary, the incompleteness of knowledge system in CN-PEP makes it inferior to others in every aspect. Meanwhile the difference in logical order of operations between CN-SEP and JP-SKS brings about discrepancy in width and depth of knowledge.

Finally, the authors analyze the presentation of concepts and principles in the complex number chapters of the four series of textbooks. Upon analysis, we find that, though there is a difference in numbers of concepts and principles among the four series of textbooks, they have similar explanations for the common contents. However, we can also see that JP-SKS does not define the concept the *algebraic (Cartesian) form* of complex numbers. Instead, it defines such a special form of complex numbers as *complex numbers.* This has somehow overlooked the other forms of complex numbers. The overgeneralization misleads students to consider what they learn as the complete content of complex number, which is not beneficial to students' understanding of complex numbers. Although both the CN-PEP and CN-SEP define complex numbers in a similar form $a + b\mathrm{i}$, the presence of the concept of "the algebraic form of complex numbers" gives students a hint that other forms of complex numbers may exist and what they are learning is only part of complex number content. It could, at least, avoid overgeneralization. The main text in SG-PPE does not introduce the division of complex numbers in algebraic form, but you can see after gaining insights into the textbooks that SG-PPE commonly uses calculators for division of complex numbers. This reduces requirement on students' operation ability to some extent even though it matches with the expected curriculum. In CN-PEP and CN-SEP textbook, the fundamental theorem of algebra is stated in reading materials; though fundamental theorem of algebra is not explicitly introduced in JP-SKS and SG-PPE, its relevant content appears in the main text of textbook, see Fig. 2-4(1) and Fig 2-4(2). Therefore the requirements of the two series of textbooks from China might not be higher in comparison with those from Japan and Singapore.

高次方程式の解の個数を，2重解は2個，3重解は3個などと数えることにすると，n次方程式はn個の解をもつことが知られている。

Fig. 2-4(1) Relevant statement of the fundamental theorem of algebra in JP-SKS text

Translation of Fig. 2-4(1) For the number of roots of equation of higher degree, if a double root counts 2, a triple root counts 3, and so on, then we know that an equation of degree n has exactly n roots.

Note | Guide | Questions

If a polynomial equation,
$f(x) = a_1x^n + a_2x^{n-1} + \ldots + a_nx + a_{n+1} = 0$,
where each a_i is a **real number** and
n is an integer with $n \geqslant 2$,
has a complex root $u + iv$, then the
conjugate $u - iv$ is also a root.

Fig. 2-4(2) Relevant statement of the fundamental theorem of algebra in SG-PPE text

(3) Summary

We highlight the following findings of the four series of textbooks in complex number content through the above quantitative and qualitative analysis.

- There are big differences in the selection of complex number content. The differences are: CN-PEP, CN-SEP and SG-PPE incline to stress knowledge about complex numbers themselves, the CN-PEP textbook includes the least content and lacks knowledge of applications and is incomplete in knowledge structure. SG-PPE pays special attention to complex number itself and has the most complete knowledge structure among the four series of textbooks. CN-SEP is in between the two series of textbooks but pays more attention to applications than CN-PEP and SG-PPE. JP-SKS focuses on applications of complex numbers though there are some issues about definition of concept.

• There are some differences among the four series of textbooks in arrangement sequence of complex number content and knowledge. With regard to the main contents and key knowledge points, SG-PPE is in obviously spiral form while the other three series of textbooks take linear approaches. As for section setup, CN-SEP textbook is of horizontal tile pattern while the other three series of textbooks are of vertical mode.

• There are obvious differences among the four series of textbooks in width and depth of knowledge and the complexity of concept map relating to complex numbers. SG-PPE surpasses the other three series of textbooks in every aspect and it is followed by CN-SEP, JP-SKS and CN-PEP in turn.

4.2.2 Content presentation

The data and analysis in this part mainly answers the second research question about the similarities and differences of each series of textbook in content presentation. We will analyze and compare from four aspects, namely, the presentation of concepts, the presentation of principles, the presentation of examples and the presentation of chapter summaries.

(1) Presentation of main concepts

We will choose a common concept and make the investigation of its presentation mode of the concepts in each series of textbooks. Presentation of concept can be divided into concept formation and conceptual understanding. The analysis indicators of concept formation include the ways of concept introduction, the existence of aside remark, the existence of figures, and the usage of IT. The analysis indicators of conceptual understanding include the existence of verbal complements or explanations, numbers of examples[b] and exercises to help understanding, the existence of aside remarks and figures, and the usage of IT.

Considering the consistency of the concepts of "complex numbers" and the "algebraic (Cartesian) form of complex numbers" in CN-PEP, CN-SEP and JP-SKS, and the lack of interpretation about the term "the

---

[b]There are two kinds of examples in CN-PEP and JP-SKS: one is easy and informal examples, usually appearing in the contexts of the interpretation of certain knowledge points; the other is formal examples, usually following the establishment of new knowledge points and aiming to help students to master and apply the knowledge.

algebraic (Cartesian) of complex numbers" in SG-PPE, we investigated these two concepts as a whole, and also choose the concept of conjugate complex numbers for analysis. The results of data coding are shown in Tab. 2-8 and Tab. 2-9.

Tab. 2-8 Presentation of complex numbers and the algebraic form of complex numbers

			CN-PEP	CN-SEP	JP-SKS	SG-PPE
Concept establishment	Way of introduction		Through the whole process of expansions of number system	To find the square root of negative numbers	To give solution to $x^2 = k$ (no matter k is positive or negative)	Given directly
	Figure	Math			1	
		Non-math		1		
	Aside remark		2	1		
	IT					Introducing graphing calculator
Conceptual understanding	Supplementary explanation		Yes	Yes	Yes	
	Informal examples		4		3	
	Formal examples		1	2	1	
	Number of exercises		2	4	1	
	Figure	Math	1		1	
		Non-Math				
	Aside remark			1	1	
	IT					

Tab. 2-9 Presentation of conjugate complex numbers

			CN-PEP	CN-SEP	JP-SKS	SG-PPE
Concept establishment	Way of introduction		Examples with the multiplication of complex numbers	Examples with the addition of complex numbers	Examples with the multiplication of complex numbers	Examples with solving equations
	Figure	Math			1	
		Non-math				
	Aside remark		1			
	IT					
Conceptual understanding	Supplementary explanation			Yes	Yes	
	Informal examples					
	Formal examples					
	Number of exercises			2	1	
	Figure	Math		1		
		Non-math				
	Aside remark		1			
	IT					

It can be seen from Tab. 2-8 and Tab. 2-9 that, during the introduction of the concept of complex numbers, CN-PEP textbook explains the whole process of the expansion of number systems, which not only reflects mathematical culture but also explains the necessity of establishing the concept. Moreover, when introducing the concept of conjugate complex numbers, CN-PEP starts from the property of the product of a pair of conjugate complex numbers and prepares for the future learning of division. There are aside remarks with the introduction of a concept to supplement certain special cases or some details to ensure the correctness and completeness of the concept. CN-PEP may explain a key concept with (informal) examples and designs formal examples and necessary exercises for concept identification in mixed circumstances, which helps students consolidate their understanding of the concept in time. Questions for "Think" and "Inquiry" are inserted between concept formation and conceptual understanding so as to stimulate students' active learning. For example, questions like "what is the relation between complex number set C and real number set R?" Therefore, in the overall teaching approach of concepts, CN-PEP follows the student-centered presentation mode upon request of the expected curriculum, that is, arranging the teaching materials from necessity to concept formation, then to question-driven conceptual understanding, and finally to concept clarification.

For concept formation, the CN-SEP textbook usually takes the formal features of concept as the starting point and explains with aside remarks, which are mostly supplementary description about the definition of the concept. This is one of the characteristics of CN-SEP. There are word explanations to support conceptual understanding after the concept is formed; and finally examples and exercises are designed to consolidate understanding of the concept and to guide the usage of the concept. The presentation of a concept generally follows the process from concept formation, to concept interpretation, then to concept usage.

JP-SKS shows another feature of presentation, i.e., the presentation with figures to impress students. It can be found after checking the presentation of some other concepts (such as square root of negative number) that, in JP-SKS, figures are often used to highlight the main mathematical features of a concept form, to impress students and to help them recognize the concept quickly after forming the concept with text, see Fig. 2-5 for example. Similar to CN-PEP, JP-SKS also starts from the necessity of establishing the concept and ends with examples and exercises to clarify the concept. The presentation mode of a concept in JP-SKS is usually as "the aim or necessity → establishing concept → enhancing concept → explaining concept → clarifying concept".

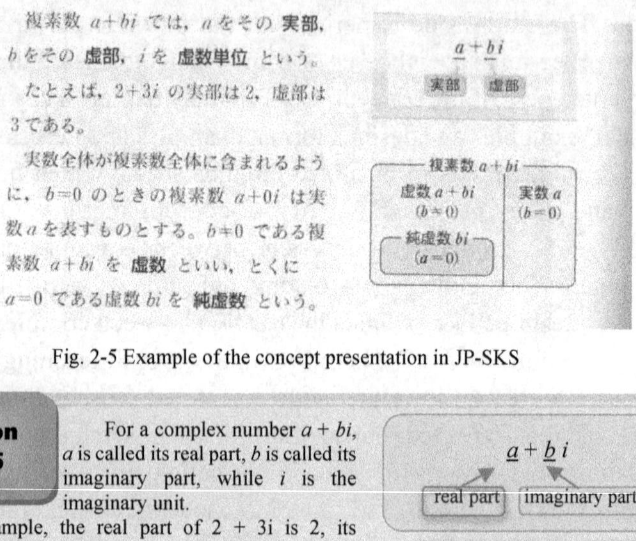

Fig. 2-5 Example of the concept presentation in JP-SKS

Translation of Fig. 2-5

For a complex number $a + bi$, a is called its real part, b is called its imaginary part, while i is the imaginary unit.

For example, the real part of 2 + 3i is 2, its imaginary part is 3.

If $b = 0$, the complex number $a + 0i$ is a real number, denoted simply by a, and in this way, real numbers, as a whole, are contained in complex numbers as a whole. If $b \neq 0$, complex number $a + bi$ is called an imaginary number. In addition, when $a = 0$, the imaginary number bi is called a purely imaginary number.

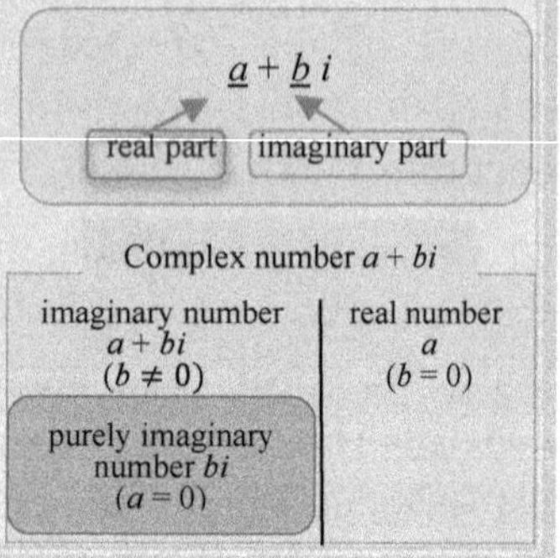

Unlike the above three series of textbooks, in SG-PPE the definition of a concept is usually given directly, like in college textbooks); and then the nature of concept is explored. It reflects the declarative process completely following the logical order.

It can be seen from the above analyses that the CN-PEP textbook has made changes in the presentation of concepts to allow students re-create the concepts they learned and to further activate their initiative in learning. Such effort is consistent with the requirements of curriculum standard. The other three series of textbooks do not fully meet the requirements of their respective curriculum standards or syllabi in this aspect.

(2) Presentation of main principles

Data shows that SG-PPE includes 9 principles in the complex number chapter, while CN-SEP, JP-SKS and CN-PEP have 8, 7 and 6, respectively. Among all the principles there are 3 included by all the four series of textbooks. We choose "the rule of subtraction of complex numbers" from the 3 common principles as a sample to analyze the presentation of principles in different textbooks.

We consider two aspects of the presentation of principles, namely, the principle introduction and the principle mastery, where the analysis indicator of principle mastery includes the way of proof of the principle, the existence of explanation of principle, the status of the usage (examples and exercises), and existence of figure, aside remark, and information technology for assistance of understanding. The result is shown in Tab. 2-10.

Tab. 2-10 Teaching of a selected principle in the four series of textbooks

				CN-PEP	CN-SEP	JP-SKS	SG-PPE
Statement of the principle				Let $z_1 = a + bi$ and $z_2 = c + di$ be two complex numbers. Then $(a + bi) - (c + di) = (a - c) + (b - d)\text{i}$			
Principle introduction				Analogy with real numbers	Inverse operation of addition	Given directly	Given directly
Principle mastery	Proof	Proof in text	Complete	—	Yes	—	—
			Partial	—	—	—	—
		Other proof		No	—	No	No
	Explanation			Yes	Yes	Yes	No
	Usage	Number of examples		1	2	1	1
		Number of exercises		2	3	2	No
	Figures			No	1	No	No
	Aside remark			No	Yes	No	No
	IT			No	No	No	No

In terms of principle introduction, JP-SKS and SG-PPE directly give the principle, while CN-PEP and CN-SEP start from the connection between old and new knowledge and build teaching and learning on students' prior knowledge.

As can be seen from Tab. 2-10, in terms of the requirements on the principle mastery, SG-PPE has the least requirements. There are neither proof of the principle, nor explanation of the principle in any forms. We find that this is not an exception in SG-PPE. The textbook does not require proving any principles in the complex number chapter. For the formulas and properties (such as the properties of operations with conjugate complex numbers), SG-PPE only makes statements, without reasoning process. Even for theorems (such as De Movire theorem), the textbook only has a short explanation before the theorem is given. On the contrary, the other three series of textbooks give explanations for most of the principles. It is particularly worth mentioning that CN-SEP textbook proves most of the principles (such as the subtraction and the division of complex numbers and the modulus of product), which enables students to fully understand the origin and the development of the principles. The CN-PEP textbook also makes corresponding statement and explanations to some theorems (such as the fundamental theorem of algebra) without formal proofs.

We can conclude from above analysis that:

There are some differences in the number of principles about complex numbers in the four series of textbooks, where SG-PPE has the most number of principles (9) and CN-PEP has the least (6). There are three principles included by all the four series of textbooks, which indicates that the four series of textbooks have basically the same requirements on the operation of complex numbers.

However, in contrast to the numbers, SG-PPE has the minimum requirements on the teaching of principles, while the CN-SEP and CN-PEP textbooks have higher requirements.

The general presentation patterns of principle are as follows: "introduction → proof → usage" in CN-PEP, "introduction → proof → explanation → usage" in CN-SEP, "given → explanation → usage" in JP-SKS, and "given → usage" in SG-PPE.

(3) Presentation of examples

There are mainly three indicators for analysis, namely, number of solutions, steps in solutions and understanding of examples. The number of solutions checks the amount of examples with multiple solutions in the textbook, steps in solutions cover the completeness of solutions, hints provided in the text for analyzing the questions and the reflections on the examples, while understanding of examples includes the use of figures, aside remarks and information technology for assistance of understanding. It should be noticed that here multiple solutions not only include the complete solving processes clearly described in the textbook but also prompts and hints given in the textbooks leading to other solutions.

After careful analysis on all the examples related to complex number in the four series of textbooks, we find that the number of examples decreases from CN-SEP, SG-PPE, JP-SKS down to CN-PEP with the numbers being 24, 23, 19 and 5, respectively. CN-PEP has the fewest examples and this matches with the previous conclusion that it includes the least amount of knowledge contents. In addition to the number of examples, the other indicators are summarized in Tab. 2-11.

Tab. 2-11 Presentation of examples in the four series of textbooks

		CN-PEP	CN-SEP	JP-SKS	SG-PPE
Number of solutions	Examples with one solution	4	22	18	20
	Examples with two solutions	1	2	1	3
	Total examples	5	24	19	23
Examples with method of analysis		2	0	2	0
Examples with additional explanation		0	0	3	0
Examples with figures	Math	0	4	1	16
	Non-math	0	0	0	0
Examples with tables		0	0	0	0
Examples with asides	Question type	0	2	0	1
	Explanatory type	1	0	9	10
	Consolidating type	0	1	0	0
Examples using IT		0	0	0	8

We have the following interesting findings:

• *On the examples with multiple solutions.* All the four series of textbooks include examples with two solutions to one problem, which shows they have the requirement in this aspect. However, after an investigation on the examples and their solution processes given in the textbooks, we find big differences between different series. Though 3 out of 23 examples in SG-PPE have two solutions, there are only hints for the second solutions, and, in fact, the solutions pointed out by the hints are essentially the same as the first solutions with only certain differences in the forms of representations (see Fig. 2-6, for example). Both CN-PEP and JP-SKS include one example (out of 5 examples and out of 19 examples, respectively) with two solutions, in which the first solution is given in detail, while the second one has a hint only. Different from SG-PPE, the second solutions in CN-PEP and JP-SKS are different in nature from the first ones. For CN-SEP, 2 out of 24 examples are given with two essentially different solutions, and all solutions are described in details. Therefore, among the four series of textbooks, CN-SEP outperforms in terms of multiple solutions to one problem and should be helpful to developing students' thinking.

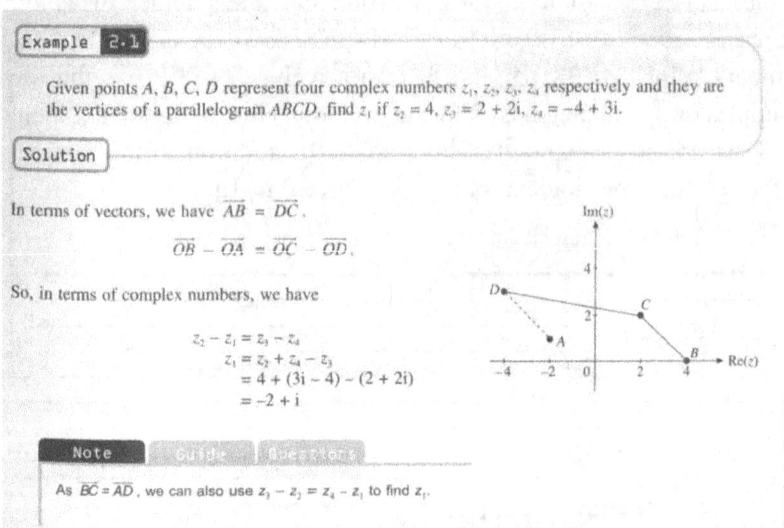

Example 2.1

Given points A, B, C, D represent four complex numbers z_1, z_2, z_3, z_4 respectively and they are the vertices of a parallelogram $ABCD$, find z_1 if $z_2 = 4$, $z_3 = 2 + 2i$, $z_4 = -4 + 3i$.

Solution

In terms of vectors, we have $\overrightarrow{AB} = \overrightarrow{DC}$.

$$\overrightarrow{OB} - \overrightarrow{OA} = \overrightarrow{OC} - \overrightarrow{OD}.$$

So, in terms of complex numbers, we have

$$\begin{aligned} z_2 - z_1 &= z_3 - z_4 \\ z_1 &= z_2 + z_4 - z_3 \\ &= 4 + (3i - 4) - (2 + 2i) \\ &= -2 + i \end{aligned}$$

Note

As $\overrightarrow{BC} = \overrightarrow{AD}$, we can also use $z_3 - z_2 = z_4 - z_1$ to find z_1.

Fig. 2-6 Example of multiple solutions to one problem in SG-PPE

• On the steps in solution. JP-SKS gives the most complete steps of solution among the four series of textbooks. It analyzes the key points with prompts for complicated examples, and provides additional remarks to the matters requiring attention in the solution after the problem is solved. There is no analysis or additional remarks in the CN-SEP and SG-PPE, which hinders students from understanding where the solution comes from and may negatively influences students' reading or self-study.

● *On the example understanding.* Over half of the examples in both JP-SKS and SG-PPE have aside remarks in the problem solving process to analyze and justify the key steps. This helps students to better understand the problem solving processes and the key steps. We also find that CN-SEP, JP-SKS and SG-PPE have 4, 1 and 16 examples, respectively, accompanied by figures. However, further analysis shows that in SG-PPE, most of the examples with figures are found in the part "relationship of complex number with geometry", where the figures are required by the problem nature and have no effect on enhancing understanding (see Fig. 2-7). The figures in JP-SKS are the reappearance of concepts used in problem solving and it only helps understanding to a limited extent. In CN-SEP, some of the figures are added for illustrating problem-solving strategy of "symbolic-graphic combination". It is vital for understanding and it also enriches the types of examples. A considerable number of examples in SG-PPE are accompanied with usage of graphic calculator. The figures in those examples are used to either check the result or to display the process. SG-PPE reduces unnecessary tedious calculation by using graphic calculator and it also demonstrates the goal of "effectively use of various mathematics tools" in *Singapore Syllabus*.

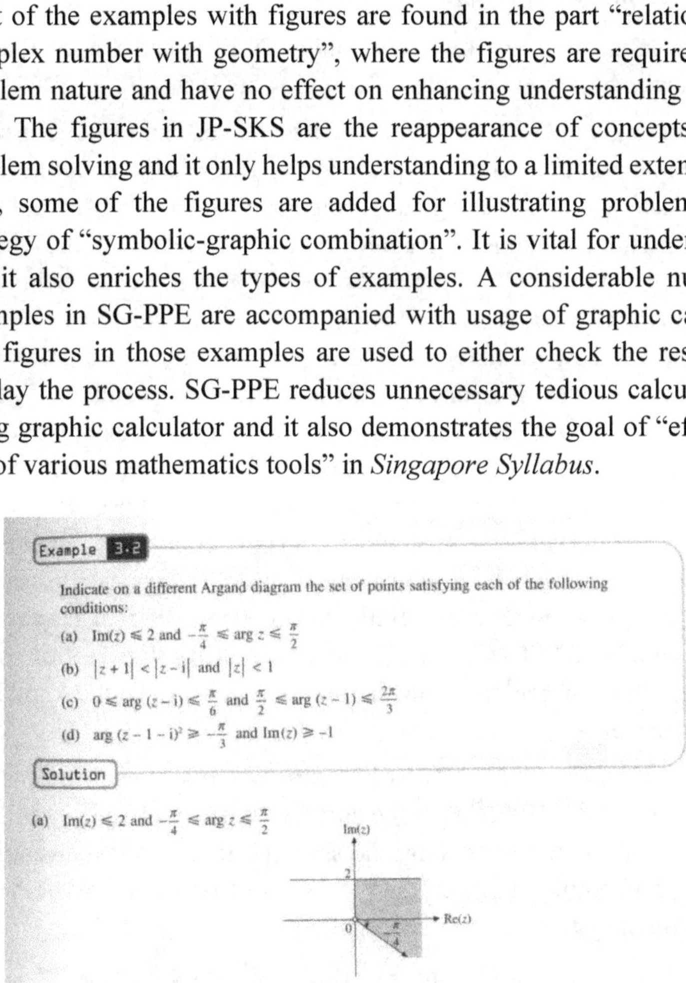

Example 3.2

Indicate on a different Argand diagram the set of points satisfying each of the following conditions:

(a) $\mathrm{Im}(z) \leqslant 2$ and $-\frac{\pi}{4} \leqslant \arg z \leqslant \frac{\pi}{2}$

(b) $|z+1| < |z-\mathrm{i}|$ and $|z| < 1$

(c) $0 \leqslant \arg(z-\mathrm{i}) \leqslant \frac{\pi}{6}$ and $\frac{\pi}{2} \leqslant \arg(z-1) \leqslant \frac{2\pi}{3}$

(d) $\arg(z-1-\mathrm{i})^3 \geqslant -\frac{\pi}{3}$ and $\mathrm{Im}(z) \geqslant -1$

Solution

(a) $\mathrm{Im}(z) \leqslant 2$ and $-\frac{\pi}{4} \leqslant \arg z \leqslant \frac{\pi}{2}$

Fig. 2-7 Figure without effect on enhancing understanding in an example in SG-PPE

Overall, with the most diversified forms, the examples in CN-SEP are the most conductive in expanding students' thought and demonstrating the connection of knowledge. Examples in JP-SKS are the most readable and those in SG-PPE are the most reflective about the idea of using modern technology.

(4) Presentation of chapter summaries

In the four series of textbooks, only CN-PEP and CN-SEP attach chapter summaries to the chapter of complex numbers.

The summary in CN-PEP includes two parts, i.e., "knowledge structure in the chapter" and "review and thinking". "Knowledge structure in the chapter" demonstrates the logical arrangement of the chapter and helps students master the general framework of knowledge in the chapter. "Review and thinking" summarizes the important ideas behind the contents in the chapter, draws an analogy between old and new knowledge, and further enlighten students on the understanding from their study experiences. Besides, it arranges some expansion tasks such as "what is your understanding of the rationality the four arithmetic operations of complex numbers", "how do you understand the consistency between the addition and subtraction of complex numbers with those of vectors", "collect historical materials about expansion from real numbers to complex numbers, and know the number system expansion process of 'integral numbers → fractions (rational numbers) → real numbers → complex numbers'". The CN-PEP summary takes a bird's view of the overall structure of contents, thus helps students to grasp the knowledge comprehensively.

The summary in the CN-SEP textbook sorts the concepts and principles learned by students in the chapter with some blanks be filled in by students. It reviews the details and asks questions such as "what is the real part of a complex number" and "what operation laws are obeyed by the addition operation of complex numbers".

Thus, it can be said that the CN-PEP textbook enhances the overall mastery of contents while the CN-SEP textbook focuses on the details of knowledge. This is consistent with the conclusion obtained from previous analysis of the section setup in these two series of textbooks.

JP-SKS and SG-PPE integrate the knowledge by merely using exercises at the end of chapters and their structures of textbooks seem somewhat incomplete.

(5) Summary

• In terms of the overall presentation of teaching contents, CN-PEP has made changes from the traditional way, trying engage students in re-creation of mathematical knowledge by using columns such as "Thinking" and "Exploration", which reflects the "exploratory" teaching mode; while CN-SEP, JP-SKS and SG-PPE still adopt the "declarative" mode. Moreover, SG-PPE basically has no requirement on proving principles; in the other extreme, CN-SEP is most strict in this aspect and requires both proofs and written explanation in the main text. The CN-PEP textbook requires students to prove most of the formulas in the chapter.

• The structure the CN-PEP and CN-SEP as textbooks are relatively complete while JP-SKS and SG-PPE lack summary of knowledge. The summary in the CN-PEP inclines to the overall view while CN-SEP focuses on details.

• As for the presentation of examples, all the four series of textbooks are intended to embody the idea of multiple solutions to one problem and expand students' thinking; however, there are differences among them. The multiple solutions in SG-PPE are only *pro forma* while that in the other three series of textbooks are substantial. CN-PEP and JP-SKS provide only the clues or hints and the CN-SEP provides multiple solutions completely. Moreover, the CN-SEP textbook enhances "symbolic-graphic combination" strategy and is more diversified in the presentation of problem solving process. JP-SKS includes analyses before the formal problem solving steps and reminds precautions after the solution, and it even has interpretations of the steps. It pays much attention to the details; thus it is the most readable. The graphic calculator is frequently used in the problem solving process of the examples in SG-PPE in order to reduce unnecessary tedious calculation and lower the complexity of operation for students. It also demonstrates the connection between algebra and geometry through graphic calculator. This is a fair reflection of the requirement of its curriculum reform.

References

[1] Alajmi A. H. (2011). How do elementary textbooks address fractions? A review of mathematics textbooks in the USA, Japan, and Kuwait. *Education Studies in Mathematics*. 79(2): 239–261.

[2] Bao, Jiansheng (2002). Comparative study on composite difficulty of mathematics Chinese and UK middle school [In Chinese]. Ph. D. Dissertation. East China Normal University.

[3] Charalambousa, C. Y., Delaney, S., Hsu, H.-Y and Mesa V. (2010). A Comparative Analysis of the Addition and Subtraction of Fractions in Textbooks from Three Countries. *Mathematical Thinking and Learning*, (12): 117–151.

[4] Chen, Yuelan (2006). Japanese new high school mathematics curricula, textbooks and college entrance examinations [In Chinese]. *Mathematics Teaching*, (1): 13–16.

[5] Chen, Yuelan (2009). The 2008 edition of the mathematical curriculum standard for the Japanese middle schools will be applied in 2009 [In Chinese]. *Shuxue Tongbao (Bulletin des Sciences Mathematics)*, 48(12): 5–8.

[6] Chen, Yuelan (2010). A Panorama of Japanese mathematical education. *School Mathematical Education Monthly*, (5): 1–4.

[7] Cheng, Guangwen (2009). Comparison and interpretation of Chinese and American mathematics curriculum standards [In Chinese]. Journal of Quanzhou Teacher's College (Natural Science), 27(6): 119–122.

[8] Cheng, Zhou (2008). Comparison of editions A and B of the high school mathematics experimental textbooks (compulsory courses) by People's Education Press [In Chinese]. *Education Science Study*, (3): 35–38.

[9] Fan, Lianghuo and Zhu, Yan (2007). Representation of problem-solving procedures: A comparative look at China, Singapore and US mathematics textbooks. *Education Studies in Mathematics*, 66(1):61–75.

[10] Fu, Yingfang and Zhou, Junhua (2007). Comparison of the middle school mathematics curriculum standards of China and UK [In Chinese]. *Mathematical Teaching Study*, (4): 2–4.

[11] Hough, S., O'Rode, N., Terman N. and Weissglass, J. (2007). Using concept map to assess change in teachers' understandings of algebra: a respectful approach. *Journal of Mathematics Teacher Education*, (10): 23–41.

[12] Jiang, Boju (2005). What has been lost for mathematics in the new curriculum standard? [In Chinese]. *Guangming Daily*, March 16.

[13] Kong, Qiping and Hu, Songlin (2002). Overview of the contents strengthened and weakened in the domain of numbers and algebra: Comparisons of the new curriculum standard with the old syllabus, I [In Chinese]. *Journal of Subject Education*, (9): 1–5.

[14] Kyungmee Park, K. and Leung, K. S. F. (2006). A Comparative Study of the Mathematics Textbooks of China, England, Japan, Korea, and the United States. *New ICMI Study Series*, 9: 227–238.

[15] Li, Qinglin (2009). Comparative study of the high school mathematics curriculum standards of US and China [In Chinese]. Master Dissertation, Huazhong Normal University.

[16] Li, Yeping (1999). An analysis of algebra content organization and presentation, and to-be solved problems in eighth-grade mathematics textbooks from Hong Kong, Mainland China, Singapore and the United States.

[17] Li, Yeping (2000). A comparison of problems that follow selected content presentations in American and Chinese mathematics textbooks. *Journal for Research in Mathematics Education*, 31(2): 234–241.

[18] Li, Yeping, Chen, Xi and An, Song (2009). Conceptualizing and organizing content for teaching and learning in selected Chinese, Japanese and US mathematics textbooks: the case of fraction division. *ZDM Mathematics Education*, 41: 809–826.

[19] Liu, Jiao (2009). Comparative study of the contents for middle schools in the domain of numbers and algebra in the curriculum standards of China and Singapore [In Chinese]. *Journal of School Mathematics*, (10): 25–27.

[20] National Office for Education Science Planning (2008). Review of the Ministry of Education's key research project "comparative study of the mathematical curricula and textbooks of US, Japan, Macau, Hong Kong, and China Mainland" [In Chinese]. *Forum on Contemporary Education*, (4): 5–7.

[21] Schmidt, W. (1995). Curriculum analysis technical report series No.1 — The TIMSS curriculum analysis: an overview of an integrated system of curriculum measurement.

[22] Schmidt, W. (1995). Curriculum analysis technical report series No.2 — Document analysis: data collection and processing.

[23] Schmidt, W. (1995). Curriculum analysis technical report series No.3 — Document analysis: generation of analysis files.

[24] Schmidt, W. (1995). Curriculum analysis technical report series No.4 — Document analysis: representative country profiles.

[25] Schmidt, W. (1995). Curriculum analysis technical report series No.5 — General topic trace mapping: data collection and processing.

[26] Schmidt, W. (1995). Curriculum analysis technical report series No.6 — Analysis of pre-selected country groups.

[27] Son, J. (2005). A Comparison of How Textbooks Teach Multiplication of Fractions and Division of Fractions in Korea and the U.S. *PME 29*, 4: 201–208.

[28] Sun, Mingfu and Gao, Weizong (2002). Comparison on the teaching objectives and curriculum contents of Chinese and Japanese high school mathematics and its enlightenment [In Chinese]. *Journal of Subject Education*, (7): 39–42.

[29] Wang, Fenping (2011). Comparative Study of the contents of complex numbers in high school mathematics textbooks of China and Britain: the case of the comparison between AQA textbook and the version A textbook of People's Education Press [In Chinese]. *Journal of Mathematics Education*, 20(3): 83–86.

[30] Xiao, Songmei (2007). Comparative study between the textbooks of Stage I Curriculum Reform and the Stage II Curriculum Reform in Shanghai [In Chinese]. Master Dissertation, Shanghai Normal University.

[31] Xin, Ying (2008). Comparative study on the algorithm contents in high school mathematics curriculum of China and Japan [In Chinese]. Master Dissertation, Northeast China Normal University.

[32] Xue, Youcai, Hu, Yue, and Ye, Saiying. (2006). A comparison on typical textbooks on calculus in US and China [In Chinese]. *Technology and Publication*, (2): 73–74.

[33] Yang, Junbo (2007). Comparative study on the middle school mathematical curricula of China and Japan [In Chinese]. Master Dissertation, Shenyang Normal University.

[34] Yang, Shuqing (2007). Comparative study of the textbooks on numbers and algebra for middle schools: the case of the comparison of the textbook by Beijing Normal University Press and the textbook by Panpac Education [In Chinese]. Master Dissertation, Northwest China Normal University.

[35] Yao, Xue (2010). Comparative study on high school mathematics textbooks of Shanghai, Hong Kong, and Singapore [In Chinese]. Master Dissertation, East China Normal University.

[36] Zhang, Weizhong and Li, Fangqi. (2009). Comparison on the characteristics of the mathematics textbooks of Singapore and China [In Chinese]. *Foreign School Education*, (2): 32–36.

[37] Zhu, Yan and Fan, Lianghuo (2006). Focus on the representation of problem types in intended curriculum: A comparison of selected mathematics textbooks from Mainland China and the United States. *International Journal of Science and Mathematics Education*, (4): 609–626.

https://doi.org/10.1142/9789814713955_0003

Chapter 3

A Comparative Study on the Organization and Presentation of Geometry Content in Textbooks of China and Germany

Lijun Ye

School of Sciences, Hangzhou Normal University
Hangzhou, Zhejiang 310036, China
yeatsylj@126.com

Macroscopic and microscopic comparison models are established based on an overall comparison of textbooks from China and Germany. By using qualitative and quantitative methods, the geometry content in the Lambacher Schweizer mathematics textbooks from Germany (DE-LSM) and that in the textbook published by People's Education Press, version A, from China (CN-PEP) are compared. Taking the subject of solid geometry as an example, we find that: DE-LSM covers a relatively narrower but more profound range of content and pays more attention to the cultivation of students' comprehensive ability, while CN-PEP covers a wider range of content and focuses on the cultivation of spatial imagination. As for the writing style, DE-LSM is well organized and more targeted at the specific contents, while CN-PEP is more systematical and attaches more importance to the integrity of mathematical knowledge; and as for the presentation of knowledge, DE-LSM emphasizes the application, while CN-PEP gives more attention to students' intuitive perception.

1. Research Background

Textbooks play important roles in schools: they provide teaching materials based on curriculum standards/syllabi and students' cognitive structure, embody the curriculum contents and teaching goals, and serve as major tools for teachers and students. Textbooks in a country carry the pedagogical thought and educational philosophy of the country. To some extent textbooks determine the learning opportunity and academic achievement of students. The curriculum reform in China and other countries put much emphasis on textbooks.

Different countries have different selections and structure arrangements for the geometry contents, due to various reasons such as teaching philosophies, goals, national cultures, etc.

Many scholars (see, for example, Fuson *et al.*, 1988 and Yeping Li, 2000) believe that textbook comparison can not only find out the similarities and differences of curricula, but also explain, to a large extent, the reasons for students' different performances in mathematics. As a Chinese proverb goes, "stones from other hills may serve to polish our jade". Though the social structures and cultural traditions vary in different countries, there are many similarities in their mathematics textbooks due to the relatively strong stability and continuity of mathematics itself. By comparing mathematics textbooks, the involved countries can mutually benefit from each other, thus can better serve the curriculum reform of mathematics in their own countries.

Geometry studies the shape, size and positional relation of figures and solids abstracted from the objects in the real world, is a branch of mathematics being axiomatized first, and is one of the earliest contents of mathematical education. Geometry curriculum is always the focus of reform in different historical periods of mathematics teaching. Study of geometry in high school, the final stage of the national basic education, is very important for both the students' professional development in the future and the development of individual mathematics thought (Lijun Ye, 2011).

With the above background, we carried out a comparative study on the teaching content and presentation of geometry in the textbooks from China and some other countries. In this chapter, we limit ourselves in the comparison of China and Germany.

2. Current Research Status

The comparative studies of school mathematics textbooks are usually from two dimensions, horizontal and vertical. The former mainly includes comparisons of textbooks from different publishers within a country or comparisons across countries; while the latter investigates the development and evolution of textbooks within a country, e.g. based on the new round of curriculum reform, many scholars start to compare the geometric contents in the new and old versions of textbooks in China.

The comparison of textbooks across countries often focuses on the content and arrangement sequence of the textbooks for the same grade, as well as the characteristics of the textbook arrangement and correlation. The most influential study in this area is the Trends in International Mathematics and Science Study (TIMSS) sponsored by the International Association for the Evaluation of Educational Achievement (IEA). In 1994, TIMSS researchers studied thousands of textbooks, curriculum guidance documents and other curriculum material from dozens of countries. It is the first time in the history that textbook is treated as an important research object on a large scale (Schmit *et al.*, 1997). The research results have been summarized by Howson (1995).

Since then, many scholars follow TIMSS and start to make various comparisons of textbooks. For example, Yeping Li (2000) has established a three-dimensional framework — mathematical features, contextual features and performance requirements — to explain the similarities and differences in the expectations to students between China and the USA. By using this framework, he made an overall comparison of all relevant problems that followed the content presentation of addition and subtraction of integers in several textbooks from China and the USA, and found that problems in American textbook varied in problem requirements and emphasized conceptual understanding more than that in Chinese textbook. Jiansheng Bao (2002) has compared the composite difficulties of school mathematics curricula of China and the UK by a pentagon model for mathematics problems, considering five factors including computation, reasoning, topic coverage, exploration, and context. Fangqi Li (2009) has compared the structures of three versions of mathematics textbooks from People's Education Press, Zhejiang Education Press and Singapore Shing Lee Publishers by focusing on "Equations".

As for the geometry contents, Hengjun Tang and Weizhong Zhang (2005) have compared the teaching contents of geometry in primary and middle schools of China and the USA from the aspects of curriculum philosophy, content selection, and implementation of curriculum. They found that the curriculum philosophies of both countries are similar. Furthermore, they compared the "similar" contents in the *Geometry by Discovery* (Wiley) and the middle school mathematics textbook published by East China Normal University Press, and found that both countries take intuitive geometry as the beginning of geometry study, while the textbook from the USA includes more profound knowledge at a wider range due to their emphasis on construction of problem situation. Aihong Yuan (2007) has compared geometry contents in three series of textbooks (published by People's Education Press, Beijing Normal University Press and Hunan Education Press) as per the National Mathematics Curriculum Standard for Compulsory Education (experimental), and also compared with the University of Chicago School Mathematics Project textbook of the USA and *Наглядная Геометрия* from Russia. Shuwen Li and Ningzhong Shi (2012) has made a comparative study on content of school geometry curricula in China and Japan. Although some comparative studies focused on the geometry contents have been done, research in this field is far from complete, especially for high school level.

3. Research Design

3.1 Research object

In recent years, due to the influence of TIMSS and PISA, Germany is undergoing a large scale curriculum reform from primary to high schools and from regular education to vocational education (Binyan Xu, 2002). This curriculum reform has gone through a tortuous process of exploration. A new kind of curriculum philosophy and new standards for different disciplines are emerging. Meanwhile, mathematics education reform in China is in full swing. Thus it is interesting to compare high school mathematics textbooks of China and Germany, and it is particularly important to analyze their differences in style, arrangement and characteristics.

The Lambacher Schweizer mathematics textbook series published by Ernst Klett Verlag is of great influence in Germany. It has been used in

Bavaria, Baden-Württemberg, North Rhine-Westphalia, Rheinland-Pfalz, Niedersachsen, Bremen, Hamburg and Berlin. Meanwhile the textbook published by People's Education Press (version A) as per the *National Mathematics Curriculum Standard for High Schools (Experimental)* (MOE of PRC, 2003) is also widely used and occupies the greatest percentage of the market of textbooks in China. Therefore, we select these two series of textbooks (abbreviated as DE-LSM and CN-PEP respectively) for our comparative study.

In Germany, children enter primary school at six and study there for 4 years. Grades five and six are the transition phase for the students during which they and their parents have enough time to select a suitable secondary school to enter. The secondary schools in Germany have three types: Hauptschule, Realschule, and Gymnasien. Since August 1992, Bavaria has started to implement a new mathematics teaching plan in secondary schools, which does not separately set up several tracks in these schools as before, but instead, from grade nine, additional mathematics and science courses are required for the students who are willing to purse mathematics or science majors in their college study. There are still foundation courses and ability courses for students from Grade 12 on (Changping Chen, 2000).

Although Grades 11 to 13 in Germany correspond to the high school stage in China, the algebra and geometry contents learned by grade ten students in Germany — such as the power, exponent, and logarithmic functions and conic sections in the required courses for math-science majors — basically equals to what high school students learn in China. Thus the author selects the geometry part of mathematics textbooks for Grades 10 to 13 in Germany and the high school textbooks in China for a comparative study.

There are three volumes related to geometry for Grades 10 to 13 in DE-LSM:

— Intuitive geometry required for grade ten students: *Lambacher Schweizer Geometrie Bayern 10*,

— Fundamental analytical geometry course: *Lambacher Schweizer Analytische Geometrie Grundkurs*, and

— Enrichment analytical geometry course: *Lambacher Schweizer Analytische Geometrie Leistungskurs*.

The latter two textbooks are used in the additional courses for the potential math-science students.

In CN-PEP textbook, geometry related textbooks include

— Two volumes from the Compulsory Courses: Math. 2 and Math. 4,
— Two volumes from the Optional Series 1: Opt. 1-1 and Opt. 1-2,
— Two volumes from the Optional Series 2: Opt. 2-1 and Opt. 2-2, and
— Two volumes from Optional Series 3 and 4: Opt. 3-3: Geometry on the Spherical Surface, and Opt. 4-1: Selected topics on Geometry Proofs.

Since the content of linear geometry in the textbooks of DE-LSM enrichment analytic geometry course is not included in the Chinese courses, it will not be analyzed in our study. CN-PEP Optional Series 1 and 2 are designed for social science students and natural science students, respectively, with contents overlapped. Therefore, in this study Optional Series 1 is excluded from the sample.

3.2 Research method

The literatures used in this study are mainly collected from various sources such as thesis, books, newspapers and websites. We sorted and analyzed all relevant materials, collected theoretical research results and coordinated the common research methods and modes used in international textbook comparative studies in recent years, and finally developed a comparative analysis framework for the current research.

Based on the framework, we then made an in-depth text analysis on the related contents of the selected textbooks. By analyzing the perspective of main texts and graphs in textbooks, we probe the characteristic of knowledge system and arrangement structure of textbooks from China and Germany in terms of textbook organization; we also evaluate the thinking level of textbooks from both countries and the differences in the presentation and background knowledge.

Finally, we employed both quantitative and qualitative ways to analyze the data. Qualitative methods include expansion of contents, extraction of key points, etc. By quantitative method, the collected data were sorted, calculated, analyzed, interpreted and explained. In addition, concept maps and composite difficulty model were also used in the research. The results obtained are visually represented in graphs. Thus we get the differences and similarities between textbooks from both countries in terms of knowledge structures, cognition levels, presentation styles and knowledge backgrounds.

3.3 Research framework

We have established the following models (Fig. 3-1) for textbook comparative study by referring to other research experiences through repeating the cycle of "discussion — experiment — modification".

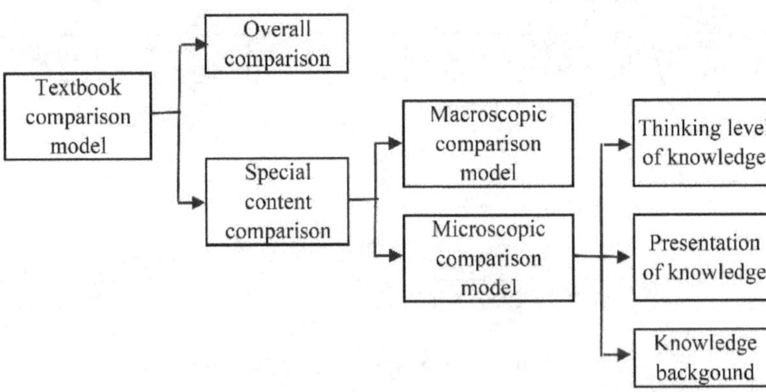

Fig. 3-1 Research Framework

Firstly we expand all contents in the textbooks of both countries vertically and compare the scope and arrangement sequence of contents as well as the arrangement structure of chapters. Hence we can track the line of thought and focus on the arrangement of mathematics textbooks from both countries.

Secondly, we select several representative segments of geometry content for more detailed comparative study. Comparison of specific content starts from sorting the arrangement sequence and comparing the correlation degree of knowledge points; common contents in different textbooks are further studied from the cognition level, presentation style and knowledge background, with the specific knowledge contents as the clue, in order to find the compiling characteristic of mathematics textbooks from both countries. Moreover, we hope to provide theoretical and practical implications and suggestions for the development of textbook in China.

4. Comparison of Geometry Content

4.1 Overall comparison

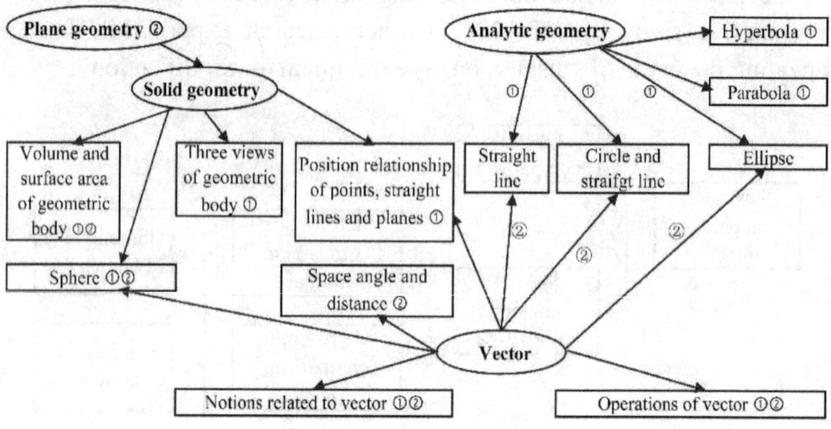

Note: ① represents CN-PEP; ② represents DE-LSM.

Fig. 3-2 Knowledge distribution in textbooks from China and Germany

4.1.1 Comparison of the content distribution

It can be seen from the Fig. 3-2 that CN-PEP covers a relatively wide scope of contents, including the surface area of a geometric body, the three views, the position relationship of points, lines and planes, the straight line, the circle and straight line, the conic section, the sphere, the vector, etc., focusing on the position relationship of points, lines and planes in space, and the straight line and curves; in which the three views, the hyperbola and parabola are not found in DE-LSM. A significant feature of DE-LSM

Tab. 3-1 Arrangement sequences of the content in textbooks from China and Germany

DE-LSM	Main content	CN-PEP	Main content
Circle	➢ Area ➢ Circumference	Solid geometric bodies	➢ Surface area and volume ➢ Three-views of a solid body
Cylinder, circular cone and sphere	➢ Surface area ➢ Volume	Position relations of points, straight lines and planes	➢ Position relations of points, lines and planes ➢ Parallel and perpendicularity of lines and planes
Vector and linear equation	➢ Vector and its coordinate representation ➢ Operation of vector coordinates ➢ Vector scalar multiplication ➢ System of linear equations ➢ Basis and dimension	Straight line and equation	➢ Inclination angle and slope ➢ Equation of straight line ➢ Intersection point of straight lines ➢ Point-point distance and point-line distance
Straight line and plane	➢ Representation of straight line and intersection point of straight lines ➢ Proofs related to plane ➢ Transversal by planes ➢ Position relations of planes ➢ Equation of plane	Circle and equation	➢ Equation of circle ➢ Straight line and circle ➢ 3 dimensional rectangular coordinate system
Length, distance and angle	➢ Length, included angle and dot product ➢ Distance from point to plane and straight line ➢ Line-line, line-plane and plane-plane angle	Plane vector	➢ Concept ➢ Linear operation ➢ Coordinates presentation ➢ Dot product
Circle and sphere	➢ Circle and straight line on the plane ➢ Sphere ➢ Sphere and straight line ➢ Sphere and plane	Conic section	➢ Curve and equation ➢ Ellipse ➢ Hyperbola ➢ Parabola
Ellipse	➢ Ellipse ➢ Area of ellipse ➢ Equation of ellipse ➢ Tangent to an ellipse ➢ Properties of an ellipse	Space vector	➢ Space vector and its operation ➢ Application of space vector in solid geometry

different from CN-PEP is that the former is permeated with the concept of vector, which is used to solve the problems in solid geometry and that concerning the connections of straight lines, circles and conic sections, focusing more on the algebraization of geometry. Moreover, though the coverage of DE-LSM is not so broad, the knowledge in each block is presented in more details with greater depth, and that is reflected in the presentation of the specific content about ellipse and circle (see Tab. 3-1).

4.1.2 Comparison of the content arrangement sequences

It can be seen from the above table that DE-LSM starts from a lower knowledge level. Students firstly study the fundamental figure "circle" before learning the "surface area and volume of a solid". Then, the topic of graphical content is paused, switching to a new topic "vector", which in turn will be a basic tool in the further study of geometry. It is followed with discussion of the topic "straight line", "plane", "intersection point", "distance" and "position relationship" by means of vectors. After that, the theme "circle" is resumed with the vector approach and, finally, it is extended to "sphere" and "ellipse". We can see that DE-LSM emphasizes the significance of vectors and such feature appears in the textbooks of other European countries.

CN-PEP is written in form of modules. Students first study the "solid bodies" to get a preliminary spatial concept. Then the "position relations of points, lines and planes in space" is discussed on the basis of spatial imagination and logical reasoning. By connecting geometry with algebra, students study "straight line and equation", "circle and equation" and finally the "vector". Meanwhile, "conic section" and "space vector" are included in the optional course textbook. The writing style of CN-PEP is largely determined by the large population and huge diversity of students in China.

4.1.3 Comparison of textbook structures

(1) Chapter structures

Each chapter of DE-LSM consists of a chapter introduction (in text or graph), sections (text and exercises), and a set of review exercises, while CN-PEP has a three-level structure, i.e. "chapter — section — subsection" structure. More precisely, in addition to the chapter introduction (in text or graph, too) at the beginning of each chapter, sections also include their own introductions. Sections are usually divided into subsections. And, in the end of each chapter, there is a chapter summary, followed by two groups of review exercises. Sometimes columns such as “Exploration and Discovery” and “Reading and Thinking” are also contained in the chapter.

The chapter structure of CN-PEP is relatively logical and hierarchical with titles and subtitles under a layer structure and knowledge is presented in the form of “module”; while in DE-LSM knowledge points in each sections are laid at the same level and there is no upper level title to grasp them. Besides, there is a chapter summary at the end of each chapter in CN-PEP which presents the knowledge in this chapter in the form of “flow diagram”, so that students can better review the content, and then do “Group A and Group B” hierarchical exercises. However, in DE-LSM the review session is named as a new lesson of “Blended Task” with over 20 exercises without hierarchy.

(2) Section structures

Sections in DE-LSM are usually consists of one or more cycles of “Knowledge Introduction — Example(s)” followed by “After-Class Exercises”, while sections in CN-PEP, as mentioned before, contain “Section Introduction”, several subsections devoting to knowledge introduction. In a subsection, mini-columns “Thinking” and “Inquiry” are often designed to give questions leading to the main text or examples in the subsection. “In-Class Exercises” are always a component of a subsection.

The arrangement of CN-PEP is of greater diversity and the modules lead by “Inquiry” and “Thinking” are designed with numerous marginal notes. This arrangement can better inspire reflection and help students learn through exploration. Moreover, CN-PEP is also designed with reading materials, which not only expands students’ scope of knowledge, but also effectively fosters students’ interest in mathematics. DE-LSM is very concise and clear, with all the contents presented to students in proper sequence, so it is like a book for students to read and study by themselves.

4.2 Comparison of the specific contents

We study four aspects: “arrangement sequence”, “cognition level”, “presentation”, and “knowledge background”.

It is generally believed that the basic requirements of the compulsory serial courses of mathematics curriculum for high schools include understanding common geometric solids, cultivating and developing students' spatial imagination, reasoning and arguing skills, abilities of communication with graphic language, and the geometrical intuition. The content of common geometric solids, which connects students' real life experience on one end and the geometric abstraction on the other, is the best starting point and the key to the learning of solid geometry. We thus select this content as the case of our comparison.

4.2.1 Macroscopic comparison

In DE-LSM and CN-PEP, the textbooks containing the content of common geometric solids are *Lambacher Schweizer Geometrie Bayern 10* and the compulsory course textbook *Mathematics 2*, respectively. Firstly, we vertically unfold the knowledge content in the related chapters and sections (Tab. 3-2).

As shown in Tab. 3-2, the knowledge about the geometric solids in DE-LSM lies in Chapter II and its preceding chapter is Chapter I "Flächeninhalt und Umfang eines Kreises" (Area and circumference of a circle), which shows that the study of circle in Germany starts relatively later (in China, the content of circle appears in middle school curriculum) and the geometric solids come immediately after studying plane figures, moving from two dimensions to three dimensions directly (and returning back to two dimensions in the following chapter). The related content of geometric solids in CN-PEP is in Chapter I of the Compulsory Course *Mathematics 2.* This chapter initiates the learning of geometry in high school mathematics course. The following chapter is on the position relationship of points, straight lines and planes in space. Both chapters belong to solid geometry, aiming to develop students' spatial imagination and basic understanding of solid geometry. (*Mathematics 2* includes only contents of geometry with the other two chapters on plane analytic geometry.)

Tab. 3-2 Arrangement sequences of the contents of common geometric solids

	DE-LSM	CN-PEP
Preceding chapter	I Flächeninhalt und Umfang eines Kreises (Area and circumference of a circle)	N/A
Title of the chapter	II Zylinder.Kegel.Kugel (Cylinder, cone, and sphere)	Chapter I Geometric Solids
Subtitles in the chaper	1 Zylinder (Cylinder) 2 Kegel (Cone) 3 Kugel (Sphere) 4 Vermischte aufgaben (Blended task)	1.1 Structure of geometric solids 1.1.1 Structure characteristics of prism/cylinder, pyramid/cone, frustum of a pyramid/cone, and sphere 1.1.2 Structure characteristics of simple combined solids 1.2 Three-views and intuitive pictures of geometric solids 1.2.1 Centre projection and parallel projection 1.2.2 Three-view of a geometric solid 1.2.3 Intuitive pictures of a geometic solid 1.3 Surface area and volume of a geometric solid 1.3.1 Surface area and volume of prism/cylinder, pyramid/cone, and frustum of a pyramid/cone 1.3.2 Volume of a ball and surface area of a sphere
Following chapter	III Trigonometrie im Dreieck (Trigonometry in triangle)	Chapter II Position relationship of points, straight lines and planes in space

As shown in Tab. 3-2, the knowledge about the geometric solids in DE-LSM lies in Chapter II and its preceding chapter is Chapter I "Flächeninhalt und Umfang eines Kreises" (Area and circumference of a circle), which shows that the study of circle in Germany starts relatively later (in China, the content of circle appears in middle school curriculum) and the geometric solids come immediately after studying plane figures, moving from two dimensions to three dimensions directly (and returning back to two dimensions in the following chapter). The related content of geometric solids in CN-PEP is in Chapter I of the Compulsory Course *Mathematics 2*. This chapter initiates the learning of geometry in high school mathematics course. The following chapter is on the position relationship of points, straight lines and planes in space. Both of these two chapters belong to solid geometry, aiming to develop students' spatial imagination and basic understanding of solid geometry. (*Mathematics 2* includes only contents of geometry with the other two chapters on plane analytic geometry.)

If we look into the chapters of geometric solids alone, we find that both textbooks organize contents in the order of "cylinder (prism) — cone (pyramid) — sphere". The differences are that CN-PEP firstly guides students to understand the characteristics of these geometric bodies before learning the surface area and volume. However, DE-LSM does not emphasize on the characteristics of geometric bodies. Instead, it mainly discusses the surface area and volume of three types of geometric bodies. Furthermore, CN-PEP introduces the relevant knowledge of frustum of a pyramid/cone and simple combined solids, and the contents of three views and intuitive pictures are also included. In summary, CN-PEP takes a more theoretical approach and contains richer material for the development of students' spatial imagination.

4.2.2 Microscopic comparison

According to Tab. 3-2, we select eight common knowledge points around the main concept "geometric solids" — the concept and structure of cylinder, the surface area of a cylinder, the volume of a cylinder, the concept and structure of cone, the surface area of a cone, the volume of a cone, the surface area of a sphere, and the volume of a ball — for comparative study through the following comparison models.

— Comparison model of teaching objectives: According to curriculum standard for high school mathematics (MOE of PRC, 2003), we divide it into three levels: to know, to understand and to master.

— Comparison model of knowledge presentations: Following the curriculum standard, we divide the knowledge presentation into five aspects: knowledge introduction, knowledge experience, rigorous proof, knowledge representation and application of the knowledge.

- ➢ Knowledge introduction: the way of introducing the learning objects;
- ➢ Knowledge experience: knowledge exploration, cooperative learning, and hands-on activities;
- ➢ Rigorous proof: logical proofs of formula, properties, or theorems;

- ➢ Knowledge representation: the way of knowledge elaboration; and
- ➢ Application of the knowledge application: examples.

— Comparison model of knowledge backgrounds: On the basis of specific content in textbooks, we divide the background levels related to knowledge points into four aspects: mathematical background, life background, scientific background, and cultural background.

- ➢ Mathematical background, which is only related to mathematics itself;
- ➢ Life background, which comes from the real world, daily lives, learning supplies, or games;
- ➢ Scientific background, which is related to science and technology;
- ➢ Cultural background, which comes from social culture, custom, etc.

(1) Comparison of teaching objectives

The teaching objectives of the eight knowledge points have the following distribution in the two editions of textbooks (Tab. 3-3).

Tab. 3-3 Comparison of teaching objectives

Level of teaching objectives	DE-LSM	CN-PEP
To know		6
To understand	2	
To master	6	2

As shown in Tab. 3-3, the level of teaching objectives about the knowledge points of geometric solids in the CN-PEP is relatively low, as students are only required to know the formulas of the surface areas and volumes. It does not elaborate these six formulas at length. However DE-LSM makes a detailed introduction to the generation of these formulas. Each formula is accompanied with 10 to 20 exercises mostly based on the real life. It requires students to understand and use these formulas in real life proficiently. Instead, CN-PEP focuses on cultivating students' spatial intuition – they are required to observe lots of solids by physical models and computer software; they also need to know the structure features of cylinder, cone, cube, sphere as well as simple combination of these solids and use these features to describe the structure of simple objects in real life.

(2) Comparison of knowledge presentations

The two editions of textbooks are consistent with their own characteristics in presenting the surface and volume formula of each geometric body. Therefore, by taking the "volume of a ball and surface area of sphere" as an example, we make comparison on the presentation of such knowledge points in the two editions of textbooks (Tab. 3-4).

Tab. 3-4 Comparison of representations of the volume formula of a ball

	DE-LSM	CN-PEP
Knowledge introduction	The question of the mass of cork ball, and the illustration of Earth	Null
Knowledge experience	Use Cavalier's principle 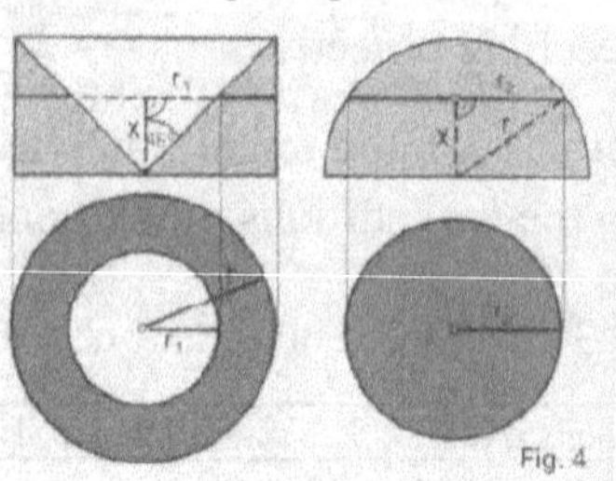	Null
Rigorous proof	Null	A remark: this formula can be proved later
Knowledge representation	Language and symbol	Function and symbol
Knowledge application	Example 1: Direct application of formula Example 2: Calculation of volume with given surface area	Applying formulas to compare the volume of a ball with that of a cylinder

DE-LSM firstly explains at the beginning of the section that though cork is a very light material, it is impossible to carry a cork ball with diameter of 1 meter, and brings out the necessity of exploring methods to solve the volume of a ball. In DE-LSM, Cavalier's principle occupies a very important position — it attempts to help students explore the generation of formula independently with the Cavalier's principle.

CN-PEP states directly that "the volume of a ball depends only on the radius *R* and is a function on the variable *R*" and gives the volume formula. There is neither intuitive explanation nor rigorous proof. It is only marked with marginal notes "this formula can be proved later". As can be seen, this knowledge occupies a very short length in the textbook and the formula is only required to be known and memorized. More detailed discussion is given in the column of Exploration and Discovery later in this chapter by using the Zu Geng's Principle (i.e., Cavalier's Principle).

Tab. 3-5 Comparison of presentation of sphere surface area formula

	DE-LSM	CN-PEP
Knowledge introduction	Null	Null
Knowledge experience	Using limit thinking 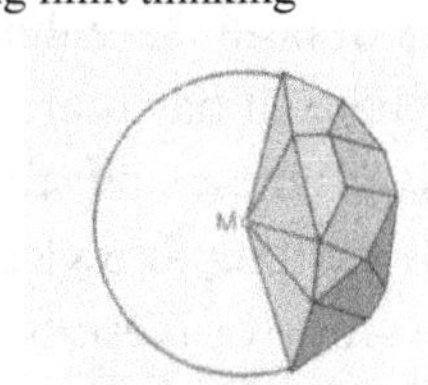	Null
Rigorous proof	Null	A remark: this formula can be proved later
Knowledge representation	Language and symbol	Function and symbol
Knowledge application	Example 1: a) Direct application of formula b) Calculation of volumes with the given surface area Example 2: Calculation of area of circle on the sphere cross-section with the given surface area	Applying formulas to compare the surface area of a sphere with the side area of a cylinder

The two versions of textbooks treat the surface area formula of a sphere in a method similar to that for the volume formula of a ball. Though CN-PEP marks it with marginal notes "his formula can be proved later", we did not find any proof of this formula in the textbook. Whereas DE-LSM uses a naive limit thinking, viewing the volume of a ball as the sum of volumes of n pyramids for large n, and then deduces the formula of surface area with the volume formula of a ball.

As can be seen, DE-LSM has explanations about both the volume formula of a ball and the surface area formula of a sphere and the reasoning processes are intuitive and easy to understand for high school students. Whereas CN-PEP only presents the formula alone without detailed explanation and these knowledge points are required only to be remembered. Although it is not difficult to memorize, it will be better if it is decorated with some mathematical culture. In practice, teachers often make some additional comments and explanations when using textbooks, which requires individual mathematical literacy of the teachers. From this perspective, CN-PEP is like a book for teachers, rather than a book for students.

Regarding the applications of knowledge, we can see that DE-LSM is richer and more flexible in the topics of both the volume formula of a ball and the surface area formula of a sphere. It has more examples than CN-PEP, and there are problems solved by using formula directly and reversely, as well as by combining with other knowledge. This shows that DE-LSM pays more attention to the study of these two knowledge points whereas CN-PEP has low requirement on them.

(3) Comparison of knowledge background

We compared the knowledge background of the eight knowledge points in the two versions of textbooks, see Tab. 3-6.

Tab. 3-6 Comparison of knowledge background

Knowledge background	DE-LSM	CN-PEP
Mathematical background	6	5
Life background	1	3
Scientific background	1	
Cultural background		

As is shown from Tab. 3-6, the backgrounds involved in the process of knowledge presentations in the two versions of textbooks are mostly mathematical background. CN-PEP takes examples of real life in describing the structure characteristics of geometric solids.

Moreover, many exercises in DE-LSM are illustrated with related pictures and it requires some common sense and scientific knowledge in the solving process. Whereas exercises in CN-PEP are interpreted with mathematicalized figures and the background is relatively weakened.

5. Conclusion

After the comparison of the content of solid geometry from the overall view to the specific contents in DE-LSM and CN-PEP, we reach the following conclusions:

(1) DE-LSM covers a narrower scope of contents but in greater depth and focuses on the cultivation of application ability, while CN-PEP covers a relatively wider scope of contents and pays more attention to the development of spatial imagination.

DE-LSM concentrates on the vector which is used throughout the whole geometry content. Spatial positional relationship, sphere, ellipse and other modules rely on the vector to solve related problems. However, CN-PEP only mentions the application of space vector in certain section in Optional 2-1. It is obvious that the geometry content in DE-LSM tends to be algebraized. CN-PEP pays more attention on the development of spatial imagination, for instance, to figure out the position relationship of point, line and plane in space through the so-called "traditional method", as well as to include the content of three-views and intuitive pictures of geometric solids.

(2) DE-LSM is compiled methodically and is targeted, while CN-PEP is more systemic and emphasizes on the integrity of knowledge.

The map of knowledge points in CN-PEP is very logical. The knowledge points appear in modules, and are linked with one another so that students can shape an integral sense of knowledge conveniently. DE-LSM focuses on knowledge flow and the practical application of knowledge. The knowledge points are explained in great detail with concise outlines which can be remembered by students easily.

Moreover, CN-PEP is more lively and vivid. For example, mini-columns "Inquiry" and "Thinking" are designed to lead knowledge modules, and numerous marginal notes provide extra explanations to the main context. In contrast, DE-LSM is concisely clear and focused.

From the context of CN-PEP, the teaching objective level of each knowledge point can be clearly seen. Depending on the level difference, some knowledge points are only briefly mentioned and some are expounded at length. Every knowledge point involved in DE-LSM is arranged in the principle of "understanding how and why".

In summary, CN-PEP is more consistent with our understanding of traditional textbook, while DE-LSM looks like a reading book for self-study.

(3) From knowledge presentation, DE-LSM focuses on the application of knowledge and CN-PEP pays more attention to students' intuitive perception.

Knowledge points in DE-LSM are provided with a lot of examples and exercises. The difficulty level of these exercises increases gradually so that students can improve from general understanding to well grasp of relevant knowledge points. DE-LSM facilitates teaching in the mode of "instructing with knowledge points – practicing with exercises". Students experience the application of knowledge in the process of independent exercises and gradually discover the extension of the knowledge in order to master the knowledge point. Whereas CN-PEP focuses on students' intuitive perception of knowledge, it makes students master the knowledge points by memory after simple representation of the knowledge and explanation by one or two examples, which is insufficient in developing students' ability of mathematical application.

References

[1] Bao, Jiansheng (2002). A comparative study on the composite difficulty of intended mathematics curriculum in Chinese and UK junior high schools, *Global Education*, 31(9): 48–52.

[2] Chen, Changping (2000). *Comparison and Study of Mathematics Education*. Shanghai: East China Normal University Press.

[3] Fuson, K. C., Stigler, J. W. and Bartsch, K. (1988). Grade placement of addition and subtraction topics in Japan, mainland China, the Soviet Union, Taiwan, and the United States. *Journal for Research in Mathematics Education*, 19(5): 449–456.

[4] Howson, A. G. (1995). Mathematics Textbooks: A Comparative Study of Grade 8 Texts (TIMSS Publication Series 3). Vancouver: Pacific Educational Press.

[5] Li, Fangqi (2009). Comparison on the Structure of Mathematics textbooks for Middle Schools from China and Singapore. Master Dissertation, Hangzhou: Zhejiang Normal University.

[6] Li, Shunwen and Shi Ningzhong (2012). Comparative study on content of Chinese and Japanese school geometry curricula. *Global Education*, 41(1): 82−85.

[7] Li, Yeping (2000). A comparison of problems that follow selected content presentations in American and Chinese mathematics textbooks. *Journal for Research in Mathematics Education*, 31(2): 234–241

[8] MOE of PRC (Ministry of Education of the People's Republic of China) (2003). *National Mathematics Curriculum Standard for High Schools (Experimental)*. Beijing: People's Education Press.

[9] Schmit, W. H., McKnight, C. C., Valverde G. A., Houang R. T. and Wiley D. E. (1997). *Many Visions, Many Aims: A Cross-national Investigation of Curricular Intentions in School Mathematics*. Boston: Kluwer Academic Publisher.

[10] Tang, Hengjun and Zhang, Weizhong (2005). Comparison of "similarity" in geometry textbook between China and America. *Journal of Mathematics Education*, 14(4): 55−58.

[11] Xu, Binyan (2002). Characteristics and development of the high school curriculum framework in Germany. *Global Education*, 31(10): 11−15.

[12] Ye, Lijun (2011). Comparative Study on the organization and presentation of geometry contents in main countries. *School Mathematics*, (1): 39−40.

[13] Yuan, Aihong (2007). Research of the domain "Space and Figures" in middle school mathematics course based on the new mathematics curriculum standard. Master Dissertation, Guangzhou: Guangzhou University.

[14] Zhou, Xuemei (2012). Comparison on reasoning and proving in geometry between Chinese and USA secondary school textbooks. Master Dissertation, Shanghai: East China Normal University.

https://doi.org/10.1142/9789814713955_0004

Chapter 4

Comparative Study on Binomial Distribution Content in High School Textbooks

Jun Li[1], Jing Cheng[2], Tingting An[3], and Dan Zhou[4]

[1]*Deakin University, Australia*
[2]*East China Normal University, Shanghai 200062, China*
[3]*Yizhuang Experimental Primary School, Beijing, China*
[4]*Shanghai Fengban Senior High School, Shanghai, China*

This chapter focuses on the comparison of the contents of statistics and probability in six series of high school textbooks in China and other countries. By taking the content of binomial distribution as an example, the six series of textbooks published by People's Education Press of China, Sukan Shuppan Publishing Company of Japan, Panpac Education of Singapore, Pearson Education of U.S., Wright Group of U.S. as well as the Cambridge University Press of UK are compared in three aspects, namely, selection of topics, design of learning tasks and presentation of content, sequentially relevant suggestions are made to the curriculum developers and textbook writers in China based on the similarities and differences. The study has found that the Chinese textbooks do not pay due attention to binomial distribution and the course needs to be deepened and extended. The Singapore's textbooks and American Wright Group textbooks cover the most topics in the closest connection with other knowledge and both attach great importance to the teaching of binomial distribution. The Chinese textbooks have good command of the cognitive level of tasks though the number of tasks is too few and activity type is monotonous, while the two series of American textbooks use a variety of activity types and it is convenient for students to learn from multiple perspectives with a wide range of activities. In terms of presentation, the Chinese textbooks value the design of knowledge introduction, despite the lack of application materials; American Pearson Education stands out in the design of statistical activities especially in using simulations; while Singapore's textbooks are unique in solving computation and internal mathematics problems with graphing calculator and give very detailed guidance for using the calculator. In general, the Chinese textbooks should be credited with its efforts in inspiring reflection and giving basic knowledge to students. It is suggested that we add a small number of topics at the time of revising the curriculum standard and textbooks, deepen the teaching and learning appropriately, use tables and graphs more often, introduce graphing calculator or software, and increase the amount of examples and exercises, in particular, those with applications to scientific context.

1. Background Introduction

In 2004, Prof. Xiru Chen, a statistician, published a long article in a mathematical teaching periodical targeting at secondary school teachers in China. Prof. Chen began with a quote from British scholar H. G. Wells that "Statistical thinking will one day be as necessary for efficient citizenship as the ability of read and write" (Xiru Chen, 2004, p. 1). He explained in detail the important role of statistics for social development and scientific and technological progress. He illustrated that contents about sampling, data, error, average, standard deviation, statistical models etc. are closely related to our real life. He argued that the significance of studying statistics lies not only in the part opposite to metaphysics (function), but also in the metaphysics, namely, in ideological cultivation. "Statistics, or the statistical idea, is a way of observing the world, and people who are trained in statistics tend to take a balanced and robust view of things without going to extremes" (Xiru Chen, 2004, p. 7).

In recent decades, "Probability and Statistics" were introduced into the primary and secondary school curriculum all over the world. In the early 21st century, China Ministry of Education has successively promulgated *Standards of Mathematics Curriculum for Compulsory Education (Experimental)* (MOE of PRC, 2001), *Standards of Mathematics Curriculum for High Schools (Experimental)* (MOE of PRC, 2003), *and Standards of Mathematics Curriculum for Compulsory Education (2011 edition)* (MOE of PRC, 2012), all of which reflect the positive response of China school mathematics curriculum to the addition of the new content, and the compulsory and optional courses of high school in China have arranged a few hours to learn probability and statistics. However, comparing with countries like Britain, Germany, America and France which have over 40-years' history of statistical education and research, popularization of probability and statistics in primary and secondary schools in China has only been slightly more than 10 years, students' performance in tests is relatively low. For instance, in 1991, in the large-scale international test IAEP2 (the Second International Assessment of Education Performance) students from mainland China ranked tenth in the test score of "data analysis, statistics and probability" while they ranked No. 1 in all the other four areas (numbers and operations, measurement, geometry, algebra and functions) (Lianghuo Fan and Yan Zhu, 2004). In PISA 2012, Shanghai students took the first place in the average score of four areas, namely, "change and relationships", "space and shape", "quantity" and "uncertainty and data", however, their performance in areas "quantity" and "uncertainty and data" was lower than in the areas "change and relationships" and "space and shape" (OECD, 2013). Yeping Li (1999) holds that the different performance of Asian and American students in TIMSS test is partially related to the different requirements in the textbooks they use, thus it should be meaningful to find the cause of relatively low performance of our students in the area of probability and statistics through the comparison of textbooks.

This study aims to know the characteristics of probability and statistics education in the content of binomial distribution in the six series of textbooks (see Tab. 4-1) from five countries, namely, UK and U.S. which introduced probability and statistics education earlier than many other countries, China and Japan which introduced probability and statistics lately and Singapore which are between the above two groups and is greatly influenced by the British curriculum. By comparing the topics, design of learning tasks as well as presentation of content of each series of textbook, we found the similarities and differences of those textbooks in various aspects such as the selection of content, the expectations of student performance and the presentation of content. After serious consideration, relevant suggestions were made to the curriculum developers and textbook writers for reference when making decisions. It must be pointed out that the education decision-making must consider the cultural background. The task of this study is not to agglomerate the practice of various countries, instead, after understanding the similarities and differences of different national practices, we hope to draw on practices of others based on the present education background in China, guide our development of probability and statistics education and reduce the time as we move forward.

Tab. 4-1 Sample Textbooks

Country	Publisher	Textbook and year of publication	Code
China	People's Education Press	Mathematics as per the National High School Curriculum Standard (Experimental): Version A, Compulsory 3 (2007 Ed.), Optional 2-3 (2009 Ed.)	CN-PEP
Japan	Sukan Shuppan Publishing Company	New Edition of Mathematics: Mathematics A (2006 Ed.), Mathematics B (2007 Ed.), Mathematics C (2007 Ed.)	JP-SKS
Singapore	Panpac Education	New Express Mathematics 4 (2008 Ed.), H2 Mathematics 2nd Volume (2006 Ed.)	SG-PPE
U.S.	Pearson Education	Prentice Hall Mathematics: Algebra 1 (2009 Ed.), Algebra 2 (2009 Ed.)	US-PHM
U.S.	Wright Group/McGrow Hill Companies	University of Chicago School Mathematics Project: Advanced Algebra (2010 Ed.), Functions, Statistics and Trigonometry (2010 Ed.)	US-SMP
UK	Cambridge University Press	School Mathematics Project AS/A2 Mathematics: Statistics 1 (2004 Ed.), Statistics 2 (2005 Ed.)	UK-SMP

2. Literature Review

The term "curriculum" has many meanings, but basically it means a plan that regulates the learning content and learning sequence, and the textbooks or other teaching material carefully designed and organized for teachers and students is an important type of curriculum (Goodlad *et al.*, 1979; Robitaille *et al.*, 1993; Valverde *et al.*, 2002). Textbook is an important resource to achieve the curriculum objectives and implement teaching (MOE of PRC, 2003), an important representation of the curriculum and a link between the curriculum and pedagogy (Pepin and Haggarty, 2001). It is even considered as one of the indices of quality of schools' educational resources that all mathematics classes in the school use the same textbook (OECD, 2013). A topic is more likely to be taught if it is included in the textbook than if it is omitted (Valverde *et al.*, 2002). The information about which topics are included and what performances are expected could be seen from the curriculum guides. Thus, textbook research would identify how the topics are organized and presented specifically in various countries. In any case, textbooks have a very big impact on student learning.

After reviewing the relevant literature in terms of textbook analysis, Pepin and Haggarty point out that "there appear to be four main areas according to which textbooks have been analyzed in terms of their content and structure: the mathematical intentions of textbooks; pedagogical intentions of textbooks; sociological contexts of textbooks; and the cultural traditions represented in textbooks" (Pepin and Haggarty, 2001, p. 160). Among them, research on mathematical intentions of textbooks often investigates "what mathematics is represented in textbooks; beliefs about the nature of mathematics that are implicit in textbooks; and the presentation of mathematical knowledge" (ibid, p. 160). They state that research on the pedagogical intentions of textbooks includes ways in which the learner is helped within the content of the text to learn the materials, within the methods included in the text, and by the rhetorical voice of the text. Some other researches focus on how the textbook itself passes its cultural traditions, social expectations and nationally advocated cultural values to students. Existing literature mostly compares textbooks from the two aspects: mathematical contents and pedagogical processing. This study and the following several researches fall into this type.

Charalambous *et al.* (2010) establish a two-dimensional (vertical and horizontal) research framework for comparison of textbooks and analyze the treatment of addition and subtraction of fractions in textbooks used in Cyprus, Ireland and Taiwan. From the horizontal analysis, they selected the topics covered with respect to the addition and subtraction of fractions and their sequencing. From the vertical analysis, they considered the fraction constructs, the worked examples, the potential cognitive demands of textbook tasks and the type of responses asked for. Based on their comparative study of one series of Cyprus textbook, two series of Ireland and two series of Taiwan textbooks, they found that there were many commonalities and similarities in the selection and arrangement of this content, but two series of textbooks from Taiwan showed obvious characteristics different from textbooks of Ireland and Cyprus. For instance, two series of textbooks from Taiwan emphasize the concept of the unit fraction; their worked examples use more diverse fraction constructs and representations than the other textbooks and they prompt multiple solutions by means of speech bubbles. Textbooks from Taiwan impose higher expectations on students' thinking for the same content. They conjectured that "textbook signature" existed and they suggested "more variation exists in the textbooks across countries rather than within the same country" (ibid, p. 146).

Houang and Schmidt (2008) reviewed how International Curriculum Analysis was used to analyze textbooks from nearly 50 nations. The TIMSS mathematics framework includes the content (mathematics topic area), performance expectation (expected student performances to complete mathematical tasks) and perspective (broad themes that might underlie many tasks). They divided the teaching content into 10 major categories such as numbers, measurement, data representation, probability and statistics. Each category contains one or more levels of specific subcategories. For instance, relative to the level 1 category "number", "whole numbers" is listed in level 2 categories, and further divided into level 3, including the "meaning", "operations" and "properties of operations". The level of detail and organization reveals the breadth and depth of content in textbooks. Each document was separated into a series of units and then each unit was subdivided into blocks. Category codes were assigned to blocks. Instead of coding page-by-page, TIMSS curriculum analysis focused on introduction, summary and exercises. As is mentioned in the paper, the textbooks used by students of six top-achieving countries in TIMSS test of Population 2 (Grades 7 and 8) are generally characterized as being of strong coherence and hierarchical sequencing, and the teaching of "data representation and analysis" takes a relatively long time.

In order to reveal why Singaporean students are more successful in mathematics than their U.S. counterparts in TIMSS 2003, Ginsberg *et al.* (2005) have conducted an exploratory study to identify major differences in their frameworks, textbooks, assessments, and teachers in primary grades. They compared Singapore's textbooks and American textbooks, both traditional and nontraditional, from textbook level (organization and content), lesson level, and problem level. They find that the appearance of Singapore's textbooks might look like as simple as American textbooks from the 1950s, but the design is very attentive. Singapore's textbooks value construction of students' concept, the topics covered in each grade are nearly half of American textbooks and the content is deeper and more systematic. It stresses the connections between knowledge, uses concrete to pictorial to abstract approach, and keep deepening and improving students' understanding as they progress. For example, how multiplication and division facts are interrelated is discussed along with the expansion of numbers throughout the years. In contrast, "traditional U.S. textbooks rarely get beyond definitions and formulas, developing only students' mechanical ability to apply mathematical concepts" and "individual topic coverage in U.S. textbooks is much shorter and less comprehensive than what is found in Singaporean texts". Ginsberg and his colleagues also indicated that U.S. textbooks were less effective than expected because "the spiral in U.S. textbooks includes significant repetition and reteaching of the same content in two or three consecutive grades" ((ibid, pp. xii–xiii). Related topics are split into different lessons, chapters in a book or even in different books and the solution strategy involved does not strongly support U.S. students' deep conceptual development. In addition, they found that Singapore' textbooks attached importance to creating problem situation and opportunity for students to review and use what they have learned by providing more practice exercises, including multistep, complex word problems and problems integrating previous knowledge. They pointed out that "if the Singapore texts are weak in any way, it is that the problems they provide, although numerous and varied, are not often based on real-world data and information, making the mathematics perhaps less applicable to real life than it should be" (ibid, p. 40). While, the U.S. places a greater emphasis on data analysis and probability and other applied mathematics and introduce topics such as geometric coordinate systems, congruence and similarity, and patterns in the early grades, whereas Singapore does not. The nontraditional textbook they examined focuses more on problem-based learning and attaches importance to the cultivation of students' communication skills in the goal of training qualified talents for the 21st century.

Yeping Li (1999) has compared the algebra part in nine series of textbooks for Grade 8 students from the United States, Hong Kong, mainland China, and Singapore, checked with the respect to the content included, key features of content organization and presentation, and the number and nature of to-be-solved (TBS) problems. Similar to the findings of Ginsburg *et al.* (2005), Yeping Li found that there was a substantial difference between U.S. and Asian textbooks. He said "the U.S. textbooks vary widely in their emphasis on algebra content. However, these textbooks are similar in the ways they dissemble content into small units, introduce content within real-world problem contexts, emphasize students' practice, and include various types of TBS problems to develop students' mathematical problem solving competency. In contrast, the Asian textbooks are similar in their content emphasis and their high mathematics requirements in algebra. Moreover, they use larger units of coherent algebra topics, introduce content as pure mathematics, and emphasize the development of students' procedural skills" (Yeping Li, 1999, p. iv).

Compared to algebra and geometry, as most countries started probability and statistics education only for a short time, there are still very few comparative studies of textbooks about probability and statistics. Focusing on the content of probability and statistics at the middle school level, Haiyue Zhang (2007) and Ping Jia (2008) compared different versions of textbooks published in China or between China and other countries. However, both of the studies only examined the worked examples and exercises in the selected textbooks in relation to the difficulty levels and problem contexts. In the last few years, some textbook comparative studies on probability and statistics at the high school level were reported. For example, Xiaoping Zhao and Xue Yao (2010) compared the similarities and differences of the content included and arrangement of probability and statistics section in middle school and high school textbooks from Singapore and Shanghai. They found that Singapore high school textbooks have more content about probability and statistics than Shanghai textbooks and unlike Singapore textbooks, the latter don not include classical probability distributions such as binomial distribution and Poisson distribution, as well as the very useful concepts such as correlation coefficient, unbiased estimator, hypothesis testing and distribution of sample means. As an integral part of students' learning, how to use graph calculator is described in great detail in Singapore textbooks. Relationships between different contents are clearly discussed, which reflect the depth of learning. Graphing calculator is widely used and understanding of related calculation content before using calculator is emphasized. There are also more exercises in Singapore textbooks than Shanghai's, which reflects statistics has a higher place in Singapore curriculum.

Yingkang Wu and Yun Gan (2013) compared the use of technology in high school mathematics textbooks from UK and China. They compared the probability and statistics content in UK-SMP and CN-PEP. They found that judging from the contents of textbooks, the scope and difficulty of the British textbooks is apparently higher than the Chinese textbooks, however, both countries attach great importance to the application and integration of technology in teaching, and their use of information technology gives priority to explorations and worked examples in problem solving. The difference between the two countries is that CN-PEP makes more use of technology to generate random numbers for the purpose of simulations and data collections, whereas UK-SMP pays slightly more attention to different hardware and software than the Chinese textbooks, and provides more detailed guidance of operations, and uses technology more often in calculations and explorations.

Tingting An (2013) has carried on a comparative study on the four series of textbooks, namely, CN-PEP, US-PHM, US-SMP and SG-PPE from the perspectives of discrete random variables and probability distributions, binomial distribution, normal distribution, and Poisson distribution. She found out that in terms of appearance, the two American series of textbooks are very thick, colorful and rich in columns while Chinese textbooks have a small number of illustrations and sidebars, being relatively simple and centering on introducing new topics by solving problems. Singapore's textbooks have almost no illustrations except graphing calculator screenshots, and they have the least number of columns and ways of introduction. However, American and Singapore's textbooks have more extensive appendices; for instance, the appendices of US-PHM involves extra practice, skills handbook, tables, properties and formulas, glossary, answers, index etc. While the appendix of Singapore's textbooks includes exploring mathematics using graphing calculator, formulas and tables, proof of least squares linear regression, review papers, and so on. In terms of contents selection, Singapore's textbooks include the most learning topics, followed by US-SMP. Some relationships between the distributions are discussed in these two textbooks but are missed in other textbooks. The Chinese textbooks include relatively less contents and US-PHM has the least. American textbooks have more and various kinds of learning activities than Chinese and Singaporean texts, which have only one activity "represent the data then draw conclusions or explore". Activities like "investigation or experimentation" and "simulations" are not arranged in China and Singapore textbooks. Except textbooks in China, the other three versions have more exercises for learners' practice. In the two America series of textbooks, exercises are even more than three times of problems in the text. Three series of textbooks from China and U.S. are consistent in the requirement for students in the main text part and exercise part, however, US-SMP cognitive requirement mainly aims to "understanding" level and this is not challenging enough for students; over 80% of exercises in the Singapore's textbooks are at "analysis" level and this requirement is too high for Chinese students. America and Singapore pay much more attention to graphic representation on the teaching of probability distributions than China. Questions in the four series of textbooks are mostly presented in the contexts of "public" and "personal", scientific context is relatively less used. As opposed to the other three editions of textbooks, the game situation is of higher proportion in the CN-PEP. Singapore's textbooks use information and communications technology (ICT) most frequently for different purposes while it is only used occasionally in China.

The three studies by Xiaoping Zhao and Xue Yao (2010), Yingkang Wu and Yun Gan (2013) and Tingting An (2013) are mostly related with this study. Judging from the existing literature, we believe that probability and statistics education in China needs improvement and the relevant teachers and educators need to learn from others' experience in teaching content and teaching approach. Even though limited by the current conditions of teachers and new technology, some of the advanced practices still cannot be implemented in our class, curriculum planning requires foresightedness, thus it is very significant to make an in-depth comparison of high school textbooks from various countries in specific content such as binomial distribution.

3. Research Methods

With the binomial distribution as an example, in this research we will describe the similarities and differences of various textbooks in terms of the selection of topics, design of learning tasks, and presentation of content based on our overall understanding of the various series of textbooks.

3.1 Research framework

For the past two years, we have obtained a preliminary framework for comparative study by first reading literatures, reflections and discussions, and used it for pre-study of some contents selected from textbooks from China, America and Singapore. On the selection of topics, we studied both the coverage of knowledge and the links between the knowledge. In terms of design of learning tasks, on one hand we focused on the variety of the learning task types designed in textbook; on the other hand we also considered the requirements of cognitive level for the learning task. In terms of presentation of the content, we examined the means of introduction in a lesson, the use of tables and graphs, ICT and context in different textbooks. Once constructing the framework, we decided to concentrate on the similarities and differences in three aspects, namely, selection of topics, design of learning tasks and presentation of content about the probability and statistics content in various series of textbooks. The research framework is shown in Tab. 4-2.

Tab. 4-2 Research Framework

Selection of topics	Design of learning tasks	Presentation of content
Amount of topics	Type of task	Introduction of key content
Links between topics	Amount of task	Use of tables and graphs
	Task level	Use of ICT
		Use of context

Following the TIMSS International Curriculum Analysis (Houang and Schmidt, 2008), teaching content is subdivided into 3 levels and the contents in Level 3 are called topics. For example, under the Level 1 category "Statistics and Probability", "binomial distribution" is a Level 2 sub-category and it is divided into Level 3 topics such as binomial distribution, binomial probability density function, mean value of a binomial distribution, and normal approximation to the binomial distribution. In addition to studying the topics included in every series of textbook, links between these topics as well as with other mathematical knowledge are also examined in our research.

As in Stein *et al.* (2007, p. 346), we define tasks as "classroom activities the purpose of which is to focus students' attention on a particular mathematical idea". A textbook is divided into the main text and after-class exercise. The main text of each lesson covers its introduction, main topics explanation, in-class practice, as well as extensions and reading materials. After-class exercise includes not only exercises after a lesson, but also questions in the exploration part and exercises in the chapter review, but excludes self-test questions, if any. Tasks can include examples, exercises, activities, extension, exploration, thinking, reading, and application of ICT, exercises in this lesson, and related exercises at the end of the chapter review and so on except the self-test questions. The tasks are divided into four types, namely, pure thinking, investigation, construct tables and graphs, and simulations; on the cognitive level, it is divided into four levels, namely, remembering, understanding, analyzing and creating (see Tab. 4-3) in order to reflect the performance expectations in different series of textbooks. Every single problem is recorded as one learning task. More precisely, an exercise with three sub problems is calculated as three learning tasks; however, in case two questions are asked in a row in one exercise, it will be recorded as one task.

Tab. 4-3 Descriptions of the cognitive level classification

Level	Description	Example
Remembering	Recognize and recall the learned definitions, propositions (axioms, law, formulas, and theorems), mathematical facts, basic concepts, algorithms and methods in a way almost identical to the textbook.	Given $x = 2, n = 5, p = 0.25$, find the probability with the binomial probability formula.
Understanding	Interpret the meaning of concepts, propositions, and mathematical structures; solve the routine problems in real-life contexts, involving one to two concepts and skills	Find the probabilities of having exactly 8 successes or at least 8 successes in 10 trials for the given probability of success p on each trial
Analyzing	Make comparison, analysis and connection, organize a few implied information in reasoning when contexts or givens are changed, compare the concepts, graphs or properties and establish meaningful connections, involving three or more concepts and skills	Examine distribution of $X \sim B\ (10, 0.8)$; think over which value of random variable X is taken to get the maximum probability
Inquiring	In the situation with absence of reference, creatively solve new problems or putting elements together to make some conjectures or construct a model independently	What is your understanding of the relationship between the hypergeometric distribution and the binomial distribution?

Finally, we compared the presentation of content that involves the means of introducing binomial distribution probability formula, the use of tables and graphs and the use of ICT in various series of textbooks. Tables and graphs are fundamental communicative tools, widely used in statistics and should be learned by every knowledgeable citizen. They make information visually and easy for communication. Integration of information technology in the teaching of probability and statistics in high school is international trend of development, however, we're still weak in this aspect. Thus we compared the frequency and the main purposes of using information technology in different textbooks and divide it into three functions, namely, calculation, graphing and exploration. Finally, like the classification of context by PISA (OECD, 2010), according to the distance of the situation from students' life, the task context in the "text" and "exercises" is divided into 6 categories, namely, context-free, game context, personal context, public context, occupational context and scientific context (Tab. 4-4) and the game context is added considering the particularity of probability knowledge. When there are multiple problems in a row under the same problem situation, the situation is recorded only once to reflect the richness of usage of situation.

Tab. 4-4 Description of task context classification

Context	Description	Example
Context-free	completely mathematical context.	only involves numbers, formulas, calculations, mathematical concepts, etc.
Game context	Entertaining or challenging problem situations due to uncertainty.	Rolling dice, flipping coins and drawing marbles etc.
Personal context	Context related to students' daily life, campus life or activities which are familiar to most of the students.	Students' height, weight and academic records etc.
Public context	Widely reported in the media and known to the public	Catalogues, athletic records, weather forecast data, elections, etc.
Occupational context	Of not high public familiarity and only known to special group of people.	Quality inspection, agricultural harvest, polygraph, etc.
Scientific context	About knowledge of a specific subject, more abstract and only professionals of one area know the details.	Running speed of an animal, temperature of the ocean, seismic data, chemical experiment data, medical data, etc.

For the above three dimensions of comparison, we have at least two researchers to encode independently according to the data coding guidelines, the inconsistency is finally agreed after discussion and consultation.

3.2 Research object

Through consultation with local mathematics educators, the research project of *Comparative Study of High School Mathematics Textbooks from Certain Countries* selected textbooks from China, Japan, Singapore, U.S., Germany, Russia, and UK for comparison. Due to the language limitation, our sub-project "Comparation of Probability and Statistics Contents" only did a comparative study on the probability and statistics content in Grades 10–12 textbooks published by People's Education Press (China)[a], Sukan Shuppan Publishing Company (Japan), Panpac Education (Singapore), Pearson Education (U.S.), Wright Group (U.S.), Cambridge University Press (UK) (see Tab. 4-1 for more information). Here, we do not distinguish compulsory or optional courses or identify which grade a textbook is applied to, mainly because the education system and school system varies among different countries. There are explicit users and using order in some countries, while some countries do not have strict requirements, thus we have no way to explain. In addition, we did not collect the corresponding exercise books for each series of textbook and the comparisons were restricted to content in textbooks. We did not take the chapters that are dedicated to permutation and combination as research object because they are not related to randomness.

Statistics education has been launched for a relatively long time in Britain and America but relatively short in China and Japan. Singapore is under the influence of East Asia culture but the education system and evaluation is British. We believe it is interesting to select these countries for comparative study from the view of drawing on the development experiences of other countries. Due to the space limitation, this chapter will focus on the topic of "binomial distribution", which is an important teaching content of statistics education in high school. This is a common

[a]In addition to the National Curriculum Standard, there is also a Shanghai local Curriculum Standard in China, and the two Standards have big differences in probability and statistics contents. We choose the most widely used textbooks rather than Shanghai textbooks compiled based on Shanghai Standard, since the latter are used only in Shanghai area.

distribution of discrete random variable. It is not difficult to list all the values of the random variable and calculate the corresponding probabilities by means of calculator. It is of low starting point compared to the normal distribution probabilities calculation, which requires integral. Thus we will take "binomial distribution" as a case and compare the above textbooks at home and abroad.

4. Research Results

In order to have a preliminary overall understanding of the six series of textbooks, we begin by looking at the basic information of every version of textbooks such as layout design, structural arrangement and length. As is shown in Tab. 4-5, from the appearance the two series of textbooks in America are the most decorated ones and rich in columns, whereas textbooks from China, Japan, Singapore and UK are relatively simple with fewer pictures. American textbooks not only put questions for thinking or tips in the sidebar like China but also put on a lot of pictures, give links to web sites, terminology explanation, answers and so on. Chinese and Japanese textbooks not only attach importance to training students' thinking and problem-solving skills, but also try to connect the teaching contents to the real world in order to stimulate students' learning enthusiasm and expand their horizons through extracurricular reading, however, the materials are not as rich as American textbooks. The UK and Singapore textbooks focus on solving mathematical problems and with obvious intention of preparing for exams. The Singapore textbooks require students to use the graphing calculator as a learning tool, which is also impressive. Two series of American textbooks not only attach importance to attracting students by text based on the real world and interesting activities and clear instructions, but also pay attention to highlighting learning objectives and preparing students' test-taking, which should be related to their hope to meet more market demand. The US-SMP sets the Big Idea column at the beginning of each lesson, although it includes only a few words, it manifests the important ideas and concepts in this lesson, for instance, the Big Idea of Lesson 13-9 of the Advanced Algebra, "Binomial and Normal Distributions" is that "As the number of trials of a binomial experiment increase, the graphs of the probabilities of each event and the relative frequencies of each event approaches a distribution called a normal distribution", which enlightens that in this lesson in addition to the specific knowledge, it requires to establish the big idea "we can use the normal distribution as a close approximation to the binomial distribution under certain conditions" in the minds of students. In the six series of textbooks, only CN-PEP does not give answers to exercises.

Pages allotted to a topic in a textbook implies the importance of the topic from one side. Textbook styles vary slightly among different countries and the pages for the "Content", "Instructions of How to Use the Book" before the body of the textbook, the "Appendix" and "Answers to Exercises" at the end of textbooks are not counted in order to make it more comparable. As page sizes, margins, languages, font sizes are different indifferent textbooks, so a direct comparison of pages does not make sense.

Tab. 4-5 Relative information of six series of textbooks

Textbook	Color	Book size	Side bar	Columns	Pages Prob. & Stat.	Pages Binom. Distr. (%)
CN-PEP	No	L	Yes, not fixed	Guiding picture and sentences, examples, practice, exercises, think about it, exploration, reading and thinking, brief summary, review questions	156	4(3%)
JP-SKS	No	S	No	Lead-in questions, examples, practice, supplementary problems, chapter review problems, investigation, extracurricular column, Answers	106	9(8%)
SG-PPE	No	L	No	Guiding diagram of main function buttons on a calculator, examples, note, questions, calculator operating instructions, guide, exercises, miscellaneous exercise, answers	274	25(9%)
US-PHM	Yes	L	Yes	Guiding picture, what you'll learn next, key vocabulary, what you'll learn, … and why, check skills you'll need, go for help, new vocabulary, example, quick check, vocabulary tip, graphing calculator hint, exercises, test preparation, mixed review, real-world connection, go online, extension, practice and problem solving, activity lab, guided problem solving, checkpoint quiz, chapter review, chapter test, test-taking tip, selected answers	143	7(5%)
US-SMP	Yes	M	Yes	Guiding picture and introduction, vocabulary, big idea, mental math, activity, example, quiz yourself, questions, review, exploration, projects, summary and vocabulary, self-test, chapter review, selected answers	347	22(6%)
UK-SMP	No	S	No	You will learn, lead-in questions, example, exercises, key points, mixed questions, test yourself, answers	244	16(7%)

Instead, we calculate the proportion of the number of pages of binomial distribution to the total number of pages of probability and statistics for each series of textbooks. Tab. 4-5 shows that the binomial distribution takes the biggest proportion in Japan, UK and Singapore textbooks, followed by two series of America textbooks, and the least are the Chinese textbooks, accounting for only 3%. Of course, this percentage is only a cursory description of binomial distribution status in the various series of textbooks, later it will be compared and discussed from multiple angles.

The main conclusions obtained will be elaborated below from the selection of topics, design of learning tasks and presentation of content in turn.

4.1 Selection of topics

Throughout the textbooks of binomial distribution in China and other countries, there are two main aspects of content, namely, knowledge of binomial distribution itself and connections between binomial distribution and other distributions. Tab. 4-6 reflects the selection of topics in six series of textbooks and here it only checks whether they are covered in main text other than exercises, questions, reviewing exercises, and test questions in the textbook.

What the binomial distribution is and how to find probabilities for a given binomial distribution are common contents in the textbooks we compared. The graph of a binomial distribution, mean and variance of a binomial distribution are covered by most textbooks. Only CN-PEP fails to show binomial distribution graphically. SG-PPE and US-SMP cover the most of the topics, up to 12, and are of the most contact with the other knowledge, up to 4, UK and Japan come in the middle, respectively 9 and 7 in total, CN-PEP and US-PHM cover the least topics, and both are 5. Although CN-PEP involves the connection between binomial distribution and two point distribution, it is only raised as a problem to be considered and not expounded like SG-PPE, JP-SKS and US-SMP, which not only explain the sufficient conditions of using the normal distribution to approximate binomial probabilities, but also design the specific examples and exercises to consolidate and strengthen such links.

Tab. 4-6 Topics covered in various textbooks

Theme	Topics	CN-PEP	JP-SKS	SG-PPE	US-PHM	US-SMP	UK-SMP
Binomial distribution	Meaning	√	√	√	√	√	√
	Assumptions	×	×	√	√	√	√
	Probability distribution	√	√	√	√	√	√
	Distribution graph	×	√	√	√	√	√
	Explore the impact of n, p changes on the binomial distribution graph	×	√	√	×	√	×
	Mode	√	×	√	×	√	×
	Mean value	√	√	√	×	√	√
	Variance	√	√	√	×	√	√
	Standard deviation	×	√	√	×	√	√
	Probability density function	×	×	√	×	√	×
	Cumulative distribution function	×	×	√	×	√	√
	Calculate by means of binomial distribution table or calculator	×	×	√	√	√	√
Subtotal		5	7	12	5	12	9
Links with other knowledge	Binomial theorem	√	√	×	√	√	√
	Two-point distribution	√	×	×	×	×	×
	Conditions for using a Poisson distribution to approximate a binomial distribution	×	×	√	×	×	√
	Use a Poisson distribution to approximate a binomial distribution	×	×	√	×	×	×
	Conditions for using a normal distribution to approximate a binomial distribution	×	√	√	×	√	×
	Use a Normal distribution to approximate a binomial distribution	×	√	√	×	√	×
	Confidence interval	×	×	×	×	√	×
Subtotal		2	3	4	1	4	2

The difference in selection of topics in each series of textbook is mainly shown in the following aspects:

— Exploring the impact of n, p changes on the binomial distribution graph,

— Expressing requirement on the calculation of probability by means of calculator or binomial table,

— Discussing mode of a binomial distribution,

— Introducing the binomial probability density function or the binomial cumulative distribution function, and

— Using a normal distribution to approximate a binomial distribution.

Compared with other textbooks, CN-PEP may consider increasing the readability and applicability from the following two aspects in order to enhance the depth of our existing probability and statistics curriculum, namely, more frequent use of distribution graph to increase students' intuitive feeling of the binomial distribution, and illustrating how to recognize situations where the binomial model applies and to solve calculation problems by means of calculator with probability density function and cumulative distribution function.

Considering the proportion of the above-mentioned binomial distribution in textbooks, we found that both indicators of SG-PPE ranked first. It indicates that these textbooks are the ones with the most profound and extensive explanation to probability and statistics among the six series of textbooks, for example, they devote great space to explore the connections between binomial distribution and normal distribution, Poisson distribution and apply them to solve problems. US-SMP includes a lot of space and extensive content, however, the proportion of the binomial distribution in the textbooks is not very large and this may be related to the large total length of textbooks that is full of illustrations and emphasizes the real situation. There are a large number of various types of exercises that relatively occupies a large space. Binomial distribution is involved in two volumes and three chapters of the textbook and it is explained in depth. The chapter of normal distribution closely follows the chapter of binomial distribution. It designs the content of using a normal distribution to approximate a binomial distribution after the introduction of how to transform any normal distribution into the standard normal distribution. Firstly through visual graphs and then by testing the hypothesis that the coin is fair, it displays the convenience brought about to use a normal distribution to approximate a binomial distribution. Some similar exercises to consolidate in practice are arranged. CN-PEP only have one section on binomial distribution in the chapter of "random variables and their probability distributions", as well as the two sections "conditional probability" and "mutually independent events". The emphasis to the binomial distribution is clearly not enough.

4.2 Design of learning tasks

As common topics among various series of high school textbooks are the binomial experiment and binomial probability distribution, to make it more comparable, we only coded and compared the learning tasks interrelated with this in various textbooks, for instance, “determine the binomial probability distribution as well as further probability calculation”. Other topics such as the mean and variance of binomial distribution, relations between binomial distribution and other distributions are not compared. Tab. 4-7 shows the numbers of tasks counted by types and cognitive levels arranged in the main text and after-class exercise of textbooks.

Tab. 4-7 Numbers of tasks counted by types and levels in selected topics

			CN-PEP	JP-SKS	SG-PPE	US-PHM	US-SMP	UK-SMP
Activity type	Pure Thinking	Main Text	10	4	2	8	7	14
		After-class exercise	4	0	7	28	54	60
	Investigation	Main text	0	0	0	0	0	0
		After-class exercise	0	0	0	1	1	0
	Construct tables and graphs	Main Text	0	1	13	5	8	0
		After-class exercise	0	0	18	11	25	0
	Simulations	Main text	0	0	0	5	0	0
		After-class exercise	0	0	0	7	1	0
Cognitive level	Remembering	Main Text	0	0	0	3	3	0
		After-class exercise	0	0	0	7	12	0
	Understanding	Main text	3	4	5	9	7	13
		After-class exercise	1	0	7	18	41	34
	Analyzing	Main Text	5	1	9	4	5	1
		After-class exercise	2	0	11	18	26	25
	Creating	Main text	2	0	1	2	0	0
		After-class exercise	1	0	7	4	2	1
Subtotal			14	5	40	65	96	74

It shows from Tab. 4-7 that the task number in Japanese and Chinese textbooks is far less than other textbooks. This difference is particularly evident in the amount of exercises. Arrange more tasks in main text rather than in after-class exercise is abnormal and this is not beneficial to students' understanding and consolidation of knowledge learned. Japanese textbook is short in length and the contents such as mean and variance of a binomial random variable, and to use a normal distribution to approximate a binomial distribution are not counted in Tab. 4-7. Thus it shows the shortage of emphasis on calculation instead of understanding in this textbook. Textbooks from China and Japan are concise in writing as a tradition, besides, it may also because statistics education is introduced in school curriculum much later in China and Japan. However, textbooks of China and UK have the problem of being monotonous in task types and only pure thinking from the four types of activities: pure thinking, conducting a survey, constructing tables and graphs and stimulations, is arranged. Other textbooks provide construct tables and graphs opportunities for students when they ask students to explore the impact of n, p changes on the binomial distribution graph or to construct tables and graphs to calculate probabilities with graphing calculators. These tasks make learning content concrete and visual, easy to understand and conform to cognitive rules, thus it is worth learning.

From the cognitive level requirement of tasks, CN-PEP arranges tasks in a reasonable sequence, tasks of remembering and inquiring level are few and tasks of understanding and analyzing take the majority. This reflects the high learning requirements of CN-PEP on the selected teaching content. Only CN-PEP and SG-PPE mainly focus on analyzing level tasks, and other textbooks give priority to understanding level tasks. However, CN-PEP sometimes lacks foreshadowing to tasks of high requirement, for example, in Elective 2-3, it requires students to think about the relation of binomial distribution and two-point distribution as well as hypergeometric distribution, where two-point distribution and hypergeometric distribution are rarely mentioned, students lack cognitive basis and it is difficult to finish tasks of such high expectations.

The tasks arranged in the two series of American textbooks have covered all four types of activities, which are helpful for students to learn from multiple angles with a wide range of activities, and such teaching conforms to the mainstream of international statistics education (Garfield *et al.*, 2008). Writers of Chinese textbooks need to learn from American textbooks in learning activities design and implementation. The section of binomial distribution in US-PHM Algebra 2 has two such kinds of exercises in succession (p. 689):

24. About 11% of the general population is left-handed. At a school with an average class size of 30, each classroom contains four left-handed desks. Does this seem adequate? Justify your answer.
25. For a group of 40 people, what is the probability that exactly three people in the group will celebrate their birthdays on a Wednesday this year?
    a) Find the probability by using the binominal probability formula.

b) Find the probability by designing and running a simulation.

c) Compare your results for parts (a) and (b). Explain any discrepancy.

Both tasks aim at active use with life background. The problems are somewhat challenging and suitable for high school students. The previous task is purely thinking task of analyzing level and students need to converge the real problem into calculation of the probability that over 4 students of a class are left-handed, then into the probability that no more than 4 students of a class are left-handed, finally make decision based on the probability that over 4 students in a class are left-handed. The first and third sub-question of the latter task is also pure thinking task of analyzing level, the second sub-question is simulation of analyzing level, and it helps to improve students' understanding of probability from multiple angles, make connection between experimental probability and theoretical probability, and develop students' awareness and ability of solving problems by modelling. Such similar tasks are very rare in both main text and after-class exercises of CN-PEP.

4.3 Presentation of content

Textbook should play an exemplary role in the aspects of didactical design of curriculum content; the introduction of big ideas should be well designed to connect students' existing physical reality and mathematical reality and stimulate students' zeal for learning. Here the processes of introducing binomial probability theorem in the six series of textbooks are compared.

JP-SKS directly gives the binomial probability theorem followed by a worked example and a relatively more difficult exercise. Then it turns to introduce the mean, variance and standard deviation of a binomial random variable. SG-PPE also directly gives the theorem and it only takes the throwing of a biased coin as an introduction before giving the probability formula. Both series of textbooks are not well designed in the introductory parts. The design in China, UK and U.S. textbooks is more interesting.

Because the related content of exclusive events and independent events as well as the combinations are learned, CN-PEP starts from repeated independent trials and gives students a primary and concrete experience of solving similar problems by finding the probability that a thumbtack tips up exactly once in three throws. Then it requires students to see whether there is any pattern and extend it to generality, get the theorem at last, and think about the connection between this theorem and the binomial theorem. Such introduction is relatively concise, however, it requires rigor reasoning in a specific example. The key to success lies in that students have mastered related previous knowledge. Thus it requires students to have a solid foundation.

UK-SMP starts from binostat (like a pinball machine) experiment, and leads students to understand the two meanings of numbers in Pascal triangle through questions, namely, it represents how many possible routes are there for the ball to drop from the starting point to the current position, and how many different ways are there to randomly take r objects out of n objects. In this way, combinations and the binomial probabilities are connected by Pascal's triangle. Students are then told that the probability formula is still true when two possible outcomes with unequal probabilities. This design requires less preparative knowledge and there is physical experiment or throwing coin to decide whether the ball falls to the left or the right on the paper. Not only students use their senses in solving the problem, but also their understanding of Pascal's triangle is improved. The downside is that the case of unequal probabilities is treated too simple.

US-SMP starts from the fixed column of big idea and points out that "In a binomial experiment, if an event E has probability p, then the probability that E occurs r times in n trials is a term in the expansion of $(p+q)^n$, where $q = 1 - p$". US-SMP firstly reviews how to find the probability of getting 2 heads in 5 tosses of a fair coin by expanded binomial expression, with the problem situation, it then explains the features of a binomial experiment and the relation between the number of outcomes and Pascal's triangle number. Next it transits from equal probability to unequal probability, explains the meaning of binominal trials and binomial probability, and gives the binomial probability theorem after examples and specific calculations. This design has relative high requirements on the preparative knowledge of students.

US-PHM firstly clarifies the learning objectives, then checks students' preparative knowledge of combinations and binomial theorem, and informs students which pages to review if necessary. These are regular columns in this textbook. Under the title of "Finding Binomial Probabilities", it firstly helps students to know what a binomial experiment is and to notice the characteristics of binomial experiments. The following Example 1 is to further understand binomial experiment, design and conduct a simulation to find the probability of getting at least two correct answers when guessing answers to three multiple-choice questions. Then it requires students to use the tree diagram to check the answer. Example 2 is a merchandising problem (unequal probabilities) and it is solved using the tree diagram. Then the binomial probability formula is given. The advantage of this design lies in that firstly it pays attention to students' cognitive differences and provides learning guidance for students who need extra help. Secondly, it focuses on the students' participation, and combines reading, observation, induction, simulation and theoretical analysis in one, which promotes students' understanding of knowledge and cultivates their ability. Though this series of textbook does not have high requirements for the depth and scope of contents, it is still worth for other textbooks to refer to in the design of statistical activities, in particular, in conducting simulations in statistics education.

Secondly, the use of statistical graphs and ICT in the parts of binomial experiment, binomial probability distribution and related problems of finding probabilities in the text of various series of textbooks are compared. Tab. 4-8 shows that CN-PEP rarely use them. As mentioned above, it does not include distribution graph in the teaching, gives little space for the calculation of probability and does not mention the impact of n, p changes on the binomial distribution graph, thus it rarely shows any opportunity in this part to use ICT. This is the direction of further improvement of CN-PEP. Although both British and Singapore's textbooks include probability calculation tasks, the difference is that UK-SMP uses binomial table while SG-PPE uses calculator. In terms of using graphing calculator, SG-PPE is the most prominent in the six series of textbooks and it turns graphing calculator into the tool of exploration, for instance, in teaching the mode of binomial distribution, CN-PEP uses inequality solving techniques, but SG-PPE adopts graphing calculator to calculate all the binomial probabilities and find the value of variable with the highest probability, which reduces the threshold of knowledge. In the calculation of binomial probabilities, it borrows graphing calculator and overcomes the calculation barrier and makes teaching more concentrated on explorations and applications. It includes the most types of probability calculation among the six series of textbooks, namely, eight types such as $\mathrm{P}(X \leq r)$, $\mathrm{P}(X > r)$, $\mathrm{P}(X \geq r)$, $\mathrm{P}(a \leq X < b)$. The numerical approach of SG-PPE is more acceptable to the public.

Tab. 4-8 Frequencies of using statistical graphs and ICT in the main text of selected topics

		CN-PEP	JP-SKS	SG-PPE	US-PHM	US-SMP	UK-SMP
Tables and graphs	Tables	0	1	1	1	2	5
	Graphs	0	1	0	3	14	2
ICT	Calculation	1	0	6	3	4	3
	Graphing	0	1	2	1	4	2
	Exploration	0	1	4	3	4	0

Finally, the authors compare the richness of using contexts in the parts of the binomial experiment, binomial probability distribution and the related problems of finding probabilities in the main text and after-class exercise of various series of textbooks.

Tab. 4-9 Numbers of contexts used in main text and after-class exercise of selected topics

		CN-PEP	JP-SKS	SG-PPE	US-PHM	US-SMP	UK-SMP
Context-free	Main Text	4	1	5	3	5	4
	After-class exercise	0	0	5	12	25	1
Game context	Main text	2	2	1	0	2	3
	After-class exercise	1	1	2	0	5	8
Personal context	Main Text	0	0	2	1	1	1
	After-class exercise	1	0	2	5	5	6
Public context	Main text	1	0	0	3	3	1
	After-class exercise	0	0	3	2	6	2
Occupational context	Main Text	0	0	0	1	0	0
	After-class exercise	1	0	4	3	1	5
Scientific context	Main text	0	0	0	0	0	0
	After-class exercise	1	0	0	0	0	3

Data from Tab. 4-9 shows that the British textbook is the richest in context types of after-class exercise, tasks of context-free or game context take the majority in various series of textbooks and they generally lack scientific context. US-SMP and UK-SMP textbooks include respectively 96 and 74 tasks in all, while it is 14 and 5 in Chinese and Japanese textbooks (see Tab. 4-7, because under the same exercise, one context appears in multi-task is only counted once in order to reflect the richness of contexts used in textbook, the total amount of task in Tab. 4-9 is not equal to that of Tab. 4-7). Thus the percentages of tasks with contexts are not low in China and Japan textbooks, which indicates that China and Japan have been aware of the value of using context, however, the textbook content is generally not substantial due to the limitation of development status of statistics education and statistics application. Statistics is highly applicable and high school textbooks can integrate more problems of scientific context.

5. Conclusions and Suggestions

The following main conclusions are obtained and suggestions are provided through our comparative study of six series of Chinese and foreign textbooks in the three aspects of the sub-category of binomial distribution, namely, selection of topics, design of learning tasks and presentation of content.

(1) CN-PEP does not give adequate attention to the binomial distribution and the related course needs to be deepened and extended.

The data shows that in the binomial distribution of the six series of textbooks, CN-PEP takes the least space and covers the least topics, which occupy only one section. In particular concrete presentations of the content such as distribution graphs and the numerical lists are missing, while such presentations could enhance students' experience of binomial distribution. Substantial connection with normal distribution and other knowledge are also missing, for instance, the characteristics of binomial distribution, which distribution can be used for approximation under which condition, are not mentioned. Examples, activities or exercises do not reflect the wide application of binomial distribution. It should use statistical graphs more frequently and introduce graphing calculator or other information technology in order to make the textbook more readable, explorative and related to applications.

(2) CN-PEP has good command of the cognitive level of tasks but various activities are needed and more attention should be paid to develop students' proficiency.

After comparison of common learning contents such as the binomial experiment, binomial probability distribution and related calculation of probability, it is found that although CN-PEP allots the fewest learning tasks, they are normally distributed to different cognitive levels, the tasks of understanding level and analyzing level are in majority and tend to be at analyzing level. It demonstrates high expectation to student learning. The shortfall is that all of these tasks are pure thinking type, there is no common activity such as building charts, let alone such activities as simulation experiments and investigation activities that are also rare in other textbooks. Knowledge learning and ability training are the core tasks of education, which need the careful design of textbook editors.

(3) CN-PEP emphasizes the design of introduction but lacks application materials.

Realistic mathematics education advocated by H. Freudenthal is widely recognized in China and CN-PEP also focuses on new knowledge introduction by connecting with students' social reality and mathematical background. In terms of introducing the binomial probability theorem, CN-PEP has a unique and concise design; however, it is more theoretically oriented and has high requirement for students' learning basis. Generally speaking, its learning tasks' design ignores the accumulation of students' manipulative experiences and rarely use visual graphs and tables; examples together with exercises are the main learning approach. Although education development is unbalanced all over the country, as a textbook, it should provide learning opportunities to students who have access to new technology. In the tasks arranged in CN-PEP, game context is dominant and the one with connection to real application is very few. This is not helpful for students to identify the value of statistical education and stimulate the enthusiasm for learning.

(4) With a simple and plain appearance, CN-PEP focuses on training students' thinking, but needs to extend the space for independent learning of students.

CN-PEP is simple in appearance and rich in internal contents. It usually emphasizes the introduction of knowledge and the training of students' thinking and problem solving, and often makes simple explanation or instruction by sidebars. It attaches importance to didactical design of mathematics contents with few signs of examination-oriented teaching. This gives a good guidance to students' learning. However, learning is ultimately a students' own behavior, CN-PEP may consider to follow the way of other textbooks, that is, on the basis of clarifying learning objectives and preparing knowledge, highlighting important concepts, presenting various representations of knowledge, providing links to further learning materials, and publishing exercise answers, give students more responsibilities and more scope for self-development. It is an important issue to be discussed as to how can textbooks better meet the need of different students in a populous country like China.

We need to analyze the results reported above objectively. For example, Singapore's textbook is prepared for students to attend A level statistics exam, who usually receive 13 or 14 years of school education before such test (Ginsburg *et al.*, 2005), thus statistics learning contents are naturally much more than high school students of China who only receive 12 years of school education. Shortage of application materials in textbook is related to the external elements such as the development and popularity degree of statistics in China and this cannot be solved by the efforts of the authors. Introduction of graphing calculator is also restricted by factors such as funding, teachers and materials.

Textbook contents do not identical to the materials taught in class and pure text research has limitations. Textbook cannot determine students' learning result itself. Pepin and Haggarty (2001) think that the writing and use of textbook reflect the culture of one country. This chapter only deals with one topic — binomial distribution, and therefore there are also some limitations in the research output. However, the main conclusions and recommendations are consistent with the final report completed by our sub-research group.

References

[1] An, Tingting (2013). A comparative study on probability distributions in high school textbooks [In Chinese]. Master Dissertation. Shanghai: East China Normal University.

[2] Charalambous, C. Y., Delaney, S., Hsu, H.-Y. and Masa, V. (2010). A comparative analysis of the addition and subtraction of fractions. *Mathematical Thinking and Learning*, 12(2): 117–151.

[3] Chen, Xiru (2004). The importance of statistics [in Chinese]. *Secondary School Mathematics Monthly*, (02), 1–2; 7.

[4] Fan, Lianghuo and Zhu, Yan (2004). How have Chinese students performed in mathematics? A perspective from large-scale international comparisons. In L. Fan, N-Y. Wong, J. Cai, and S. Li. (eds.), *How Chinese learn mathematics: perspectives from insiders* Singapore: World Scientific, pp. 3–26.

[5] Garfield, J., Ben-Zvi, D., Chance, B., Medina, E., Roseth, C. and Zieffler, A. (2008). Learning to reason about distribution. In J. Garfield and D. Ben-Zvi (eds.), *Developing students' statistical reasoning: Connecting research and teaching practice*. Dordrecht: Kluwer Academic Publishers, pp. 165–186.

[6] Ginsburg, A., Leinwand, S., Anstrom, T. and Pollock, E. (2005). What the United States Can Learn from Singapore's World-class Mathematics System (and What Singapore Can Learn from the United States): An Exploratory Study. Washington, DC: American Institutes for Research.

[7] Goodlad, J. I., Klein, M. F. and Tye, K. A. (1979). The domains of curriculum and their study. In: J. I. Goodlad *et al.* (eds.): *Curriculum Inquiry: The Study of Curriculum Practice*. New York: McGraw-Hill Book Company, pp. 43–76.

[8] Houang R. T. and Schmidt W. H. (2008). TIMSS International Curriculum Analysis and Measuring Educational Opportunities. In *3rd IEA International Research Conference*, Chinese Taipei. Retrieved from http://www.iea.nl/sites/default/files/irc/IRC2008_Houang_Schmidt.pdf

[9] Jia, Ping (2008). Comparison of Chinese and American middle school mathematics textbooks — focusing on statistics and probability [In Chinese]. Master's Thesis. Shenyang: Liaoning Normal University.

[10] Li, Yeping (1999). An analysis of algebra content, content organization and presentation, and to-be-solved problems in eight-grade mathematics textbooks from Hong Kong, Mainland China, Singapore, and the United States. Doctoral dissertation. Pittsburgh: University of Pittsburgh.

[11] MOE of PRC (Ministry of Education of the People's Republic of China) (2001). *Standards of Mathematics Curriculum for Compulsory Education (Experimental)* [In Chinese]. Beijing: Beijing Normal University Press.

[12] MOE of PRC (Ministry of Education of the People's Republic of China) (2003). *National Standards of Mathematics Curriculum for High Schools (Experimental)* [In Chinese]. Beijing: People's Education Press.

[13] MOE of PRC (Ministry of Education of the People's Republic of China) (2012). *Mathematics Curriculum Standards for Compulsory Education (2011 edition)* [In Chinese]. Beijing: Beijing Normal University Press.

[14] OECD (Organisation for Economic Co-operation and Development) (2010). *PISA 2009 Results: What Students Know and Can Do: Student Performance in Reading, Mathematics and Science (Volume I)*. OECD Publishing.

[15] OECD (Organisation for Economic Co-operation and Development) (2013). *PISA 2012 Assessment and Analytical Framework: Mathematics, Reading, Science, Problem Solving and Financial Literacy*. OECD Publishing.

[16] Pepin, B. and Haggarty, L. (2001). Mathematics textbooks and their use in English, French and German classrooms: A way to understand teaching and learning cultures. *Zentralblatt für Didaktik der Mathematik*, 33(5), 158–175.

[17] Robitaille, D. F., Schmidt, W. H., Raizen, S., McKnight, C., Britton, E. and Nicol, C. (1993). *Curriculum Frameworks for Mathematics and Science* (TIMSS Monograph No. 1). Vancouver: Pacific Educational Press.

[18] Stein, M. K., Remillard, J. and Smith, M. S. (2007). How curriculum influences student learning. In F. Lester (ed.), *Second Handbook of Research on Mathematics Teaching and Learning*. Greenwich, CT: Information Age Publishing.

[19] Valverde, G. A., Bianchi, L. J., Wolfe, R. G., Schmidt, W. H. and Houang, R. T. (2002). *According to the Book: Using TIMSS to Investigate the Translation of Policy into Practice through the World of Textbooks*. Dordrecht: Kluwer Academic Publishers.

[20] Wu, Yingkang and Gan, Yun (2013). A comparative study of applying new technology in statistics education in high school textbooks used in China and Britain [In Chinese]. *Secondary School Mathematics Monthly*, (1): 41–45.

[21] Zhang, Haiyue (2007). Comparative experimental study of "statistics and probability" in two mathematics textbooks compiled as per the national mathematics curriculum standards — "Shandong Edition" and "Bejing Normal University Edition" (Grade 7). Master's Thesis. Jinan: Shangdong Normal University.

[22] Zhao, Xiaoping and Yao, Xue (2010). A comparative study of statistics textbooks used in Shanghai and Singapore high schools [In Chinese]. *Mathematical Teaching*, (2): 9–11.

https://doi.org/10.1142/9789814713955_0005

Chapter 5

Comparative Study on Calculus in High School Mathematics Textbooks

Xiaoping Zhao*, Guangxiang Zhang, Ming Ni, Hua Huang, Hua Wang, Zhengwu Long, Xuefen Gao, Jiao Liu

Department of Mathematics, East China Normal University
Shanghai 200241, China
**xpzhao@math.ecnu.edu.cn*

In order to steadily put forward the reform in the teaching of calculus in high schools and carefully construct high-school calculus course, we aim to obtain rich information and useful ideas by investigating high school calculus textbooks from other countries, which can form a basis for reviewing our calculus-teaching reform over the last decades. This is an efficient way for us to look back at the experiences and adjust strategies to construct a more appropriate calculus course for high schools in China.

1. Overview

1.1 Significance of the study

It has become a trend for calculus to be a course taught in high schools. In particular, most countries with a well-developed economy and education system include calculus in their high school curricula. In China, from the 1978 *Full Time Secondary School Mathematics Syllabus (tentative version)* to the 2003 *National Mathematics Curriculum Standards for High Schools (Experimental)*, there has been calculus content in all versions of the mathematics syllabi/standards. However, over more than thirty years there have echoed the questions: Why study calculus in high schools? What should students learn and how? On these questions, the mathematics education researchers, mathematicians and mathematics teachers in high schools have different opinions. The sharp controversies lead to different appeals for reform and result in worries and hesitations in the implementation of reform.

Calculus was originally a college course. Its knowledge structure was built in the 19th century based on the strict foundation of theories of limit and real numbers. The theories of limit and real numbers are highly rigorous and abstract; calculus requires deep fundamental knowledge and mathematical methods which are difficult for high school students. In addition, calculus is a prerequisite course for college students in science, engineering and economics majors. These majors differ in the required depth and breadth of calculus knowledge, and their calculus courses are thus often independently constructed. Thus, calculus as a high school course should fit the common cognitive features of high school students and at the same time, as well take into account their different needs after graduating from high schools. Hence, how to appropriately select teaching contents, grasp teaching requirement, and attain teaching goals catering high school students within the limited amount of time become important issues to discuss.

In order to stably promote the reform in teaching calculus in high schools and discreetly construct the high-school calculus course, we hope to obtain rich information and beneficial ideas by investigating calculus textbooks for high school students in other countries, which can form a basis for reviewing our decades of calculus-teaching reform. This is an efficient shortcut for us to summarize experiences and adjust strategies to construct a more suitable calculus course for high schools.

1.2 Basic information about sample textbooks

In this study, mathematics textbooks of high schools from six countries, namely China, Japan, Singapore, Germany, France and Russia have been selected as samples to be analyzed. These countries have diverse cultural traditions and good achievements in mathematics teaching. Moreover, in the 21st century, these countries have carried out large-scale mathematics curriculum reforms, and published a variety of mathematics textbooks for schools, which reflect their reform intentions and specific arrangements while retaining their respective traditions in mathematics education. Differences exist among the course design and education systems in different countries. Some countries have a nationwide curriculum standard or syllabus, some have multiple local standards, and others only have recommended curriculum standards from academic institutions. Therefore, the content selection and the course requirements of high school textbooks are subjected to different levels of regulation in different countries. In order to ensure the validity of this study, we first investigate the background of each sample textbook.

(1) In the high school phase of China, a three-year school system is typically adopted. The currently used mathematics textbooks were written in accordance with the *National Mathematics Curriculum Standards for High Schools (Experimental)* issued in 2003, and they consist of compulsory courses and optional courses. The compulsory courses must be learned by all high school students while the optional courses are further divided into four series, and students aiming at humanities/social sciences, science/engineering/economics, and arts/sports in the future are respectively assigned different series among them. The course requirements for science/engineering/economics students are higher, and the number of such students is also the largest, so the textbooks used by these students will be considered in this study, including five textbooks for the compulsory courses, three textbooks from Optional 2 (optional courses assigned to these students), and parts of the topics from Optional 3 and 4 (selected at students' will). Calculus content is designed in the second volume of textbooks of the Optional 2 (Optional 2-2 for short). The selected textbook, *High School Mathematics Textbook* (Edition A) published by the People's Education Press, is the most extensively used one throughout China.

(2) Japan adopts a three-year schooling system for high schools, and the mathematics textbooks selected in this study were written in accordance with the *Mathematics Study Guidance for High Schools* issued in 1999 by the Monbusho (Ministry of Education), including Mathematic Foundation, Mathematics I, Mathematics II, Mathematics III, Mathematics A, Mathematics B, and Mathematics C, seven books in total. Mathematic Foundation and Mathematics I are required for all students and the rest are either optional or partially required. Mathematics II includes a wide range of important mathematical topics and most Japanese students choose "Math II/Math B Exam Paper" in their College Entrance Examination, and in the entrance examinations of the most prestigious

universities, even contents from Mathematics III and Mathematics C are frequently involved as well (Yuelan Chen, 2006). Hence, it is necessary to conduct a comprehensive study of all seven Japanese high school mathematics textbooks, especially Mathematics II and Mathematics III, which contain the calculus content. The selected Japanese textbooks in this study are the ones published by the Sukan Shuppan Publishing Company, which boast of a large market share in Japan[a].

(3) The schooling system of basic education in Singapore is different from that of China; Singaporean students study for four years (special or express programs) or five years (regular program) in secondary schools before taking a "General Certificate of Education Ordinary Level" test, (O-level Test). Those who pass the O-level Test could enter a two-year or three-year higher schools or junior college, after which they need to pass a "General Certificate of Education Advanced Level" test (A-level Test) to obtain the qualification to enter a university. The higher school or junior college students are of similar age to Grade 11 to 12 students in Chinese high schools. The courses they learn are of three types — H1, H2 and H3. According to the A-level examination syllabus issued by Singapore Examinations and Assessment Board in 2013, there are two levels of requirements on mathematics, namely H1 and H2. Both levels include calculus, with the H2 requirement being stringent and fairly similar to the teaching requirement of calculus content in the O-level "Additional Mathematics" course as specified by the *Secondary Mathematics Syllabus* issued by the Singapore Ministry of Education in 2006. In this study, the textbooks entitled *H2 Mathematics 1 — A Comprehensive Guide*, Volumes 1 & 2 published by the Singapore Panpac Education Press are selected as high school textbook samples from Singapore.

(4) There are three types of secondary schools in Germany: Gymnasium, Hauptschule and Realschule. The Gymnasium which teaches arts and sciences in more depth, is divided into two periods: the first period

[a]Japan has issued a new mathematical "Study Guidance" since 2009, in which the original seven subjects were revised and integrated into six textbooks, namely, Mathematics I, Mathematics II, Mathematics III, Mathematics A, Mathematics B and Flexible Mathematics Application, and the calculus content were also re-integrated. This study started relatively earlier, therefore the 1999 "Study Guidance" edition textbooks were selected as samples.

is mainly meant for children at the age of 10–15 or 10–16 (slight differences exist among the states), corresponding to the higher grades of primary school in addition to the middle school period in China; the second period — in either a regular high school or a vocational high school is meant for students at the age of 15–18 or 16–19. Students from regular high school can enter universities after graduation, corresponding to students in Chinese high schools (Dayuan Jiang, 2005). The federal states of Germany enjoy a high level of autonomy in education and each state has its own syllabus and textbooks, among which Bavaria holds a long-term crown in educational quality. Since 2007, Bavaria has launched a comprehensive reform and revision over its mathematics syllabus (Binyan Xu, 2009), which is expected to embody the direction and characteristics of the reformation of school mathematics courses in Germany. Therefore, in this study the *Mathematics Textbooks* Volume 10–12 used in Bavaria (published by the Ernst Klett Verlag) are selected as sampled high school textbooks from Germany.

(5) The currently used mathematics syllabus in France has been in effect since 2000. It was developed by a subordinate math science and technology group in CNP (le conseil national des programmes). This syllabus reflects the latest reform orientation of mathematics teaching in France, which specifies unified regulations on course content, teaching requirements and teaching hours for the students between middle school and grade one in high school. Starting from Grade Two in high schools, the students are grouped into natural sciences, literature and economics majors. Although mathematics is still the required course for all majors, the focus of content and teaching requirements vary among different majors (Shuping Pu, 2012). There are many series of mathematics textbooks for high school students currently used in France. This study selected two series of textbooks published by Editions Belin and Hachette Education, respectively. Both series of textbooks were written after the France mathematics curriculum reform in 2000 and applicable to majors of natural sciences. In addition, the two series are widely used and fairly representative among mathematics textbooks in France.

(6) The currently implemented mathematics syllabus in Russia was issued by the Education Ministry of Russian Federation in 2003, in which

Grades 5–9 are the period of incomplete secondary education and Grades 10–11 are the period of secondary general education. The two periods constitute the complete secondary education. The graduates from Grade 11 of regular schools can apply for colleges and universities. Therefore, the Grades 10–11 students correspond to those who study in high schools in China. During this period, the teaching plan can be classified into two types: "basic level" and "professional level (higher level)". The Ministry of Education of Russian issued the No. 349 Decree in 2007 and promulgated the recommended list of textbooks to the educational organizations. The whole series of mathematics textbooks for Grades 10–11 published by the Moscow Education Press (Просвещение Издательство), includes *Algebra and Elements of Analysis* (Алгебра и начала анализа) for Grade 10, Algebra and Elements of Analysis for Grade 11, and Geometry for Grades 10–11. This is the series under the Moscow University School Mathematics Project and the only series of mathematics textbooks applicable to both "basic level" and "professional level" students, where content meant for "professional level" students are marked with asterisks. Teaching suggestions for each level are also given at the end of each textbook. As V. A. Sadovnich (В. А. Садовнич), an academician of Russian Academy of Sciences and the president of Moscow University, pointed out in the preface of the textbooks: "This series of textbooks enable school students to obtain a good basic education, and help them develop a proper perspective over basic scientific knowledge." Therefore, this series of textbooks were selected as the sample textbooks from Russia.

In summary, the high schools of the above six countries are either two-year or three-year long; the mathematics textbooks from all the surveyed countries have both "higher requirement" (for natural sciences majors) and "lower requirement" (for arts, social science majors) versions. All the sample textbooks selected in this study are the "higher requirement" version and are all published after the educational reforms in these centuries.

All six series of textbooks selected contain the content of calculus, and the details are listed in the following Tab. 5-1.

Tab. 5-1 Sample textbooks

Country	Publisher	Textbook (publication year)	Code
China	People's Education Press	Optional 2-2 (2007) (Version A)	CN-PEP
Japan	Sukan Shuppan	New Math II (2007) & III (2004)	JP-SKS
Singapore	Panpac Education	H2 Mathematics Vol. 1	SG-PEH
Germany	Ernst Klett Verlag	Lambacher Schweizer Mathematik für Gymnasien, 10, 11 and 12 (2008)	DE-LBS
France	Editions Belin Hachette Education	Math 1^{re}S (2005) Déclic Maths Terminale S enseignement obligatoire (2002)	FR-EDB FR-HLT
Russia	Просвещение Издательство	Алгебра и начала анализа, 11 (2006)	RU-MGU

US textbooks are not selected in this study, because calculus courses in US are mainly taught in AP (Advanced Placement) program which run by the US College Board for those students who apply for specific majors in colleges and intend to obtain replaced credits. The AP courses have their own textbooks on specified topics, e.g., *Calculus* by Ron Larson *et al.* and *Calculus: Concepts and Applications* by Paul A. Foerster. These textbooks are usually over 1,000 pages and can cover almost all of single variable calculus normally learned in university. Perhaps due to the existence of these AP courses, calculus content in regular US high school mathematics textbooks is insignificant and not representative of the actual amount of calculus studied in the schools. For example, in the school mathematics textbooks published by the Chicago University Press (under the University of Chicago School Mathematics Project, or UCSMP for short), only the so-called "Precalculus" is covered, while the high school mathematics textbooks published by Prentice Hall contains no calculus content at all. In this situation, it is inappropriate to compare either the regular high school textbooks or the AP calculus materials with the textbooks from other countries.

1.3 Issues to be discussed

Through the comparison of textbooks from various countries, we hope to arrive at a better understanding of the following questions:

(1) How the teaching objectives of high school calculus are positioned of various countries and what positive role they are expected to play for

students. This will help us to set up the premise "why should high school students learn calculus".

(2) What content of calculus are chosen in high school textbooks of various countries, whether such content can reflect the key ideas and the value of applications of calculus, how to connect up properly with other mathematical content in high school textbooks, and how much calculus content can a high school textbook accommodate. It will help us to further clarify the basic question "what content of calculus must be learned and can be learned by high school students", providing us with ideas and materials to build a more rational and scientific high school calculus course.

(3) How the textbooks of various countries find a compromise between the theoretical rigor of the classic calculus and the limited cognitive abilities of high school students. It will help us to find a balanced strategy between the scientificness and instructiveness in the high school calculus course, making calculus easy to learn and encouraging students' enthusiasm on one hand, and keeping mathematics as systematic and rigorous as possible on the other hand.

1.4 Research approach

In this study, both quantitative and qualitative analyses are adopted.

(1) With quantitative approach, the textbooks will be analyzed and compared in terms of:

- weight of calculus in high school mathematics;
- the breadth of calculus knowledge;
- the density and distribution of calculus material in the textbooks.

Certain case studies are involved.

(2) The qualitative analysis mainly concerns the integrity, logicality, rigorousness and conciseness of the knowledge structure in the textbooks, in particular:

- appropriateness and completeness of the knowledge structure;
- the rigorousness of the deduction process;
- the balance between simplicity and logical rigor.

Case studies are involved as well.

The study only analyzes the main text of the textbooks, generally reviewing the whole text and then carefully reviewing from one mathematical topic to another. Therefore, the common topics in the textbooks and unique features of each series of textbooks could be summarized.

2. Comparison of Calculus Content

2.1 Weight comparison of calculus in high school textbooks

The school system and textbook structure are not alike among the surveyed countries. Different countries may have two-year or three-year school systems, or have two or three different levels of teaching requirements. The courses are divided into the compulsory, optional or extensive types. Therefore, the situation is complicated. In this study, the proportion of pages used by calculus content over an entire textbook series is taken as the basis to describe the weight of calculus in high school textbooks, as is shown in Tab. 5-2. The calculus contents counted in Tab. 5-2 include limit of sequence, limit and continuity of functions, and differentiation and integration. For detailed information, please refer to Section 2.2.

Tab. 5-2 Weight statistics of calculus in high school mathematics textbooks

Textbooks	Total pages of textbooks	Pages on calculus	Percentage	Textbooks counted as the totality
CN-PEP	1004	67	7%	Compulsory 1~5 and Optional 2-1, 2-2, 2-3
JP-SKS	1060	197	19%	Mathematic Foundation; New Mathematics I, II, III, New Mathematics A, B and C
SG-PEH	598	102	17%	H2 Mathematics Vol. 1 and Vol. 2
DE-LBS	552	229	41%	Lambacher Schweizer Mathematik für Gymnasien 10, 11 and 12
FR-EBD /HLT	1038	253	24%	Math 1reS, Maths 2e, and Maths Terminale S enseignement obligatoire
RU-MGU	1186	202	17%	Алгебра и начала анализа.10 & 11, Геометрия 10-11

As can be seen from Tab. 5-2, there is a significant difference in the weight of calculus content in the high school mathematics textbooks. DE-LBS and FR-EBD have a greater weight, particularly DE-LBS, whose calculus weight reaches 41% of the high school mathematics. JP-SKS,

RU-MGU and SG-PEH are all close to a 20% in proportion. The calculus takes the minimum weight in China, only at approximate 7%.

It must be mentioned that some textbooks (such as DE-LBS, FR-EBD, and RU-MGU) give a deeper review of functions before the formal introduction of calculus, studying properties of functions from the viewpoint of limit and continuity, which makes it difficult for us to detach precisely calculus content from the review of functions. In Tab. 5-2, such content is also counted in the calculus pages, so the weights calculated from the data are not precise, yet they could still interpret the weights to some extent.

2.2 Content range comparison of calculus

The calculus content selected by the six series of textbooks all belongs to single-variable calculus, and mainly consists of derivatives, integrals and their applications. Since the topic of limit and continuity are not only the basis of modern calculus, but also the main indicators of the high school calculus teaching requirements, such content in the high school textbooks are also classified as calculus content. There is some coverage of differential equations in the textbooks of some surveyed countries, and we categorize such content as the applications of calculus. Therefore, in this study, we roughly divide the content scope of calculus into "limit and continuity", "derivatives" and "integrals", involving six knowledge points in total (see Tab. 5-3). Any content of such knowledge points contained in the textbooks, regardless of its quantity or depth are all listed as a solid square "■" in the table. The blank square "□" indicates a missing knowledge point. (We will use these marks in the sequel.)

Tab. 5-3 Content range of calculus in the sample textbooks

Textbook	Limit and continuity			Derivative	Integral	
	Limits of sequence	Limits of function	Continuous function		Indefinite	Definite
CN-PEP	■	■	■	■	■	■
JP-SKS	■	■	■	■	■	■
SG-PEH	□	■	□	■	■	■
DE-LBS	■	■	□	■	■	■
FR-EBD	■	■	■	■	■	■
RU-MGU	■	■	■	■	■	■

From Tab. 5-3 we see that the major topics of calculus covered by the textbooks of various countries are very similar. They differ from each other only in the part of "limit and continuity", which indicates that all the surveyed countries mostly agree on the scope of calculus to be taught in high schools.

2.3 Comparison of the specific knowledge points being covered

To facilitate a more comprehensive and in-depth view of the calculus content in the sample textbooks, the chapter and section titles for calculus content in various textbooks are listed below.

CN-PEP

Chapter 1 Calculus and its applications

Section 1 Rates of change and derivative

Section 2 Computation of derivatives

Section 3 Applications of derivatives in the study of functions

Section 4 Examples of optimization problems in daily life

Section 5 Definite integral

Section 6 Fundamental theorem of calculus

Section 7 Simple applications of definite integrals

JP-SKS, New Math II

Chapter 6 Differentiation and integration

Section 1 Differential coefficients and derived function

Section 2 Changes of values of functions

Section 3 Integration

JP-SKS, New Math III

Chapter 2 Limits

Section 1 Limits of number sequences

Section 2 Limits of functions

Chapter 3 Differentiation

Section 1 Derived functions

Section 2 Derived functions of elementary functions

Chapter 4 Applications of differentiation

Section 1 Applications of derived functions

Section 2 Other applications

Chapter 5 Integration and its applications

Section 1 Indefinite integrals

Section 2 Definite integrals

Section 3 Applications of integration

SG-PEH

Chapter 5 Calculus

Section 1 Differentiation

Section 2 Mclaughlin series

Section 3 Integration

Section 4 Definite integrals

Section 5 Differential equations

DE-LBS, 10

Chapter 6 Properties of functions and their graphs

Section 4 Limit at infinity

DE-LBS, 11

Chapter 1 Graphs of rational functions

Section 1 Behaviors around the gaps of the domains

Section 2 Behaviors at infinity

Section 3 Relations between graphs and factors

Chapter 2 Local and global differentials

Section 1 Difference quotient and average rate of change

Section 2 Differential quotient and local rate of change

Section 3 Differentiability

Section 4 Derived functions

Section 5 Primitive functions

Section 6 Derivative of $x \to x^n$ $(n \in \boldsymbol{Z})$

Section 7 Rules for sum and scalar multiple

Section 8 Rules for product and quotient

Chapter 3 Applications of derivative

Section 1 Monotonicity

Section 2 Extreme point and extremum

Section 3 Conditions for extrema

Section 4 Explorations of rational functions

Section 5 The Newton method

Chapter 5 Other rules of derivation

Section 1 Derivatives of sine and cosine functions

Section 2 Inverse functions

Section 3 Composite functions

Section 4 Derivatives of composite functions

Section 5 Power functions with rational exponents and their derivatives

Chapter 6 Natural Exponential function and natural logarithmic function

Section 1 Natural Exponential function and its derivative

Section 2 Natural logarithmic function and its derivative

Section 3 Derivatives of complicated functions

Section 4 Exponential functions and exponential equations
Section 5 Exponential and logarithmic functions and their graphs
Chapter 8 Applications of differentiation
Section 1 Polynomial functions in the real situations
Section 2 Extreme value problems
Section 3 Complicated extremum problems
Section 4 Functions with parameters
Section 5 Determination of function expressions
Section 6 Functional adaptations

DE-LBS, 12

Chapter 1 Integration
Section 1 Local rates of change and total amount of change
Section 2 Integral
Section 3 Integral as area; integral function
Section 4 Fundamental theorem of calculus
Section 5 Primitive functions
Section 6 Properties of primitive functions and integrals
Section 7 Computation of areas with integrals
Section 8 Infinitely extensive areas
Chapter 2 Further properties of functions and their graphs
Section 1 Second derivatives
Section 2 Bending of graphs
Section 3 Turning point, type of extrema
Chapter 6 Applications of differentiation and integration
Section 1 Exponential increase
Section 2 Determination of doubling time and half-life period
Section 3 Review of extremum problems
Section 4 Exploration of complicated functions

FR-EDB

Chapter 3 Derivation and applications
Section 1 Derived number — tangent
Section 2 Derived functions
Section 3 Applications of derivation
Chapter 4 Asymptotic behavior — limits
Section 1 Asymptotes of a curve
Section 2 Limits of reference
Section 3 Operations on limits

FR-HLT

Chapter 1 Functions — variations and continuity
Section 1 Review of derivatives
Section 2 Variation of a function

Section 3 Continuity
Section 4 Properties of continuous functions
Chapter 2 Limits of sequences and functions
Section 1 Limits of sequences
Section 2 Limits of a function at $+\infty$ *or* $-\infty$
Section 3 Limit of a function at a real number a
Section 4 Limits and orders
Section 5 Limits and operations
Section 6 Theorem of intermediate value over open intervals
Chapter 3 Derived and primitive functions
Section 1 Complements on derivation
Section 2 Primitive functions
Chapter 4 Exponential functions
Section 1 Functions verifying $f' = kf$
Section 2 Study on exponential function
Section 3 Differential equations $y' = ay + b$
Chapter 5 Logarithmic and exponential functions
Section 1 Function ln
Section 2 Study on ln
Section 3 Derived and primitive functions
Section 4 Powers of a strictly positive number
Chapter 6 Sequences and recurrence
Section 1 Recurrence
Section 2 Behavior of a numeric sequence
Section 3 Adjacent sequences
Section 4 Convergence of monotonic sequences
Chapter 7 Integral
Section 1 Notion of integral of a positive function
Section 2 Integral of a function of arbitrary sign
Section 3 Integral and primitive function
Section 4 Properties of integrals
Section 5 Integration by parts

RU-MGU

Chapter 1 Functions, derivatives and integrals
Section 1 Functions and their graphs
Section 2 Limits of functions and continuity
Section 3 Inverse functions
Section 4 Derivative
Section 5 Applications of derivatives
Section 6 Primitive function and integral

2.4 Comparison of several typical topics

Although the major topics of calculus covered by the high school textbooks in surveyed country are almost the same, the detailed coverages of each topic are very different. In order to see the differences, we will study certain selected typical topics from the fundamental topics in calculus (such as limit and continuity), the skills of operations (such as deduction of derivative formulas and skills in integral calculation), and applications (such as applications of derivatives and of definite integrals).

2.4.1 Comparison of the coverage of limit and continuity

From the viewpoint of mathematical logic, notions of limit and continuity serve as the foundation of calculus. However, vigorous treatment of limit and continuity is too abstract for high school students and might be a huge barrier in their learning of calculus. It is challenging to write a textbook of calculus which can balance the logical rigorousness and students' cognitive level. In addition, the breadth of content, the depth of theory and the writing style of the whole calculus content are greatly affected by the comprehensiveness of the treatment of limit and continuity in each series of textbooks. In the following we will make detailed comparison and analysis.

First we note all the textbooks avoid the use of the "$\varepsilon - N$" or "$\varepsilon - \delta$" language, which helps to decrease the difficulty in calculus course. This seems to be a consensus among the textbook writers in all the surveyed countries. Even with such consensus, the concepts of limit and continuity are put across quite differently. Some textbooks take a comparatively rigorous approach: in certain textbooks the limit process is presented with the help of a gradually-approaching table of values or the graph of a function, elaborating the concept of limit intuitively. In this way the knowledge points related to limit and continuity could be presented relatively completely. Other textbooks avoid mentioning limit and continuity, or describe them as vaguely as possible. Hence the corresponding knowledge points are relatively few. Some related typical knowledge points concerning limit and continuity and their appearance in the surveyed textbooks are listed in Tab. 5-4.

Tab. 5-4 Knowledge points related to limit and continuity

Knowledge points	CN-PEP	JP-SKS	SG-PEH	DE-LBS	FR-EDB	RU-MGU
Limit introduced before derivatives	□	■	□	■	□	■
$\lim_{x\to x_0} f(x) = a$	■	■	■	■	■	■
$\lim_{x\to x_0} f(x) = \infty$	□	■	□	■	■	■
$\lim_{x\to\infty} f(x) = a$	□	■	□	■	■	■
$\lim_{x\to\infty} f(x) = \infty$	□	■	□	■	■	■
Operational rules of limits	□	■	□	□	■	■
Analytical properties of limits	□	■	□	□	■	■
One-sided limit	□	■	□	■	■	■
Methods of evaluating limit	□	■	□	■	■	■
Relationship between continuity and limit	□	■	□	□	■	■

As can be seen from Tab. 5-4, JP-SKS, FR-EDB, and RU-MGU contain more knowledge points on limit and continuity while CN-PEP and SG-PEH contain very few. The different situations will be briefly introduced below with the textbook contexts.

(1) Cases with less content of limit and continuity

Example 1. In CN-PEP, the term "limit" and the notation "lim" are first introduced in the definition of derivative with three steps towards the introduction.

Step 1. Calculation and listing. Around $t = 2$, calculate the average speed $\bar{v} = \frac{h(2+\Delta t)-h(2)}{\Delta t}$ for Δt having values 0.01, 0.001, 0.0001, ... and $-0.01, -0.001, -0.0001$, ..., and list the results in a table.

Step 2. Observing the trend. By observing the table, $\bar{v}$ is discovered to be closer and closer to the value -13.1.

Step 3. "Limit" coming on stage. The observation leads to the notion of "limit", and an expression with "lim" is given:

> To facilitate expression, we use $\lim_{\Delta t\to 0}\frac{h(2+\Delta t)-h(2)}{\Delta t} = -13.1$ to represent "around $t = 2$, when Δt approaches to 0, the average speed $\bar{v}$ approaches to a definite value -13.1".

Meanwhile, an aside remark is given, explaining the meaning of the term "limit":

> The definite value -13.1 is called the *limit* of $\frac{h(2+\Delta t)-h(2)}{\Delta t}$ when Δt approaches to 0.

Here the notion of limit is given through the observation of only one specific example, with the sole aim of allowing the definition of derivatives to be given. Since the definition, properties and operations of limits are not further discussed, it is difficult to deduce the derivatives of all elementary functions and the derivation rules from the definition of derivatives. Therefore, the textbook has no choice but to list out most of the derivative formulas and derivation rules directly rather than by deduction.

Also, in CN-PEP, the notion of "continuous function" first appeared in the form of an intuitive description:

> If the graph of function $y = f(x)$ in an interval $[a, b]$ is a continuous curve, it must have maximum and minimum values.

Based on this description, the term "continuous function" starts to be used in the following pages to elaborate the fundamental theorem of calculus:

> If $f(x)$ is a continuous function in the interval $[a, b]$ and $F'(x) = f(x)$, then $\int_a^b f(x)dx = F(b) - F(a)$.

In fact, the textbook did not give a definition of continuous function. The abuse of the term "continuous function" in the following text only relies on the intuitive fact that graphs of elementary functions in their domains are continuous curves.

Example 2. In the SG-PEH, the concept of limit and continuity is mentioned even less. When introducing the definition of derivatives, the textbook simply makes Δx smaller and smaller so that the secant slope $\frac{f(x+\Delta x)-f(x)}{\Delta x}$ approaches a definite value. The textbook directly describes this kind of phenomenon with the term and notation of limit, and present the definition of the first derivative of $y = f(x)$ as $\frac{dy}{dx} = f'(x) = \lim_{\Delta x \to 0} \frac{f(x+\Delta x)-f(x)}{\Delta x}$.

Prior to this definition, neither the notion of limit nor the notation "lim" are mentioned in the textbook. Then with the above definition, the textbook demonstrates the method of derivation with $y = x^2$ and $y = \sin x$ as examples. Then the textbook contains no further material of limit and continuous functions. Although the notions of limit and continuity are necessary for the construction of a deductive system of calculus, this series of textbooks requires students to do calculation of derivatives and integrals by using graphing calculators, overpassing the theoretical necessity.

(2) Examples with more content on limit and continuity

Example 3. There are two cycles of calculus in JP-SKS, which are presented in the sample textbooks *New Math II* and *New Math III* respectively.

In *New Math II*, "limit" is processed in a way similar to that of CN-PEP. Students first observe a list of values from one single example to feel the phenomenon of limit. Then the textbook presents the language and notation of limit and uses them to define derivatives. The derivatives of $f(x) = 2$, $f(x) = x$ and $f(x) = x^3$ are deduced from the definition, and the derivative of $f(x) = x^n$ is obtained by an incomplete induction. The remainder of the textbook involves only the calculus of polynomial functions.

In *New Math III*, the content of limit and continuity are more completely presented. Before Chapter 3 on differentiation, the entire Chapter 2 is dedicated to the concepts of limit and continuity. The chapter and section titles and the main knowledge points in this chapter are as follows:

Chapter 2 Limits
- Section 1 Limits of sequences
  1. Limits of sequences
     - A Concept of limits of sequences
     - B Convergent sequences
     - C Properties of limits of sequences (1)
     - D Properties of limits of sequences (2)
  2. Limits of infinite geometric sequences
     - A Limits of sequences $\{r_n\}$
     - B Applications of limits of sequences $\{r_n\}$
     - C Limits of inductively defined sequences
  3. Infinite geometric series

A Convergence and divergence of infinite series
B Infinite geometric series
C Infinite geometric series of point motion
D Infinite geometric series on similar figures
E Properties of infinite series

Section 2 Limits of functions

4. Limits of functions (1)
   A Properties of limits of functions
   B Evaluation of limits
   C The case of the limit being infinity
   D One-sided limits
5. Limits of functions (2)
   A Limits at $x \to \infty$, $x \to -\infty$
   B Limits of exponential and logarithmic functions
6. Limits of trigonometric functions
   A Limits of trigonometric functions
   B $\lim_{x\to 0}\frac{\sin x}{x} = 1$
7. Continuity of functions
   A Continuity of functions
   B Continuous functions over an interval
   C Properties of continuous functions over closed intervals

From the list we see that *New Math III* contains not only the concepts of limits of sequences and functions, rules for arithmetic operations of limits, but also other important properties of limits. For example, Section 1 has a subsection "properties of limits of sequences (2)" including properties of limits of sequences like

Property 5: If $a_n \leq b_n$ for all *n*, and if $\lim_{n\to\infty} a_n = a$, $\lim_{n\to\infty} b_n = b$, then $a \leq b$.

Property 6: If $a_n \leq b_n$ for all *n*, and if $\lim_{n\to\infty} a_n = \infty$, then $\lim_{n\to\infty} b_n = \infty$.

Property 7: If $a_n \leq c_n \leq b_n$ for all *n*, and *if* $\lim_{n\to\infty} a_n = \lim_{n\to\infty} b_n = a$, then $\lim_{n\to\infty} c_n = a$.

In Section 2, the subsection "limit of trigonometric functions" includes the following properties of limit of functions:

Suppose $\lim_{x\to a} f(x) = \alpha$ and $\lim_{x\to a} g(x) = \beta$.

Property 5: If $f(x) \leq g(x)$ for all *x* in a neighborhood of *a*, then $\alpha \leq \beta$.

Property 6: If $f(x) \leq h(x) \leq g(x)$, for all x in a neighborhood of a, and if $\alpha = \beta$, then $\lim_{x \to a} h(x) = \alpha$.

These properties not only mathematically present the relationship between the process and the results of limits, but also provide powerful tools for the rigorous proofs of important results that followed, such as the proofs of $\lim_{x \to 0} \frac{\sin x}{x} = 1$ and $\lim_{x \to 0} x \sin \frac{1}{x} = 0$.

New Math III also presents the definition of continuous functions in an explicit and orderly manner, which paves the way for the instruction of subsequent material.

(i) The definition of function continuity at a point is given:

> Suppose function $f(x)$ is defined in a neighborhood of $x = a$, and $\lim_{x \to a} f(x)$ exists; if $\lim_{x \to a} f(x) = f(a)$, then we say that $f(x)$ is continuous at $x = a$.

(ii) The definition of left (or right) continuity of functions at a point is given (the definition of left and right limit has been given previously when defining limits):

> Suppose function $f(x)$ is defined in a certain left (or right) neighborhood of $x = a$, and $\lim_{x \to a^{+0}} f(x)$ (or $\lim_{x \to a^{-0}} f(x)$) exists. If $\lim_{x \to a^{+0}} f(x) = f(a)$ (or $\lim_{x \to a^{-0}} f(x) = f(a)$), then we say $f(x)$ is left (or right) continuous at $x = a$.

(iii) The definition of function continuity over various intervals (such as (a, b), $[a, b]$ or (a, ∞)) is given.

(iv) The continuity of elementary functions over their defining domains is discussed.

(v) The two important theorems involving continuous functions over a closed interval, namely the extremum theorem and the intermediate value theorem, are given (the two theorems are not proved rigorously in the textbooks, instead, they are only illustrated with graphs).

It can be seen that the content of limit and continuity in JP-SKS, particularly the second circle, are comparatively richer, thus guaranteeing that the presentation of subsequent content could maintain the traditional deductive style.

Example 4. FR-EBD/HLT also divide the calculus content into two cycles, one in FR-EDB, the 2nd grade textbook for high school, including Chapter 3 "derivative and its application" and Chapter 4 "asymptotic behavior — limit". The second cycle is placed in FR-HLT, the 3rd grade textbook for high school, involving 6 chapters of content. Such design is somewhat similar to that of the JP-SKS. But compared to JP-SKS, the calculus content in FR-EDB/HLT is more systematic and employs a more rigorous style, which is evident even from the first cycle in FR-EDB, for instance:

(i) The derivative concept is expressed using a generic function $f(x)$. Most high school calculus textbooks take a specific function as example to explain the concept of derivative, but in FR-EDB, a more abstract approach of using a generic function $f(x)$ is employed from the very beginning to elaborate the definition of a derivative, i.e., to express limit problem with generic function $f(x)$.

(ii) The definition of derivative is presented in two limit forms. When the average rate of change of $f(x)$ around x_0 is expressed both the forms $\frac{f(x)-f(x_0)}{x-x_0}$ and $\frac{f(x_0+h)-f(x_0)}{h}$ are used, so the derivative concept has two limit expressions:

$$f'(x_0) = \lim_{x \to x_0} \frac{f(x)-f(x_0)}{x-x_0} \text{ and } f'(x_0) = \lim_{h \to 0} \frac{f(x_0+h)-f(x_0)}{h}.$$

(iii) Counterexamples that limit does not exist are given. Although Chapter 3 of FR-EDB is only the first time that the limit concept is encountered by the students, three typical counterexamples (see Fig. 10-1) are already given in the textbook, which undoubtedly enables students to have a more accurate understanding of the limit concept and thus derivative concept.

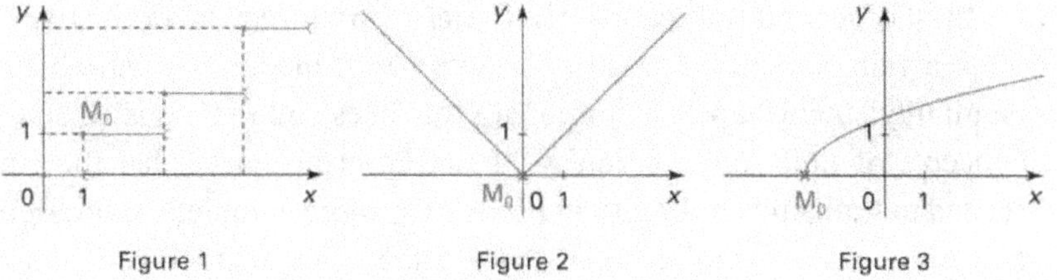

Fig. 5-1 Counterexamples in p. 77, FR-EDB (1)

(iv) The "limit" concept is generalized to involve infinity. In Chapter 3 of FR-EDB, the limit concept of a function at a point x_0 inside the defined domain is given, and then in Section 1 of Chapter 4, limit phenomena including one-sided cases or infinite cases are introduced, for instance:

$$\lim_{\substack{x\to x_0\\x>x_0}} f(x) = +\infty, \qquad \lim_{\substack{x\to x_0\\x>x_0}} f(x) = -\infty,$$

$$\lim_{\substack{x\to x_0\\x<x_0}} f(x) = +\infty, \qquad \lim_{\substack{x\to x_0\\x<x_0}} f(x) = -\infty;$$

$$\lim_{x\to+\infty} f(x) = l, \qquad \lim_{x\to-\infty} f(x) = l,$$

$$\lim_{x\to+\infty} f(x) = +\infty, \qquad \lim_{x\to+\infty} f(x) = -\infty,$$

$$\lim_{x\to-\infty} f(x) = +\infty, \qquad \lim_{x\to+\infty} f(x) = -\infty.$$

This section also explains the limit relation of $f(x)$ and $y = ax + b$ gives their relationship of $f(x)$ and $y = ax + b$ in terms of limits when the graph of function $f(x)$ has non-horizontal asymptote $y = ax + b$: $\lim_{x\to+\infty} [f(x) - (ax + b)] = 0$ or $\lim_{x\to-\infty} [f(x) - (ax + b)] = 0$, thus allowing the asymptote to be computed.

(v) Results related to the computation of limits are given in a more comprehensive manner. Limits of sums, differences, products, quotients and reciprocals are presented as theorems in Section 3, Chapter 4 of FR-EDB (see Fig. 10-2) so as to facilitate computation of limits in subsequent content.

The indeterminate forms of limit, such as $\infty - \infty$, $0 \times \infty$, $\frac{\infty}{\infty}$, and $\frac{0}{0}$, are introduced in the margins as notes, so as to help students to grasp the limit operation more accurately.

In the second cycle, FR-HLT starts to introduce derivatives and integrals in Chapter 3, while Chapter 1 "function — variables and continuity" and Chapter 2 "limits of sequences and functions" discuss the concepts of limit and continuity. It can be clearly seen that the content related to limit and continuity in FR-HLT is more complete, more rigorous and more widely applied than that of the other textbooks. Taking the "intermediate value theorem" for example, Section 4 "the properties of

THÉORÈME 2

Si $\lim_{x \to \bullet} u(x) =$	ℓ	ℓ	ℓ	$+\infty$	$-\infty$	$+\infty$
et $\lim_{x \to \bullet} v(x) =$	ℓ'	$+\infty$	$-\infty$	$+\infty$	$-\infty$	$-\infty$
alors $\lim_{x \to \bullet} (u+v)(x) =$	$\ell + \ell'$	$+\infty$	$-\infty$	$+\infty$	$-\infty$	?

THÉORÈME 3

Si $\lim_{x \to \bullet} u(x) =$	ℓ	$\ell > 0$	$\ell < 0$	$\ell > 0$	$\ell < 0$	0	$+\infty$	$-\infty$	$-\infty$
et $\lim_{x \to \bullet} v(x) =$	ℓ'	$+\infty$	$+\infty$	$-\infty$	$-\infty$	$+\infty$ ou $-\infty$	$+\infty$	$-\infty$	$+\infty$
alors $\lim_{x \to \bullet} (u \times v)(x) =$	$\ell \times \ell'$	$+\infty$	$-\infty$	$-\infty$	$+\infty$	?	$+\infty$	$+\infty$	$-\infty$

THÉORÈME 4

Si $\lim_{x \to \bullet} u(x) =$	$\ell \neq 0$	0 (par valeurs supérieures à 0)	0 (par valeurs inférieures à 0)	$+\infty$	$-\infty$
alors $\lim_{x \to \bullet} \left(\frac{1}{u}\right)(x) =$	$\frac{1}{\ell}$	$+\infty$	$-\infty$	0	0

THÉORÈME 5

Si $\lim_{x \to \bullet} u(x) =$	ℓ	$\ell \neq 0$	$\ell \neq 0$	$\ell \neq 0$	0
et $\lim_{x \to \bullet} v(x) =$	$\ell' \neq 0$	0	$+\infty$	$-\infty$	0
alors $\lim_{x \to \bullet} \left(\frac{u}{v}\right)(x) =$	$\frac{\ell}{\ell'}$	$+\infty$ ou $-\infty$	0	0	?

Si $\lim_{x \to \bullet} u(x) =$	$+\infty$	$+\infty$	$-\infty$	$-\infty$	$+\infty$ ou $-\infty$
et $\lim_{x \to \bullet} v(x) =$	$\ell' > 0$	$\ell' < 0$	$\ell' > 0$	$\ell' < 0$	$+\infty$ ou $-\infty$
alors $\lim_{x \to \bullet} \left(\frac{u}{v}\right)(x) =$	$+\infty$	$-\infty$	$-\infty$	$+\infty$	?

Fig. 5-2 Limits of operations of functions in p. 124 FR-EDB (1)

continuous functions" of Chapter 1 in FR-HLT gives the following description:

> **Theorem 1.4.1** (intermediate value theorem). Suppose f is a continuous function defined in interval I, and $a, b \in I$. For any given real number k between $f(a)$ and $f(b)$, there exists a real number c between a and b such that $f(c) = k$.

Such a description is commonly seen in most textbooks. For example, New Math III of JP-SKS explains the intermediate value theorem in a similar way, only with I assumed to be a closed interval $[a, b]$. However, FR-HLT further extends this theorem. In Section 6 "intermediate value theorem on the open interval" of Chapter 2, there is a proposition as follows:

> **Proposition 2.6.1** (intermediate value theorem in open interval or half open interval). Let a, b, α and β denote either real numbers or $+\infty$ or $-\infty$. If f is a continuous and strictly monotonic function on (a, b) such that
> $$\operatorname*{Lim}_{x \to a} f(x) = \alpha \text{ and } \lim_{x \to b} f(x) = \beta,$$
> then for every real number k strictly included between α and β, the equation $f(x) = k$ admits a unique solution in the open interval (a, b).

Compared to the original intermediate value theorem, this proposition weakened the constraints on interval endpoints. It also allows the interval and the value range to be infinite, making it applicable to a wide range of problems.

The property that every monotonic and bounded sequence possess a limit is given in the Chapter 6 "sequences and mathematical induction" of FR-HLT.

Example 5. Section 2 in Chapter 1 of RU-MGU is specifically dedicated to the concepts of limit and continuity of functions, which serve as prerequisite knowledge before learning calculus. The content it contains is listed below:

Section 2 Limits of functions and continuity

- 2.1 The concept of limits of functions
- 2.2 One-sided limits
- 2.3 Properties of limits of functions
- 2.4 The concept of continuity of functions

2.5 Continuity of elementary functions
2.6 Discontinuous functions

In RU-MGU, the continuity of functions is described in more detail than that of other textbooks, such as in Section 2, the relationship between continuity and limits is given with proof; in Section 3, the relationship between continuity and derivability is also given with proof. In addition, the continuity of elementary functions and composite functions, properties of continuous functions over a closed interval, and examples of various kinds of discontinuous functions are given as well. Some are elaborated through rigorous proof and others are illustrated by graphs. In "2.3 Properties of Function Limits", not only the limits of sums, products and quotients are given, but also other simple practical limit operation properties are summarized to facilitate the computation of limits, e.g., "if $\lim_{x \to a} f(x) = 0$, then $\lim_{x \to a} \frac{1}{f(x)} = \infty$"; "if $\lim_{x \to a} f(x) = \infty$, then $\lim_{x \to a} \frac{1}{f(x)} = 0$", "$\lim_{x \to a} [f(x)]^2 = \left[\lim_{x \to a} f(x)\right]^2$" and so on, which can be used directly as formulas.

In short, the three series of textbooks from Japan, France and Russia all give relatively complete elaboration on the topic of limit and continuity, and provide the necessary basis and tools for the derivation of subsequent calculus knowledge.

(3) Other examples

The limit and continuity content in DE-LBS is less difficult than that of the Japan, France and Russia counterparts, but still far more comprehensive than the Chinese and Singaporean textbooks. This series of textbooks is more focused on the integration of the "limit and continuity" knowledge, ideas and methods into practical usage, with lower requirements on rigorous mathematical language. Specifically:

(1) There is no chapter or section named by "limit" or "continuity" in the textbooks. Yet in the content of "Properties of function and its graphs", the limit concept is used very appropriately to discuss the properties of functions, in which multiple types of function limits are involved, including one-sided limits and its relationship with ordinary two-sided limits.

(2) Much limit computation is done in the textbooks, but the computation rules have not been explicitly stated in the text. Instead, they are only given as a side note in the form of annotations.

(3) The concept of continuous functions is not formally mentioned in the text of the books, but an introduction of the concept is provided in the reading materials.

2.4.2 Comparison of methods in obtaining derivative formulas

The number of derivative formulas varies in the textbooks of different countries and the obtaining methods differ as well. By the amount of deduction required, we divide the methods into the following three levels:

Level-A: formula obtained by deduction according to the definition and the existing rules and properties;

Level-B: formula obtained by incomplete induction after some partial deduction or explanation;

Level-C: formula given directly without reasoning or explanation.

Here we take the addition rule of derivatives

$$[(u+v)(x)]' = u'(x) + v'(x)$$

as an example to illustrate the three levels of formula obtaining methods.

RU-MGU gives a proof of this rule on Page 92, which is briefly described as follows:

Let $f(x) = u(x) + v(x)$. Then

$$\Delta f = f(x+\Delta x) - f(x) = [u(x+\Delta x) + v(x+\Delta x)] - [u(x) + v(x)]$$
$$= [u(x+\Delta x) - u(x)] + [v(x+\Delta x) - v(x)] = \Delta u + \Delta v.$$

Therefore,

$$\frac{\Delta f}{\Delta x} = \frac{\Delta u + \Delta v}{\Delta x} = \frac{\Delta u}{\Delta x} + \frac{\Delta v}{\Delta x}.$$

Since

$$\lim_{\Delta x \to 0} \frac{\Delta u}{\Delta x} = u'(x), \quad \lim_{\Delta x \to 0} \frac{\Delta v}{\Delta x} = v'(x),$$

we obtain that

$$f'(x) = \lim_{\Delta x \to 0} \frac{\Delta f}{\Delta x} = \lim_{\Delta x \to 0} \left(\frac{\Delta u}{\Delta x} + \frac{\Delta v}{\Delta x}\right)$$
$$= \lim_{\Delta x \to 0} \frac{\Delta u}{\Delta x} + \lim_{\Delta x \to 0} \frac{\Delta v}{\Delta x} = u'(x) + v'(x).$$

This proof is obtained strictly by using the definition of derivative and the addition rule of limits, so it can be categorized as *Level-A*.

The formula is obtained in New Math II of JP-SKS as follows:

Calculate the derivative function of function $y = x^3 + x^2$.

$$\begin{aligned} y'(x) &= \lim_{h\to 0}\frac{[(x+h)^3+(x+h)^2]-(x^3+x^2)}{h} \\ &= \lim_{h\to 0}\left[\frac{(x+h)^3-x^3}{h}+\frac{(x+h)^2-x^2}{h}\right] \\ &= \lim_{h\to 0}[(3x^2+3xh+h^2)+(2x+h)] \\ &= 3x^2+2x. \end{aligned}$$

Let

$$f(x) = x^3, g(x) = x^2, \text{ i.e.}, y = f(x) + g(x),$$

since $f'(x) = 3x^2$, $g'(x) = 2x$, we see that the addition rule

$$y' = f'(x) + g'(x) \text{ for } y = f(x) + g(x)$$

is satisfied.

This is a conclusion drawn from a specific example, and can be categorized as *Level-B*.

CN-PEP gives the addition rule directly without providing any proof or explanation. The original text is as follows:

The following "differentiation rules" can help us compute the derivative of the sum, difference, product or quotient of two functions.

导数运算法则

1. $[f(x)\pm g(x)]'=f'(x)\pm g'(x)$;
2. $[f(x)\cdot g(x)]'=f'(x)g(x)+f(x)g'(x)$;
3. $\left[\frac{f(x)}{g(x)}\right]'=\frac{f'(x)g(x)-f(x)g'(x)}{[g(x)]^2}\ (g(x)\neq 0)$.

We can classify this method of obtaining the addition rule as *Level-C*.

Based on the above classification, we selected some typical derivative formulas, observed the way of obtaining such formulas in the surveyed textbooks in various countries and listed the observations in Tab. 5-5. The blank square "□" indicates that the series of textbooks does not present the corresponding formula.

Quantitatively, CN-PEP has the fewest derivative formulas and SG-PEH has the most. The textbooks from Japan and Russia also have a relatively large number of derivative formulas. In terms of the level of deduction required in obtaining the formulas, the textbooks rank from low to high as CN-PEP, SG-PEH, JP-SKS, FR-EDB, DE-LBS, and RU-MGU. Below we will choose some textbooks with lower and higher deduction requirements for detailed study.

Tab. 5-5 Comparison of the levels of deducing derivative formulas among the surveyed textbooks

Derivative formulas	CN-PEP	JP-SKS	SG-PEH	DE-LBS	FR-EDB	RU-MGU
Sum, product and quotient rules	C	A	C	B	B	A
$(x^\alpha)' = \alpha x^{\alpha-1}$	B	A	C	B	B	A
$(e^x)' = e^x$	C	A	C	B	A	A
$(\ln x)' = \frac{1}{x}$	C	A	C	B	A	A
$(\sin x)' = \cos\ x$	C	A	A	B	A	A
$(\tan x)' = \sec^2 x$	□	A	C	B	A	A
$(\arcsin x)' = \frac{1}{\sqrt{1-x^2}}$	□	□	C	□	□	A
$(\arctan x)' = \frac{1}{1+x^2}$	□	□	C	□	□	A
Derivatives of composite functions	B	A	B	B	A	A
Second and higher-order derivatives	□	A	A	□	A	A
Derivative of implicit functions	□	B	B	□	□	□
Derivative of inverse functions	□	B	□	B	□	A
Derivative of parametric functions	□	B	B	□	□	□

(1) Cases of low deduction requirement in derivative formulas

The derivative formulas are derived in CN-PEP as follows:

(a) The derivative formulas of functions $y = c$ (c is a constant), $y = x$, $y = x^2$, $y = \frac{1}{x}$ and $y = \sqrt{x}$ are deduced according to the derivative definition (Level-A).

(b) The derivative formulas of other basic elementary functions (8 in total) and the sum, product and quotient rules are directly listed out (Level-C).

(c) The differentiation rule of composite functions is obtained through a specific example $[\ln(3x+2)]' = \frac{3}{3x+2}$ (Level-B).

(d) Based on the above formulas and rules, the derivative formulas of other elementary functions (mostly polynomial functions) are deduced (Level-A).

Steps of deriving the derivative formulas in SG-PEH:

(a) The derivative formulas of $y = x^2$ and $y = \sin x$ are deduced according to the derivative definition (Level-A).

(b) The derivative formulas of other basic elementary functions (12 in total) and the sum, product and quotient rules are directly listed out (Level-C).

(c) The differentiation rule of composite functions and the methods for finding the derivatives of implicit functions and parametric functions are directly listed out (Level-C).

(d) A detailed description of how to calculate derivatives with graphing calculators is presented.

(2) Cases of high deduction requirement in derivative formulas

The three series of textbooks from Russia, Japan and France present a relatively complete set of derivative formulas and rules, along with proofs for the most part. In addition, the order of the numerous proofs and the preparatory groundwork have been designed carefully (the details will be discussed in Section 3 of this Paper), not only following the logical order of the mathematics, but also in a form suitable for high school students to learn.

It is necessary to mention that while JP-SKS (1) and (2) in Tab. 5-5 do not specifically give the derivatives of arc-sin functions and arc-tangent functions, New Math III of JP-SKS (2) does not cover the general topic of "derivatives of inverse functions", and the method of finding the derivatives of an inverse function is given at Level-B. It just omitted the special cases of inverse trigonometric functions.

2.4.3 Comparison of the application of derivatives

We now list the main applications of derivatives in the textbooks of various countries, as shown in Tab. 5-6.

Tab. 5-6 Statistics of applications of derivatives in all textbooks

Applications of derivatives	CN-PEP	JP-SKS	SG-PEH	DE-LBS	FR-EDB	RU-MGU
Monotonicity of functions	■	■	■	■	■	■
Extremal and extremum values	■	■	■	■	■	■
Concavity and convexity	□	■	■	■	□	■
Equations of tangent and normal lines of curves	□	■	■	□	□	■
Approximate calculation	□	■	■	■	□	■
Proof of equations and inequalities	□	■	□	□	□	□
Taylor formula	□	□	■	□	□	■
Intermediate value theorems	□	■	□	□	□	■

As can be seen from Tab. 5-6, on the application of derivatives, there are both similarities and differences among the textbooks from various countries:

(1) Shared features

The applications of derivatives in the textbooks of various countries mainly focus on mathematical applications. The use of derivatives to determine the monotonicity of a function and to obtain the extremal and extremum values of functions is shared by all the surveyed textbooks. Moreover, when calculating the extremal and extremum of functions, all the textbooks have chosen a number of problems with real-life backgrounds.

(2) Difference in the ranges of applications

JP-SKS and RU-MGU give a wide range of applications, while CN-PEP and FR-EDB give relatively few.

(3) Mean value theorem with different requirements for students

Even on the shared application topics, there are differences among the textbooks of the surveyed countries. For example, the mean value theorem exists in both RU-MGU and JP-SKS, but there are significant differences. There are two mean value theorems in the Russian textbooks — Rolle mean value theorem and Lagrange mean value theorem, while JP-SKS only offers Lagrange mean value theorem. The two mean value theorems are both proved in the Russian textbooks, while the theorem in the Japanese is only illustrated with a graph of the function. Although the Rolle mean value theorem is only a special case of Lagrange mean value theorem (when $f(a) = f(b)$), it can serve as a bridge towards Lagrange mean value theorem and make it significantly less abstract. Thus the mean value theorem in RU-MGU will make it easier for students to have an accurate intuition.

(4) Taylor formula with different requirements for students

Regarding the Taylor formula, the part in SG-PEH is titled as "Maclaurin series", while the corresponding part in RU-MGU is titled "Taylor series". But in fact, they only discussed the case of $x_0 = 0$, i.e., they only discussed the application of Maclaurin series in polynomial expansion and approximate calculation. The largest difference between this part of the two textbooks lies in the way of presenting the formulas: SG-PEH directly presents the Maclaurin formula; on the other hand, RU-MGU, based on the intermediate value theorems, derives the Maclaurin formula via a number of generalizations, induction, supposition and construction steps, during which the properties of the remainder is also explained. The Russian approach will not only reinforce the internal connection of knowledge, but also convince the students of the necessity and reasonableness of Maclaurin formula.

(5) Applications of derivatives in the proof of equations and inequalities

In JP-SKS (2), there is a section dedicated specifically to the proof of equations or inequalities with derivatives (or mean value theorem). Although the number of examples presented is not large, the conclusions drawn are fairly nice, and the proof methods are simple and concise. For example, it is illustrated that algebraic proportions such as "when $x > 0$, $e^x > 1 + x$" and "when $x > 0$, $\ln(x + 1) < x$" can be effectively proved using derivatives. Other textbooks do not choose this application, probably due to the difficulties and high flexibility of such propositions.

2.4.4 Comparison of integration methods

Integration requires a relatively high level of transformation skills. When we examine the textbooks of the six countries, the introduced integration methods are as shown in Tab. 5-7.

Tab. 5-7 Statistics of the main methods of integration in the surveyed textbooks

Methods of integration	CN-PEP	JP-SKS	SG-PEH	DE-LBS	FR-EDB	RU-MGU
Definition of definite integral	■	□	■	■	■	■
Inverse operation of derivatives	■	■	■	■	■	■
Methods of substitution	□	■	■	■	■	■
Integration by parts	□	■	■	□	■	■
Integration of fractional functions	□	■	■	□	□	□
Trigonometric substitution	□	■	■	□	□	□
Integrand of form $f\left(\sqrt{a^2 \pm x^2}\right)$ or $\frac{1}{a^2 \pm x^2}$	□	■	■	□	□	□

The "definition of definite integral" in Tab. 5-7 refers to the method to obtain definite integral via partition, summation and taking the limit (i.e., the definition of the Riemann integral). Since a definite integral is defined by the "fundamental theorem of calculus" in JP-SKS, they do not include the "definition of definite integral" method. The "inverse operation of derivatives" in Tab. 5-7 only refers to the method of finding the antiderivative directly from the basic derivative formulas and computing definite integrals using it. In the textbooks, the antiderivative formulas are usually just reversed derivative formulae which can be used directly.

Regarding the integration methods in the textbooks of the aforementioned countries, we noted the following points:

(1) The integration methods emphasized by all the surveyed countries

"Definition of definite integral", "inverse operation of derivatives" and "methods of substitution" are the most basic integration methods. These methods are introduced as the main methods in the vast majority of the textbooks. The integration by parts is also given much attention by most of the textbooks.

(2) Techniques of integration in Singapore textbooks

SG-PEH caught us by surprise. In the preceding derivative part of this textbook series, only the derivatives of functions $y = \sin x$ and $y = x^2$ are deduced. The derivatives of other functions are directly listed out rather than presented after deduction, which may be due to the low coverage of limit in the textbook. Meanwhile, the textbook lays more emphasis on the use of the capability of taking derivatives in graphing calculators to solve problems. This appears to indicate that the Singaporean textbook pays less attention to calculation and transformation skills. Yet surprisingly, this series of textbooks have higher requirement of integration skills than the other surveyed textbooks, as illustrated by the following cases:

(a) A large quantity of integral formulas. In the "Inverse Operation of Derivatives" part, SG-PEH gives more integral formulas than the other textbooks. For example, formulas such as

$$\int \sec x \, dx = \ln|\sec x + \tan x| + C$$

$$\int \sec^2 x \, dx = \tan x + C, \quad \text{and}$$

$$\int \frac{dx}{x^2 - a^2} = \frac{1}{2a} \ln \left|\frac{x-a}{x+a}\right| + C$$

are not introduced in other textbooks. However, SG-PEH directly quotes them as formulas that can be used directly.

(b) A wide range of integration methods in the section of "techniques of integration" of SG-PEH. There are totally 9 sub-headings listed out, including:

I) Standard integration formulas (inverse operation of derivatives);

II) Integration of functions of the form $[f(x)]^n \cdot f'(x)$ or $f'(x) \cdot e^{f(x)}$;

III) Integration of rational functions (fractional function with both numerators and denominators are polynomials);

IV) Integration of trigonometric functions;

V) Integration of functions of the forms $\frac{1}{\sqrt{k^2-x^2}}$, $\frac{1}{\sqrt{a^2-(x-b)^2}}$, $\frac{1}{k^2+x^2}$ and $\frac{1}{k^2+(x-b)^2}$;

VI) Integration by decomposing numerators, such as $\int \frac{3-x}{\sqrt{2x-x^2+5}}\mathrm{d}x = \int \frac{\frac{1}{2}(2-2x)+2}{\sqrt{2x-x^2+5}}\mathrm{d}x = \frac{1}{2}\int \frac{(2x-x^2+5)'}{\sqrt{2x-x^2+5}}\mathrm{d}x + 2\int \frac{1}{\sqrt{2x-x^2+5}}\mathrm{d}x = \cdots$;

VII) Integration of functions of the forms $\frac{1}{k^2-x^2}$ and $\frac{1}{a^2-(x-b)^2}$;

VIII) Integration by substitution; and

IX) Integration by parts.

(c) More example questions on integration. In the section of "techniques of integration" section of SG-PEH, there are totally 11 examples and 35 questions asked over the topic of integrals. By quantity, it is far more than the other surveyed textbooks and also more than the computational problems in the other sections of the Singaporean textbooks.

2.4.5 Comparison of the application of definite integrals

The application situation of definite integrals in the surveyed series of textbooks is shown in Tab. 5-8. It mainly involves the calculation of the trapezoid area of trapezoids with curved sides as well as other planar regions. It also involves the calculation of volumes of solids such as rotary bodies, the simple first or second order differential equations, and the application of integrals in physics or other subjects.

Tab. 5-8 Statistic of applications of definite integral in each textbook

	CN-PEP	JP-SKS	SG-PEH	DE-LBS	FR-EDB	RU-MGU
Areas of curved trapezoids	■	■	■	■	■	■
Areas of planar regions	■	■	■	■	■	■
Volumes of rotary body	□	■	■	□	■	■
Volumes of other solids	□	■	□	□	□	□
Work	■	□	□	□	□	■
Distance	■	■	□	□	□	□
Differential Equation	□	■	■	□	□	■
(In)equalities of infinite series	□	■	□	□	■	□
Others		arc length of curve				barycenter, other force problem

As can be seen from Tab. 5-8, the textbooks from Japan, Singapore and Russia contain content on the solution of differential equations (see Tab. 5-9). Although the types of differential equations differ among these surveyed textbooks, their solution methods are similar in requiring the use of approaches in solving indefinite integrals. As can be seen from Tab. 5-9, SG-PEH discusses the widest range of differential equations.

Tab. 5-9 Statistics of types of the differential equations in JP-SKS, SG-PEH, and RU-MGUs

Type	JP-SKS	SG-PEH	RU-MGU
$y'(x) = a$	■	■	■
$y'(x) = ax + b$	□	■	■
$y'(x) = f(x)$	□	■	■
$y'(x) = g(y)$	■	■	■
$y'(x) = f(x)g(y)$	□	■	□
$y''(x) = a$	■	■	■
$y''(x) = f(x)$	□	■	■

2.5 Summary

Summarizing above analyses, the following preliminary conclusions can be arrived at:

(1) The content range of calculus in the surveyed textbooks of the six countries all include the derivatives of single-variable functions, indefinite integrals, definite integrals and their applications.

(2) As for the weight of calculus in high school mathematics curriculum, there is a large difference among the surveyed countries. According to the statistics, DE-LBS has the biggest weight of 41%; CN-PEP has the smallest weight of only 7%.

(3) The specific content and presentation styles of the surveyed textbooks vary significantly among the six countries. The Japanese, French and Russian textbooks cover a large selection of concepts, methods, reasoning and applications. Their content structure and rigorous elaboration approaches are also more in line with the characteristics of mathematics as a subject and the study habit of high school students. The Germany textbooks pay more attention to the infiltration and preparation foreshadowing of mathematical ways of thinking. However, the main text does not give much thought over the systematization of knowledge and rigorousness in expression. Instead, they rely more on the illustrations of function graphs and the auxiliary columns. The Singaporean textbooks

emphasize the use of calculus as a tool without much introduction of the mathematical ways of thinking and the relationship among the knowledge points. Moreover, they rely much on graphing calculators. The Chinese textbooks present comparatively less content of calculus, and the requirement is also lower.

(4) The limit content of calculus textbooks all avoids the usage the setting of $\varepsilon - N$ and $\varepsilon - \delta$, but they all adopt appropriate approaches to present the concept of limits in an accurate way. Among which the Japanese, French and Russian textbooks all set up special chapters or sections to elaborate the concepts and properties of limits and continuous functions in a detailed manner, which ensures a basic level of mathematical rigor in subsequent content.

(5) The original skill requirements on calculations of derivatives and integrations specified in the textbooks of the surveyed countries vary significantly. The majority of rules for computing derivatives and basic derivative formulas in the Japanese, French and Russian textbooks are presented after rigorous deduction, whereas such deduction is scarce in the Singaporean and Chinese textbooks, with most rules and formulas given directly due to the weak foundation of limit in the Chinese and Singaporean textbooks. The Singaporean textbooks lay more emphasis on integration skills (for this part, the limit knowledge is not required, instead, the transformation skills of algebraic expressions are more desirable), while the textbooks from other countries only focus on several basic techniques of integration.

(6) In terms of the applications of calculus, the surveyed countries mainly focus on the internal applications in mathematics (even if there are some practical problems, it is relatively easy to translate them into mathematical problems). Derivatives are mainly used to study the monotonicity and extreme values of functions. Integrals are mainly used to calculate volumes and areas. These features are shared by the various series of textbooks, but are also good at achieving good teaching results. There are also applications in approximate calculation, algebraic proof, differential equations, and applications in physics such as displacement, work and center of gravity, to name a few. The selection differs among different textbooks.

3. Comparison of Completeness and Rigorousness in Logical Deduction

Limited by the knowledge foundation of high school students and their class hours, high school calculus, which used to be studied in college, has inevitably to be cut down and restructured when fitted into high school textbooks. As can be seen from the analysis in the second part, the selected six series of textbooks in this study have all largely cut down or simplified the content compared to their original calculus curriculum. In mathematical textbooks, where rigorous logic has always been emphasized, it is thus important to make up the logical gap when content is deleted or cut down and to maintain the logical rationality when content is being simplified. All textbooks from the six countries have their own strategies for doing this.

Below we choose several specific topics as examples to investigate the practices of each country and their resultant effects.

3.1 About the logical order in deduction

At the college stage, since students have already obtained their own preliminary ability to determine logical rationality, the university textbook writers can sometimes, provided that the logic remains sound, change the presenting order of knowledge so as to ensure the conciseness of text elaboration. For example, the derivative formula "$(x^{\alpha})' = \alpha x^{\alpha-1}$, α is real" is presented in the *Advanced Mathematics* textbook by Shanghai Jiaotong University as follows:

Step 1: In Section 3 "basic derivative table" a proof of $(x^n)' = nx^{n-1}$, where n is a positive integer, is given.

Proof (sketch): According to the definition of derivative, we have

$$\begin{aligned}\frac{\Delta y}{\Delta x} &= \frac{(x+\Delta x)^n - x^n}{\Delta x} = \cdots \\ &= C_n^1 x^{n-1} + C_n^2 x^{n-2}\Delta x + \cdots + C_n^n \Delta x^{n-1}.\end{aligned}$$

Therefore,

$$(x^n)' = \lim_{\Delta x \to 0} C_n^1 x^{n-1} = nx^{n-1}.$$

Step 2: The following explanation is given immediately after Step 1:

The derivative formula $(x^n)' = nx^{n-1}$ holds not only for positive integer n, but also for any real number n. Its proof will be given in Section 5. At this moment we assume this conclusion for the time being in the calculations like $\left(\frac{1}{x}\right)' = (x^{-1})' = -\frac{1}{x^2}, \left(\frac{1}{x^3}\right)' = \cdots = -\frac{3}{x^4}, \left(\sqrt[5]{x^2}\right)' = \cdots = \frac{2}{5\sqrt[5]{x^3}}, \left(x^{-\sqrt{2}}\right)' = \ldots = -\sqrt{2}x^{-\sqrt{2}-1}$, etc.

Step 3: In Section 5 "derivatives of composite functions", the general result $(x^\alpha)' = \alpha x^{\alpha-1}$ for real number α is proved as an application of the derivative formula of composite functions.

Proof (sketch): We have $y = x^\alpha = \mathrm{e}^{\alpha \ln x}$. Let $u = \alpha \ln x$, by the rules for computing derivatives of composite functions,

$$(x^\alpha)' = (\mathrm{e}^u)'_u \cdot (\alpha \ln x)'_x = \cdots = \alpha x^{\alpha-1}.$$

Normally, high school textbooks do not give a formula in such a way as "its proof will be given later, here we borrow this conclusion for the time being", since the unproved formulas may disturb the students' newly-constructed logical thought. Therefore, the deduction sequence needs to be designed more carefully, first ensuring logical soundness and then making the presentation as concise as possible. Below we will discuss the different ways of introducing formula $(x^\alpha)' = \alpha x^{\alpha-1}$ in the sample textbooks of the surveyed countries.

3.1.1 Complete and rigorous deduction

Taking JP-SKS as an example, since JP-SKS (1) only discusses the calculus of polynomial functions, it will be sufficient to consider $(x^\alpha)' = \alpha x^{\alpha-1}$ for natural numbers α. This can be done based on Step 1 in the above-mentioned *Advanced Mathematics* textbook, without the dilemma caused by the sequence of proofs. Here we mainly analyze the treatment in JP-SKS (2), summarizing the procedure of the proof in Tab. 5-10. Note that each step of the proof is rigorous, the whole proof involves the use of the definition of derivatives, derivatives of product and quotient, derivatives of inverse functions, derivatives of implicit functions, derivatives of composite functions, and derivatives of logarithmic functions. Moreover, mathematical induction is used therein.

Tab. 5-10 Proof of $(x^{\alpha})' = \alpha x^{\alpha-1}$ in JP-SKS (2)

Step	α	Formula(s) proved	Knowledge used	Methods or process
1	1,2,3	$x' = 1$, $(x^2)' = x$, $(x^3)' = 3x^2$	definition of derivative	direct use of the definition of derivative
2	4	$(x^4)' = 4x^3$	derivative of product	$(x^4)' = (x^3 \cdot x)'$
3	Positive integer n	$(x^n)' = nx^{n-1}$	mathematical induction	derivative of product, mathematical induction
4	negative integer $-n$	$(x^{-n})' = -nx^{-n-1}$	derivative of quotient	$(x^{-n})' = \left(\frac{1}{x^n}\right)'$
5	rational number $\frac{1}{n}$	$\left(x^{\frac{1}{n}}\right)' = \frac{1}{n}x^{\frac{1}{n}-1}$	derivative of inverse function	$x = y^n \Rightarrow \frac{dx}{dy} = ny^{n-1} = nx^{\frac{n-1}{n}}$
6	rational number $\frac{m}{n}$	$\left(x^{\frac{m}{n}}\right)' = \frac{m}{n}x^{\frac{m}{n}-1}$	derivative of composite function	$y = u^m, u = x^{\frac{1}{n}} \Rightarrow \frac{du}{dx} = \frac{1}{n}x^{\frac{1}{n}-1}$, $\frac{dy}{du} = mu^{m-1} = mx^{\frac{m-1}{n}}$
7	real number α	$(x^{\alpha})' = \alpha x^{\alpha-1}$	derivative of implicit function	$\ln y = \alpha \ln x \Rightarrow \frac{y'}{y} = \frac{\alpha}{x} \Rightarrow y' = \frac{\alpha y}{x} = \alpha x^{\alpha-1}$

As can be seen from Tab. 5-10, the complete proof of formula $(x^{\alpha})' = \alpha x^{\alpha-1}$ requires the overall consideration of all the previous results that are needed. A reasonable logical sequence shall be designed for it so as to demonstrate the knowledge system more completely, rigorously and concisely.

The treatment in RU-MGU resembles that of JP-SKS (2), yet somewhat simplified and divided into four steps; its proof process is also very rigorous:

Step 1 proves the $\alpha = 1$ case according to the definition of derivative (corresponding to Step 1 in Tab. 5-10);

Step 2 proves the result by mathematical induction for any positive integer α (corresponding to Steps 2 and 3 in Tab. 5-10);

Step 3 proves the result by quotient rule for negative integer α (corresponding to Step 4 in Tab. 5-10); and

Step 4 proves the result by derivation rule of composite functions for real number α (corresponding to Step 7 in Tab. 5-10), omitting Steps 5 and 6 dealing with rational α.

It is worth noticing that in JP-SKS (2), Step 7 transforms the function $y = x^{\alpha}$ into logarithmic form $\ln y = \alpha \ln x$ before further proceeding with the proof. Different from this, Step 4 in RU-MGU transforms $y = x^{\alpha}$ into exponential form $y = \mathrm{e}^{\alpha \ln x}$ for further proof, relying on derivative formula of exponential function and the derivative rule of composite functions to compute $(\mathrm{e}^{u})' = \mathrm{e}^{u}u'$; specifically, taking the derivatives of both sides of $x^{\alpha} = \mathrm{e}^{\alpha \ln x}$ yields

$$(x^{\alpha})' = \mathrm{e}^{\alpha \ln x} \cdot \frac{\alpha}{x} = \alpha x^{\alpha - 1}.$$

The two methods reach the same goal by different means, and both are in line with logical sequences.

3.1.2 Examples and incomplete induction

In CN-PEP, derivative functions of $y = c$, $y = x$, $y = x^2$, $y = \frac{1}{x}$ and $y = \sqrt{x}$ are deduced from the definition of derivatives and incomplete induction then arrives at the formula $(x^n)' = nx^{n-1}$, where n is allowed to be a rational number, although no proof is given for general rational n. Instead, the textbook notes that the formula, as one of the derivative formulas for elementary functions, can be used by students directly. Such presentation, though lacking regular general proof, could enable students to experience the pattern of derivative formulas for this type of functions after observing the deduction process for $n = 0, 1, 2, -1$ and $\frac{1}{2}$. Thus they could consciously acknowledge the validity of the formula. Although such presentation has no ordering problems, the method is obviously not rigorous.

3.1.3 Conclusions given without proof

In SG-PEH, $(x^2)' = 2x$ is deduced from the definition of derivatives. Then the derivative formula $(x^n)' = nx^{n-1}$ is given without proof, where n is noted to be a constant (according to the context, herein n is meant to be a constant real number). Formulas obtained in this way cannot be viewed even as a result of generalization or (incomplete) induction, only as one given directly. By doing so, the students lack the opportunity to deduce formulas (derivative formulas of power functions are of lower difficulty compared to other derivative formulas) and thus students are deprived of the chance of further understanding the definition of derivative and accumulating experience in computing derivatives. In fact, SG-PEH introduces a wide range of areas in which graphing calculators can be used. The powerful graphing and numerical computation capabilities of graphing calculators make the transformation skill between a function and its derivative/antiderivative much less important. Hence, the rationality in the logical sequence of derivative formulas is paid less attention.

3.2 On the rigorousness of deduction

The Fundamental Theorem of Calculus is the bridge between differentials and integrals, it transforms the tedious and lengthy partition, summation and limit process in definite integration into a simple formula $\int_a^b f(x)\mathrm{d}x = F(b) - F(a)$, based on the antiderivative. It is an important and non-ignorable key point in the content of calculus in the textbooks. But some basic knowledge involved in the proof of the Fundamental Theorem of Calculus is, more often than not, simplified or omitted in the calculus textbooks for high schools. For example, the "mean value theorem of integrals" is necessary for a rigorous proof of the Fundamental Theorem of Calculus, but it may well have been omitted in the simplified calculus textbooks. However, the proof of the mean value theorem requires some properties of definite integrals and continuous functions, which may again have been omitted in the textbook. If the mean value theorem of integrals is to be proved rigorously, no such omissions will be possible, preventing simplification. Yet if the textbook has to be simplified and its difficulty reduced, the Fundamental Theorem of Calculus cannot be proved rigorously. The author reviewed the corresponding measures in each sample book in order to find their approaches in solving such dilemma:

3.2.1 Mostly rigorous mathematical proof

Taking RU-MGU as an example, its proof procedures are as follows:

Step 1: Define the integral upper limit function, $\Phi(u) = \int_a^u f(x)\mathrm{d}x$.

Step 2: Prove $\Phi'(u) = f(u)$ in the following way:

Let $h > 0$ (or $h < 0$). Since $\Phi(u+h) - \Phi(u) = \int_u^{u+h} f(x)\mathrm{d}x$, we have $\Phi'(u) = \lim_{h\to 0} \frac{\Phi(u+h)-\Phi(u)}{h} = \frac{\int_u^{u+h} f(x)\mathrm{d}x}{h}$.

Let M and m be the maximum and minimum values of $f(x)$ in the interval$[u, u+h]$ (or$[u+h, u]$), then $mh \le \int_u^{u+h} f(x)\mathrm{d}x \le Mh$, i.e., $m \le \frac{\int_u^{u+h} f(x)\mathrm{d}x}{h} \le M$. When $h \to 0$, we have $M \to f(u)$ and $m \to f(u)$. It follows that $\frac{\int_u^{u+h} f(x)\mathrm{d}x}{h} \to f(u)$, i.e., $\Phi'(u) = f(u)$.

Step 3: Prove $\int_a^b f(x)\mathrm{d}x = F(b) - F(a)$, for any function $F(x)$ with $F'(x) = f(x)$.

Both $\Phi(x)$ and $F(x)$ are both antiderivatives of $f(x)$, so $\Phi(x) = F(x) + C$, where C is a constant. Because $\Phi(a) = \int_a^a f(x)\mathrm{d}x = 0$, we obtain that $C = F(a)$, and $\int_a^b f(x)\mathrm{d}x = F(b) + C = F(b) - F(a)$. This completes the proof.

It should be said that the above-mentioned proof is fairly complete, but still not entirely rigorous or desirably stringent. For example, in Step 2, why should function $f(x)$ have maximum M and minimum m in interval $[u, u+h]$? Why should inequality $mh \le \int_u^{u+h} f(x)\mathrm{d}x \le Mh$ hold? Why does $\frac{\int_u^{u+h} f(x)\mathrm{d}x}{h} \to f(u)$ when $M \to f(u)$ and $m \to f(u)$? The properties of continuous functions and definite integrals involved here are likely to have been omitted in the past lessons, or only briefly mentioned, but not specified as theorems or properties to serve as the basis of inference. RU-MGU does not add any relevant properties specifically for the proof of the Fundamental Theorem of Calculus. Instead, a picture is added herein to help the students who have not gone through strict training understand the rationality of the facts introduced without noticing any defects (just like how the lack of an axiom of continuity in Euclid's *Elements* went unnoticed for over 2000 years).

On the whole, the proof of the Fundamental Theorem of Calculus in DE-LBS and FR-EDB is complete in logic but also not without flaws (somewhat more severe than RU-MGU), especially for the proof of Step 2. For instance, FR-EDB only uses a single example $\ln x = \int_1^x \frac{1}{t}\mathrm{d}t$ to explain that the integral upper limit function $\Phi(u) = \int_a^x f(t)\mathrm{d}t$ is an antiderivative of integrand $f(x)$.

3.2.2 Introducing by examples and generalization

Taking CN-PEP as an example, the Fundamental Theorem of Calculus is introduced in the following way:

Step 1: Reinvestigate a specific familiar example. Recall the known relationship$v(t) = s'(t)$ between displacement function $s(t)$ and velocity function $v(t)$, and calculate the overall displacement in the time interval $[a, b]$ as $s(b) - s(a) = \int_a^b v(t)\mathrm{d}t = \int_a^b s'(t)\mathrm{d}t$.

Step 2: Introduce the general formula. For a continuous function $f(x)$ in $[a, b]$, if a function $F(x)$ satisfying $F'(x) = f(x)$, then

$$\int_a^b f(x)\mathrm{d}x = F(b) - F(a).$$

This approach is acceptable by students. Since they are familiar with the relationship between displacement function and velocity function, they could accept the equation given in Step 1. But students still cannot really understand the generalization in Step 2, so they could only memorize and apply it as a formula.

The strategies adopted in SG-PEH resemble that of CN-PEP. The example in Step 1 is replaced by function $f(x) = x^2$ and its antiderivative $F(x) = \frac{1}{3}x^3$. Then the formula for computing definite integrals is introduced. SG-PEH does not even mention that the formula is the Fundamental Theorem of Calculus or Newton-Leibniz formula.

3.2.3 Changing the definition of definite integral

In JP-SKS, the Fundamental Theorem of Calculus is used as the definition of definite integral:

> Suppose that a continuous function $f(x)$ in interval $[a, b]$ has $F(x)$ as its antiderivative, then we call $\int_a^b f(x)dx = F(x)|_a^b = F(b) - F(a)$ the *definite integral* of $f(x)$ in $[a, b]$.

Then it explains that when $f(x) \geq 0$, the geometric meaning of $\int_a^b f(x)dx$ is the area of the trapezoid with a curved edge and with x axis as the base.

Such treatment avoids the need to prove the fundamental theorem of calculus as well as the cumbersome details in defining the definite integral via partition, summation and taking the limit. In logic, it is also logically sound and meanwhile simplifies the textbook considerably. However, we believe that changing the primary meaning of definite integrals is risky. There is no error in logic sequence or defect in concept that have been discovered in this series of textbooks. But, there is no guarantee that students will not encounter conceptual conflicts in other occasions in the future.

3.3 Strategies for providing the required preparatory knowledge for deduction

As can be seen from the above discussion, it would be very difficult, or impossible to ensure absolute rigor in high school calculus textbooks. The rigorous proof of some important propositions requires a large amount of preparative knowledge as a foundation. It would require careful selection, elaborate design and artful compilation by the writers to select the necessary basic content and important examples, such that the course framework can be kept reasonably complete and concise, and meanwhile good groundwork can be laid for the deduction of important knowledge points.

We found that the approaches adopted in some textbooks are worth learning. For example, RU-MGU, as discussed in the previous analysis, covers a wide range of knowledge and its theoretical rigorousness in theoretical requirement also tops the surveyed textbooks. Yet the calculus proportion in this textbook is not high (accounting for 17% of the entire high school mathematics course, see Tab. 5-2), so it manages to be fairly comprehensive and maintain a concise structure at the same time. Herein we take the process of deriving $(a^x)' = a^x \ln a$ in Section 4.5 "derivatives of elementary functions" as an example to demonstrate its thoughtful "foreshadowing" strategy.

The main text of the proof: In Section 4.5, RU-MGU starts from the definition of derivatives to obtain

$$\frac{\Delta y}{\Delta x} = \frac{a^{x+\Delta x} - a^x}{\Delta x} = \frac{a^x(a^{\Delta x} - 1)}{\Delta x} = \frac{a^x(\mathrm{e}^{\Delta x \ln a} - 1)}{\Delta x}$$
$$= a^x \ln a \cdot \frac{\mathrm{e}^{\Delta x \ln a} - 1}{\Delta x \ln a}.$$

Set $\alpha = \Delta x \ln a$. Since $\alpha \to 0$ when $\Delta x \to 0$, we have

$$\lim_{\Delta x \to 0} \frac{\Delta y}{\Delta x} = \lim_{\alpha \to 0} \left(a^x \ln a \cdot \frac{\mathrm{e}^{\alpha} - 1}{\alpha}\right) = a^x \ln a \cdot \lim_{\alpha \to 0} \frac{\mathrm{e}^{\alpha} - 1}{\alpha}.$$

Here puts a remark: Referring to Section 2.4, we see that $\lim_{\alpha \to 0} \frac{\mathrm{e}^{\alpha}-1}{\alpha} = 1$. Therefore, $(a^x)' = \lim_{\Delta x \to 0} \frac{\Delta y}{\Delta x} = a^x \ln a$.

Foreshadow 1: In order to have a complete proof, we have to track back to Section 2.4 "the concept of continuous function" to look for a proof of $\lim_{\alpha \to 0} \frac{\mathrm{e}^{\alpha}-1}{\alpha} = 1$. The key step of the proof is to transform $\lim_{\alpha \to 0} \frac{\mathrm{e}^{\alpha}-1}{\alpha}$ into $\lim_{t \to 0} \frac{\ln(t+1)}{t}$.

The transformation is realized as follows: Let $\mathrm{e}^{\alpha} - 1 = t$. Then $\alpha = \ln(t+1)$ and $\frac{\mathrm{e}^{\alpha}-1}{\alpha} = \frac{t}{\ln(t+1)}$. Since $\alpha \to 0$ when $t \to 0$, to prove $\lim_{\alpha \to 0} \frac{\mathrm{e}^{\alpha}-1}{\alpha} = 1$, it suffices to show $\lim_{t \to 0} \frac{\ln(t+1)}{t} = 1$.

Foreshadow 2: The proof of $\lim_{t \to 0} \frac{\ln(t+1)}{t} = 1$ appears in the previous page. According to the properties of continuous functions, $\lim_{t \to 0} \frac{\ln(t+1)}{t} = \lim_{t \to 0} \ln(1+t)^{\frac{1}{t}} = \ln \lim_{t \to 0} (1+t)^{\frac{1}{t}}$. So we only need to prove $\lim_{t \to 0} (1+t)^{\frac{1}{t}} = \mathrm{e}$.

Foreshadow 3: Now we further track to Section 2.2 "one-sided limit", where $\lim_{t \to 0} (1+t)^{\frac{1}{t}} = \mathrm{e}$ is proved and labeled as "important limit". During the proof, the textbook gives a remark again: Please refer to the proof of $\lim_{n \to \infty} \left(1 + \frac{1}{n}\right)^n = \mathrm{e}$ in Section 4.7 of Vol. 10 (the previously mentioned Sections 4.7, 2.4 and 2.2 are from Vol. 11).

Foreshadow 4: Finally, we track to Section 4.7 "powers with irrational exponents" in Vol. 10, where we see the proof of the sequence limit $\lim_{n\to\infty}\left(1+\frac{1}{n}\right)^n = \mathrm{e}$, the key point here is using the properties of monotonic bounded sequences.

As can be seen, the textbook only presents the main text of the proof for this formula $(a^x)' = a^x \ln a$, which is very concise and simple. Yet the foreshadowing procedure is rather long and well scheduled. If the textbook lacks the foreshadows of $\lim_{n\to\infty}\left(1+\frac{1}{n}\right)^n = \mathrm{e}$, $\lim_{t\to 0}(1+t)^{\frac{1}{t}} = \mathrm{e}$, $\lim_{t\to 0}\frac{\ln(t+1)}{t} = 1$ and $\lim_{\alpha\to 0}\frac{\mathrm{e}^{\alpha}-1}{\alpha} = 1$, then the proof of derivative formula $(a^x)' = a^x \ln a$ will be extremely lengthy and difficult. Intermediate results like $\lim_{t\to 0}\frac{\ln(t+1)}{t} = 1$ and $\lim_{\alpha\to 0}\frac{\mathrm{e}^{\alpha}-1}{\alpha} = 1$ are normally not emphasized as theorems or properties. Even if the students have learned the results, they still have difficulties to apply them. The elaborately designed examples (or exercises) in the previous text not only manage to explain and consolidate the new knowledge, but also foreshadow the following content and break down the difficulties therein. With explicit hints, this series of textbooks gives clear tracking clues for students to find the relevant foreshadows, and string the involved knowledge together. By doing so, the proof of the formula is kept rigorous, the content in the textbook is made more closely connected, and the presentation also becomes more concise, making it a good example for Chinese textbook writers to learn from.

3.4 Summary

As the high school calculus course necessarily involves cutting down or simplifying the content in classic calculus textbooks, the deduction chain is inevitably affected. Different countries have different standards and practices for maintaining the most basic deductive style of the math course and restructuring the knowledge system of high school calculus.

(1) The Russian and Japanese textbooks can be used as reference if a relatively high standard is desired; there the course content arrangement mainly follows the logical order of mathematics deduction, and complete

proofs are given for important theorems and formulas whenever possible. Some textbooks get conclusions through examples, induction and verification; in fact, such methods are more or less used in all the surveyed textbooks. Though there are flaws, this process could still allow the students to understand the correctness and reasonableness of the results. Some textbooks give up deductive reasoning entirely on some occasions, and instead present the conclusions directly, which makes the textbooks more like manuals of formulas.

(2) Rigorous proof of certain important conclusions (such as the fundamental theorem of calculus) would require the supplementation of a large amount of knowledge to handle the details. As long as the correctness of these details is intuitive, some textbooks will make some compromise in the preciseness by combining strategies of observation, reasoning and interpretation, so as to maintain the integrity of thinking in the proving process as far as possible. Thus the textbooks can reflect the deductive features. There are also some textbooks that give up proving those difficult knowledge points, only providing the conclusions directly.

(3) Some textbooks have a better overall design. They present a reasonable decomposition of some important and difficult knowledge, and make careful preliminary groundwork. They also provide leads and tracking clues for relevant materials to simplify the difficult points in a naturally simplified solution.

4. Conclusions and Recommendations

4.1 On the main objectives of high school calculus

After examining the high school calculus textbooks of the surveyed countries, we note that their primary teaching objective does not seem to be the preparation for college calculus courses. After all, the teaching content in the high school textbook has covered the majority of college calculus courses, and there is no need for students to learn the same topics twice. High school calculus courses are also not designed to provide mathematical tools for studying science and engineering, or economics majors in university; in the textbooks, the real-world scenarios that calculus can be applied to are very sparse, and the value of calculus in other fields is not clearly shown. If the purpose of high school calculus is to study the properties of functions, then why are the properties of functions so intensively studied before students learn calculus (monotonicity of functions and extremes, etc.)? Why not learn these properties after the course of calculus? In our selected textbooks, only the German Textbook discusses the monotonicity of function and extreme after the learning of derivatives.

We also believe that, high school calculus is a course whose main role is to improve the mathematical cultivation of the high school students. It is designed to make the students experience a course with exquisite elements and advanced thinking methods, such as variables, movement, infinities, limits, and dialectics. This course is a product of modern mathematics development and provides advanced mathematical tools in many applied areas. It should become part of the common foundation needed by future talents of different fields. All of these characteristics should be the main purpose for calculus courses in high school.

4.2 On the content of high school calculus

All the sample textbooks of high school calculus courses reach a consensus on the scope of the content, which should include the main concepts, methods and application of single-variable derivatives, indefinite integrals and the definite integrals based on the limit concept; in the limit part $\varepsilon - N$ and $\varepsilon - \delta$ languages are avoided.

Judging from the status quo of high school education in China, for students intending to study science and engineering in the future, we should develop the teaching of calculus to a higher level. Specifically, we can make the teaching of limit and continuous function closer to the higher standard of the six series of textbooks, which is to make comprehensive comparison of limit and the characteristics of continuous functions. This will lay a solid foundation for subsequent learning, and will make students experience the main ways of thinking in calculus. In the part of derivatives, we recommend students master all derivatives of basic elementary functions, the derivation methods of composite functions, inverse functions, and implicit functions. In the integral part, we recommend students master the basic integral formulas and the method of integration by substitution and integration by parts. We also recommend students know the definition of definite integrals, and that they learn to calculate the definite integral with the fundamental theorem of calculus. In the application part, we recommend students master the various applications in mathematics itself, and that they know as many applications in other fields as possible.

4.3 On the class hours

By referring to the weight of high school calculus in other countries, we recommend an increase in the weight of high school calculus in China's mathematics curriculum to 15%~20%, or about 50~65 hours.

4.4 On the style of textbooks

According to the learning experience of the six series of textbooks, we feel that the more rigorous the textbooks are, the deeper the students will understand calculus ways of thinking. Therefore, we recommend that high school calculus textbooks in China pursue a well-integrated knowledge structure and rigorous deduction and reasoning as far as possible. When encountering the dilemma of "simplified course" and "rigorous logic", some compromise can be made regarding the latter. Rigorous proving is allowed to be replaced by reasoning that is scientifically correct, but with defects in rigorousness. Correct intuition is allowed to be regarded as truth. The conclusions of examples and exercises can be used as the basis of reasoning. The key is to keep the compromise at an acceptable level, and preserve the deductive style of mathematics in most of the high school calculus textbooks. In this regard, the Russian and Japanese textbooks are worthy of our reference.

4.5 On the use of calculators

Among the six series of textbooks investigated, the three series from Singapore, Germany, and France made use of graphing calculators. Singaporean textbooks have used calculators most and seem to "rely on" them. If the calculator contents were deleted, the text content would become incomplete. Calculators in the German and French textbooks are a "secondary" tool, mainly for drawing and numerical calculation. Textbooks of other three countries have not mentioned the use of graphical calculators. We recommend the Chinese calculus textbook to follow the practice of France and Germany, we could make graphing calculators a complementary tool to reduce the tedium in plotting graphs and in numerical calculation, but they should not undermine the training of deductive reasoning, basic calculation, and transformation skills. In particular, when computing derivatives and antiderivatives, there is no need to use a graphical calculator. When students perform numerical calculations, observe the trends of value change, plot function graphs and examine their features, they can be encouraged to use calculators.

References

[1] Chen, Yuelan (2006). Japanese new high school mathematics curricula, textbooks and college entrance examinations [In Chinese]. *Mathematics Teaching*, (1): 13–16.

[2] Department of Applied Mathematics, Shanghai Jiaotong University (1994). *Advanced Mathematics*. Shanghai: Shanghai Jiaotong University Press.

[3] Ho, F. *et al.* (2006). *H2 Mathematics Volume 1*. Singpore: Panpac Education.

[4] Никольский, С. М., Потапов, М. К. Решетников, Н. Н. Шевкин, А. В. (2006). *Алгебра и Начала Анализа. 11 Класс*: *Учебник для Общеобразовательных Учреждений*/-5-е изд. -М.: Просвещение.

[5] Jacob, N. *et al.* (2005). *Math 1reS*. Paris: Editions Belin.

[6] Jean-Paul Beltramone *et al.* (2002). *Décuc Maths Terminale S enseignement obligatoire*. Paris: Hachette Education.

[7] Jiang Dayuan (2005). Basic facts about the education structure in Germany [In Chinese]. *Forum of Vocation Education*, 2005(3): 62–64.

[8] Masanori Ohya (2004). *New Mathematics Textbooks III*. Tokyo: Suken Shuppan.

[9] Masanori Ohya (2007). *New Mathematics Textbooks II*. Tokyo: Suken Shuppan.

[10] METI (1999). *Mathematics Study Guidance for High Schools*. Tokyo: Tech Solid Press.

[11] Liu Shaoxue *et al.* (2007). *Standard Experiment Textbooks of Senior High School Curriculum: Mathematics Option 2-2* [In Chinese]. Beijing: People's Education Press, 2007.

[12] MOE of PRC (Ministry of Education of the People's Republic of China) (2003). *National Mathematics Curriculum Standard for High Schools (Experimental).* [In Chinese]. Beijing: People's Education Press.

[13] Singapore Examinations and Assessment Board (2013). *GCE A-Level Syllabus Examined in 2014: Mathematics Higher 1 (Syllabus 8864), Mathematics Higher 2 (Syllabus 9740) and Mathematics Higher 3 (Syllabus 9824)*. Singapore Examinations and Assessment Board Official Document.

[14] Singapore Ministry of Education Curriculum Planning and Development Division (2006). *Secondary Mathematics Syllabuses*. Year of implementation: from 2007. Singapore Ministry of Education Official Document.

[15] Schmid, A. *et al.* (2008). *Mathematik für Gymnasien*. Stuttgart Leipzig: Ernst Klett Verlag.

[16] Pu Shuping (2012). Characteristics of French high school mathematics textbooks and their inspiration [In Chinese]. *Foreign Primary and Secondary Education*, (8): 53–59.

[17] Xu Binyan (2009). Reform trends of the high school mathematics curriculum in Germany [In Chinese]. *High School Mathematics Monthly*, (2): 1–4.

https://doi.org/10.1142/9789814713955_0006

Chapter 6

Comparison of Contents Design in Mathematical Inquiry in High School Textbooks

Binyan Xu

School of Educational Sciences, East China Normal University
Shanghai 200062, China
byxu@kcx.ecnu.edu.cn

This chapter proposes five elements for analysis of inquiry contents in high school mathematics textbooks, namely situation expression, problem statement, partnership, activity type, and context interaction; Their sub-indexes are described in detail to construct an analytical framework for the inquiry contents in high school mathematics textbooks. Based on the framework, the chapter presents an analysis of seven high school textbooks from China, Germany, Japan, UK, USA, France, and Singapore, and compares their characteristics of design in the mathematical inquiry. The result suggests that the German textbooks (DE-EDM) focus on the creation of abundant real situations and the design of the activities for peer cooperation; the textbooks published by the People's Education Press of China (Version A, CN-PEP) emphasize the design of open-ended problems and usually express problems with interrogative sentences; the textbooks by Sukan Shuppan Publishing Company, Japan (JP-SKS) offer plenty of mathematical reading materials for students; the UK textbooks (UK-SMP) emphasize on the design of teamwork cooperation situations; the US textbooks (US-PHM) pay attention to the display of the role of the inquiry column in the textbooks as well as the application and development of knowledge in the inquiry; the French textbooks (FR-EDB) pay special attention to students' inquiry into information technology and the application of information technology in mathematics; and Singapore's textbooks (SG-PPE) attach importance to the cultivating of students' capacity of speculation and explanation.

1. Background

1.1 International comparative study on textbooks

The study of mathematics textbooks from the perspective of international comparison helps to understand the characteristics of mathematics textbooks and the influence of textbooks on mathematics classroom teaching activities in different countries. For example, Lianghuo Fan and Yan Zhu (2007) analyzed the representation of problem-solving in mathematics textbooks for middle schools in China, USA and Singapore based on the Pólya's four-stage problem-solving model. They found that the Chinese and Singapore's middle school textbooks present more on how to carry out the plan in the solution to problems while the US textbooks focus more on the problem-solving steps in Pólya's model. Furthermore, they found that Singapore's textbook developers aimed to regard problem-solving as independent mathematical contents and the developers of the US textbooks intended to provide the lists of problem-solving strategies for readers' easy reference (Lianghuo Fan and Yan Zhu, 2007, p. 71). Park and Leung (2006) selected mathematics textbooks used in Grade 8 of China, UK, South Korea, Japan and USA and conducted an exploratory and comparative study from five dimensions — textbook development and publication policies, choice of contents, the role of textbooks in teaching and learning, physical appearance of textbooks, and characteristics of the contents. It is found that the textbooks in western countries focus on guiding students to understand the role of mathematics in the practical life; while Chinese textbooks convey mathematical concepts by means of a relatively economical method which does not produce a significant effect in arousing students' interest in learning (Park and Leung, 2006, pp. 237–238). Chinese researches focus more on investigating the foreign mathematics textbooks and propose enlightenments on Chinese textbooks design in the end of their studies (see, for example, Yufeng Guo, 2006; Shuping Pu, 2012; Libao Wu *et al.*, 2013). Some Chinese researchers are also pursuing the international comparative studies of mathematics textbooks. For example, Yingfang Fu and Weizhong Zhang (2007) made a comparison of the situation culture of word problems in the *Practice Book*, an English middle school mathematics textbook, and *Mathematics, a Middle School*

Textbook published by the Beijing Normal University Press of China. The study shows that there is some difference in the ratio and handling method of situations of word problems due to different cultural backgrounds of the countries, and that in Chinese textbooks less attention is paid to the extensive cultural phenomenon in the problems. Wenyu Zhang and Hailun Fu (2011) made a comparative study of the primary school mathematics textbooks from Singapore and China and found that there were similarities and differences in the contents arrangement and compiling features of the two sets of textbooks. The similarities lied in students' understanding of concepts and skillful application of basic methods. However, the Singapore's textbooks better reflect the idea of placing solution of mathematical problem solving in the center of the curriculum. The Singapore's textbooks present some unconventional and challenging problems and the compilation of textbooks is based on integrated mathematical activities. The literature analysis shows that the international comparative study on high school mathematics textbooks is still relatively scarce in China, and the existing researches mainly focused on concepts, techniques, or examples and exercises. This chapter selects the contents of mathematical inquiry in seven series of textbooks from seven countries (Tab. 6-1) as the object of study. The reason for such selection is based on the national research project (Jianpan Wang, 2011) on the one hand, and on the other hand based on the practical demand for mathematical education reform in China.

Tab. 6-1 Sample Textbooks

Country	**Publisher and Year**	**Textbook**	**Code**
China	People's Education Press, 2009	Mathematics as per the National High School Mathematics Curriculum Standard (Experimental): Version A, Compulsory Course Math 1-5	CN-PEP
USA	Pearson Education, 2009	Prentice Hall Mathematics: Pre-Algebra, Algebra 1, Algebra 2, Geometry	US-PHM
UK	Cambridge University Press, 2007	Textbooks of School Mathematics Project: Core1, Core2, MM1, MS1	UK-SMP
Germany	Schroedel Verlag, 2008	Elemente der Mathematik, 10	DE-EDM
France	Éditions Belin, 2010	Math 1^{re}s, Maths 2e, Math 3e, Math 4e, Maths 5e, Maths 6^{e}	FR-EDB
Japan	Sukan Shuppan Publishing Company, 2008	New Edition of Mathematics: Foundations of Mathematics, Math I, Math A,	JP-SKS
Singapore	Panpac Education, 2008	New Express Mathematics 3	SG-PPE

The *High School Mathematics Curriculum Standard (Experimental)* of China (MOE of PRC, 2003; hereinafter referred to as *Curriculum Standard*) issued in 2003 proposes that high school mathematics textbooks should offer various inquiry activities, encourage students to study independently and experience mathematical inquiry or creation process. To begin with, we make a brief summary of the studies concerning mathematical inquiry before proposing research questions.

1.2 Studies concerning mathematical inquiry

While studying how reading promotes students' mathematical inquiry, Siegel *et al.* (1998) first recalled Peirce and Dewey's understanding of "inquiry as the process of settling doubt and fixing belief within a community", and quoted the viewpoint of Lakatos (1976) that involving students in mathematical inquiry means inviting them to experience and appreciate the first-hand ambiguity, nonlinearity and "conscious guessing". Such view of inquiry is especially essential to mathematical education. Ball and Bass (2003) carried out an empirical study of the relationship between mathematical problem reasoning and mathematical inquiry ability, and pointed out that the training of mathematical reasoning is significant in improving mathematical inquiry ability. They claimed that "mathematical reasoning comprises a set of practices and norms that are collective, not merely individual or idiosyncratic", and that "mathematical reasoning can serve as an instrument of inquiry in discovering and exploring new ideas .... Mathematical reasoning functions centrally in justifying or proving mathematical claims". J. Volmink believed that "inquiry and investigation should occupy a central place in the school mathematics curriculum" (see Segel, 1998, p. 283).

In addition, it is found according to the text analysis of curriculum documents that many countries have abundant understandings of the connotation of "mathematical inquiry". The Singapore's mathematics syllabus issued in 2000 emphasized that inquiry activities should be vigorously carried out to help students to pursue development in the job-seeking awareness. In 1989, the French mathematics syllabus for middle schools was modified and the following provisions were specified: As for teaching activities, in addition to some temporary projects, some projects of longer durations should be developed to offer students with situated inquiry; and the *Guideline for Middle School Mathematical Learning* issued and implemented by Japan in 2008 proposed that, "we should enhance foundations, study mathematical contents by means of experiment, investigation and inquiry, etc., cultivate the expression, thinking and judging abilities and flexibly apply knowledge and abilities among situations" (Yuelan Chen, 2010).

All countries pay much attention to mathematical inquiry in the curriculum standards/syllabi. In Germany, mathematical inquiry has rich connotation and emphasizes that students should experience the process of discovering, solving and further generating certain external or internal mathematical problem situation (Leuders, 2003). From the perspective of international comparative study, the evaluation and comparison on students' creative learning methods are penetrated into some large-scale international evaluation projects. For example, the mathematical evaluation framework in the latest Trends in International Mathematics and Science Study (TIMSS-11) proposed that, "non-routine problems are problems that are very likely to be unfamiliar to students, … transfer of knowledge and skills to new situations,… problems requiring reasoning may do in different ways, … reasoning involves the ability to observe and make conjectures, to make logical deductions based on specific assumptions and rules, and to justify results" (Mullis *et al.*, 2009). The *Curriculum Standards* of China claims that "mathematical inquiry … refers to a process of students to independently inquire and study a particular mathematical problem. Such process includes observing and analyzing mathematical facts, proposing meaningful mathematical problems, guessing and inquiring proper mathematical rules or conclusions and giving interpretations or justifications."

Through a series of document analysis, we believe that mathematical inquiry could be defined as follows: By focusing on a problem situation or mathematical problem, students observe, analyze and estimate mathematical facts, propose meaningful mathematical problems, conjecture and verify mathematical conclusions or rules, or give interpretations or justifications, and, furthermore, rethink conclusions or generate a new round of problems.

2. Contents of Mathematical Inquiry in Textbooks

In the *Curriculum Standards* of China, mathematical inquiry as a newly added part attracts much attention. The *Curriculum Standards* propose that high school mathematics curriculum should arrange diversified forms of inquiry activities, encourage students to get involved in active learning and experience the process of mathematical discovery and creation. The learning activities or materials in relation to mathematical inquiry are offered in the textbooks based on the *Curriculum Standards*. In UK, The School Mathematics Project textbook UK-SMP published by the Cambridge University Press provides rich development materials in order to intrigue students to further specify mathematical thoughts and skills in the exploration. In particular, UK-SMP arranges the column of "discussion questions", which helps students to create situations of discussing, exploring and illustrating the mathematical thoughts by means of group activities. In USA, the textbook US-PHM has a distinguished character in providing learning resources for students to explore the solution to open-ended mathematical problems through the design of mathematical activities. For example, in *Algebra I* and *Algebra II*, the explicit column of "Activity Lab" is designed to cultivate students' estimation and prediction abilities as well as critical thinking skills. The "Activity Lab" also intrigues students to actively perform mathematical inquiry activities and increases students' experience in mathematical inquiry. Other US textbooks also obviously reflect enough attention on mathematical inquiry. For example, in the geometry textbook of the textbook series US-SMP under the University of Chicago School Mathematics Project (UCSMP, a project initiated in 1983), one characteristic is with real world as orientation. In the textbook, relevant situations of music and building are created, aiming to help students to fully explore the mathematics embraced in these situations. The textbook also vigorously incorporates technology and provides modern technical tools and thoughts for students to explore mathematics.

According to the comparative analysis of mathematical inquiry contents in the textbooks of different countries, we try to answer the following questions: To what extent does the contents of the textbooks reflect "mathematical inquiry"? What are the features of the contents setting? The chapter also describes the differences and similarities among different textbooks, and further analyzes organization and presentation of the mathematical inquiry contents in the mathematics textbooks of China. It is quite important to make reference to the international experience as a well-known Chinese saying says that "stones from other hills may serve to polish the jade". The conclusions obtained in comparative studies will help us to modify and improve the Chinese textbooks, as well as to enhance the domestic textbook studies.

3. Analytical Index System

Mathematical inquiry requires environment. We will analyze the situations presented in the mathematical inquiry contents and the mathematical information involved in situations to evaluate whether the situations are meaningful and challenging for the students. Mathematical inquiry also requires methods. We will analyze the question types adopted in the problem statements and the forms of activities used in the learning progress so as to understand whether students experience the methods of observation, analysis, conjecture, verification, justification or reflection,

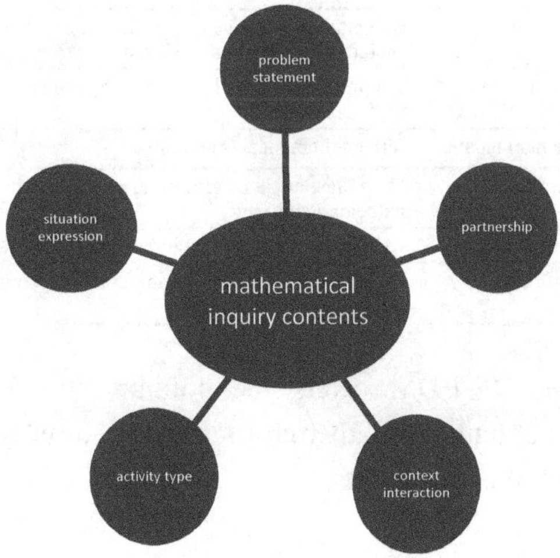

Fig. 6-1 Five-index framework for analysis of mathematical inquiry contents

etc. Moreover, the contents of textbooks must be coherent and there must be interactions between the mathematical inquiry contents and their contexts. Therefore, we need to further analyze the role of mathematical inquiry contents in textbooks. Based on these discussions, we propose five elements in the analysis of textbook inquiry contents: situation expression, problem statement, partnership, activity type and context interaction. The five-index analytical framework is shown as Fig. 6-1.

Below, we present a detailed description of the five indexes with sub-indexes.

3.1 Situation expression

"Situation expression" means the presentation of the situation in the textbook when organizing and delivering the mathematical inquiry contents. Through the investigation of textbooks both in China and abroad inquiry contents usually appears with real-world situation, virtual situation or pure mathematical situation. The presented situation contains relevant mathematical information, which may be complete, redundant and insufficient. See Tab. 6-2 for specific meanings.

Tab. 6-2 Meanings of sub-indexes of the index "situation expression"

Situation type	Real-world	True situations derived from daily life and external real world (including nature, arts, P.E. and humanity, etc.)
	Virtual	Imaginary situations based on actual life or story plots in literacy or sci-fi works
	Pure mathematics	Pure mathematical problem
Completeness of mathematical information	Complete (exactly self-contained)	All mathematical conditions are just used to solve the problem of mathematical inquiry
	Redundant	Redundant mathematical information/conditions are involved
	Insufficient	The mathematical information/conditions provided are not sufficient to solve the problems

For example, DE-EDM offers the column "Im Blickpunkt" ("In Focus") for the contents of mathematical inquiry, in which the following real-world situation is presented:

Example 1. An "Im Blickpunkt" in DE-EDM (Fig. 6-2).

Im Blickpunkt

Spiralen

In der Natur kommen oft Spiralen vor. Seit Jahrtausenden sind sie auch zur künstlerischen Gestaltung verwendet worden.

1. Erläutere, warum eine Spirale kein Graph einer Funktion ist.

Fig. 6-2 A capture of page 54, DE-EDM 10

> **Translation of Fig. 6-2**
>
> In focus
>
> Spirals
>
> There are many spirals in nature, and they are used for artistic design for thousands of years.
>
> 1. Explain why a spiral is not the graph of any function.
>
> ...

The situation used in the text is originated in the nature and in the world of artists. It connects the mathematics with natural world and guides students to observe and guess from the perspective of mathematics. Note that no specific mathematical information/ conditions are obviously provided in the problem, requiring students to extract necessary mathematical elements to answer the problem.

Also, in the very beginning of the introduction of differential calculus, DE-EDM designs a learning block called "Änderungen Beschreiben" ("changes to describe"). The following situation is an example in the block. After reading it, students are requested to draw the graph of the function of the glider's height on time.

Example 2. "Flying a glider" from DE-EDM (Fig. 6-3).

Segelflug

Konrad ist mit seinem Onkel das erste Mal in einem Segelflugzeug geflogen. Für die Schülerzeitung schreibt er einen Bericht.

„Das Segelflugzeug hat keinen eigenen Motor, deshalb ging es hoch mit einer Seilwinde. Das Variometer im Cockpit zeigte eine Steiggeschwindigkeit von 5 m/s an. Wir wurden kräftig in die Sessel gedrückt. Etwa 1 Minute nach dem Abheben klinkte das Schleppseil aus, und der Pilot machte sich auf die Suche nach einer guten Thermik. Das dauerte etwa 1 Minute, in denen wir 50 Meter an Höhe verloren. In dem Aufwind ging es schnell wieder aufwärts, weit aufwärts: innerhalb von 8 Minuten stiegen wir auf 1 200 m Höhe.

Das war vielleicht eine Aussicht! Unser Dorf war sehr klein unter uns zu sehen, und die Schule gerade noch zu erkennen. In der Ferne am Horizont konnte man das Meer sehen, zu dem wir im Auto immer mindestens zwei Stunden unterwegs sind!

Etwa 10 Minuten lang konnten wir diese Aussicht genießen. In dieser Zeit verloren wir natürlich an Höhe: das Variometer zeigte dabei eine konstante Sinkgeschwindigkeit von etwa 1 m/s an. Dann sorgte der Pilot dafür, dass wir zügig nach unten kamen. In etwa dreieinhalb Minuten gingen wir in der Nähe eines Berghanges auf 200 m herunter. Dort konnten wir uns mithilfe des Hangaufwindes in gleichbleibender Höhe einen guten Einflug zum Landeplatz suchen. Nach 2 Minuten hatten wir uns entschieden und den Landeanflug eingeleitet. Die Erde schien in einem schnellen Tempo auf uns zu zu stürzen. Das Variometer zeigte eine Sinkgeschwindigkeit von 4 m/s an. Eine halbe Stunde nach dem Start hatte uns die Erde wieder."

- Stelle den Segelflug als Graph der Funktion *Zeit → Höhe* grafisch dar. Über welchem Zeitabschnitt ist die Steiggeschwindigkeit am höchsten? Wann ist die Sinkgeschwindigkeit am höchsten? Woran kann man das am Graphen erkennen?
- Berechne für jeden angegebenen Zeitabschnitt die durchschnittliche Steig- bzw. Sinkgeschwindigkeit. Vergleiche mit dem vorher gezeichneten Graphen der Funktion *Zeit → Höhe*.
- Skizziere nun den Graphen der Funktion *Zeit → Steig- bzw. Sinkgeschwindigkeit*. Vergleiche mit dem vorher gezeichneten Graphen zu der Funktion *Zeit → Höhe*. Welche Zusammenhänge kannst du erkennen?

Fig. 6-3 A capture from pages 134–135, DE-EDM 10

There is rich mathematical information in the report. To draw the graph of the function, students need to screen necessary data from rich information. This example gives a situation with redundant mathematical information.

Translation of Fig. 6-3

Glider

Konrad flew a glider for the first time with his uncle. He reported his experience in the Student Newspaper.

> The glider has no engine of its own, therefore it takes off by a rope winch. The variometer in the cockpit showed the climbing speed being 5m/s (with a picture in the textbook). We were pushed into the armchairs. About 1 minute later the tow rope loosed, and the pilot was searching after a good thermal lift. This took about one minute, in which we lost 50m on height. In the upwind, it quickly rebounded, going far upwards: within 8 minutes, we climbed at the altitude of 1 200m.
>
> How splendid the view was! Our village was very small to see below us, and the school just barely recognizable. In the distance on the horizon you could see the sea, to which we always drive at least two hours in the car!
>
> We were enjoying the view for about 10 minutes. During this time, we lost on height of course: the variometer showed a constant velocity of about 1m/s. Then, the pilot made a quick descending, and in about three and a half minutes, we had lower down about 200m approaching to a hill slope. There, we were able to use the upslope wind to keep in a stable level, looking for a good place to land. Two minutes later, we began the landing process. The earth seemed fall to us at a rapid pace. The variometer showed a rate of descent of 4m/s. Half an hour after the start, we reached the ground again.

- Draw the graph of the function of the glider's height on time. What time does the glider have the fastest climbing speed? What time does the glider has the quickest fall velocity?
- Calculate for every specified period the rate of ascend or descend, comparing with the previously drawn graph of the function of height on time.
- Now sketch the graph of the function of rate of ascend or descend on time, and compare it with the previously obtained graph of height on time. What connection do you see?

The *Algebra 2* of US-PHM provides an explicit column of mathematical inquiry as "Critical Thinking" in which a linear programing problem is shown as follows:

Example 3. Maximum Profit (Fig. 6-4).

23. Critical Thinking Sometimes two corners of a graph both yield the maximum profit. In this case, many other points may also yield the maximum profit. Evaluate the profit formula $P = x + 2y$ for the graph shown. Find four points that yield the maximum profit.

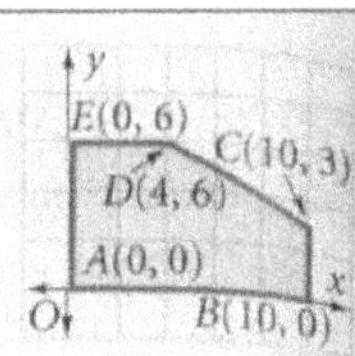

Fig. 6-4 A capture of page 144, US-PHM, Algebra 2

3.2 Problem statement

According to the pre-analysis of textbooks, we found that textbooks normally use declarative sentences or interrogative sentences to raise problems. Thus, the sub-indexes of the index "problem statement" should include, on one hand, the sentence type used for the question raising, declarative or interrogative; and, on the other hand, the openness of the conclusion or the problem-solving process must also in our consideration. See Tab. 6-3 for the detailed description of the indexes.

Tab. 6-3 Meanings of sub-indexes of the index "problem statement"

Sentence type	Declarative	Direct statement of mathematical facts, tasks or activity requirements; a full stop usually follows in the end of the sentence.	
	Interrogative	The enquiry or rhetorical method is used to show the requirements of mathematical tasks or activities and question mark usually follows in the end of the sentence.	
Question type	Closed type	Both the conclusion and process are unique.	
	Open type	Open-ended	No default conclusion; students' inquiry conclusions may be diversified.
		Open process	The methods or strategies adopted by students to solve the problem are not specified and the students' inquiry process is diversified.

For example, there is a column (named as "Detective" in our study) in DE-EDM with the left picture in Fig. 6-5 as its flag.

Example 4. A "Detective" problem in DE-EDM (Fig. 6-5).

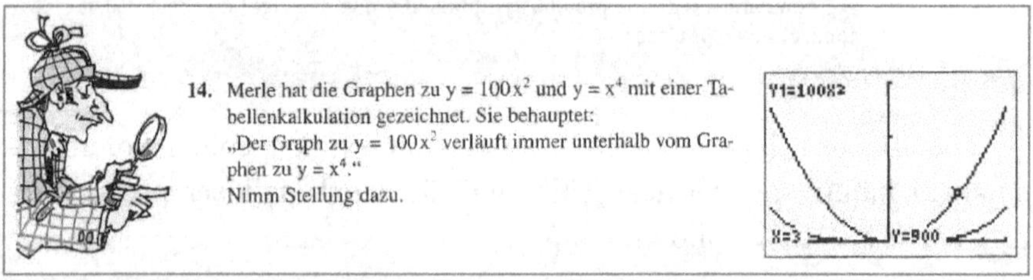

Fig. 6-5 A capture of page 68, DE-EDM 10

A declarative sentence is used here. Students are requested to express their opinions towards a mathematical viewpoint with no proofs required. Therefore, students may arrive at different conclusions.

Translation of Fig. 6-5

14. Merle used a graphing calculator to draw the graphs of $y = 100x^2$ and $y = x^4$. She claimed:

"The graph of $y = x^4$ always runs below to the graph of $y = 100x^2$".

Give your opinion.

In addition, no prefixed approach to the question is given. This means that the exploration process is open.

A mathematical inquiry column "Exploration Task" is provided in the SG-PPE of Singapore, from which we quote an example as follows.

Example 5. An "Exploration Task" in SG-PPE (Fig. 6-6)

> **Exploration Tasks**
>
> 1. Search the internet or other resources for information on how to make a simple clinometer. Use your clinometer and your knowledge of trigonometric ratios to find the following.
>   (a) The heights of the five tallest buildings in the Shenton Way area.
>   (b) The heights of Mount Faber or Bukit Timah Hill.
>
> In your report, state clearly the assumptions and limitations of your findings.

Fig. 6-6 A capture of page 275, SG-PPE, New Express Math 3

Similarly, declarative sentences are used here and students are asked to give answers. However, the method to get answers is not specified. The process for students' exploration is diversified.

3.3 Partnership

This index refers to whether students work individually, with a partner, or as a team in some inquiry contents. According to its connotations, mathematical inquiry also demands for students' spirit of cooperation. Thus, the analysis of this index is crucial for interpreting in organization and presentation methods of mathematical inquiry contents. The sub-indexes of partnership are described in Tab. 6-4:

Tab. 6-4 Meanings of the sub-indexes of the index "partnership"

Individual	Activity without explicit requirement on cooperation
Partner	Activity requiring the cooperation of two persons
Team	Activity requiring team work with three or more members

For example, the Chinese high school textbook CN-PEP arranges an explicit mathematical inquiry column "Internship Task".

Example 6. "Internship Task" in CN-PEP (Fig. 6-7).

实习作业

走进微积分

目的

通过收集微积分创立的时代背景和历史意义的有关材料，体会微积分在数学思想史和科学思想史上的价值.

过程

1. 微积分（differential and integral calculus）是研究函数的微分（differential）、积分（integral）以及有关概念和应用的数学分支. 微积分中的基本概念是极限、导数、积分等. 导数、定积分都是微积分中的核心概念.

◆ 请同学们收集介绍"微积分"的有关书籍，了解微积分的研究对象，以及微积分的基本概念.

2. 微积分的创立不仅是数学思想史上的里程碑，也是科学思想史上的里程碑.

◆ 请同学们收集历史上对微积分创立和发展的一些重要评价.

3. 微积分的创立具有悠久的历史渊源.

◆ 请同学们收集历史上我国和古代欧洲有关微积分思想的一些代表性的工作.

4. 微积分的创立有深刻的时代背景，当时的技术对数学提出了许多要求.

◆ 请同学们通过收集从欧洲文艺复兴到 17 世纪期间的社会、经济状况、科学发展、贸易、航运等情况，了解它们对数学的要求.

5. 牛顿（Newton，1642—1727）和莱布尼茨（Leibniz，1646—1716）是微积分的创立者.

牛顿

莱布尼茨

◆请同学们收集牛顿和莱布尼茨的生平资料，以及他们创立微积分时所做的开创性工作.

思考

微积分的创立说明了什么？它的历史意义是什么？你有什么体会？

要求

以小组为单位，按上述问题分头收集材料，最后汇总. 每组写一个实习报告，各组分头交流.

主要参考书目：

[1] 中国大百科全书总编辑委员会《数学》编辑委员会，中国大百科全书出版社编辑部. 中国大百科全书：数学卷. 北京：中国大百科全书出版社，1992.

[2] 莫里斯·克莱因. 古今数学思想（第二册）. 朱学贤，申又枨，叶其孝，等译. 上海：上海科技出版社，2002.

[3] 张顺燕. 数学的源与流. 北京：高等教育出版社，2003.

[4] 李文林. 文明之光：图说数学史. 济南：山东教育出版社，2005.

Fig. 6-7 A capture of pages 61–62, CN-PEP, Opt. 2-2

Translation of Fig. 6-7

Getting into Differential and Integral Calculus

Objective:

Experience the value of differential and integral calculus (calculus for short) in the history of mathematical thoughts and of scientific thoughts through collecting the materials about the time background and historical significance of the establishment of calculus.

Process:

1. Calculus is a mathematical branch studying differential and integral of functions and related notions together with their applications. The fundamental concepts in calculus include limit, derivative and integral, with derivative and definite integral as the cores.
   - ◆ Collect and read some introductory books on calculus, understanding its studying objects and its basic concepts.
2. The establishment of calculus is a milestone not only in the history of mathematical thoughts but also in the history of scientific thoughts.
   - ◆ Collect important historical comments and opinions about the establishment and development of calculus.
3. The establishment of calculus has a long historical trace.
   - ◆ Collect historical works related to the idea of calculus in ancient China and Europe.
4. The establishment of calculus deeply rooted in the era when the development of technology has highly demand on mathematics.
   - ◆ Collect the information on status of society, economy, science and technology, trading, and shipping, from European Renaissance to the 17th century, and understand their demand on mathematics.
5. Newton (1623-1727) and Leibniz (1646-1716) are the founders of calculus
   - ◆ Collect reading materials about their life and their creative contributions to the establishment of calculus.

Thinking:

What are the implications of the establishment of calculus? What is the historical significance? And, what is your experience with this study?

Requirements:

Students are divided into groups according to the above themes, each group in charge of collecting materials with one theme. Then put all materials together, based on which each group is required to write a report and then communicate with other groups.

Main References:

[1] Mathematics Sub-committee of Encyclopedia of China Editing Committee. *Encyclopedia of China: Mathematics Volume*. Bejing: Encyclopedia of China Publishing House, 1992.

[2] M. Kline. *Mathematical Thought from Ancient to Modern Times (Volume 2)*. Chinese translation by Zhu Xuexian, Shen Youcheng, Ye Qixiao *et al*. Shanghai: Shanghai Scientific & Technical Publishers, 2002.

[3] Zhang Shunyan. *Sources and Flows in Mathematics*. Beijing: Higher Education Press, 2003.

[4] Li Wenlin. *The Glory of Civilization: A Pictorial History of Mathematics*. Jinan: Shandong Education Press, 2005.

It is clearly specified that students solve problems through team work; while the following inquiry activity designed in the DE-EDM is explicitly specified as a partner activity (Partnerarbeit).

Example 7. A partner activity in DE-EDM (Fig. 6-8).

Nimm

Partnerarbeit: Eine nicht zu geringe Anzahl Streichhölzer wird auf den Tisch geschüttet. Gespielt wird zu zweit. Die beiden Spieler ziehen abwechselnd bei jedem Zug mindestens 1 Holz, aber weniger als die Hälfte der noch vorhandenen Hölzer – so lange das geht.
Wer abräumen muss, d.h. das letzte Holz nehmen muss, verliert.

- Spielt das Spiel einige Male miteinander. Was fällt euch auf?
- Könnt ihr eine Strategie herausfinden, die zum Gewinn des Spieles führt?
- Erläutert und begründet die Aussage des nebenstehenden Artikels.

Das 2-Personen-Streichholzspiel „Nimm" weist charakteristische Verlustzahlen auf: 3, 6, 12, ...
Wenn man am Zug ist, kann man von diesen aus nicht gewinnen.

Fig. 6-8 A capture of page 63, DE-EDM 10

Translation of Fig. 6-8

Taking out

Partner activity: Some matches is sprinkled on the table, and the game is played by two persons. The players take out matches alternatively, with at least one match but less than half of the remaining matches being took out each turn.
The person who takes the last match loses the game.

- Play the game with each other a few times. What do you find?
- Could you find a strategy that will lead to winning the game?
- Explain and justify the statement in the adjoining box.

The 2-person-strike game "taking out" has characteristic loss number: 3, 6, 12, ... If you met a number in the series, you cannot win.

3.4 Activity type

"Activity type" refers to the contents and form of a mathematical inquiry activity. Through the pre-analysis of textbooks, we classify the mathematical activities as problem-solving, writing, project, reading, and experiment. Through these activities, the textbooks encourage students to conduct multiple-perspective mathematical inquiries. The meanings of the sub-indexes of the activity type index are described in Tab. 6-5.

Tab. 6-5 Meanings of the sub-indexes of the index "activity type"

Problem-Solving	Verification and reflection	Verification or reflection on the existing problem-solving process or known results of yourself or others
	Calculation and proving	Conduct calculation, proofs or figure construction in the mathematical sense by using mathematical formulas, theorems etc.
	Conjecture and interpretation	Conjecture, judgment or interpretation on the possible problem-solving process or conclusions according to the problem situations.
Experiment	IT type	IT devices (such as graphing calculator, geometric sketchpad, and software) are used to tentatively explore problems and experience the mathematization process.
	Science type	Design or participate in small-scale experiments in the fields of physics, chemistry or biology, observing and analyzing the mathematics in the experiments
	Daily life type	Daily activities (games, sports, etc.) are organized to experience the discovery of mathematical rules in activities.
Writing	On concept	Text reports or exhibition boards are used to express mathematical concept
	On conclusion	Text reports or exhibition boards are used to present inquiry conclusion
	On process	Text reports or exhibition boards are used to present the process and method of problem inquiry
Project	Text works	Conduct deep inquiry on a theme with text works as the output
	Real objects	Conduct deep inquiry on a theme with material objects as the output
	Electronic works	Conduct deep inquiry on a theme with electronic or computerized works as the output
Reading	With questions	Reading text and answer questions.
	Without questions	Reading text, without the requirement of answering questions.

Example 4 above requires that students make verification and reflection (diagnosis and reflection) on the conclusions concerning functional graphs; Example 6 requires students to prepare reports and write essays on mathematical concepts and process; Example 7 is a game activity, which allows students to observe and analyze the mathematical models of the game rule.

Let us consider the following

Example 8. A problem in the column "Lernfeld" of DE-EDM (Fig. 6-9).

Genug abgekühlt

Die Abkühlung einer erhitzten Flüssigkeit soll untersucht werden. Dabei wird die Temperaturabnahme von der Ausgangstemperatur und der Umgebungstemperatur abhängen.

- Miss die Umgebungstemperatur. Führe die Messungen des Abkühlungsprozesses durch und betrachte die Entwicklung der Temperaturunterschiede zwischen Getränke- und Umgebungstemperatur. Übertrage die Tabelle in dein Heft und trage die Messergebnisse ein.

Zeitpunkt n	Getränketemperatur g	Umgebungstemperatur u	Temperaturunterschied
1			
2			
3			
n			

- Stelle den Vorgang grafisch dar.
- Gib für die Temperaturunterschiede einen Term zur Berechnung an.
- Bestimme die Zeit, nach der der Abkühlungsvorgang praktisch abgeschlossen ist.

Fig. 6-9 A capture of page 62, DE-EDM 10

Translation of Fig. 6-9

Cooling fully

The cooling process of a heated fluid is investigated. The temperature decreasing depends on the starting temperature and the ambient temperature.

- Measure the ambient temperature. Observe the cooling process by measuring the change of the temperature differences between the drink and the ambient temperature. Copy the table to your notebook and carry on the measurement results.

No. (n)	Fluid temperature (g)	Ambient temperature (u)	Temperature difference

- Show the process graphically.
- Give a term for calculation of temperature difference.
- Determine the time when the cooling process is practically completed.

This inquiry requires that students conduct experiments and observe temperature changes so that they can experience the process of seeking for mathematical rules from physics phenomena.

3.5 Context interaction

The inquiry contents in mathematics textbooks is not independent. Instead, it plays different roles in different contexts, mainly including introduction (of new knowledge), transition (from the preceding to the following), summary, enhancement and deepening, and application and extension. The analysis of the index can be used to probe into the overall intention of the textbooks in different countries in arranging the contents of inquiry. The meanings of the sub-indexes of the index "context interaction" are described in Tab. 6-6.

Tab. 6-6 Meanings of sub-indexes of the index "context interaction"

Introduction	The activity only relates to the following chapter/section or knowledge points
Transition	The activity relates to the preceding and following chapters/sections or knowledge points
Summary	The activity only relates to the preceding units or knowledge points
Enhancement & deepening	The activity aims to train and deepen the learned contents
Application & extension	The activity aims to apply or extend the present knowledge to other mathematics context or to other subjects

The experimental activities in Example 8 involve the contents of "exponential growth" in the following section. Therefore, the contents has the characteristic of "introduction (of new knowledge)".

The above analytical indexes construct an analytical index system integrating complete mathematical inquiry contents, as shown in Fig. 6-10.

Based on the analytical index system, we made a comparative study of the presentation means and organization forms of the mathematical inquiry contents in the high school textbooks of relevant countries.

4. Process of Study

4.1 Selection of textbooks and mathematical inquiry contents

This study is one of the sub-topics in the *Comparative Study of High School Mathematics Textbooks of Certain Countries*, an educational key

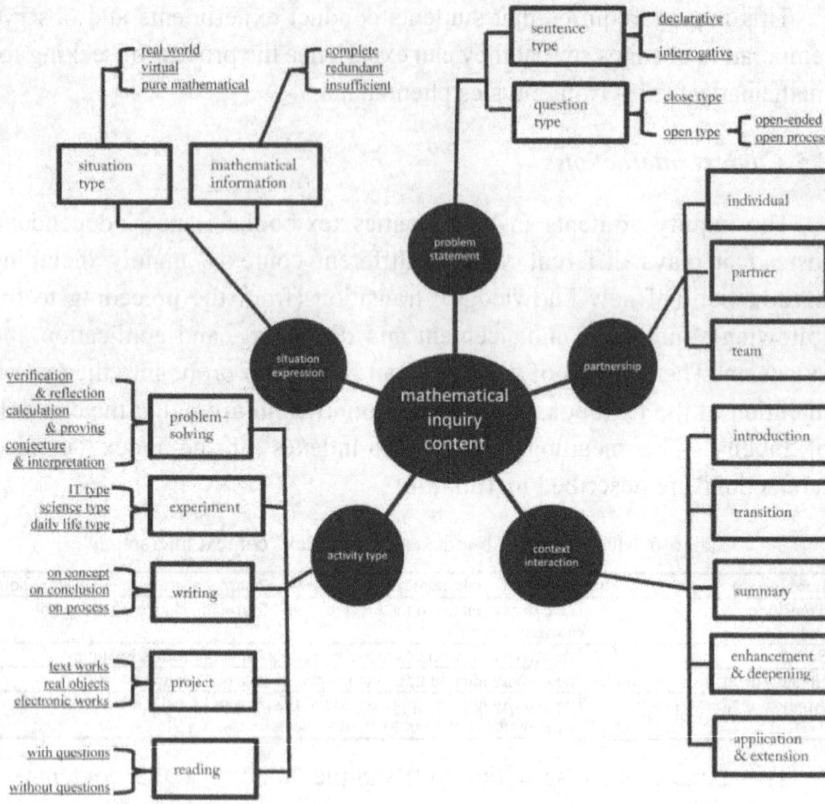

Fig. 6-10 Analytical index system involving mathematical inquiry contents

project under the "Eleventh Five-Year Plan" of the National of Social Science Foundation of China. Jianpan Wang (2011) offered a general criterion for textbook selection. This study confined our sights within USA, UK, Germany, France, Japan, and Singapore, and the chosen textbooks were recommended by experts in these countries. We also chose CN-PEP for comparison. Due to the increasing attention to mathematical inquiry in the mathematics curriculum reforms of all these countries, all the chosen textbooks have columns in relation to mathematical inquiry, such as "Detective" in DE-EDM, "Critical Thinking" in US-PHM and "Inquiry and Finding" in CN-PEP. This study focuses on these explicit columns and examine the design of mathematical inquiry contents in textbooks through encoding, statistics and analysis. The information of textbooks studies and relevant inquiry column information are summarized in Tab. 6-7.

Tab. 6-7 Inquiry columns in selected textbooks

Code	Inquiry columns
CN-PEP	"Thinking", "Reading and Thinking", "Applying IT", "Internship Tasks", "Inquiry", "Observation", and "Inquiry and Discovery"
US-PHM	"Practice and Problem-Solving", "Error Analysis", "Reasoning", "Open Ended", "Activity Lab", "Writing", "Critical Thinking", "Design", and "Make a Conjecture"
UK-SMP	"Discussion Questions"
DE-EDM	"Detective", "Projekt", "Lösung" "Im Blickpunkt", "Auf den Punkt gebracht", "Zum Selbsternen", "Lernfeld", "Partnerarbeit", and "Teamarbeit"
FR-EDB	"Activité", "Communiquer", "(Exercices) Approfondir", "Avec un Tableur", "Travaux Pratiques"
JP-SKS	"Column", "Study", and "Supplemented Questions"
SG-PPE	"In-class Activity", "Exploration Task", "Project Task", "Thinking Time"

4.2 Reliability test of the encoding system

We did reliability test after formulating the analytical index system. To do this, we randomly chose one column from each of the seven textbooks, and let two researchers encode the same contents. In this first attempt, the consistency of the encoding was not satisfactory. Later, through detailed discussions our team had a better understanding of the index system. Based on it, we redid the contents selection and encoding comparison, and finally reached a high consistency. After this, we proceeded to encode the inquiry columns in the surveyed textbooks based on the index system.

5. Comparative Analysis and Enlightenments

As mentioned above, the analytical index system for the inquiry contents comprises five indexes. Our analysis is conducted from the five aspects.

5.1 Situation expression of mathematical inquiry contents

5.1.1 Situation types

Statistics shows that all the seven textbooks prefer pure mathematical situations for mathematical inquiry. More precisely, in all inquiry situations the proportion of pure mathematical situations is as high as 91.18% in JP-SKS, 81.85% and 67.34% in the CN-PEP and FR-EDB, respectively, and nearly 60% in DE-EDM, UK-SMP and US-PHM. The mathematical inquiry column in the SG-PPE uses not only pure mathematical situations (40.74%) but also virtual mathematical situations (48.77%). This SG-PPE textbook fabricates question situations based on some real-world situations for students to explore mathematics.

The data also shows that all the textbooks also pay attention to real, challenging and interesting question situations so as to motivate students to actively participate in inquiry activities. For example, 32.66%, 27.08% and 23.73% of the mathematical inquiry situations in FR-EDB, DE-EDM and US-PHM, respectively, are real-world situations; this proportion is 10.32% in CN-PEP (Fig. 6-11).

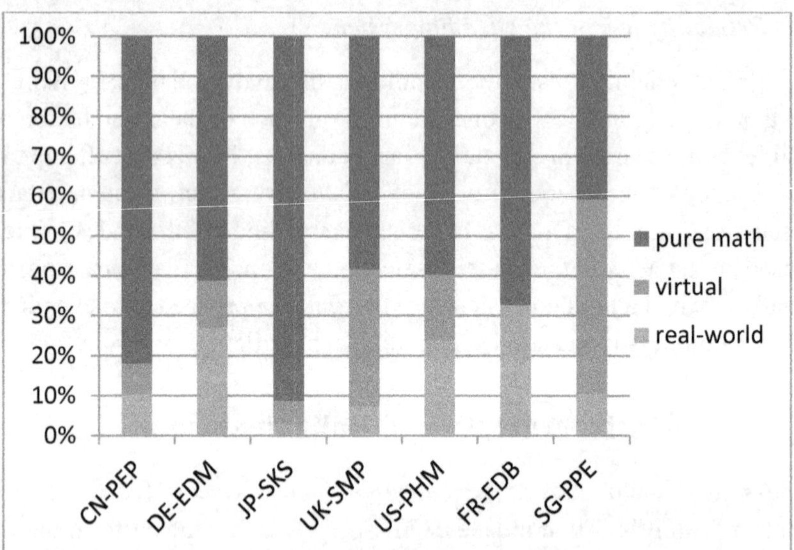

Fig. 6-11 Distribution of mathematical inquiry situations

It should be mentioned that FR-EDB contains the highest percentage of real-world situations, so the students in France will have more opportunities to conduct inquiry with real-world background. On the other hand, JP-SKS focuses on the inquiry with pure mathematical situations, and SG-PPE contains the largest proportion of virtual situation.

5.1.2 Mathematical information

The mathematical information presented in the mathematical inquiry situations also has different features in different textbooks. The statistics shows (Fig. 6-12) that more than 95% of the inquiry situations in CN-PEP, JP-SKS and US-PHM provide students with complete mathematical information and students only need to consider the conditions or information given in problems. While the mathematical information contained in half of the inquiry situations of DE-EDM is not in correspondence with the mathematical conditions necessary for problem-solving; 47.92% of the situations do not provide sufficient information so that the students need to reasonably supplement conditions according to the tasks. Such feature is also reflected in SG-PPE. 26.71% of the inquiry situations provide insufficient mathematical information. DE-EDM also has one unique feature not reflected in the other textbooks — 6.25% of the inquiry problems contains redundant information, and students need to learn how to analyze and screen these mathematical conditions.

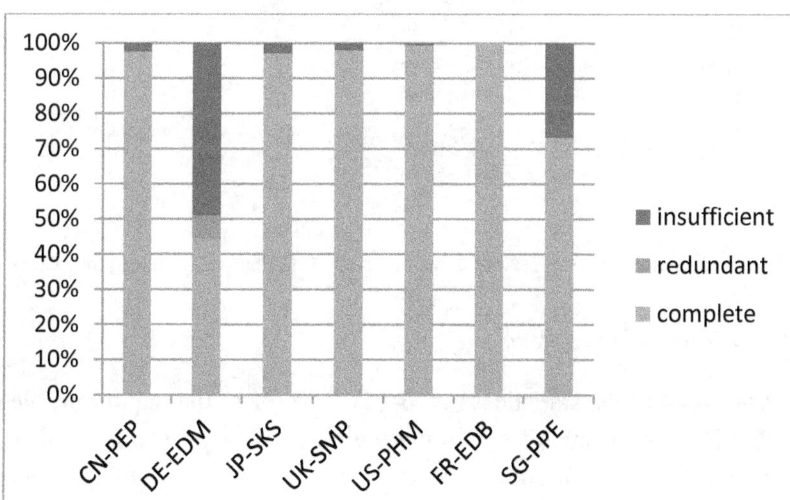

Fig. 6-12 Completeness of mathematical information presented in inquiry situations

5.2 Problem statement of mathematical inquiry contents

5.2.1 Sentence types of statements

The data on this sub-index are shown in Fig. 6-13. We found that the CN-PEP prefers to use interrogative sentences in the statements of mathematical inquiry problems, with the proportion as high as 82.87%, ranged number one in the textbooks from the seven countries. This shows that the CN-PEP use "question marks" to activate students' consciousness of mathematical thinking and exploring, and then turn it to the inquiry actions. As for the other textbooks, more than 60% of the statements uses declarative sentences. Of course, the extent of sentence types in triggering students' mathematical inquiry action is quite complex and the problem type presented through these sentence types is also contributory.

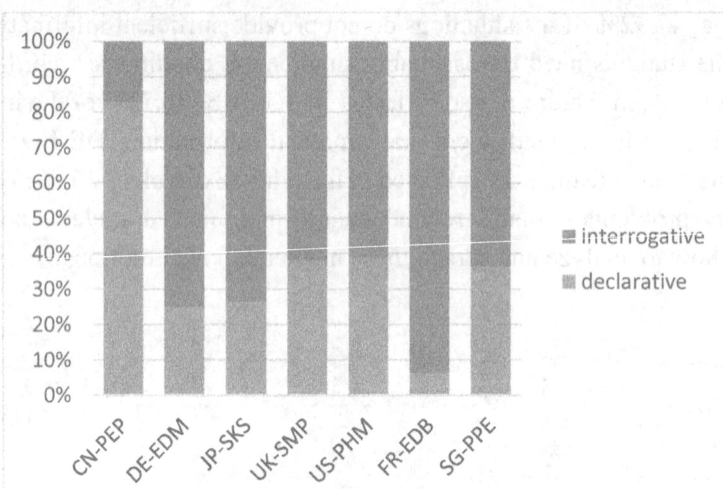

Fig. 6-13 Sentence types used in the statements of mathematical inquiry problems

5.2.2 Openness of problems

According to the statistics (Fig. 6-14), 83.85% of the inquiry problems in DE-EDM are of open type. In particular, 26.69% problems emphasize open conclusions, or open-ended. When dealing with open-ended problems, students may make bold assumptions on possible conclusions and then do careful analysis, interpretation or verification. Or, they may make reasonable inductive inferences to obtain possible conclusions and then go forward to give proofs for the conclusions. The remainder of the problems of open type (54.16% in all problems) focuses on open process. For such problems students may take different approaches to the conclusions. They may conduct various experiments, do different calculations, verifications or demonstrations, etc. The large proportion of the problems of open type shows that DE-EDM emphasizes the creation of a free learning environment for mathematical inquiry. As for the percentage of problems of open type, CN-PEP and US-PHM are next only to DE-EDM, with percentages of 59.35% and 50.69% respectively. These data show that CN-PEP and US-PHM also make great efforts to create a rich environment for cultivating students' mathematical inquiry ability, providing students opportunities to face uncertain in real world and in mathematical world, encouraging them to make bold induction and assumption, and guiding them to rational reasoning.

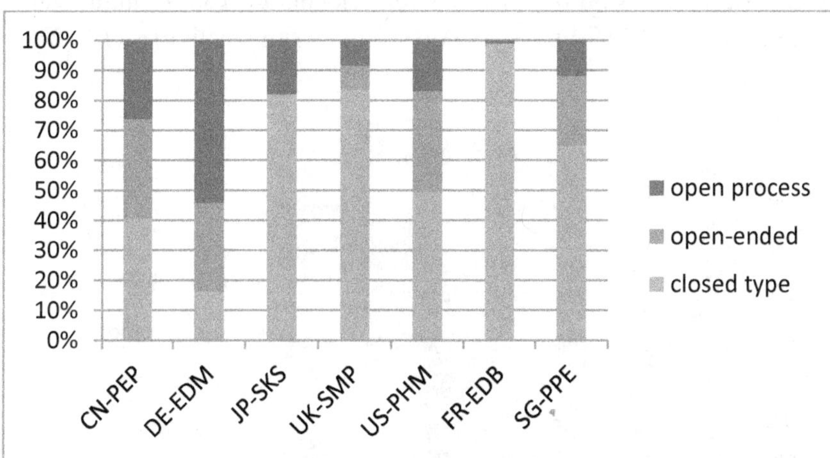

Fig. 6-14 Types of mathematical inquiry problem statement in textbooks

5.3 Activity type of mathematical inquiry

Both situation expression and problem statement of mathematical inquiry activities create an environment for students to make inquiry, while the activity type and its presentation contain the requirements of an activity, which provides suggestions for students to make inquiry. After encoding textbooks (Fig. 6-15), we found that except for JP-SKS, the problem-solving activities dominate in the mathematical inquiry in SG-PPE, UK-SMP, CN-PEP, US-PHM, DE-EDM, and FR-EDB, with percentages of 99.35%, 98.21%, 86.57%, 76.42%, 72.97% , and 72.76% respectively. Meanwhile, JP-SKS also pays special attention to reading activities (55.88%) to intrigue students' awareness to inquiry. Besides the problem-solving activities, CN-PEP, DE-EDM, US-PHM and FR-EDB organize rich inquiry activities, including experiment, writing, project and reading. Different activities with different requirements play different roles in cultivating students' mathematical inquiry ability. In particular, 13.51% of the activities in DE-EDM are experiments. This means that the textbook makes efforts to highlight students' experience of mathematics in the process of "doing mathematics", to enhance students' sense of mathematics in interesting activities, and to develop students' ability to usie information technology in mathematical inquiry. CN-PEP focuses more on the provision of resources for students to read, to think and to explore. And, US-PHM pays attention to the writing activities (13.32%).

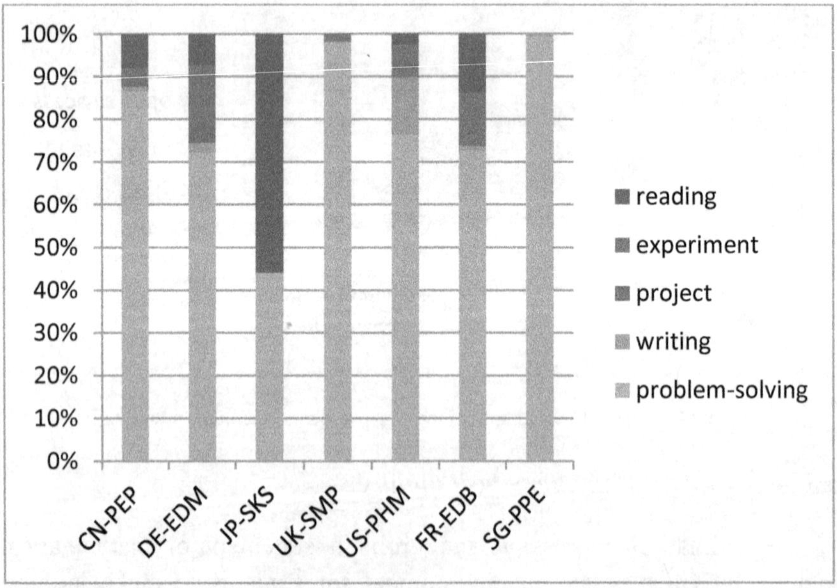

Fig. 6-15 Types of mathematical inquiry activities

Using the index framework, we also found that while paying attention to individual inquiry activities, GE-EDM also emphasizes the cultivation of students' cooperative and teamwork spirits. UK-SMP pays special attention to students' teamwork awareness with all the problems in the inquiry column highlighting team activities. CN-PEP textbooks rarely specify partnership or teamwork requirements in activities with only 1.07% of the problems being clearly marked as "teamwork".

5.4 Features of mathematical inquiry activities

5.4.1 Features of problem-solving activity

In view of the fact that problem-solving dominates in textbooks, we offer a classified comparison of the requirements for problem-solving activities and obtain the following statistical data (Fig. 6-16).

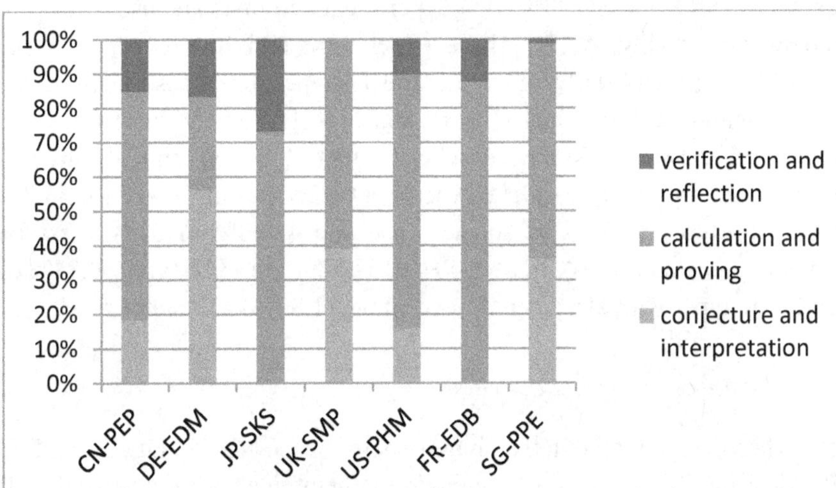

Fig. 6-16 Features of problem-solving activities in mathematical inquiry contents

Except for DE-EDM, calculations and proofs in the mathematically strict sense dominate in the inquiry-problem solving activities. The calculation and proof activities of FR-BM, US-PHM, JP-SKS, CN-PEP and SG-PPE accounts for 87.61%, 73.98%, 73.33%, 66.53% and 62.50% of all problem-solving activities. While emphasizing calculation and proof, the six sets of textbooks also pay attention to students' ability of conjecture, interpretation and simulation. For example, 39.09% and 36.18% of the problem-solving activities in UK-SMP and SG-PPE, highlight conjecture and interpretation abilities. Also, problem-solving activities also focus on enhancing students' ability of verification and reflection; 26.67%, 15.10% and 12.39% of problem-solving activities in JP-SKS, CN-PEP and FR-EDB contribute to this aspect. More than half (56.17%) of the problem-solving activities designed by DE-EDM highlight students' ability to conjecture and interpret. In summary, the problem-solving activities in textbooks have different focal points in the cultivating students' mathematical inquiry ability.

5.4.2 Features of experiment activities

While analyzing the activity type of inquiry in textbooks, we found that except JP-SKS and SG-PPE, rich experiment activities are also offered in the textbooks (Fig. 6-17). These experiment activities cover IT type, scientific type and daily life type. The IT type dominates 100% of the experimental activities in UK-SMP and FR-EDB and 90% in CN-PEP. DE-EDM and US-PHM have a richer content of the experiment activities organized. For example, in the experiment activities of DE-EDM, 43.3% are of IT type, 13.34% of science type and 43.33% of daily life type (mostly game activities). In t US-PHM, 57.72% are of IT type, 40.94% of daily life type, with also a small percentage (1.34%) of science type.

5.4.3 Features of writing activities

When design mathematical inquiry activities, CN-PEP, DE-EDM, US-PHM, FR-EDB and SG-PPE include mathematical writing activities. It can be seen from further analysis that these textbooks have different focal points in writing activities (Fig. 6-18). 100% of the CN-PEP textbooks require that students write on the mathematical processes, FR-EDB emphasizes the writing on mathematical conclusions, and SG-PPE pays attention to the writing on concepts. The DE-EDM emphasized the diversity of writing activities with 20%, 40% and 40% of the writing activities targeting at concepts, conclusions and processes respectively.

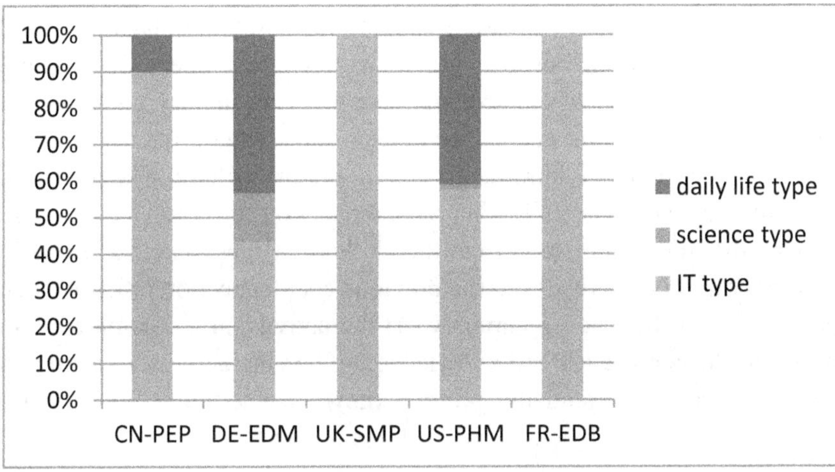

Fig. 6-17 Features of experiment activities in the mathematical inquiry contents

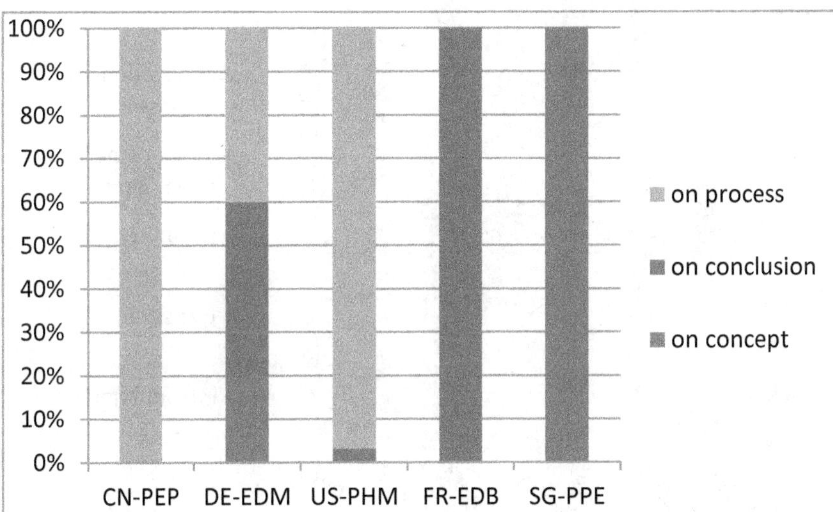

Fig. 6-18 Features of writing activities in the mathematical inquiry contents

5.5 Context interaction

By "context interaction" we mean the interaction between inquiry columns and the context in the textbook. Through investigation one can find that, in terms of the setting of inquiry columns, each textbook not only has its own features in situation expressions, problem statements and activity requirements, but also has its own context interaction design. There are five types of context interaction: 1) an inquiry column only relates to the following chapter/section or knowledge point, playing the role of new knowledge introduction; 2) an inquiry column relates to both the previous and the following chapter/section or knowledge point, playing the role of transition from the old knowledges to new ones; 3) an inquiry column only relates to the previous chapter/section or knowledge point, playing the role of summary; 4) the activity in a column aims to train and deepen the learned knowledge; and 5) the activity in a column aims to apply or extend the present knowledge to other mathematics context or to other subject.

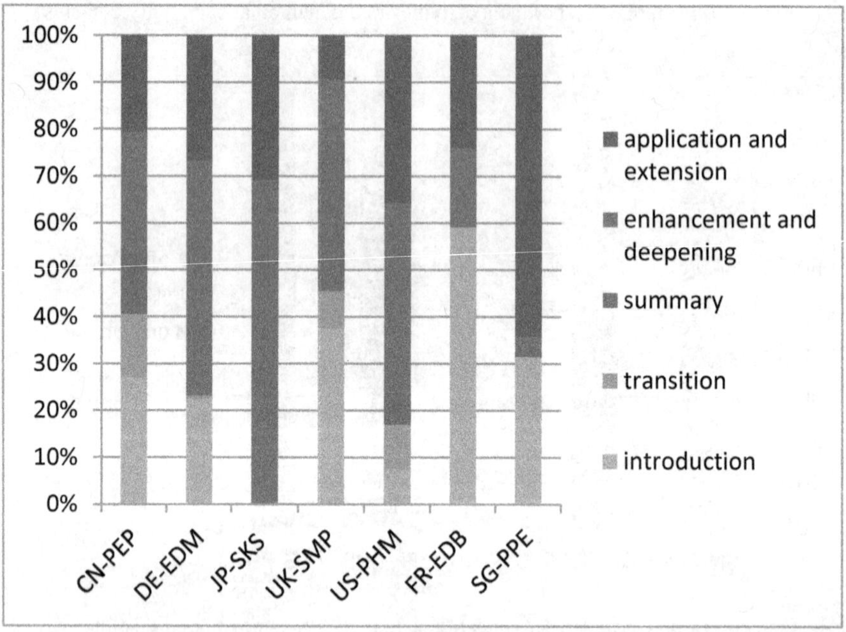

Fig. 6-19 Features of Context interaction of the mathematical inquiry contents

The statistics (Fig. 6-19) shows that Only CN-PEP, UK-SMP and US-PHM have all the five types of context interaction. CN-PEP makes a relatively balanced distribution of the five types, 27.05% being introduction, 13.52% transition, 88% summary, 24.91% enhancement and deepening, and 20.64% application and extension. On one hand, CN-PEP often designs inquiry contents while introduces new mathematics contents, intriguing students to get involved in new knowledge learning during the activities of thinking, analysis and experiment. On the other hand, it also focuses on knowledge enhancement, deepening, application and extension, so inquiry columns may also occur in exercises or the end of a chapter. This feature is also found in DE-EDM, JP-SKS and US-PHM — the three sets of textbooks emphasize the introduction through inquiry problems as well as the enhancement, deepening and applications of knowledge. The column design of DE-EDM pays attention on the introduction of new knowledge rather than summaries of the old knowledge, 22.46% of the inquiry columns being introduction and none for summary. Normally, the inquiry contents of JP-SKS is neither a new knowledge introduction nor a summary of the previous chapter/section. Instead, it highlights the enhancement and deepening of knowledge, so most of the mathematical inquiry contents (61.67%) is arranged in exercises. The inquiry contents of FR-EDB has close interaction with the following chapter/section (59.12% new knowledge introduction), while 64.20% of inquiry columns in SG-PPE devotes to application and extension. These two sets of textbooks do not place the inquiry contents between chapters or knowledge points, i.e. they have no columns for transition. Although the inquiry columns of UK-SMP may appear in various positions, most of them serve as new knowledge introduction (37.56%) or old knowledge summaries (31.92%).

6. Summary

6.1 Features of mathematical inquiry contents design

According to the above data and analysis, the high school textbooks in the seven countries have their own features in the organization and presentation of mathematical inquiry contents. DE-EDM creates real situations for students. Through the design of experiment activities and problems with real backgrounds it gives students opportunities to learn how to collect, identify and process mathematical information, and how to extract mathematical problems. Also, it pays much attention to students' abilities of reflection, conjecture and interpretation. CN-PEP has less percentage of real situations than DE-EDM but higher percentage than JP-SKS and UK-SMP. Also, CN-PEP has a high percentage in the design of open-ended problems. It shows that CN-PEP makes efforts to create real-world and open problem situations so that students may observe, analyze and guess mathematical facts and may experience the process of proposing meaningful mathematical problems. Of course, more mathematical activities need to be designed, thus offering students with more opportunities to develop their innovation awareness and practice ability. JP-SKS contains less mathematical inquiry contents but it has the following character: rich reading materials being offered in order to extend students' vision through reading, or to drive students to discover, propose and solve mathematical problems from the materials. UK-SMP focuses more on the cultivation of students' abilities of assumption, reasoning and interpretation. US-PHM emphasizes the interaction of inquiry columns with the main text of the textbook. The columns are used for introduction of new knowledge, summary of previous knowledge from the inquiry view, as well as applications and extensions of knowledge through inquiry. US-PHM also focuses on the promotion of students' thinking about process. FR-EDB lays emphasis on students' inquiry into related information technology or applications of information technology in mathematics. Influenced by the problem-solving thought of Singapore, SG-PPE pays much attention to the cultivating students' abilities of conjecture and interpretation in the design of inquiry contents.

6.2 Comparison on the explicit features of mathematical inquiry

As can be seen from the study, textbooks in the seven countries have their different features in the design of mathematical inquiry contents, which is helpful for mutual reference. In order to highlight the essence of mathematical inquiry, we select five indexes that directly embody the nature of inquiry for comparison. The indexes are: the proportion of problems of real world situations, the proportion of problems with insufficient mathematical information, the proportion of problems of open type (either open-ended or open process), the proportion of activities of type other than problem-solving, and the proportion of activities of problem-solving requiring conjecture and interpretation in all activities of problem-solving. We get the following statistical chart (Fig. 6-20):

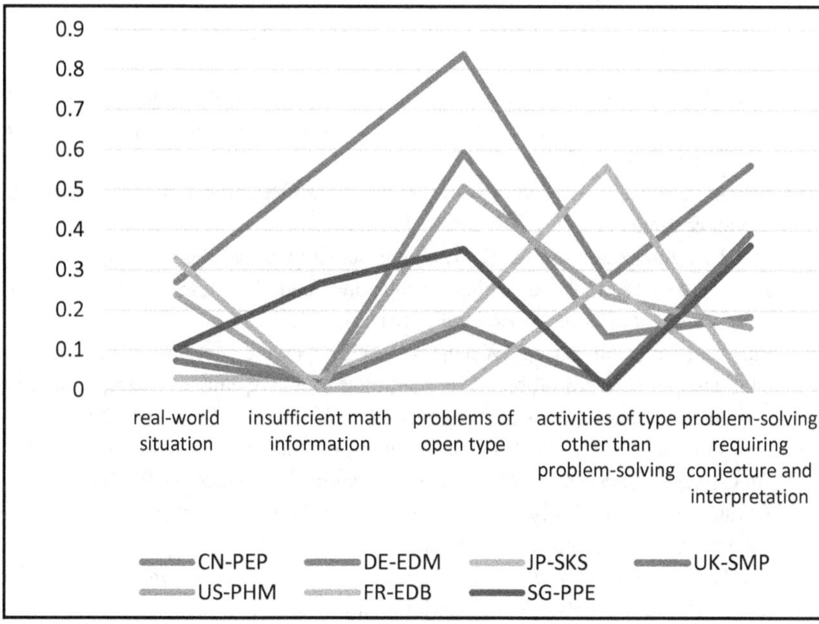

Fig. 6-20 Comparison of indexes of the features of mathematical inquiry

The findings show that among the five indexes, three indexes of DE-EDM are higher than the other textbooks and the other two indexes are also high, showing that DE-EDM embodies most of the essence of inquiry in the inquiry column designed. UK-SMP and JP-SKS have relatively low indexes: UK-SMP only has a high proportion in "activities requiring conjecture and interpretation in all activities of problem-solving", and JP-SKS only occupies a top position in "activities of type other than problem-solving". This shows that the inquiry columns of UK-SMP and JP-SKS do not fully reflect the essence of inquiry. Except for the low proportion of "problems with insufficient mathematical information", CN-PEP has moderate results in other indexes. It shows that the design of inquiry columns in CN-PEP does preferably reflect the essence of inquiry.

References

[1] Ball, D. L. and Bass, H. (2003). Making mathematics reasonable in school. In J. Kilpatrick, W. G. Martin, and D. Schifter (eds). *A Research Companion to Principals and Standards for School Mathematics.* Reston, VA: National Council of Teacher of Mathematics. pp. 27–44.

[2] Chen, Yuelan (2010) The frame structure and the analysis of the latest (2008) edition Japanese *Guidelines for Mathematical Learning* [In Chinese]. *Primary and Secondary Schooling Abroad*, (3): 40–49.

[3] Ernest, P. (1991). *The Philosophy of Mathematics Education.* London: Falmer Press, Taylor & Francis Inc.

[4] Fan, Lianghuo and Zhu, Yan (2007). Representation of problem-solving procedures: A comparative look at China, Singapore, and US mathematics textbooks. *Educational Studies in Mathematics*, 66(1): 61–75.

[5] Fu, Yingfang and Zhang, Weizhong (2007). On the cultural characteristics of situations in the word problems in middles school mathematics textbooks of China and UK [In Chinese]. *Primary and Secondary Schooling Abroad*, (2): 29–32.

[6] Guo, Yufeng (2006). Analysis and thinking on Mathscape textbooks of Australia [In Chinese]. *Curriculum, Teaching Material and Method*, 26(1): 92–96.

[7] Lakatos, I. (1976). *Proofs and Refutation: The Logic of Mathematical Discovery.* Cambridge, UK: Cambridge University Press.

[8] Leuders, T. (2003). *Mathematik-Didaktik: Praxishandbuch für die Sekundarstufe I und II.* Berlin: Cornelsen Verlag Scriptor GmbH & Co.

[9] MOE of PRC (Ministry of Education of the People's Republic of China) (2003). *Mathematics Curriculum Standards for High Schools (Experimental)* [In Chinese]. Beijing: People's Education Press.

[10] Mullis, I. V. S. Martin M. E., Ruddock, G. J., O'Sullivan, C. Y. and Preuschoff, C. (2009). *TIMSS 2011 Assessment Frameworks.* Chestnut Hill, MA: TIMSS & PIRLS International Study Center, 42–46.

[11] Park, K. and Leung, F. K. S. (2006). A comparison of the mathematics textbooks in China, Hong Kong, Japan, Korea, United Kingdom, and the United States, Chapter 2–5 in F.K.S. Leung, K.D. Graf and F.J. Lopez-Real (eds.). *Mathematics Education in Different Cultural Traditions: A Comparative Study of East Asia and the West, the 13th ICMI Study*, pp. 227–238. New York: Springer-Verlag.

[12] Pu, Shuping (2012). Features and implications of mathematics textbooks in French secondary schools [In Chinese]. *Primary and Secondary Schooling Abroad*, (8): 53–59.

[13] Siegel M., Borasi R. and Fonzi J. (1998) Supporting students' mathematical inquiries through reading. *Journal for Research in Mathematics Education*, 29(4), 378–413.

[14] Wang, Jianpan (2011). Comparative study on high school mathematics textbooks in certain countries [In Chinese]. *Curriculum, Teaching Material and Method*, 31(7): 105–106.

[15] Wu, Libao, Song, Weifang and Yang, Fan (2013). Characteristics of the arrangement structure of the USA IM mathematics textbooks and its inspiration [In Chinese]. *Primary and Secondary Schooling Abroad*, (8): 60–64.

[16] Zhang, Wenyu and Fu, Hailun (2011). A comparative study on primary school mathematics textbooks between Singapore and China [In Chinese]. *Studies in Foreign Education*, 38(7): 36–39; 82.

https://doi.org/10.1142/9789814713955_0007

Chapter 7

Comparison of Difficulty Levels of Examples in High School Mathematics Textbooks

Jiansheng Bao[1] and Jianpan Wang[2]

Department of Mathematics, East China Normal University
Shanghai 200241, China
[1]*jsbao@math.ecnu.edu.cn* and [2]*jpwang@admin.ecnu.edu.cn*

Examples in mathematics textbooks play a multi-functional role in mathematics teaching and learning, and the difficulty of examples reflects the difficulty of mathematics textbooks. With six series of high school mathematics textbooks from China, the United States, France, Russia, and Australia as samples, this chapter uses a composite difficulty model to analyze the difficulty levels of examples in the textbooks with respect to 5 factors — the problem context, the mathematical cognition, the computation complexity, the reasoning difficulty, and the topic coverage. The results compare difficulty characteristic of examples in the sample textbooks, which may provide references for the textbook compilation in China.

In view of the overloaded schoolwork of school students in China, the *National Outline for Medium and Long Term Education Reform and Development (2010−2020)* (MOE of PRC, 2010) has required "the content of textbooks shall be adjusted and the difficulty of the curriculum shall be designed in a scientific way". As a result, the difficulty level of textbooks for school students has become a hot topic in international comparison studies. The research group on "International Comparative Study of the Difficulty Level of School Science/mathematics Textbooks", led by Zhenguo Yuan, then the director of China Institute of Education Sciences, has carried out a large-scale international comparison on difficulty levels of textbooks for primary, middle and high schools, respectively. The empirical research arrived at the following conclusions: Among the ten sample countries, the school mathematics textbooks are ranked in the middle in terms of difficulty level, with the primary school textbooks ranked number 4, and textbooks for secondary schools ranked number 3 (Qinghuan Wang, 2014). The announcement of this finding soon attracted a wide attention, and afterwards Xinhua News Network as well as other network, newspapers and periodicals made series of relevant reports.

Obviously, there are a bunch of affecting factors to levels of difficulty, and the research conclusions also depend on the definition of the levels and on the tools used in the assessment. As our contribution to topics of levels of difficulty, this chapter presents part of a national research project conducted by us titled *Comparative Study on High School Mathematics Textbooks in Certain Countries* (see Jianpan Wang, 2011). We will show the analytic results on the composite difficulty level of examples in the mathematical textbooks with a pentagon framework.

Examples play an important role in mathematics teaching and learning. Firstly, examples are the media or concrete means for students to understand the rather abstract mathematical concepts, principles, thoughts and methods, and thus the example-based teaching has become a basic form of mathematics teaching; secondly, examples reflect the basic requirements of a textbook in problem solving, one of the main activities in mathematics learning. In this sense, the difficulty level of examples, to some extent, reflects the difficulty of textbooks.

It should be noted that our aim is not to provide a ranking for countries, but to investigate the features and levels of difficulty factors in textbooks of various countries from the perspective of the examples designed in textbooks.

1. Research Design and Tools

1.1 Sample textbooks

In international comparative studies on mathematics textbooks, the selection of sample textbooks is somehow a difficult task. On the one hand, how widely the textbooks are adopted shall be taken into account, and on the other hand, whether the textbooks conform to the mainstream thinking of curriculum reform in its own country shall also be considered. However, the publication of textbooks in many countries is simply a commercial activity, and the selection of textbooks lacks guidance at the national level; therefore, it is not easy to select a set of representative textbook samples. Our research group, taking into account of both factors mentioned above, together with the publication time of textbooks as well as the convenience in obtaining the textbooks, finally selected the following high school mathematics textbook samples for the comparison, see Tab. 7-1.

Tab. 7-1 Sample Textbooks

Country	Publisher and Year	Textbook	Code
China	People's Education Press, 2009	Mathematics as per the National High School Mathematics Curriculum Standard (Experimental): Version A, Compulsory Course Math 1-5, Optional Course Opt. 2-1, 2-2 and 2-3	CN-PEP
China	Shanghai Education Press, 2007	High School Mathematics Textbooks as per the Shanghai School Mathematics Curriculum Standard: Two volumes for Senior Grade 1, two volumes for Senior Grade 2 and one volume for Senior Grade 3; Extensive II for science students	CN-SEP
USA	Pearson Education, 2009	Prentice Hall Mathematics: Algebra 1, Algebra 2, Geometry	US-PHM
France*	Éditions Belin, 2010 Hachette Livre, 2010	Math 1^{re}s, Maths 2e Déclic Maths Terminale S	FR-EDB FR-HLT
Russia	Просвещение Издательство (Education Press), 2006	МГУ-Школа: Алгебра и начала анализа, класс10, класс 11 (Algebra and Preliminary Analysis, Grade 10, Grade 11); Геометрия, классы10-11 (Geometry, Grades 10-11)	RU-MGU
Australia	Haese & Harris Publications, 2004	International Baccalaureate Diploma Programme, Mathematics for the International Student: Mathematical Studies SL & HL	AU-IBM

*We only found the 1998 edition of the French Belin textbook Math Terminale STT (polytechnic Senior 3 student oriented), which, however, is outdated; and so the Hachette Déclic Maths Terminale S is used as a substitute.

In this study, the examples in textbooks refer to those formally presented in the texts which are given full solutions or analyses leading to solutions. Since each difficulty index of mathematical problems is analyzed according to the solving process, the given solutions to examples, therefore, may ensure the objectivity of analysis to a certain degree. When identifying the specific examples in each series of textbooks, the following rules are adopted:

(1) Examples in CN-PEP and CN-SEP

According to the Chinese traditions of textbook design, we take the mathematical problems marked with "例" or "例题" as "examples" in our investigation. Fig. 7-1 and Fig. 7-2 show some typical examples.

例 2 物理学中的玻意耳定律 $p=\frac{k}{V}$（k 为正常数）告诉我们，对于一定量的气体，当其体积 V 减小时，压强 p 将增大. 试用函数的单调性证明之.

分析：按题意，只要证明函数 $p=\frac{k}{V}$ 在区间$(0, +\infty)$上是减函数即可.

证明：根据单调性的定义，设 V_1，V_2 是定义域$(0, +\infty)$上的任意两个实数，且 $V_1<V_2$，则

$$p(V_1)-p(V_2)=\frac{k}{V_1}-\frac{k}{V_2}=k\frac{V_2-V_1}{V_1V_2}.$$

由 V_1，$V_2\in(0, +\infty)$，得 $V_1V_2>0$；

由 $V_1<V_2$，得 $V_2-V_1>0$.

又 $k>0$，于是

$$p(V_1)-p(V_2)>0,$$

即

$$p(V_1)>p(V_2).$$

所以，函数 $p=\frac{k}{V}$，$V\in(0, +\infty)$是减函数. 也就是说，当体积 V 减小时，压强 p 将增大.

Fig. 7-1 A capture of p. 29, CN-PEP, Compulsory Course textbook Math1

Translation of Fig. 7-1

Example 2. The Boyle's Law $p=\frac{k}{V}$ (k is a positive constant) in physics says that the pressure of a given mass of a gas would increase when its volume is decreased. Try to use the monotonicity of a function to prove it.

Analysis: It suffices to show that the function $p=\frac{k}{V}$ is a decreasing function in the interval $(0,+\infty)$.

Proof: To obtain the monotonicity, arbitrarily choose two real numbers V_1 and V_2 in the domain $(0,+\infty)$ with $V_1<V_2$. Then

$$p(V_1)-p(V_2)=\frac{k}{V_1}-\frac{k}{V_2}=k\frac{V_2-V_1}{V_1V_2}.$$

We obtain $V_1V_2>0$ from $V_1,V_2\in(0,+\infty)$;

And obtain $V_2-V_1>0$ from $V_1<V_2$.

Also, we have $k>0$. It follows that

$$p(V_1)-p(V_2)>0,$$

i.e.

$$p(V_1)>p(V_2).$$

This means that the function $p=\frac{k}{V}$ is a decreasing function in the interval $(0,+\infty)$. That is, the pressure p would increase if the volume V is decreased.

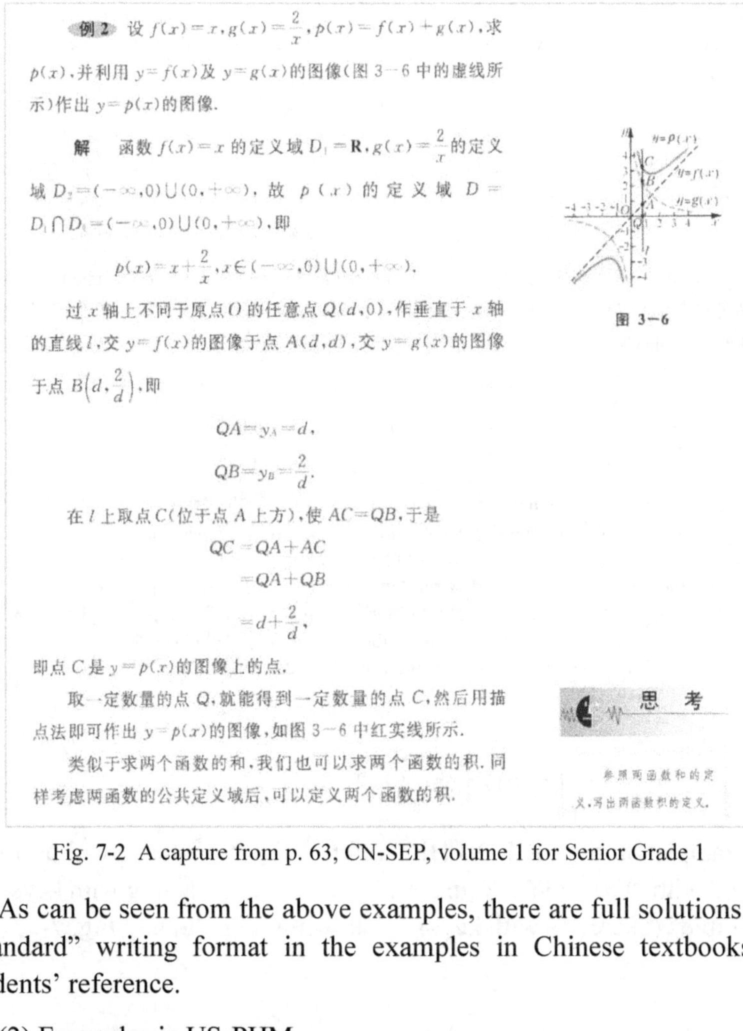

例2 设 $f(x)=x$，$g(x)=\frac{2}{x}$，$p(x)=f(x)+g(x)$，求 $p(x)$，并利用 $y=f(x)$ 及 $y=g(x)$ 的图像（图 3－6 中的虚线所示）作出 $y=p(x)$ 的图像.

解 函数 $f(x)=x$ 的定义域 $D_1=\mathbf{R}$，$g(x)=\frac{2}{x}$ 的定义域 $D_2=(-\infty,0)\cup(0,+\infty)$，故 $p(x)$ 的定义域 $D=D_1\cap D_2=(-\infty,0)\cup(0,+\infty)$，即

$$p(x)=x+\frac{2}{x},x\in(-\infty,0)\cup(0,+\infty).$$

过 x 轴上不同于原点 O 的任意点 $Q(d,0)$，作垂直于 x 轴的直线 l，交 $y=f(x)$ 的图像于点 $A(d,d)$，交 $y=g(x)$ 的图像于点 $B\left(d,\frac{2}{d}\right)$，即

$$QA=y_A=d,$$
$$QB=y_B=\frac{2}{d}.$$

在 l 上取点 C（位于点 A 上方），使 $AC=QB$，于是

$$QC=QA+AC=QA+QB=d+\frac{2}{d},$$

即点 C 是 $y=p(x)$ 的图像上的点.

取一定数量的点 Q，就能得到一定数量的点 C，然后用描点法即可作出 $y=p(x)$ 的图像，如图 3－6 中红实线所示.

类似于求两个函数的和，我们也可以求两个函数的积. 同样考虑两函数的公共定义域后，可以定义两个函数的积.

Fig. 7-2 A capture from p. 63, CN-SEP, volume 1 for Senior Grade 1

As can be seen from the above examples, there are full solutions with "standard" writing format in the examples in Chinese textbooks for students' reference.

(2) Examples in US-PHM

For US-PHM, all mathematical problems marked with "Example" are reckoned as an analysis object, as shown in Fig. 7-3.

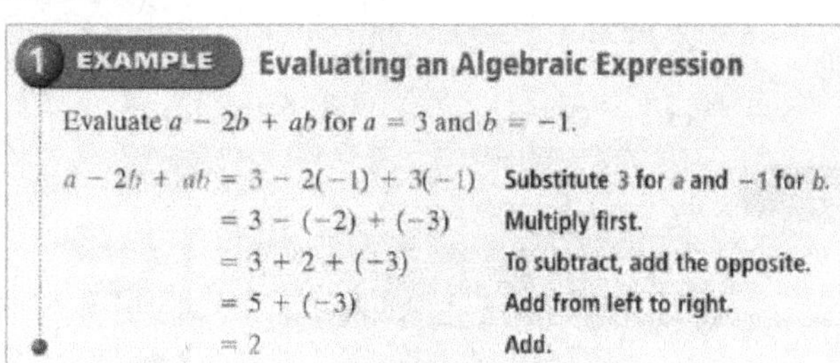

1 EXAMPLE Evaluating an Algebraic Expression

Evaluate $a-2b+ab$ for $a=3$ and $b=-1$.

$a-2b+ab=3-2(-1)+3(-1)$	Substitute 3 for *a* and −1 for *b*.
$=3-(-2)+(-3)$	Multiply first.
$=3+2+(-3)$	To subtract, add the opposite.
$=5+(-3)$	Add from left to right.
$=2$	Add.

Fig. 7-3 A capture from p. 12, US-PHM, Algebra 2

Translation of Fig. 7-2

Example 2. Let $f(x) = x$, $g(x) = \frac{2}{x}$, and $p(x) = f(x) + g(x)$. Find $p(x)$, and draw the graph of $y = p(x)$ based on the graphs of $y = f(x)$ and $y = g(x)$ (as shown by dash lines in Fig. 3-6).

Solution: The domain of function $f(x) = x$ is $D_1 = \mathbf{R}$, and that of function $g(x) = \frac{2}{x}$ is $D_2 = (-\infty, 0) \cup (0, +\infty)$. Thus, the domain of $p(x)$ is $D = D_1 \cap D_2 = (-\infty, 0) \cup (0, +\infty)$. That is

$$p(x) = x + \frac{2}{x_2}, \qquad x \in (-\infty, 0) \cup (0, +\infty).$$

Through any point $Q\ (d, 0)$ other than the origin O, draw the line l perpendicular to the x-axis. The intersection points of l with the graphs of $y = f(x)$ and $y = g(x)$ are $A\ (d, d)$ and $B\left(d, \frac{2}{d}\right)$, respectively. That is

$$QA = y_A = d, \qquad QB = y_B = \frac{2}{d}.$$

Given a point C on line l above the point A with $AC = QB$, then

$$QC = QA + AC = QA + QB = d + \frac{2}{d}.$$

Therefore, C is a point on the graph of $y = p(x)$.

Choose a number of points Q, we obtain a number of points C. Then draw the graph of $y = p(x)$ by point-by-point manner, as shown by red real line in Fig. 3-6.

We can also find the product of two functions in the similar way to find the sum: the domain of the product is also obtained by identifying the common part of the domains of the two functions.

> Thinking:
> With the definition of the sum of two functions as reference, write down the definition of the product of two functions.

(3) Examples in FR-EDB and FR-HLT

Examples in the FR-EDB/HLT refer to those mathematical problems marked with "Example", "Capaceté" (competence training with keys) and "Exemple" (exercises with keys), as shown in Fig. 7-4 and Fig. 7-5.

***Exemple* :**

OAB est un triangle équilatéral direct,

donc : $(\overrightarrow{OA}, \overrightarrow{OB}) = \frac{\pi}{3}\ [2\pi]$.

D'où : $\cos(\overrightarrow{OA}, \overrightarrow{OB}) = \cos\left(\frac{\pi}{3}\right) = \frac{1}{2}$

et $\sin(\overrightarrow{OA}, \overrightarrow{OB}) = \sin\left(\frac{\pi}{3}\right) = \frac{\sqrt{3}}{2}$.

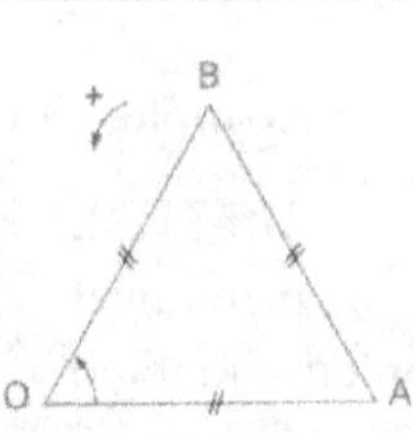

Fig. 7-4 A capture of p. 251, FR-EDB, Math 1res

Translation of Fig. 7-4

Example:
OAB is an equilateral triangle,
so: $(\overrightarrow{OA}, \overrightarrow{OB}) = \frac{\pi}{3} \quad [2\pi]$.

Thus: $\cos(\overrightarrow{OA}, \overrightarrow{OB}) = \cos\left(\frac{\pi}{3}\right) = \frac{1}{2}$

and $\sin(\overrightarrow{OA}, \overrightarrow{OB}) = \sin\left(\frac{\pi}{3}\right) = \frac{\sqrt{3}}{2}$.

Capacité 4 **Donner le sens de variation d'une fonction définie par sa courbe**

Énoncé

→ Voir aussi exercice corrigé 25 p. 57

La courbe suivante représente une fonction *f* définie sur [-3 ; 8].

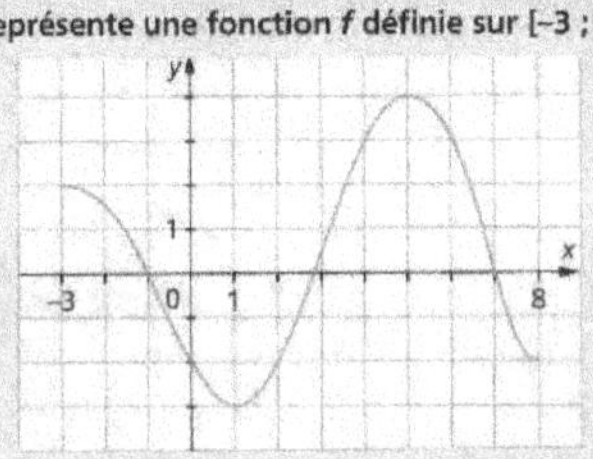

a/ Décrire les variations de *f*.
b/ Dresser le tableau de variations de *f*.

Solution

a/ La fonction *f* est décroissante sur [-3 ; 1], croissante sur [1 ; 5] et décroissante sur [5 ; 8].
b/ Le tableau de variations de *f* est le suivant :

x	-3	1	5	8
Variations de *f*	2 ↘	-3 ↗	4 ↘	-2

Sur l'axe des abscisses
Sur l'axe des ordonnées

Fig. 7-5 A capture of p. 49, FR-EDB, Maths 2^{e}

Translation of Fig. 7-5

Capacity 4: Describe the variation of the function by its curve.
Problem
The following curve represents a function *f* on the interval $[-3; 8]$.
(Figure omitted)

See also Exercise 23, p. 57

a/ Describe the variations of *f*
b/ Draw the table of the variations of *f*.

Solution

a/ The function *f* is monotonically decreasing on $[-3; 1]$, monotonically increasing on $[1; 5]$, and monotonically decreasing on $[5; 8]$.
b/ The table of the variations of *f* is as follows:

x	−3	1	5	8
Variations of *f*	2 ↘	−3 ↗	4 ↘	−2

On the abscissa axis
On the ordinate axis

Such mathematics problems are presented with questions and answers and thus have similarity with examples in Chinese textbooks.

Except for the above-mentioned two types of "examples", there is a column titled "Exercices Résolus" (Exercises Resolved) in the French textbooks, in which problems are usually accompanied with solutions and hints on methods if necessary, as shown in Fig. 7-6. We also consider such problems as examples.

exercices résolus

1 Calculs de limites de fonctions

Déterminer, si elle existe, la limite des fonctions suivantes :

a) $f : x \mapsto \dfrac{x^2+4x}{3x-1}$ en $+\infty$; b) $g : x \mapsto \sqrt{x^2-2x}-x$ en $+\infty$;

c) $k : x \mapsto \dfrac{\cos x - \sin x}{4x-\pi}$ en $\dfrac{\pi}{4}$; d) $m : x \mapsto \dfrac{x}{2-\cos x}$ en $-\infty$.

• Méthodes

Si les théorèmes sur les limites (somme, produit, quotient, composée) ne permettent pas de conclure, on peut :

- repérer les termes prépondérants dans l'expression algébrique de la fonction ou utiliser un résultat remarquable ;
- utiliser un changement de variable ; par exemple $\lim\limits_{x\to a} f(x) = \lim\limits_{h\to 0} f(a+h)$;
- comparer la fonction f à une fonction dont on connaît la limite et utiliser un théorème de comparaison.
- conjecturer la limite ℓ de la fonction f, puis montrer que $\lim\limits_{x\to a} |f(x)-\ell| = 0$.

a) En $+\infty$, x^2 « l'emporte » sur $4x$ et $3x$ « l'emporte » sur -1 donc $f(x)$ à la même limite que $\dfrac{x^2}{3x}$.

On obtient donc $\lim\limits_{x\to+\infty} \dfrac{x^2+4x}{3x-1} = \lim\limits_{x\to+\infty} \dfrac{x}{3} = +\infty$.

b) À l'aide d'une calculatrice ou d'un tableur, on conjecture que la limite de g en $+\infty$ est -1.

On a $|g(x)-(-1)| = |\sqrt{x^2-2x}-(x-1)|$
et en utilisant l'expression conjuguée :

$$|g(x)-(-1)| = \frac{1}{|\sqrt{x^2-2x}+x-1|}.$$

Comme x^2 « l'emporte » sur $-2x$,

$$\lim_{x\to+\infty} (x^2-2x) = +\infty$$

et, par composée, $\lim\limits_{x\to+\infty} \sqrt{x^2-2x} = +\infty$.

Donc, par somme :

$$\lim_{x\to+\infty} (\sqrt{x^2-2x}+x-1) = +\infty,$$

donc $\lim\limits_{x\to+\infty} |g(x)-(-1)| = 0$.

En conséquence, $\lim\limits_{x\to+\infty} g(x) = -1$.

c) On pose $x = \dfrac{\pi}{4}+h$. On obtient, en utilisant les formules d'addition :

$$k\left(\frac{\pi}{4}+h\right) = \frac{\cos\left(\frac{\pi}{4}+h\right) - \sin\left(\frac{\pi}{4}+h\right)}{4h}$$

$$= \frac{-\sqrt{2}\sin h}{4h} = -\frac{\sqrt{2}}{4}\left(\frac{\sin h}{h}\right).$$

Comme $\lim\limits_{h\to 0} \dfrac{\sin h}{h} = 1$, on obtient :

$\lim\limits_{h\to 0} k\left(\dfrac{\pi}{4}+h\right) = -\dfrac{\sqrt{2}}{4}$, donc $\lim\limits_{x\to\frac{\pi}{4}} k(x) = -\dfrac{\sqrt{2}}{4}$.

d) Pour tout réel x, $-1 \leqslant \cos x \leqslant 1$, donc :

$$1 \leqslant 2-\cos x \leqslant 3.$$

La fonction m est donc définie sur $\mathbb{R}$.
Pour tout réel x négatif, on a :

$$\frac{1}{3} \leqslant \frac{1}{2-\cos x} \leqslant 1, \quad \text{d'où} \quad \frac{1}{3}x \geqslant m(x).$$

Comme $\lim\limits_{x\to-\infty} \dfrac{1}{3}x = -\infty$, on en déduit que :

$\lim\limits_{x\to-\infty} m(x) = -\infty$ par un théorème de comparaison.

Fig. 7-6 Capture of p. 44, FR-HLT

Translation of Fig. 7-6

1 Calculation of function limits

Determine, if there exist, the limits of the following functions:

a) $f\colon x \mapsto \frac{x^2+4x}{3x-1}$ at $+\infty$; **b)** $g\colon x \mapsto \sqrt{x^2-2x} - x$ at $+\infty$;

c) $k\colon x \mapsto \frac{\cos x - \sin x}{4x-\pi}$ at $\frac{\pi}{4}$; **d)** $m\colon x \mapsto \frac{x}{2-\cos x}$ at $-\infty$.

Methods

If the theorems on limits (sum, product, quotient, and composite) do not apply, one can try

- To identify the dominant terms in the algebraic expression of the function or to use a well-known result;
- To use the change of variable, for example $\lim_{x\to a} f(x) = \lim_{h\to 0} f(a+h)$;
- To compare the function f to a function whose limit is conceded and to use the theorem of comparison;
- To conjecture the limit l of the function f, then to show that $\lim_{x\to a}|f(x)-l| = 0$.

a) At $+\infty$, x^2 "wins" $4x$ and $3x$ "wins" -1, so $f(x)$ has the same limit as $\frac{x^2}{3x}$.

Thus $\lim_{x\to+\infty} \frac{x^2+4x}{3x-1} = \lim_{x\to+\infty} \frac{x}{3} = +\infty$.

b) Using a calculator on a spreadsheet, one conjectures that the limit of g at $+\infty$ is -1.

We have

$|g(x)-(-1)| = \left|\sqrt{x^2-2x} - (x-1)\right|$,

and use the conjugated expression:

$$|g(x)-(-1)| = \frac{1}{\left|\sqrt{x^2-2x} + x - 1\right|}.$$

Since x^2"wins" $-2x$,

$$\lim_{x\to+\infty} (x^2-2x) = +\infty;$$

take square root, $\lim_{x\to+\infty} \sqrt{x^2-2x} = +\infty$.

By sum, $\lim_{x\to+\infty} (\sqrt{x^2-2x} + x - 1) = +\infty$,

so $\lim_{x\to+\infty} |g(x)-(-1)| = 0$.

In short, $\lim_{x\to+\infty} g(x) = -1$.

c) Let $x = \frac{\pi}{4} + h$. Using the addition formulas, we obtain:

$$k\left(\frac{\pi}{4}+h\right) = \frac{\cos\left(\frac{\pi}{4}+h\right) - \sin\left(\frac{\pi}{4}+h\right)}{4h}$$

$$= \frac{-\sqrt{2}\sin h}{4h} = \frac{\sqrt{2}}{4}\left(\frac{\sin h}{h}\right).$$

As we get $\lim_{h\to 0} \frac{\sin h}{h} = 1$, we obtain:

$\lim_{h\to 0} k\left(\frac{\pi}{4}+h\right) = -\frac{\sqrt{2}}{4}$, so $\lim_{x\to\frac{\pi}{4}} k(x) = -\frac{\sqrt{2}}{4}$.

d) For any real x, $-1 \le \cos x \le 1$, then:

$$1 \le 2 - \cos x \le 3.$$

The function m is, therefore, defined on $\mathbb{R}$.

For any negative real number x, we have

$\frac{1}{3} \le \frac{1}{2-\cos x} \le 1$ from where $\frac{1}{3}x \ge m(x)$.

Since $\lim_{x\to-\infty} \frac{1}{3}x = -\infty$, we deduce that:

$\lim_{x\to-\infty} m(x) = -\infty$ by the theorem of comparison.

(4) Examples in RU-MGU

In our study, samples are those mathematical problems marked with "Пример" (example). For example:

ПРИМЕР 2. Вычислим приближенно интеграл $\int_0^{\frac{\pi}{2}} \sin x\, dx$.

Для этого разобьем отрезок $\left[0;\ \frac{\pi}{2}\right]$ на 10 равных частей точками $0<\frac{\pi}{20}<\frac{2\pi}{20}<\ldots<\frac{9\pi}{20}$ и вычислим приближенно интеграл $I=\int_0^{\frac{\pi}{2}} \sin x\, dx$ по формуле (5), воспользовавшись таблицами значений синуса:

$$I\approx\left(\sin\frac{\pi}{20}+\sin\frac{2\pi}{20}+\ldots+\sin\frac{9\pi}{20}+\frac{\sin 0+\sin\frac{\pi}{2}}{2}\right)\cdot\frac{\pi}{20}\approx$$

$$\approx(\sin 9^\circ+\sin 18^\circ+\ldots+\sin 81^\circ+0,5)\cdot 0,1571\approx 0,998.$$

Полученный результат отличается от точного, равного 1, не больше чем на 0,2%.

Fig. 7-7 Capture of p. 174, RU-MGU, Алгебра и начала анализа класс 11

Translation of Fig. 7-7

Example 2. Calculate the integral approximately $\int_0^{\frac{\pi}{2}} \sin x\, dx$.

To do that, we divide the interval $\left[0;\frac{\pi}{2}\right]$ into ten equal parts by points $0<\frac{\pi}{20}<\frac{2\pi}{20}<\cdots<\frac{9\pi}{20}$, and calculate the integral $I=\int_0^{\frac{\pi}{2}} \sin x\, dx$ approximately by formula (5), using the value table of sine:

$$I\approx\left(\sin\frac{\pi}{20}+\sin\frac{2\pi}{20}+\cdots+\sin\frac{9\pi}{20}+\frac{\sin 0+\sin\frac{\pi}{2}}{2}\right)\cdot\frac{\pi}{20}$$

$$\approx(\sin 9^\circ+\sin 18^\circ+\cdots+\sin 81^\circ+0.5)\cdot 0.1571\approx 0.998.$$

The obtained result is different from the exact value, which equals to 1, with an error not larger than 0.2%.

Note: the content in this chapter is the approximate calculation of definite integral. In the beginning, it introduces how to divide the intervals in a detailed manner, then how to make approximation for integrals, and at last, an intuitive explanation is presented by using the areas of trapezoids with curved edges.

(5) Examples in AU-IBM

Examples are those mathematical problems marked with “Example”, as shown in Fig. 7-8.

Example 15

Let $A = \{1, 2, 3, 4\}$ and define a relation R by: $xRy \Leftrightarrow x + y$ is even.

a Show that R is an equivalence relation. b Find the equivalence classes.

a *Reflexive:* $x + x = 2x$
But $2x$ is even for all $x \in A$ so, xRx for all $x \in A$

Symmetric: If xRy then $x + y$ is even.
Now $x + y = y + x$ for all $x, y \in A$
$\Rightarrow$ $y + x$ is also even $\Rightarrow$ yRx also
i.e., if xRy, then yRx

Transitive: Suppose xRy and yRz
Then $x + y$ is even and $y + z$ is even.
i.e., $x + y = 2m$ and $y + z = 2n$ where $m, n \in \mathbb{Z}$
$\Rightarrow$ $x + y + y + z = 2m + 2n$
$\Rightarrow$ $x + 2y + z = 2m + 2n$
$\Rightarrow$ $x + z = 2m + 2n - 2y$
$\Rightarrow$ $x + z = 2(m + n - y)$

But as $m, n, y \in \mathbb{Z}$ $m + n - y \in \mathbb{Z}$ also
$\therefore$ $x + z$ is even i.e., if xRy and yRz then xRz

Fig. 7-8 Capture of p. 125, AU-IBM, Mathematical Studies HL

From Tab. 7-2, US-PHM gives the most examples, and the next is AU-IBM and then RU-MGU; there is no much difference in the two series of Chinese textbooks; and FR-EDB/HLT provides the least examples, which are only about one quarter of those in US-PHM or one third in AU-IBM. It is definitely that the statistical data is only on the text of the textbooks. While in actual teaching, the Chinese teachers, according to the specific requirements of class, will use a large number of examples from supplemented teaching materials.

Tab. 7-2 Total number of examples in sample textbooks

Country	Textbook Code	Number of examples
China	CN-PEP	458
	CN-SEP	548
USA	US-PHM	1,155
France	FR-EDB/HLT	301
Russia	RU-MGU	634
Australia	AU-IBM	907

1.2 Model of overall difficulty

In order to facilitate the comparison of textbook examples, Jiansheng Bao (2002) built a "pentagon model", which is often used to compare the overall difficulty of mathematical problems in recent years such as Lili Hu (2008), Wenjun Gao and Jiansheng Bao (2009), Jian Miao (2009), Weizhong Zhang and Lihong Huang (2009), Songyu Jin (2012), Yuming Zhao (2013).

Based on the original model, the research group firstly makes a further improvement on the difficulty factors, levels, index system as well as coding scheme, and conducts a verification over the content validity, structure validity and reliability of the relevant analytical tools with expert authentication and statistical analysis. Secondly, the basic units and scope of difficulty analysis are determined so as to ensure the comparability and fairness between different countries. With the established index system, all the textbooks are carefully reviewed to collect relative statistical data, and exceptional cases are also interpreted. Finally, the statistical data are analyzed from teaching perspective and background factors to obtain conclusions.

The revised model is shown in Tab. 7-3, including five factors, namely Context, Mathematical cognition, Computation, Reasoning and Topic coverage. Among them, the factor of Mathematical cognition, which replaces the original factor Discovery in Jiansheng Bao (2002), is adopted from the refined results of Qingpu Experiment (Dan Liu *et al.*, 2007). The other four difficulty factors, their levels, and the secondary indexes still follow Jiansheng Bao (2002).

Tab. 7-3 Difficulty factors and levels of mathematical examples

Level / Factor	1	2	3	4
A: Context	None (A_1)	Personal life (A_2)	Public affair (A_3)	Scientific situation (A_4)
B: Mathematical cognition	Operational memory (B_1)	Conceptional memory (B_2)	Explaining comprehension (B_3)	Exploring comprehension (B_4)
C: Computation	None (C_1)	Number computation (C_2)	Simple symbolic computation (C_3)	Complex symbolic computation (C_4)
D: Reasoning	None (D_1)	Simple reasoning (D_2)	Complex reasoning (D_3)	
E: Topic coverage	Single topic (E_1)	Two topics (E_2)	Three and above (E_3)	

Here, we give detailed descriptions of the factor of Mathematical cognition, involving its levels and secondary indexes.

Level B_1, Operational memory: be able to do basic calculation or routine operations of basic elements on the mathematical problems (including basic construction of geometric figures), adhering to the required procedure or methods of the textbooks. The secondary indexes include:

— Numerical calculation and simple symbolic operation, such as reduction of fractions to a common denominator, rationalization of denominators, and simple factorization;

— Solving linear equations and quadratic equations with one unknown, and simultaneous linear equations with two unknowns, all with numerical coefficients; and

— Routine geometric construction, such as bisecting an angle, or drawing the perpendicular bisector of a line segment.

Level B_2, Conceptional memory: be able to recall or retell the definitions, concepts, propositions, rules, expressions or mathematical facts in consistence with the textbooks or almost in the same way; or present the mathematical facts, terminologies or basic concepts learned. The secondary indexes include:

— Memorizing the basic mathematical facts, such as 144 is the square of 12, {3, 4, 5} is a Pythagorean triple, 17 is a prime number;

— Identifying the basic measurement units, such as knowing that 1 meter equals to 100 centimeters, 1 square meter equals to 100 square decimeters; 1 kilogram equals to 1,000 grams;

— Memorizing the basic rules, formulas, relations and procedures, such as the formulas of triangle area, the relation of opposite numbers, rooting formula of quadratic equation with one unknown; and

— Understanding the meaning of basic concepts, memorizing the mathematical concepts and their related notation, definitions, and basic properties, such as knowing isosceles triangles having two equal sides and angles, and the graph of a linear function being a straight line.

Level B_3, Explaining comprehension: not only to retell, but be also able to understand the concepts, principles, rules and the meaning of mathematical structures, thus be able to solve regular problems according to the textbook examples; be able to transform problems into various forms; be able to compare and analyze the similarities and differences of conditions and types of various variations of ordinary problems; and be able to track the reasoning processes according to the given relations. The secondary indexes include:

— Explaining the complex or compound concepts, propositions, principles, rules and mathematical structures in their own words;
— Establishing relations useful for the development of meaningful understanding among various forms of expression;
— Transforming of question forms, i.e., to transform a question from one form into another form, in particular to a familiar form;
— Making type identification, and suitably selecting mathematical knowledge and methods according to the question types, so that to solve problems flexibly with procedural knowledge;
— Reading, interpreting and solving regular questions, i.e., being able to read and understand the questions, and to solve or explain a question according to the given examples;
— Being able to complete the reasoning process based on the understanding of given relations, and being able to conduct inductive reasoning and informal deductive reasoning; and
— Making comparison and establishing relations, being able to make comparison among some concepts, figures, properties and solutions when the context or conditions of the problems changed, and being able to establish meaningful relations between variations as well.

Level B_4, Exploring comprehension: be able to analyze and creatively solve problems that are unseen and non-routine, to integrate the analytical processes as a whole, and to make value judgement for the solving procedures or the way approaching to the solution of problems.

— Discovering and forming appropriate mathematical problems, i.e. recognizing mathematical elements, relations or structures in various contexts, and then posing appropriate mathematical problems;

- Solving non-routine and open-ended mathematical problems;
- Making mathematical conjectures and building models, i.e. analyzing the main relationship or the key procedures between the conditions and conclusions, and proposing assumptions or preliminary mathematical models;
- Specializing and generalizing, i.e. combining all the deconstructed elements and their relations integrally, and extending or specializing the existing mathematical concepts, procedures, properties and propositions as are required by the models; and
- Mathematical reasoning and proving, i.e. drawing conclusions, giving rigorous proof, making the true/false judgement based on logical reasoning, and making the judgements of superior/inferior or complex/simple based on value standards.

Based on the above framework, we analysed each example with respect to the five difficulty factors, and evaluate the level of each factor. For a series of textbooks, we calculate the weighted mean values of a factor i for all the examples by the following formula

$$d_i = \frac{1}{n}\sum_j n_{ij}d_{ij} \quad (i = 1,2,3,4,5;\ j = 1,2,\ldots)$$

where d_{ij} refers to the j level weight of the i difficulty factor, and, in this study, we directly use the rank as the weight, i.e., $d_{ij} = j$; the number n_{ij} indicates the number of examples at the j level with respect to the ith difficulty factor in this series of textbooks; and n is the total number of samples (examples) in this series of textbooks. Obviously, $n = \sum_j n_{ij}$, for any i.

$$n = \sum_j n_{ij}, \qquad \text{for any } i.$$

2. Results of the Study

The data collected from the six series of textbooks with the weighted mean values are shown in Tab. 7-4.

Tab. 7-4 Quantitative indicators of difficulty factors

Factor	Level	Questions	CN-PEP	CN-SEP	FR-EDB/HLT	US-PHM	AU-IBM	RU-MGU
Background	A_1	Number	346	463	263	799	750	601
		Percentage	75.55%	84.49%	87.38%	69.12%	82.69%	94.79%
	A_2	Number	48	28	24	159	50	15
		Percentage	10.48%	5.11%	7.97%	13.75%	5.51%	2.37%
	A_3	Number	30	47	12	172	32	9
		Percentage	6.55%	8.58%	3.99%	14.88%	3.53%	1.42%
	A_4	Number	34	10	2	26	75	9
		Percentage	7.42%	1.82%	0.66%	2.25%	8.27%	1.42%
	Weighted mean		1.46	1.28	1.18	1.50	1.37	1.10
Mathematical cognition	B_1	Number	39	88	90	171	81	119
		Percentage	8.52%	14.96%	29.9%	14.81%	8.93%	18.77%
	B_2	Number	126	115	10	394	347	210
		Percentage	27.51%	19.89%	3.32%	34.11%	38.26%	33.12%
	B_3	Number	205	325	187	570	398	245
		Percentage	44.76%	58.03%	62.13%	49.35%	43.88%	38.64%
	B_4	Number	88	20	14	20	81	60
		Percentage	44.76%	58.03%	62.13%	1.73%	8.93%	9.46%
	Weighted mean		2.75	2.57	2.42	2.38	1.71	2.39
Computation	C_1	Number	124	94	100	316	101	90
		Percentage	27.07%	17.15%	33.22%	27.36%	11.10%	14.20%
	C_2	Number	131	135	66	392	118	114
		Percentage	28.60%	24.64%	21.93%	33.94%	13.08%	17.98%
	C_3	Number	139	246	70	362	272	208
		Percentage	47.82%	59.85%	41.20%	31.34%	30.00%	32.81%
	C_4	Number	64	73	65	85	416	222
		Percentage	13.97%	13.32%	21.59%	7.36%	45.82%	35.02%
	Weighted mean		2.31	2.54	2.33	2.19	3.11	2.89
Inference	D_1	Number	162	196	113	759	375	326
		Percentage	35.37%	35.77%	37.54%	65.71%	41.32%	51.42%
	D_2	Number	219	328	124	387	423	220
		Percentage	47.82%	59.85%	41.20%	33.51%	46.59%	34.70%
	D_3	Number	77	24	64	9	109	88
		Percentage	16.81%	4.38%	21.26%	0.78%	12.09%	13.88%
	Weighted mean		1.81	1.69	1.84	1.35	1.71	1.63
Knowledge combination	E_1	Number	173	312	79	927	454	250
		Percentage	37.77%	56.93%	26.25%	80.26%	50.06%	39.43%
	E_2	Number	145	196	148	212	346	194
		Percentage	31.66%	35.77%	49.17%	18.35%	38.15%	30.60%
	E_3	Number	140	40	74	16	107	190
		Percentage	30.57%	7.30%	24.58%	1.39%	11.80%	29.97%
	Weighted mean		1.93	1.50	1.98	1.21	1.62	1.91

Next, for each difficulty factor, we compare the difficulty level distributions of examples for the six series of textbooks, and do quantitative analysis and qualitative description on the difficulty features. Finally, we compare the six in their composite difficulty level.

2.1 Differences in context factor

The statistical results on context levels are shown in Fig. 7-9.

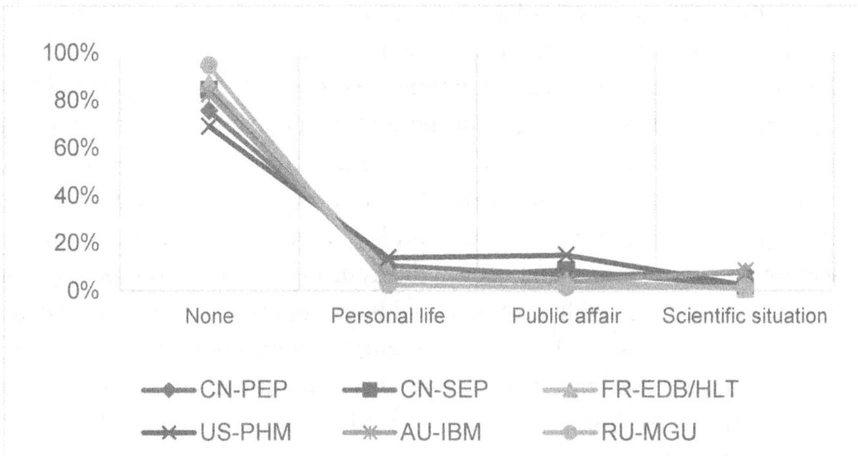

Fig. 7-9 A comparison of context levels

The above figure clearly reflects the similar horizontal distribution of six series of textbooks in terms of context level. Examples without any context account for the vast majority. For example, in CN-SEP Volume Two for Grade Senior Two, there are only five examples with contexts among all 90 examples. Overall, the percentage of examples without any context in RU-MGU, FR-EDB/HLT, CN-SEP, and AU-IBM is slightly higher than those in the other two. US-PHM has more examples with contexts, among which nearly 15% are associated with personal life and public affair but with rather low percentage of examples in scientific situation.

It is undeniable that the study has certain weaknesses by only taking examples as the study object to measure the context among the textbooks. In fact, in the textbooks, many questions with context information are not presented in the formal examples. Most textbooks have the inclination to use scenario examples in the introduction and/or inquiry and modeling part. For instance, in CN-PEP, scenario examples are often provided in the columns of “observation”, “inquiry” and “thinking”, while in FR-EDB/HLT, they are often in the chapter introductions and columns, such as “communiqure”.

The role context information plays in mathematics examples varies in different countries.

In the Chinese textbooks, those examples are usually designed as application questions in order to apply mathematical concepts, skills and principles to the practical situations. Situational (or contextualized) examples are generally placed at the end of the sections. Sometimes, it appears as a practical application after examples without contexts. Sometimes, it is a “big” and “difficult” question, requiring comprehensive understanding, exploring or even building models. For example, in the section of sine theorem and cosine theorem in CN-SEP Volume Two of Senior Grade One, eight examples without contexts are first solved by using the theorems, and then two examples with contexts are led out, which illustrate the importance of the theorems in real life, especially in measurement. In Compulsory Course Math 5 of CN-PEP, the same contents are divided into two sections, one section devoting to an introduction of theorems with five examples without contexts, and the other section presenting actual applications of the theorems in measurement. The second type of example setting can be seen in the section of logarithm function, Compulsory Course Math 1 of CN-PEP. In the section, after the introduction of the concept and operation laws of logarithm and four examples without contexts, two situational questions are set up, one explaining the magnitude of the earthquake, and the other one calculating the era of Han Tomb at Mawangdui by mathematical modeling. A similar setting of examples for the same contents appears in CN-SEP, Volume Two of Senior Grade One, where five examples without contexts are served as the direct application of the concepts and operation laws, followed by two situational questions about the GNP and the magnitude of earthquake respectively, from which the learners can understand the significance of logarithmic function and can extract the needed mathematical knowledge from real life. In addition to the two types, there are some sections with different example settings. For instance, in the section of “Probability of Independent Events Intersection” in

Extensive II for Science Students, Senior Grade Three, CN-SEP, six examples related to the public affair or personal life are used for illustration, no examples without contexts. The same content in CN-PEP is treated in the same way. This is mainly because the knowledge of probability and statistics heavily relies on situations, and we are used to using dice, poker cards, and defect rate of products as situational examples to facilitate learners to understand. Other chapters/sections on applications usually provide more situational examples.

While in US-PHM, context information of examples generally has the following functions. (1) Examples are used to introduce knowledge to be learned, appearing at the very beginning of a chapter/section. (2) Examples provide source material to extract mathematical questions for exploration or modelling, usually placed at the end of a chapter/section. (3) Examples serve as direct applications of concepts, theorems or rules, often closely following the introduction of concepts. (4) Examples promote learners to understand mathematical concepts in connections with practical situations, look for answers fitting in the situations, and make reasonable explanations; such examples have various forms, and appear in different places. The first two functions are similar to those in the Chinese textbooks. Here we focus on the latter two.

For example, Section 1-4 in US-PHM *Algebra 1*, which is on functions, contains four examples, all related to production and personal life. The first two require learners to write down function expressions, and the third one relates memory cards and their prices with which learners could understand independent and dependent quantities as well as give a reasonable explanation on the function relation. The fourth question is also related to personal life, asking students to find the relationship between the working time and the income for a baby-sitter, to give the reasonable domain and range of the situation. Through these examples with personal life context, learners could understand the meanings of functions, variables, domain and range, and are able to give suitable answers and reasonable explanations for the answers. Section 10-5 in the textbook *Geometry*, on trigonometry and area, contains three examples, among which the latter two are the practical application of the area formulas of triangles. In Section 6-7 of *Algebra 2*, about permutations and combinations, each of the definitions of factorial, number of permutations, and number of combinations is followed by one or more practical applications. Through above analyses, we can see that US-PHM lays emphasis on the connections of mathematical concepts with and reasonable explanations of the real situations, which allows students to fully appreciate the importance of mathematics in everyday life and to broaden and deepen their understanding of mathematics through comprehending its practical significance rather than dull exercises within mathematics.

In FR-EDB/HLT, traditional contents, such as "Number", "Algebra" and "Geometry", generally contain only pure mathematics questions, while examples with context information are mainly in the module of "Probability and Statistics", closely related to the features of the content. Among them, a lot of examples are related to social life, such as French population, world population, French birth rate, population growth rate, the number of visitors to France, temperature, prison, and desert; Some are related to entertainment life, such as horse racing, card games, lottery tickets, and ball games; Some are related to daily life, such as phone bills, family size, the number of poultry, and the number of fireplaces and rooms. Examples related to economic life and school life are less used, and the least used are examples associated with occupational life. Though few examples are related to scientific contexts, some of them are in fact quite interesting. For example, in Maths 2^{e} of FR-EDB, there is an example about the ideal gas equation (as shown in Fig. 7-10).

In RU-MGU, besides some application questions in topic introduction and exercises, there are few examples in the main text being designed with context information. Most examples are still purely mathematical.

As one series of the textbooks for the International Baccalaureate Program, AU-IBM pay much attention to basic skills training. However, there are plenty of examples associated with real context in the textbooks, with particular focus on interaction with other disciplines, modeling and science-technological situations. Examples related to biology, physics, and chemistry account for a large proportion in the applied questions and most of these questions need mathematical models during the process of solving.

Capacité 6 **Traduire le lien entre deux quantités, en identifiant la variable**

Énoncé

→ Voir aussi exercice corrigé 40 p. 34

L'équation d'état du gaz parfait est : PV = nRT avec P, la pression en pascal (Pa) ; V, le volume en m^3 ; n nombre de moles (quantité de matière en mol) ; R = 8,314 $Pa \cdot m^3 \cdot K^{-1} \cdot mol^{-1}$, la constante du gaz parfait ; et T la température en kelvin (K).

1. On suppose que P = $1{,}013 \times 10^5$ Pa et T = 273,15 K.
a/ Exprimer le nombre n de moles en fonction du volume V.
b/ Préciser la fonction modélisant cette situation.

2. On observe la pression en fonction de la température avec V et n fixés.
a/ Quel paramètre doit-on faire varier pour l'expérience ? En déduire la nouvelle variable associée à cette situation.
b/ Si l'on dresse un tableau de résultats de l'expérience, que met-on dans la 1re ligne (ou colonne) du tableau ? Et dans la 2de ?
c/ Si l'on représente graphiquement les résultats, quelle grandeur place-t-on en abscisses ? Et en ordonnées ?

Fig. 7-10 A capture of p. 25, FR-EDB, Mathe 2e, p. 25

Translation of Fig. 7-10

Capacity 6. Explore the link between two quantities, by identifying the variables.

Problem

See also Exercise 40, p. 34

The state equation of an ideal gas is PV = nRT, where P is the pressure in pascal (Pa), V the volume in m^3, n the number of moles (the amount of material in mol), R = 8.314 $Pa \cdot m^3 \cdot K^{-1} \cdot mol^{-1}$, which is the constant of the ideal gas, and T is the temperature in kelvin (K).

1. Suppose P = 1.013×10^5 Pa and T = 273.15 K.
a/ Express the number n of moles as a function of the volume V.
b/ Precisely describe the function modeling in this situation

2. For fixed V and n, observe the pressure as a function of the temperature.
a/ Which parameters should be varied for the experiment?
b/ If a table of results of the experiment is drawn up, what do we put in the first row (or column) of the table? And what in the second?
c/ If the results are graphically represented, what is the magnitude of the abscissa? And what the other axis?

2.2 Differences in mathematical cognition factor

Statistical results on mathematical cognitive levels are shown in Fig. 7-11.

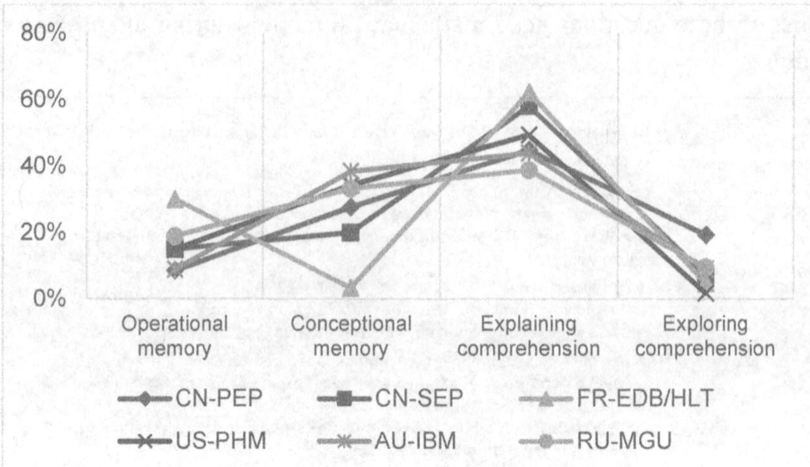

Fig. 7-11 A comparison of mathematical cognitive levels

On the whole, examples at the level of "explaining comprehension" in all mathematical textbooks account for the largest part, followed by examples at the "conceptional memory" level (except for the French textbooks), with examples at the levels of "operational memory" and "exploring comprehension" the fewest, which also reflects the high school textbooks put greater emphasis on learner's understanding of knowledge, rather than simple computation and memory. In particular, the lowest level of "operational memory" is significantly less used in high school textbooks than that in middle school. The distribution of mathematical cognitive levels for the six series of textbooks can be found from the figure. In the two series of Chinese textbooks, only about 35% of the examples are at the two lower levels (on memory), while in US-PHM, this percentage is nearly 50%. Examples at the level of "exploring comprehension" in CN-PEP are about 20%, which is substantially higher than US-PHM with a percentage of 1.73%.

After comparing the specific examples in each textbook, we can see that the Chinese textbooks put more emphasis on high-level examples. Many examples have more variations, integrated with multiple knowledge points, even requiring the practice of modeling. In US-PHM, concepts are directly and widely applied to a broader sense, without any requirement of transferring knowledge and deepening understanding. For example, for the content of equations of straight lines, CN-SEP (Chapter 11) contains 8, 2, 16, and 1 examples at the four levels from low to high, respectively, with the corresponding percentage 30%, 7%, 59%, and 4%; CN-PEP (Chapter 6), has 0, 9, 10, and 2 examples with the percentage 0%, 43%, 48%, and 9%; while US-PHM (Chapter 6) has 1, 19, 13, and 0 examples with the percentage 3%, 58%, 39%, and 0%. In the Chinese textbooks, except for some examples that require learners to use equations in, such as, "point-slope" form and general form to solve the questions by direct substitution, there are many more examples requiring learners' deeper understanding, with particular emphasis on integration with vector and graphic knowledge. Here is an example of this kind in CN-SEP (p. 9, vol. 2 of Senior Grade 2).

例 6 已知在$\triangle ABC$中，$\angle BAC=90°$，点 B、C 的坐标分别为(4,2)、(2,8)，向量 $\vec{d}=(3,2)$，且 $\vec{d}$ 与 AC 边平行，求$\triangle ABC$的两条直角边所在直线的方程.

解 如图 11－7，因为 AC 与向量 $\vec{d}=(3,2)$平行，所以，直线 AC 的点方向式方程是

$$\frac{x-2}{3}=\frac{y-8}{2}.$$

又因为 $AB\perp AC$，所以 $\vec{d}=(3,2)$是直线 AB 的法向量.

所以，直线 AB 的点法向式方程是

$$3(x-4)+2(y-2)=0.$$

因此，$\triangle ABC$的两条直角边所在直线的方程分别是

$$\frac{x-2}{3}=\frac{y-8}{2}\text{和 }3(x-4)+2(y-2)=0.$$

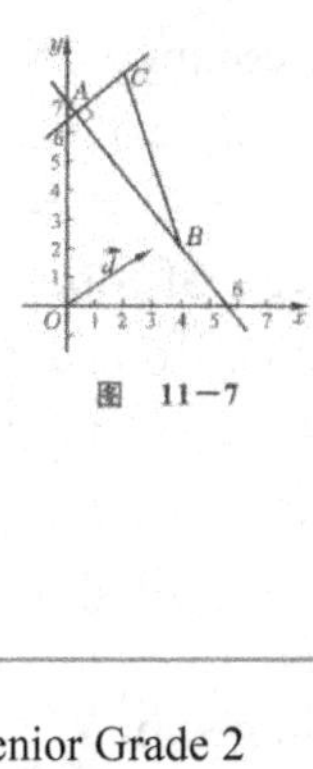

图 11－7

Fig. 7-12 A capture from p. 9, CN-SEP, vol. 2 of Senior Grade 2

This question integrates multiple knowledge points, such as equations of straight lines, vector parallel and perpendicularity, and triangle, belonging to the level of explaining comprehension in the cognitive dimension. In US-PHM, many examples in this part are set to find the intercept or slope of a straight line by a direct substitution, or to draw the graph of a linear equation in the general form, or to draw the line through a given point parallel to a given straight line. Those examples are direct application of formulas, without requirements of in-depth understanding, so they belong to the levels of operational and conceptional memory.

Translation of Fig. 7-12

Example 6. In a triangle ΔABC, $\angle BAC = 90^\circ$, the coordinates of B and C are (4,2) and (2,8), respectively, vector $\vec{d} = (3,2)$, and $\vec{d}$ parallels to line AC. Find the equations of the two straight lines on which the right-angle sides of ΔABC lie.

Fig. 11-7 omitted

Solution: As in fig. 11-7, AC parallels to $\vec{d}$, so the "point-direction" form of the equation of AC is

$$\frac{x-2}{3} = \frac{y-8}{2}.$$

Also, since $AB \perp AC$, $\vec{d} = (3,2)$ must be the normal vector of AB. Thus, the "point-normal" form of the equation of AB is

$$3(x-4) + 2(y-2) = 0.$$

Therefore, the equations of the straight lines on which the right-angle sides of ΔABC lie are

$$\frac{x-2}{3} = \frac{y-8}{2} \text{ and } 3(x-4) + 2(y-2) = 0,$$

respectively.

Another reason for the low mathematical cognitive level in US-PHM lays on the fact that many contents there are learned by Chinese students in their compulsory education stage, and these contents are relatively easy with mainly calculations and simple reasonings involved. For example, in its Chapter 2, *Algebra 1* (rational numbers), the number of examples at the level of operational memory, conceptional memory, explaining comprehension, and exploring comprehension is 23, 10, 4, and 0 with percentage 62%, 27%, 11%, and 0%, respectively. Here is another example: in Chapter 5, *Algebra 2* (quadratic equations and functions), the number of examples at the four levels is 10, 17, 18, and 0, with percentage 22%, 38%, 40%, and 0%, respectively. Similar situations can also be found in geometry textbook, such as Chapter 3 (parallel and perpendicular lines), where the number of examples at the four levels is 4, 19, 12, and 0, with percentage 12%, 54%, 34%, and 0%, respectively. These contents also appear in Chinese middle school textbooks. But US-PHM textbooks pay more attention to the direct use of the knowledge, such as mixed arithmetic operations, opposite numbers and the absolute values of rational numbers, solving quadratic equations by the rooting formula, identifying opposite angles and corresponding angles in accordance with the definition, and so on. Those questions are generally categorized as the first two mathematical cognitive levels, thus making the low-level of this part significantly higher than the latter two levels, which shall have an impact on the final results.

The proportion of examples at the forth level (exploring comprehension) in FR-EDB/HLT is lower than that in the two editions of Chinese textbooks, especially lower by nearly 15 percentage points than that in CN-PEP. This data certainly cannot cover the whole high school curriculum in France, because 1) we lack the data in the extended textbooks for math-science students, and 2) the parts "activités" and "travaux dirigés", which contain many problems at the fourth level, are not taken in to our account. On the other hand, there are a lot of probing questions in the columns of exploration and thinking in the two Chinese editions of textbooks, which are not in our consideration. However, "formal" examples, as the most important part of the textbook, have a pleasing performance in the total number and the distribution among levels have a pleasing performance. It, at least, reflects the progress of curriculum reform.

In the Russian textbook RU-MGU, it takes a relatively "balanced" distribution at the four levels with proportions 18.77%, 33.12%, 38.64%, and 9.46%, from low level to high level. The percentage of examples at the "exploring comprehension" level is high, second only to that in CN-PEP, and on the other extreme, the percentage at the "operational memory", RU-MGU also occupies the second place, next to FR-EDB/HLT. However, at the level of "explaining comprehension", though the percentage is as high as 38.64%, it is the lowest among the textbooks. The fact that more than half of the examples are at the "memory" levels shows that RU-BGU pays greatest attention to basic skill training and basic concept understanding, and, meanwhile, it still provides a good chance for students to develop mathematical ability, in particular, about one tenth of the examples are designed for students' exploration.

Compared with CN-SEP and CN-PEP, AU-IBM puts more emphasis on cultivating students' mathematics basics, such as operational steps, understanding of concepts, and mastery of mathematical formulas. But it is worth mentioning that, the exploration questions in AU-IBM are not the applied questions commonly seen, neither those which can be solved by common strategies. The questions there are not superficial, but are in some depth to inspire students to do mathematics, to explore and to discover. Though the proportion of such questions in AU-IBM is small, many of them are innovative mathematical questions.

2.3 Differences in computation factor

Statistical results of computation levels are shown in Fig. 7-13.

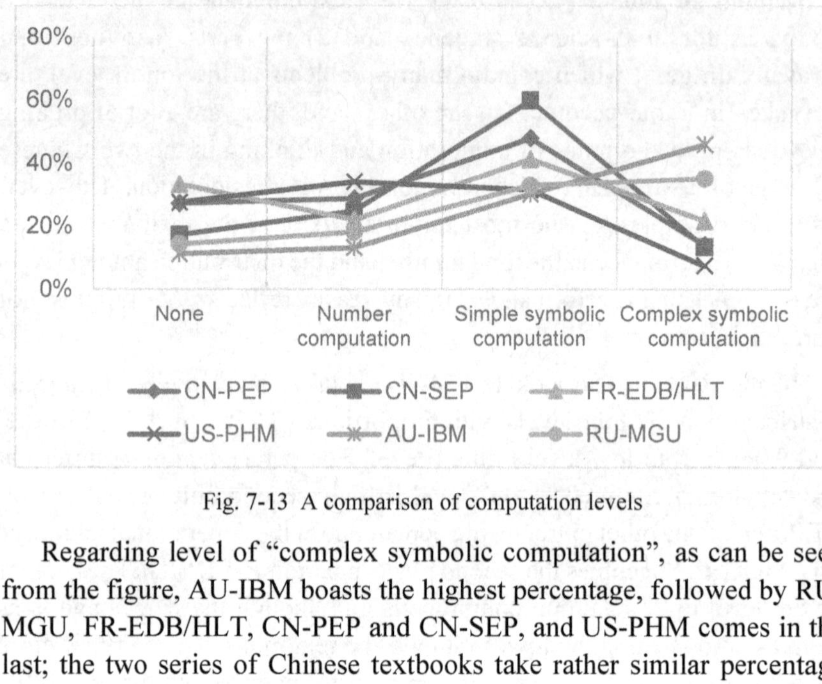

Fig. 7-13 A comparison of computation levels

Regarding level of "complex symbolic computation", as can be seen from the figure, AU-IBM boasts the highest percentage, followed by RU-MGU, FR-EDB/HLT, CN-PEP and CN-SEP, and US-PHM comes in the last; the two series of Chinese textbooks take rather similar percentage (13.32% and 13.50%, respectively), significantly lower than the RU-MGU (35.02%). For "simple symbolic computation", the two series of Chinese textbooks take the highest percentage, while AU-IBM takes the lowest.

As for the level distribution of computation factor within each series of the textbook, the four levels in RU-MGU increase progressively in proportion, with the first two levels close in percentage and the remaining two close to each other as well. In both CN-PEP and CN-SEP, the first three levels increase progressively (the first two levels of CN-PEP are much closer to each other), with simple symbolic computation the highest, and the level of complex symbolic computation has the lowest percentage among all levels.

As can be seen, the emphasis in different versions of textbooks is very different: although CN-PEP and CN-SEP have the similar strategy in level distribution, CN-SEP pays more attention to simple symbolic computation (59.85%); AU-IBM lays more emphasis on two levels of symbolic computation, complex symbolic computation (45.82%) in particular, and the non-computation and number computation take relatively small proportion. Though the overall operation level in RU-MGU is not high, its complex symbolic operation takes high percentage (35.02%).

With regard to the contents of algebra and function in FR-EDB/HLT, there are few examples at the level of number computation, and the other three levels share almost identical proportions. The complex symbolic computation is significantly higher than the two Chinese series, which is because FR-EDB/HLT has more contents and higher requirements in "calculus". For example, sandwich theorem in limit, the mean value theorem for continuous functions, and the notion of the primitive function are all included. As for the contents of geometry, the FR-EDB/HLT give more examples with no computation. This is because most of the examples are concerning geometric proof, geometric construction, and computer-aided construction, etc. (Fig. 7-14).

Capacité 1 **Démontrer que deux droites sont perpendiculaires**

→ *Voir aussi exercice corrigé 3 p. 206*

Énoncé

On considère un triangle ABC quelconque inscrit dans un cercle Γ de centre O.
Le point D est diamétralement opposé au point A sur le cercle Γ et il est distinct de B et C. Le point I est le milieu du segment [AB].
Démontrer que les droites suivantes sont perpendiculaires :
a/ (AB) et (BD) ; b/ (OI) et (AB).

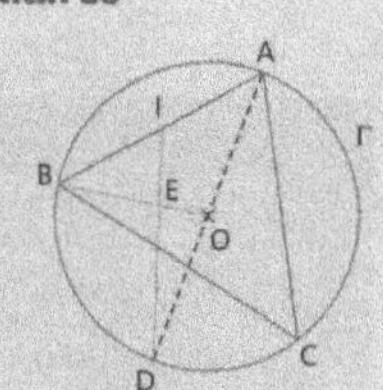

Fig. 7-14 Capture of p. 202, FR-EDB, maths 2e

Translation of Fig. 7-14

See also Exercise 3, p. 206

Capacity 1: Demonstrate that two lines are perpendicular

Problem

Consider a triangle ABC inscribed in a circle Γ with center O.
Point D is the opposite end of the diameter through A, and D is different from B and C. Point I is the midpoint of segment [AB].
Show that the following pairs of lines are perpendicular
a/ (AB) and (BD); b/ (OI) and (AB).

Figure omitted

The biggest difference between AU-IBM and the Chinese textbooks lies in the fact that the two series of Chinese textbooks place more emphasis on Euclidean geometry and the logical relations, whose related contents can be summarized as follows: straight lines and planes in space, conic sections, simple geometric solids, propositions and commonly used logical terms. This design shows that the Chinese textbooks pay much attention to geometry contents and introduction of geometry contents with geometric approach and logical reasoning. This is not the design idea of AU-IBM. Instead, AU-IBM seems prefer to solving geometric problems with algebraic methods, e.g., some examples on equations of straight lines or planes are included in the textbooks, while the solutions are all related to algebra. Such difference in philosophy will result in very different distributions of computational levels.

2.4 Difference in reasoning factor

The statistical results in terms of reasoning levels are shown in Fig. 7-15 below.

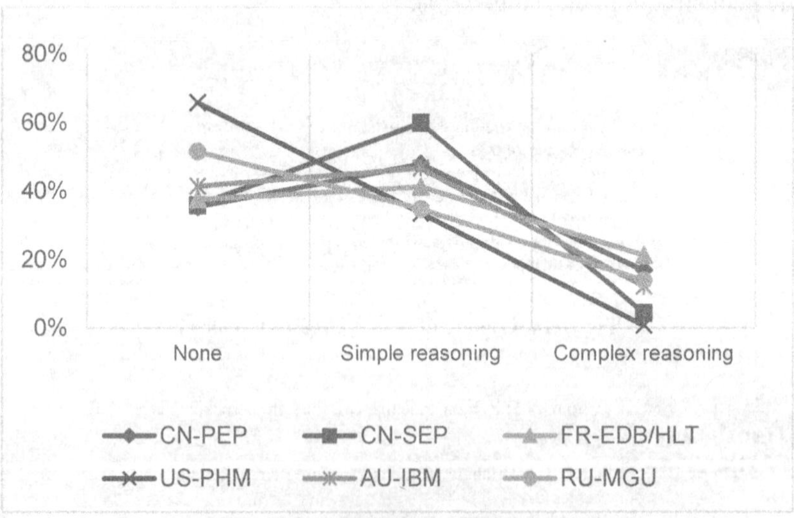

Fig. 7-15 A comparison on the reasoning levels

It is easy for us to find in the figure that both the US and Russian textbooks are almost linearly decreasing from no reasoning to complex reasoning, with two lines intersected nearly at the midpoint — simple reasoning. Nevertheless, the slope of the two lines are quite different, and US-PHM has much higher percentage at the no reasoning level and lower percentage (almost 0) at the complex reasoning level. Therefore, RU-MGU obviously exceeds US-PHM at the overall reasoning level.

The other four series of textbooks have a similar trend and are in the form of "reverse tick" — "simple reasoning" occupies the largest proportion among examples. CN-PEP and CN-SEP almost share the identical percentage in the examples of "no reasoning" (around 35%). The percentage of CN-PEP at the level of "complex reasoning" is 12.43%, much higher than that of CN-SEP (4.38%). Seen from this, the overall reasoning level of CN-PEP is higher than that of CN-SEP. However, FR-EDB/HLT has even higher percentage of examples involving "complex reasoning" than the two series of Chinese textbooks, though the three series have almost the same proportion in examples without requirement on reasoning. This certainly shows that the French textbooks have higher overall reasoning level. Among all the four series, AU-IBM takes the highest percentage at the "no reasoning" level, and occupies the third position at the other two levels. So, its overall reasoning level lies in the middle.

If we go into knowledge domains to have a more detailed investigation, we can find that the examples at the complex reasoning level in CN-PEP are concentrated in Euclidean geometry (plane and solid), functions and modeling, statistics and probability, vectors, sequences, and conic sections. Different from the middle school mathematics, the reasoning and proofs in high school textbooks are not limited to geometry, but also have higher requirements for algebra and functions, such as finding the domain of composition functions and the judgement of the monotonicity of functions. There are a number of examples in this area requiring reasoning and proofs.

The mathematical reasoning in US-PHM mainly appears in geometry. Seldom examples concerning reasoning are observed in the domains of algebra and functions, and statistics and probability. Even in the domain of geometry, many problems are only focused on some simple construction of graphs, such as drawing an angle identical with the known one. There is almost no problem requiring rigorous mathematical proof.

The percentage of examples in FR-EDB/HLT is far greater than both CN-PEP and CN-SEP in terms of "complex reasoning" level, the reason for which may be attributed to one type of special example, i.e., problems in the column "Exercices Résolus". These problems are accompanied with solutions in a standard format, and feature high difficulty involving more reasoning steps. They are often classified as problems at the level of "complex reasoning".

The mathematical reasoning level of examples in the Russian textbooks is only at the middle level among all sample textbooks, which is more or less beyond our expectation; for it can be seen that high requirements on logic reasoning is the Russia's tradition in mathematical education. However, if we go beyond the examples and give a glance to the main text, we can find the reason for the statistics. The Russian textbooks pay much attention to the systematicness and rigorousness of mathematics presented. In the text, theorems are precisely stated, with complete proofs whenever possible. If a proof cannot be provided within the scope of school mathematics, the text usually gives a clear remark. For example, in the *Geometry* textbook of RU-MGU, almost two full pages are devoted to the statement and proof of Euler Theorem on Polyhedron (see pages 62−63 of the textbook). The reader will also be referred to pages 63−64 of the *Algebra* textbook for Grade 11 where one can see the intermediate value theorem of a continuous function with a declaration of "no proof". In a textbook of this characteristic, too many examples with "complex reasoning" seem redundant.

The lower percentage of the level "complex reasoning" of AU-IBM is mainly due to the coverage of the contents together with the preference, as we mentioned before, of solving geometry problems with algebraic approaches. The examples with complex reasoning appear with the knowledge of polynomials, number sequences, complex numbers, mathematical induction, permutations and combinations, probability and statistics, vectors, calculus and applications, etc.

2.5 Difference in topic coverage

The statistical results in terms of topic coverage levels are shown in Fig. 7-16.

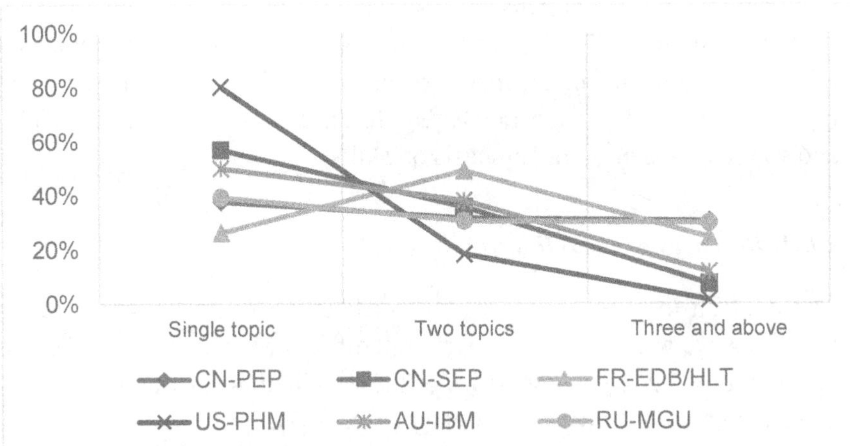

Fig. 7-16 A comparison of the topic coverage levels

In general, the topic coverage of examples in US-PHM is the lowest; the results of CN-SEP and AU-IBM are similar; RU-MGU and CN-PEP are almost identical, both at the highest level; and FR-EDB/HLT has a high topic coverage level as well.

More precisely, only 7.30% of the examples in CN-SEP involve three or more topics, and over 90% of the examples contain one or two topics, in particular, about 56.93% with single topic. The poor appearance of CN-SEP in the topic coverage factor is beyond our expectation. About half (49.17%) of the examples in FR-EDB/HLT involve two topics, while examples at the other two levels almost occupy a quarter, respectively. The percentage of examples at the "two topics" level in French textbooks ranks top 1 among all textbooks. The differences of CN-PEP and RU-MGU in terms of "two topics" level and "three and above" level are only 1.06 and 0.6 percentage points, respectively. Both of these two series of textbooks have a high proportion of examples at the "three and above" level.

One of the reasons why CN-PEP has a high level in topic coverage is that it pays attention to teaching with variation. The examples are generally comprehensive ones, which involve multiple topics and require diversified methods to solve rather than simple one-step solution or routine repetition of similar methods. However, CN-SEP has low percentage in the examples involving three or above topics, which may be related to the requirements in the Shanghai Mathematical Curriculum Standards that, "Materials should reflect the consistency with learning requirements and relevance to students' metal characters and ability level, have fundamental, typical, diversified and acceptable features, and avoid excessive difficulty and complexness or extra emphasis on skills".

2.6 Differences of overall difficulty levels

Based on the comparison of five difficulty factors of the six series textbooks, we will synthesize the overall difficulties with the model given in subsection 1.2. By the statistics of the weighted mean value in Tab. 7-4, we obtain the following radar chart (Fig. 7-17).

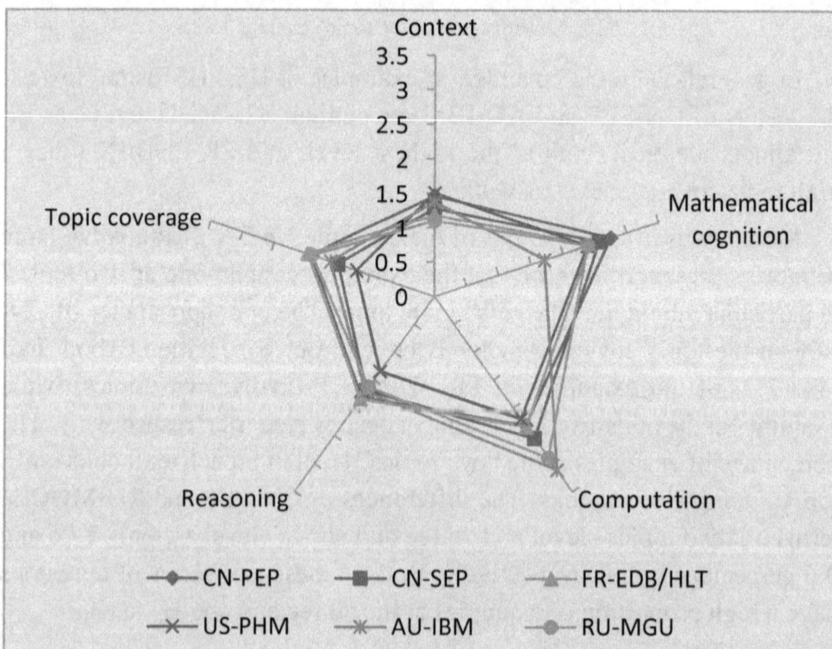

Fig. 7-17 A comparison of overall difficulty levels

As can be seen from the figure, from the mathematical cognition factor of the examples, the two Chinese series of textbooks rank top two, with CN-PEP higher than CN-SEP, followed by FR-EDB/HLT, RU-MGU, US-PHM and AU-IBM, successively. From the context factor, US-PHM ranks number 1, followed by CN-PEP, AU-IBM, CN-SEP, FR-EDB/HLT, and RU-MGU, among which the last two series still keep the traditional features of mathematics textbooks, typically confined in the internal mathematics, and less related to the reality. For the reasoning factor, FR-EDB/HLT and CN-PEP are the highest, followed by AU-IBM, CN-SEP, RU-MGU and US-PHM. With respect to computation factor, AU-IBM and RU-MGU are the highest, then CN-SEP, FR-EDB/HLT, CN-PEP, and US-PHM. In terms of topic coverage factor, FR-EDB/HLT, CN-PEP, and RU-MGU are the highest with almost each example covering two topics, then AU-IBM, CN-SEP and US-PHM with most of the examples involving one topic. Yet it is somewhat surprising that CN-SEP is not high in the level of topic coverage.

In addition, all the six pentagons take a declining trend in the context dimension, which indicates that the actual situation level of the six series of textbooks are all relatively low. This is a striking difference between textbooks of middle schools and high schools, cf. (Jiansheng Bao, 2002), indicating that in high schools, with the mathematical theory study deepening, the textbooks are designed to pay more attention to mathematics itself (except for some newly introduced topics like probability and statistics), and their relations to the real life are also fading.

References

[1] Bao, Jiansheng (2002). Comparison of the comprehensive difficulty levels of mathematics curriculum in junior high schools between China and the UK [In Chinese]. Doctoral Dissertation. Shanghai: East China Normal University.

[2] Gao, Wenjun and Bao, Jiansheng (2009). Comparison of exercises in mathematical cognition levels in China and United States — with quadratic equations and functions as examples [In Chinese]. *Journal of Mathematics Education*, 18(4): 57–60.

[3] Hu, Lili (2008). Comparative study of the difficulty levels of mathematics textbooks in junior high schools between China and the USA [In Chinese]. Master Thesis. Shanghai: East China Normal University.

[4] Jin, Songyu (2012). Comparative study on the mathematics textbooks of junior high school between China and Korea [In Chinese]. Master Thesis. Yanji, Jilin: Yanbian University.

[5] Liu, Dan, Yang, Yudong and He, Zhenzhen (2007). The taxonomic basis of teaching objective measurement — The new century action of Qingpu Experiment, V [In Chinese]. *Shanghai Educational Research* (11), 43–46.

[6] Miao, Jian (2009). Study on the comparison of Grade 7 mathematics textbooks published by People's Education Press and Jiangsu Science Press [In Chinese]. Master Thesis. Fuzhou: Fujian Normal University.

[7] MOE of PRC (Ministry of Education of the People's Republic of China) (2010). *National Outline for Medium and Long-term Education Reform and Development (2010–2020)* [In Chinese]. Beijing: People's Publishing House.

[8] Wang, Jianpan (2011). Comparative study on high school mathematics textbooks in certain countries [In Chinese]. *Curriculum, Teaching Material and Method*, 31(7): 105–106.

[9] Wang, Qinghuan (2014). The difficulty level of Chinese school textbooks is over-rated [In Chinese]. *Guangming Daily*, May 6th, 2014, Page 6.

[10] Zhang, Weizhong & Huang, Lihong (2009). Comparison and analysis of the difficulty of "triangles" of the new mathematics textbooks [In Chinese]. *Journal of Mathematics Education*, 18(4): 61–64.

[11] Zhao, Yuming (2013). A comparison research of difficulty of algebra exercise of junior middle school mathematics textbooks [In Chinese]. Master Thesis. Shijiazhuang: Hebei Normal University.

https://doi.org/10.1142/9789814713955_0008

Chapter 8

A Comparative Study on Example and Exercise Design in High School Mathematics Textbooks

Guanghua Shao, Ping Lu, Jianping Wang and Weili Lu

College of Teacher Education, Ningbo University
Ningbo, Zhejiang 315211, China

This chapter compares the designs of examples and exercises in senior high school mathematics textbooks from China, U.S. and Germany. By constructing basic designing principles of examples and exercises in mathematics textbooks, we established a comparison framework and relevant coding system for analysis. The study focuses on the content of "function", and makes the comparison from both external and internal aspects. The external aspect mainly involves the quantity, types and materials of examples and exercises. The internal aspect concerning examples covers their exemplarity, typicality, fundamentality and treatment methods; the internal aspect concerning exercises involves hierarchy, variation and coverage. Comparison results show the commonalities and differences in the textbooks from the three countries. Based on comparative analysis, this chapter also offers suggestions to optimize senior high school mathematics textbooks' examples and exercises design.

1. Research Background

As the internationalization of mathematics education grows, international mathematical exchange and collaborative projects are increasing across the world. Comparative studies of mathematics education have been receiving more and more attention, of which the mathematics textbook is one of the focuses. D. F. Robitaille has pointed out: "in every country, textbooks would have a significant impact on teaching and learning of mathematics, so it is an important field for researchers to understand the differences in the contents and methods of the textbooks of different countries" (Howson, 1995). The disciplinary characteristics of mathematics imply that the central issue of mathematics, namely mathematical problems (Halmos, 1982), is indispensable in mathematical research. Since the major forms of mathematical problems in the textbook are examples and exercises, the study of examples and exercises has always been a research hotspot in mathematics education.

The research on examples has mainly focused on two aspects. The first is empirical research by psychology researchers on the role of examples in the effect of learning, and the second is case analysis made by mathematics educators and researchers on the design of teaching using examples. Overseas studies on examples and exercises are numerous. For instance, B. H. Ross (1987, 1991 and 1997) studied the examples or samples from various aspects, and put forward new ideas on sample structures, the role of the samples and the improvement of examples. Kalyuga *et al.* (2001) carried out the comparative study between problem solving study and example study. R. Atkinson *et al.* (2000) conducted an in-depth study on the sample-learning theory. Xinming Zhu *et al.* (1987) proved by experiments the effectiveness of sample learning in class teaching. Guanghua Shao (2003) conducted theoretical and empirical studies on mathematical sample learning. Risheng Lai (2005) studied the influence of sample presentation and its cognitive mode on high school students' scores. Jinjie Ren and Haidong Lu (2009) made a study of the influence of sample-learning mode and self-explanation on transfer learning among college students' mathematics learning. Zaiping Dai (1996) classified examples into three types, namely, concept-oriented, principle-oriented, and problem-solving oriented. Xiaoming Wang (2007) studied samples from the perspective of teaching methods and deemed that samples provide learners with the expert solutions for their learning by imitation.

Researchers have also studied exercises from different perspectives. V. A. Oganesian (В. А. Оганесян) (1984) made significant studies in the structure of exercises. He considered that a mathematical exercise would consist of four elements {Y, O, P, Z}, where Y represents the conditions, O stands for the background knowledge, P denotes the problem solutions, and Z refers to the conclusion. According to the four elements mentioned above, he classified exercises into four types: standard exercises, training exercises, exploratory exercises, problem exercises. By element analysis, such classification helps teachers to know the basic structure and difficulty of an exercise. J. Sweller *et al.* (1982, 1988, 2006) conducted an relatively in-depth study on the mathematical problems (problem solving). Mathematics problem-solving has become a hotspot in international mathematics education research since the 1980s, for the new educational ideas and concepts represent a new way of teaching. In April 1980, National Council of Teachers in Mathematics released *An Agenda for Action*, which explicitly proposed problem-solving as the core of school teaching. Lingyuan Gu (1981) put forward the idea of "conceptual variation" and "procedural variation", and classified the variations of exercises into horizontal variations and vertical variations; horizontal variation is characterized mainly by the changes in the apparent forms, and the vertical variation mainly reflects changes of structural characteristics. Shanghe Cai (1990) conducted a study on the exercise characteristics of junior high school textbooks. Zaiping Dai (1996) conducted a relatively in-depth and comprehensive study on mathematical exercise theory and analyzed the features of exercises, examples and review exercises; based on their functions he even classified such exercises into scaffolding, basic, communication, analysis, integrated, thinking and inquiry types. Guanghua Shao (1997) discussed the formulation principles of four types of problems (examples, exercises, practice questions and review questions). Zengru Luo (2002) divided the mathematical problems into practice-type exercises and investigation-type exercises based on known or unknown conclusions. Based on the five factors of mathematical problems — reasoning, topic coverage, cognitive demand, computation and context — Jiansheng Bao (2002) established a composite difficulty model of mathematical problems. Xuhua Sun *et al.* (2006) proposed concepts of horizontal and vertical variation on the basis of structure of mathematics problems.

The existing literature indicates that the research on examples or exercises consists mostly of the research on psychological development of mathematics learning, hence little research has been done in the design of example designs in mathematics textbooks. As for research on exercises, layout or design of examples in math textbooks is rarely covered despite numerous general studies conducted. This paper discusses the characteristics and patterns of example designs based on a comparative study of selected math textbooks of three countries, China, the United States and Germany in the hope of achieving some beneficial enlightenment.

2. Research Methodology

2.1 Definitions of basic concepts

2.1.1 Examples

In psychological studies, examples are also called samples; they are entities which can illustrate or show more abstract concepts or relative specific principles and are the models which are able to show similar things or can be imitated (Binyan Xu, 2007). The examples in the math textbooks usually refer to math problems that further explain the content or method of mathematical learning with more detailed answers to questions or instructions. They are either permeated with some mathematical methods, or reflect a mathematical thinking, or provide some important conclusions. Freudenthal deems that examples are the ladder and keys towards the general problems and algebraic relations, which means that the methods contained in the examples can be applied to answer the more general question of learning and comply with the teaching concept of "teaching for transfer". Guanghua Shao (2011) points out that mathematical samples are explanatory, typical and exemplary, and the examples help us with a better understanding of new concepts, new knowledge and in revealing the effectiveness of the new approach. Moreover, samples can play a key role in supplementing and explaining concepts, principles, formulas, etc.; they help learners to clarify concepts, reinforce memory, and facilitate learners to know the principles or use formulas to solve problems; furthermore, selection, layout, quantity, difficulty and sequence of examples would affect learning of mathematics.

2.1.2 Exercises

Exercises usually refer to math problems affiliated to the text and examples of math textbooks, and they are an indispensable part of math textbooks. From the perspective of denotative extension, exercises include those problems tagged as "practice", "exercises", "homework", "test questions", etc. The functions of exercises are mainly embodied in knowledge and education. In the aspect of knowledge, students can obtain systematic math knowledge and form the necessary skills through math exercise; as for educational function, quality education through math learning covers both intelligence and non-intellectual emotions, attitudes and values; the math exercise is an important tool for training mathematics thinking, and plays an important role in cultivating and enhancing students' mathematical thinking and mathematical consciousness, etc., which are featured by flexibility, accuracy and criticality.

2.1.3 Design of examples and exercises

In view of the concept and function, the exercise and example design mainly include extrinsic and intrinsic structural design. Extrinsic structure refers to the extrinsic attributes of exercises and examples, and is also known as surface structure; the intrinsic structure is the essential attribute of exercises and examples, and is also called deep structure.

The extrinsic structural design mainly refers to the quantity, types and materials of examples and exercises, which are the most basic aspects we should consider in formulating examples and exercises. Firstly, the quantity of examples and exercises are normally discussed with a combination of knowledge points. The knowledge directly presented in the textbooks is limited, and the corresponding supporting exercises and examples are also limited. On a limited scale, we should rationally allocate relevant knowledge and examples and exercises so as to achieve high-quality and efficient teaching objectives. Secondly, since there are different types of knowledge points, there should be different types of designs of exercises and examples. The design of exercises and examples must have a clear purpose, either to deepen understanding of a concept, or to master a problem-solving approach, or to be skillful in the application of knowledge, etc. On the basis of previous research by experts and scholars, we make a general classification of exercises: concept- and principle-oriented, approach-oriented and application-oriented. In combination with the most intuitive presentation of exercises, we may also divide exercises into true-false questions, multiple choice questions, blank-filling questions, short-answer questions, detailed answer-questions, proof questions and construction questions, etc. Finally, the materials of examples and exercises essentially derive from three aspects: first, pure mathematical problems with no background; second, mathematical problems in daily life; third, math problems concerning other disciplines (such as the natural sciences, humanities and social sciences, etc.). The materials of examples and exercises should be drawn not only to include the internal transfer of mathematics (transfer of knowledge between the points), but also include inter-scene or interdisciplinary migrations.

Intrinsic structural design of examples mainly reflects the exemplary, typical, basic and treatment aspects of such examples. Firstly, in respect of the exemplary design, the main approach for students to learn a skill relies on observation, imitation, practice, memory and proficiency, among which imitation is particularly important. As a sample, examples not only exhibit an expert's problem-solving ideas, problem-solving process and standard solution handling, but also play a main role in providing learners with imitation approach. This requires that examples should serve as a model in guiding learners towards the application of new knowledge and problem solving; meanwhile, examples should be used to further explain the principle of mathematical concepts and, to a certain extent, to visualize abstract thinking. Secondly, as for the examples of typical design, through classroom teaching, mathematics teachers can help students master the solving methods of math problems of the same type, namely, examples play an analogical role in solving similar math problems. Different ways of learning such as self-exploration and self-learning are highly advocated in the new curriculum, and it is particularly important that examples should be typical in self-directed learning. Thirdly, as for the basic design of examples, new knowledge inevitably starts from or is connected with old knowledge. The role of examples in textbooks means that they are bound to bring students a certain degree of cognitive load. In order to make new knowledge easily accepted and assimilated, there must be some basic examples. Finally, as for the ways of example treatment, that is, the main content or presentation of examples, they include "questions + answers", "questions + analysis + answers", "questions + answers + check", "questions + answers + major step instructions", "questions + answers + major step instructions + check", "questions + answers + summary" and "questions + analysis + answers + summary".

From the perspective of functioning, examples mainly play a role in transfer and development, and in strengthening the new concept, new method and new skills. The essential attributes of exercises include hierarchy, variations, topic coverage and so on. Firstly, as for hierarchy of exercises design, organizing knowledge into a system is a premise for quick and accurate grasping, applying and transferring. Thus exercises should be conducted in an orderly manner for hierarchical design. Elementary exercises, mid-range exercises, and comprehensive problems are reasonably matched; the various knowledge and skills are organized into a system according to the degree of difficulty. The thinking level will be arranged from low to high levels, which would be more consistent with students' cognitive rules. Secondly, as for exercises' variation design, variation training is an important approach for students to build a complete, reasonable knowledge base, which helps students to deepen their understanding of the nature of the problem and to form a good cognitive structure. In combination with Lingyuan Gu's theory of exercises (Lingyuan Gu, 1981), the study of variations of the exercises is divided into vertical variation type and horizontal variation type. Finally, the coverage of design exercises, some exercises are highly integrated, which are mainly used to consolidate new knowledge and combine the existing knowledge, skills and ways of thinking; some exercises even introduce new math concepts, and show new problem-solving methods with a high mobility scalability.

2.2 Main research methods

This paper mainly uses text analysis, and analyzes the extrinsic and intrinsic structure of the examples and exercises in math textbooks of three countries, Germany, China and the USA. We use a quantitative analysis method on the quantity, type and material of such exercises so as to establish a clear understanding of the surface characteristics of exercises and examples; we use a qualitative analysis approach to make a comparative analysis of the internal structures of examples and exercises.

2.2.1 Research framework

Based on the basic theory regarding the extrinsic and intrinsic design of examples and exercises in high school mathematics textbooks, we construct a comparative research framework as shown in Fig. 8-1:

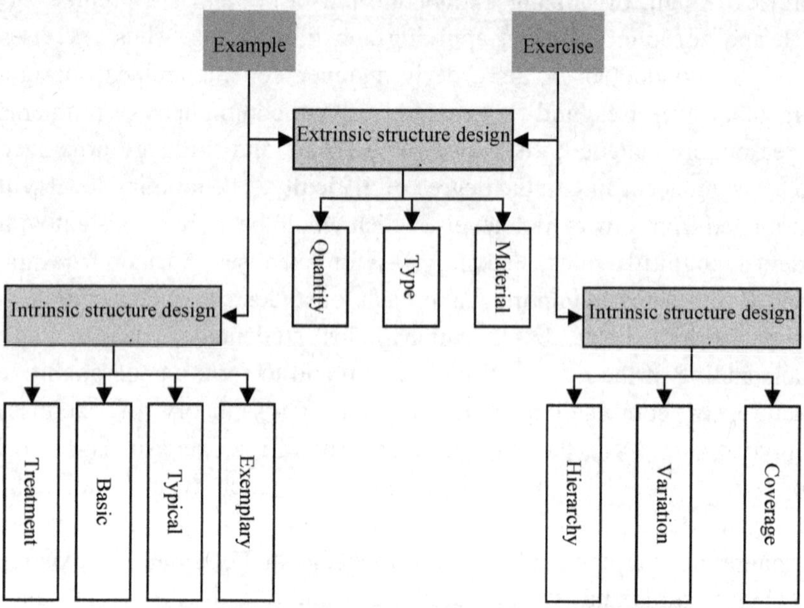

Fig. 8-1 Comparative research framework of examples and exercises

2.2.2 Coding system

On the basis of the basic principles regarding the design of examples and exercises and the comparative research framework, we construct the coding system in relation to the comparative research framework of example and exercise design as shown in Tab. 8-1 and Tab. 8-2.

Tab. 8-1 Comparative Study Coding System regarding Extrinsic Design of Examples and Exercises

Comparison Dimension	Comparison Object	Primary category	Sub-category
Extrinsic structure	Examples	Quantity	Quantity of big questions, quantity of sub-questions
		Type	Concept and principle-oriented, approach-oriented and application-oriented
		Material	Pure math questions, life background questions and disciplinary questions
	Exercises	Quantity	Quantity of big questions, quantity of sub-questions
		Type	True-false questions, multi-choice questions, blank-filling questions, short-answer questions, solution-answer questions, proof questions and construction questions
		Material	Pure math questions, life background questions and discipline background questions

2.2.3 Research text

(1) Textbook selection

Whether there are "many textbooks on one curriculum standards" or "multi-curriculum standards with many books", one country tends to publish textbooks of many different versions during the same period. We mainly consider two aspects of contemporaneity and universality of textbooks selected in our study. The ordinary high school curriculum standard experimental textbooks published by People's Education Press (in Mainland China) are the most widely used high school mathematics textbooks in Mainland China. As for textbooks in the United States, the series high school mathematics textbooks by publishers McDougal Littell is particularly distinctive and reflects a new conception of the American curriculum reform and development. For Bavaria, the largest and oldest state in Germany, its curriculum reform has always been at the forefront; the publishing house Ernst Klett Verlag has a long history of publishing mathematics textbooks, and its LS edition textbook is widely used in Bavaria. We use the examples and exercises of these popular textbooks in the three countries to reveal the typical characteristics. The details of the three versions of the textbooks selected are shown in Tab. 8-3.

Tab. 8-2 Comparative Study Coding System regarding Extrinsic Design of Examples and Exercises

Comparison dimension	Comparison object	Primary categories	Sub categories	Code	Symbol
Intrinsic structure	Examples	Exemplary	1. Knowledge use demonstration	Mathematical formulas, rules, premise of theorem application, format	
			2. Mathematical Methods demonstration	Conventional problem solving steps	
			3. Mathematical Thinking demonstration	Symbolic-graphic combination, Equivalence transformation, function and equation in conjunction with specific problems	
		Typical	1. Targeted	Application-oriented with regard to the new knowledge	
			2. Representative	Generality as to Structural problems (mathematical relationship)	
			3. Enlightening	The problem is easy to be extended, varied, expanded	
		Basic	1.Relatedknowledge	Number of the topics involved	
			2. Calculation operation	Computational complexity and steps	
			3. Level of thinking	Comprehensiveness, transference and other capacity requirements	
		Treatment	1. Questions + answers		S
			2. Questions + analysis + answers		AS
			3. Questions + answers + check		SC
			4. Questions + answers + major step instructions		SI
			5. Questions + answers + major step instructions + check		SIC
			6. Questions + answers + summary		SG
			7. Questions +answers + analysis +summary		ASG
	Exercises	Hierarchy	1. Imitation	Basic overview and basic structure are the same.	MF
			2. Variation training	Basic overview or basic structure undergoes some change.	BS
			3.Integrated application	Integration with previous knowledge and methods	ZH
			4. Open exploration	Conclusion or the solution process is not unique	KF
		Variation	1. Horizontal variation	The surface characteristics of a problem undergo a change	BS1
			2. Vertical variation	The main structural features of a problem undergo a change.	BS2
		Coverage	1. Knowledge	Knowledge in the text and examples.	FG1
				The introduction of new theorems, formulas, etc.	FG2
			2. Skills	Skills in the text and examples.	FG3
				The introduction of new skills	FG4
			3. Way of thinking	The way of thinking in the text and examples	FG5
				The newly introduced problem-solving method	FG6

Tab. 8-3 Details of the Textbooks of the Three Versions Country

	Book title	Main editors	Publisher	Publication time
China	Mathematics Compulsory I,	Peiling Qian, Jiayue Zhang *et al.*	People's Education Press,	2007, 2nd Edition
U.S.	Algebra 2: Concepts and Skills	Ron Larson, Laurie Boswell, Timothy D. Kanold, Lee Stiff	McDougal Littell	2008
Germany	Lambacher Schweizer-LS Algebra Bayern 10	August Schmid, Ralf Gleiß, Edward König etc.	Ernst Klett Schulbuchverlag	Approved in 2007

(2) Selection of Contents for Comparison

As an important mathematical model, function is used to describe the variation of the objective world; the concepts and methods of functions runs throughout the high school mathematics curriculum. Students learn exponential functions, logarithmic functions and other specific basic elementary functions; students combine the theory with practical problems, experience the process and methods by using the concept of function to construct models, learn the importance of functions in math and other subjects, and gain the initial ideas to understand and use the functions to solve the simple problems in social life. Therefore, model learning of exponential and logarithmic functions is very important. This paper will discuss the comparison of exponential and logarithmic functions in the three textbooks.

Layout of examples and exercises corresponds to knowledge points, and an in-depth analysis must be conducted on the premise that the knowledge points remain relatively consistent. After the statistical analysis, the comparison of the knowledge points regarding the exponential and logarithmic functions in the three textbooks are shown in Tab. 8-4 (namely, there are 18, 19 and 18 knowledge points in Chinese textbook, US textbook and German textbook respectively, which shows a higher knowledge consistency). Due to the need for comparative analysis, we omit the statistics for different contents in each textbook, such as Neper algorithm, logarithmic scale, geometric progression in the German textbook.

Tab. 8-4 Comparisons on the Knowledge Points

Chinese textbook (Exponential and Logarithmic Functions)	American textbook (Chapter VIII "Exponential and Logarithmic Functions")	Germany textbook (III. Exponential and Logarithmic Functions)
Actual background of exponential function model, concept, graph and property of exponential function, computer drawing, solution of real problems by exponential function; definition of logarithmic function, common logarithm, natural logarithm, relation between exponent and logarithm, operation property of logarithm, base change formula, graph and property of logarithmic function, computer drawing, inverse function; linear growth model, inverse function growth model, logarithmic function growth model and case application of function model	Definition of growth exponent; graph drawing of growth exponent; exponential falling function and its graph; application of exponential growth functions; application of exponential falling functions; solution of exponential function relation expression; definition of exponential function; definition of logarithmic function; conversion of logarithmic to exponential expression; estimate of logarithmic expression; graph drawing of logarithmic function; properties of quotient and product for logarithms; Simplification of logarithmic expressions by the properties of quotient and product; take the logarithms with the same base at both sides with identical power; take exponents at both sides; application of logarithms in the actual life; computer drawing	Linear growth, exponential growth, growth factor, exponential decrease; definition, graph and property of exponential function; exponential function growth model; definition, graph and property of logarithmic function, relation between logarithmic and exponential functions, estimate of logarithms; operation property of logarithms (formulas), base change formula; definition of logarithmic equation and solving method, definition of logarithmic equation and solving method

3. Comparison Results

3.1 Comparisons on the design of examples

3.1.1 External comparison

According to the basic principles of external example design, the external design comparative analysis is made mainly from three aspects, namely example quantity, type and material.

(1) Example quantity comparison

Examples in the Chinese textbook are mainly marked with "例" in the main text of the textbook; those in the German textbook are marked with "Beispiel (Beispiele)" and located between the concept description of the main text and class exercises; those in the American textbook are also denoted by "EXAMPLE". There is only one question for each example in the American textbook, while there are several sub-questions for each example in the Chinese and German textbooks. Therefore, we count not only the numbers of examples (as big questions), but also the number of sub-questions in examples. The statistical result is as follows (Fig. 8-2).

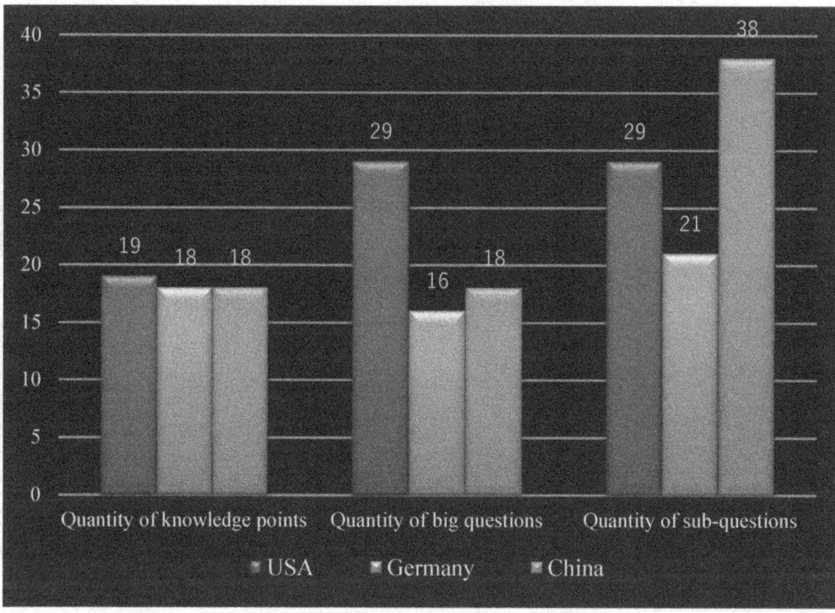

Fig. 8-2 Example quantity comparison

As shown in Fig. 8-2, there is no difference in the statistics of the quantity of big questions and the quantity of sub-questions in the United States. The United States ranks top with 29 in the quantity of big questions; China becomes the first in the statistics for the quantity of sub-questions; and Germany has kept the third place in these two statistics. The ratio between the quantity of big questions and that of relevant knowledge points in China and Germany remains consistent, 1:1, while the ratio between the quantity of sub-questions and that of relevant knowledge points greatly differs — around 1.1:1 for Germany, over 1.5:1 for America and definitely more than 2:1 for China. In fact, Chinese examples focus on horizontal variations.

(2) Example type comparison

The examples in the three textbooks are counted according the following types including concept and principle-oriented, approach-oriented and application-oriented. The quantity and percentage are shown in Fig. 8-3 below.

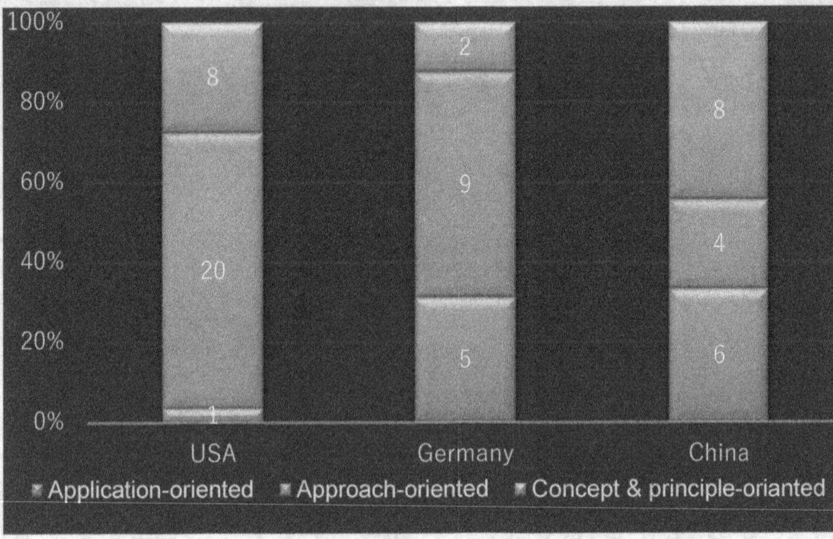

Fig. 8-3 Example type comparison

It can be seen in Fig. 8-3 that the three textbooks have a great difference in the type of examples. The American textbook focuses on approach-oriented examples, with a percentage near 70%, and there is only one concept and principle-oriented example. The German textbook also emphasizes approach-oriented examples and such examples mainly appear in Section 7, "Exponential and Logarithmic Equations", in which the four conditions for solving the two equations are given; most concept and principle-oriented examples appear in the concept parts such as exponential and logarithmic functions, etc., and reflect the applicability, and most application-oriented examples appear in the "Exponential Function Growth Model". Compared with the textbooks of Germany and the United States, the Chinese textbook gives priority to application-oriented examples — in particular, the growth models of different types of functions are presented in the form of application-oriented examples — the approach-oriented samples only occupy a small percentage in the Chinese textbook and only some methodology is shown in the application-oriented examples.

(3) Example material comparison

In terms of the source for question materials, the three sets of textbooks are counted according to the following three aspects: pure math questions, life background questions and discipline background questions. The specific data is shown in Fig. 8-4.

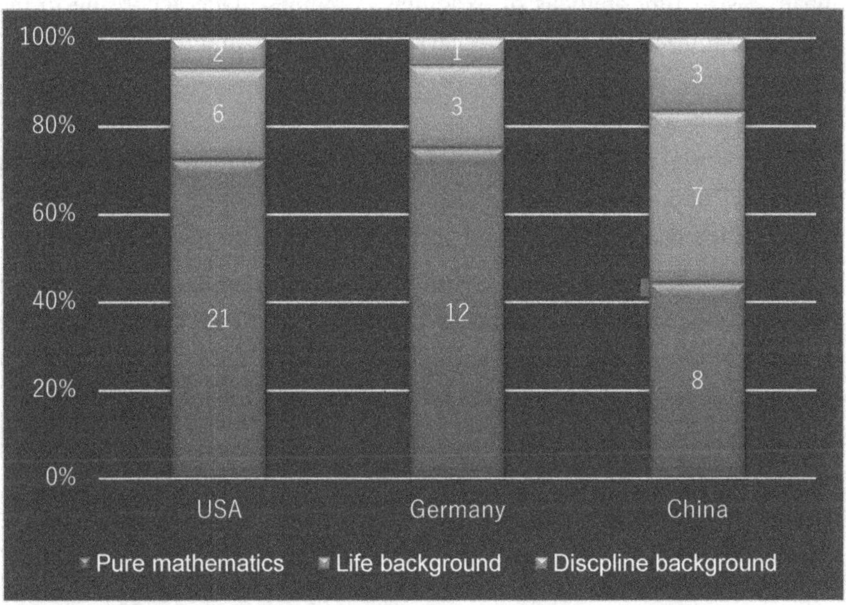

Fig. 8-4 Example material comparison

As shown in Fig. 8-4, all the three types of questions indicated above are provided in the examples of the three sets of textbooks. The ratios between pure math questions and background questions in the United States, Germany and China are 1:2.625, 1:3 and 1:0.8 respectively. The textbooks in the United States and Germany focus on pure math questions and pay attention to exponential and logarithmic operations, as well as solving exponential function equations and logarithmic function equations. In contrast, the Chinese textbook focuses on the interaction with actual background in example design, with a strong applicability. In fact, all three textbooks tend to be derived from real-life background. Due to the fact that the quantity relations in the exponential and logarithmic function models are very useful in both daily life and disciplinary research, background questions are evidently essential.

3.1.2 Internal comparison

According to the basic principles of internal design for examples, the intrinsic structure analysis of examples is mainly made according to the following four categories: exemplary, typical, basic and treatment.

(1) Exemplary comparison of examples

Exponential and logarithmic equations are required to be solved in the studies of exponential and logarithmic functions. The three textbooks reflect different exemplary characters in relevant examples.

The US textbook pays attention to the exemplary character of mathematical methods and process. "Question type" or "solving tips" are given in front of the stem and explanations are offered on the right of the solving procedures. Checking is also required, as shown in Section 8.6.

EXAMPLE 3 Take a Common Logarithm of Each Side

Solve $10^{3x-1} = 19$.

SOLUTION

$10^{3x-1} = 19$	Write original equation
$\log 10^{3x-1} = \log 19$	Take common logarithm of each side
$3x - 1 = \log 19$	$\log 10^x = x$
$3x = 1 + \log 19$	Add 1 to each side
$x = \dfrac{1 + \log 19}{3}$	Divide each side by 3
$x \approx 0.760$	Use a calculator

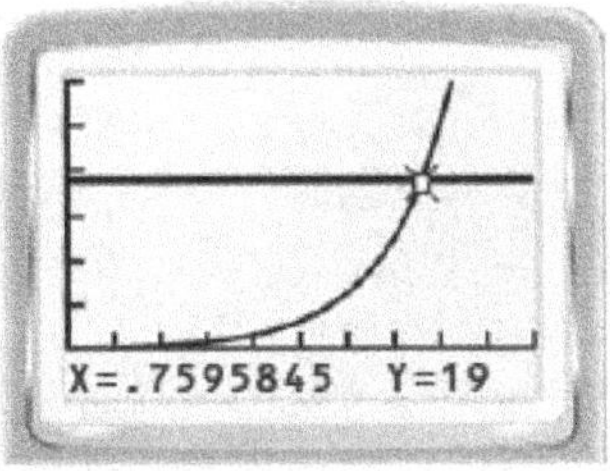

CHECK You can check the solution by substituting it into the original equation. Or, you can check the solution graphically by graphing each side of the original equation as a function.

$$y_1 = 10^{3x-1} \text{ and } y_2 = 19,$$

The two graphs intersect when $x \approx 0.760$.√

The German textbooks focus on the exemplary character of mathematical methods and thoughts. In Section 7 of Chapter III, Page 63, four examples are progressive, one after another, and the exemplary and representative characters have the effect of inferring others from one fact. The variable in example 1 only appears in the antilogarithm, with one side of the equality sign being a variable and the other being a constant. In example 2, there are variables in both sides of the equation and such variables only appear in the antilogarithm; the equation could be solved by transposition and application of operation nature of logarithms. Example 3 contains a compound variable of logarithmic and exponential functions and one side of the equation is a variable. Example 4 contains a compound variable and both sides of the equation are variables. Additionally, the idea of variable substitution is adopted. Although lacking the solving procedure explanations in the examples of the US textbook, the level, hierarchy, method and process are absolutely clear.

EXAMPLE 1

$$\begin{aligned} \log_5(2x+1) &= 3 \\ 5^{\log_5(2x+1)} &= 5^3 \\ 2x+1 &= 5^3 \\ x &= 62 \\ L &= \{62\} \end{aligned}$$

Determining the domain

$$\begin{aligned} 2x+1 &> 0 \\ x &> -0.5 \\ D &= (-0.5, +\infty) \end{aligned}$$

EXAMPLE 2

$$\begin{aligned} \lg(5x+2) &= 2 + \lg(1-x) \\ \lg(5x+2) - \lg(1-x) &= 2 \\ \lg\frac{5x+2}{1-x} &= 2 \\ \frac{5x+2}{1-x} &= 10^2 \\ 5x+2 &= 100(1-x) \end{aligned}$$

Determining the domain

$$5x+2>0 \text{ and } 1-x>0$$
$$x>-0.4 \text{ and } x<1$$
$$D = (-0.4, 1)$$

$$105x = 98$$
$$x = \frac{14}{15}$$
$$L = \left\{\frac{14}{15}\right\}$$

EXAMPLE 3

$$x^{\lg x} = 1000 \qquad D = \mathbf{R}^{+}$$
$$\lg x^{\lg x} = \lg 10^4$$
$$\lg x \cdot \lg x = 4$$
$$\lg x = 2 \text{ or } \lg x = -2$$
$$x = 100 \text{ or } x = 0.01$$
$$L = \{0.01, 100\}$$

EXAMPLE 4

$$x^{\lg x} = x^3 \qquad D = \mathbf{R}^{+}$$
$$\lg x^{\lg x} = \lg x^3$$
$$\lg x \cdot \lg x = 3 \lg x$$

Substitution: $u := \lg x$

$$u^2 = 3u$$

Thus, $u^2 - 3u = u(u-3) = 0$

$$\Rightarrow u = 0 \text{ or } u = 3$$

Back-substitution: $\lg x = 0$ or $\lg x = 3$

So $x = 1$ or $x = 1000$

$$L = \{1, 1000\}$$

No examples of strong exemplary character of exponential and logarithmic equations are specifically given in Chinese textbooks, and only Example 2, "Calculate the value of x in the following equations" in Section 2.2.1, has the taste of solving the logarithmic equation. But such operations would appear in exercise books and examination papers in large amounts. Also, variable substitution is a thinking method frequently used in the solving exponential functions. There have not been exemplary examples of such thinking method but these are shown to students in class through teachers' lesson preparations. In this regard, Chinese textbooks are slightly deficient.

(2) Typical comparison of examples

Typical examples should have clear or targeted themes and mainly target at the application of new knowledge. As for typical examples, other things could be inferred from one fact or the examples, namely being transferable and representative and the example questions are of common structure. Typical examples should be able to stimulate thinking and enlighten others, and it should be easy for the questions to be extended, varied or expanded. Quantitative analysis is done on the examples of the three sets of textbooks according to the standards mentioned above, and specific explanations are made in the form of examples.

Typical examples in the American textbooks focus on enlightenment and humanity and are skilled at guiding students' independent discovery, stimulating their thoughts and cultivating their ability of active exploration. In Section 8.4, "Exponential and Logarithmic Functions", Example 2 is mainly used to enhance the understanding of logarithm through the interpretation of the relationship between exponent and logarithm, which is different from the explanations to the major procedures in the process of mathematical problem solving. Instead, some prompts for analysis and guide are added.

EXAMPLE 2: Evaluate Logarithmic Expressions

Evaluate the expressions.

a. $\log_4 64$ b. $\log_4 2$ c. $\log_{1/3} 9$

SOLUTION To help you find the value of $\log_b y$, ask yourself what power of b gives you y.

a.	$4^? = 64$	What power of 4 gives 64?
	$4^3 = 64$	Guess, check, and revise
	$\log_4 64 = 3$	Definition of $\log_b y$
b.	$4^? = 2$	What power of 4 gives 2?
	$4^{1/2} = 2$	Guess, check, and revise
	$\log_4 2 = \frac{1}{2}$	Definition of $\log_b y$
c.	$\left(\frac{1}{3}\right)^? = 9$	What power of $\frac{1}{3}$ gives 9?
	$\left(\frac{1}{3}\right)^{-2} = 9$	Guess, check, and revise.
	$\log_{1/3} 9 = -2$	Definition of $\log_b y$

Typical examples in the German textbooks are highly targeted. In particular, some concept and principle-oriented examples give examples and interpretations based on new concepts. The difference between exponential and logarithmic functions is explained through examples upon the introduction of the concept of the logarithmic function in the text: "Example: $x \mapsto 2^x$, $x \mapsto \left(\frac{3}{4}\right)^x$ and $x \mapsto (\sqrt{2})^x$ are exponential functions; while $x \mapsto x^2$, $x \mapsto x^{\frac{1}{2}}$ and $x \mapsto x^{\sqrt{2}}$ are power functions". These examples are highly targeted but have no structures of questions. Similar situations appeared in Chinese textbooks. However, they are not presented as formal examples but rather occur upon the introduction of the concept of logarithm: "For example, because $\frac{18}{13} = 1.01^x$, x is the logarithm of $\frac{18}{13}$ to the base 1.01, which is denoted by $x = \log_{1.01} \frac{18}{13}$; also, because $4^2 = 16$, the logarithm of 16 to the base 4 is 2, which is denoted by $\log_4 16 = 2$." The above two are explanations of the concept of logarithm. Additionally, there is a large similarity between Chinese and German textbooks in deepening the understanding of logarithm through the mutual conversion between exponential expressions and logarithmic expressions, which greatly differs from American textbooks, as shown in the following examples.

In Chinese textbook:

EXAMPLES:

a) Because $2^3 = 8$, one has $3 = \log_2 8$ (reading logarithm of 8).

b) Because $5^{-2} = 0.04$, one has $\log_5 0.04 = -2$.

c) $\log_2 16 = 4$, then $2^4 = 16$; $\log_2 0.5 = -1$, then $2^{-1} = 0.5$; $\log_2 1 = 0$, then $2^0 = 1$.

d) From Fig. 2 (omitted) one can deduce that

$$\log_2 2.5 = 1.3219\cdots, \quad 0.5 = \log_2 1.4142\cdots,$$
$$\log_2 0.5 = -1, \quad \log_2 2 = 1.$$

In particular, the exponent-logarithm relation expressed in the formats of "because..., one has..." or "..., then..." is quite clear and highly targeted. The relation is simple, easy to understand, and easy to imitate, which is beneficial to concept understanding.

The mathematics textbooks in Germany focus more on enlightenment and transfer ability. The exponential function is quite a useful function model and the exponential function growth model is observed in many problems concerning natural growth process and life. Section 3 in the German textbook introduces the exponential function growth model, of which two examples are quite classical. Here Example 1 is chosen for illustration purposes.

EXAMPLE 1:

At present, a forest has a stock of 40,000 fm (m^3), with annual growth rate of 3%, which is an exponential growth trend. How large amount will be added to the stock in the following 4.5 years (assuming the trees were not cut down and there was no obvious change to the growth conditions)?

What are known?	Stock $c = 40{,}000$ (fm) Annual growth rate 3%
↓	
What to find?	Growth in 4.5 years
↓	
Suppose the growth law $x \mapsto c \cdot a^x$	Growth factor $a = 1 + 3\% = 1.03$ Thus $x \mapsto 40{,}000 \cdot 1.03^x$ (in the year x)
↓	
Substitution	$4.5 \mapsto 40{,}000 \cdot 1.03^{4.5} \approx 45{,}700$
↓	
Get the result	Increase 5,700 fm after 4.5 years

The flow chart for solving the problem is shown on the left. The process "What are known? → What to find → What to suppose? → Substitution for calculation → Get the result" is in accord with the process in Polya's steps of problem solving. The specific methods in each step of the procedure are provided on the right, which are clear and simple, and provide students with the ideas to get the answers. Almost all problems of the same type can be solved by using the above procedures.

The treatment of typical examples for exponential growth model in the Chinese textbook differs greatly from that of German textbooks, with multiple purposes, as shown in Example 8:

EXAMPLE 8 At the end of 1999, China has a population of about 13 *yi* (= hundred million). If the annual average growth rate could be confined within 1%, what population will China have at most after 20 years (accurate to *yi*)?

SOLUTION Suppose the average growth rate in the future is 1% and the population in China after x years is y *yi*.

At the end of 1999, the population is about 13 *yi*;

After 1 year (i.e., in 2000), the figure is

$$13 + 13 \times 1\% = 13 \times (1 + 1\%) \ (yi);$$

After 2 years (i.e., in 2001), the figure is

$$13 \times (1 + 1\%) + 13 \times (1 + 1\%) \times 1\% = 13 \times (1 + 1\%)^2 \ (yi);$$

After 3 years (i.e., in 2002), the figure is

$$13 \times (1 + 1\%)^2 + 13 \times (1 + 1\%)^2 \times 1\% = 13 \times (1 + 1\%)^3 \ (yi);$$

… … … …

In this way, one can see that, after x years, the population figure will be

$$y = 13 \times (1 + 1\%)^x = 13 \times 1.01^x.$$

So, when $x = 20, \ y = 13 \times 1.01^{20} \approx 16 \ (yi)$.

The setting of this example has two purposes: the first is to deduce the exponential growth model, and the second is to serve as an example of application of the function model. A simple example can not only introduce the deduction process of the exponential growth model but also lay a foundation for solving similar problems, which is quite meaningful. Although the analytical idea for problem solving is not shown in a flow chart like in the German textbook, it is useful to let student think about the idea, and students of higher caliber are able to summarize the idea. Thus, it can only be said that the three sets of textbooks are different in both arrangement of learning and functions of examples.

(3) Basic nature comparison of examples

According to the basic design principle of examples, the examples in the textbooks of China and Germany are counted as follows.[a]

As shown in Fig. 8-5, the two editions of textbooks differ greatly in the basic nature of examples. As for the knowledge points embraced in exercises, the ratio between the examples with a single knowledge point and those with multiple knowledge points is 1:0.6 in the Chinese textbook and 3:1 in the German textbook, which shows that the examples in the German textbook are relatively pure and simple and highly targeted, while the examples in the Chinese textbook emphasize comprehensive exemplary character. As for solving steps of examples, the ratio for examples with single step and those with multiple steps is 2:1 in the German textbook and 8:1 in the Chinese textbook, which shows that

[a]Due to the obviously basic character of examples in the American textbook, this study presents it only by introduction of cases.

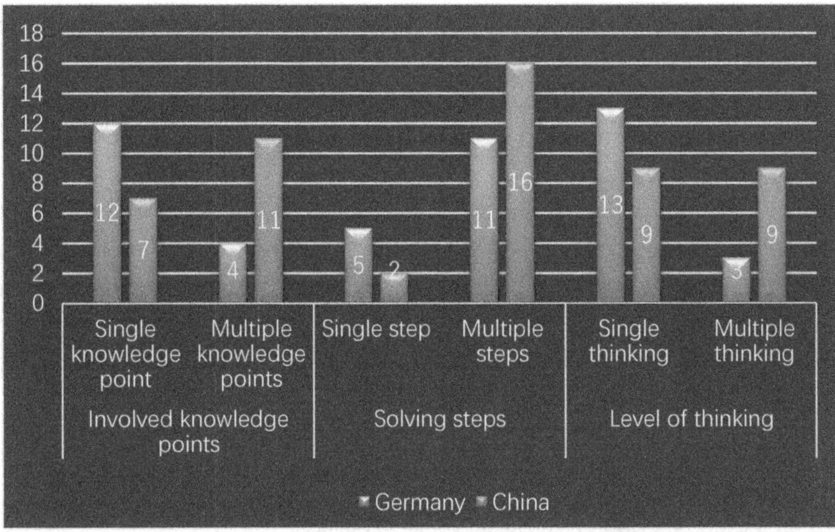

Fig. 8-5 Basic nature comparison of examples

Chinese examples emphasize the presentation of the solution process for each step. Most of the single-step examples in the German textbook are concept-principle oriented, without the difference in quantity of steps, placing them into single-step examples. In the study, the level of thinking is measured by thinking amount. Single-thinking example refers to a lower level of thinking question with only one answer or a solving method and "simple and direct" as the core. Multiple-thinking example refers to a higher level of thinking question with a variety of answers or solving methods, which is more complex. As for the thinking amount of examples, the ratio between single-thinking and multiple-thinking examples in the German textbook is around 4:1 and 1:1 in the Chinese textbook, which shows that the two sets of textbooks focus on the basic character of examples and the examples of German textbooks are much easier to understand and involve less thinking amount. The thinking amount of single-knowledge examples is less than that of multiple-knowledge examples, but it is not necessarily less in the number of solving steps, showing that examples should not only guarantee the amount of knowledge but also the detailed process of the solving process.

The examples in the American textbook demonstrate obviously basic nature. It can be observed from the exemplary and typical characters of examples that there will be solving tips before the stem of each example and the action for each step in the solving process will be directly given, which shows that they are direct, simple and involve single knowledge points and multiple steps. Also, the examples in the American textbook follow the principle of "step-by-step", from shallow to deep level and with distinguished hierarchy. They have the taste of the above-mentioned German exemplary examples 1, 2, 3 and 4, while seemingly lower in the thinking level than German ones, as shown in the following examples (pp. 448–451, solving process omitted):

EXAMPLE 1 Solve using equal powers property
Solve the equation: a. $4^2 = 4^{9x-5}$; b. $2^{3x} = 4^{x+1}$.

EXAMPLE 2 Take a common logarithm of each side
Solve $3^x = 5$.

EXAMPLE 3 Take a common logarithm of each side
Solve $10^{3x-1} = 19$.

EXAMPLE 4 Solve a logarithmic equation
Solve $\log_7(4x - 3) = \log_7(x + 6)$.

EXAMPLE 5 Exponentiate each side
Solve $\log_2(3x + 1)$.

EXAMPLE 6 Check for extraneous solutions
Solve $\log 10x + \log(x - 3) = 2$. Check for extraneous solutions.

(4) Treatment comparison of examples

In the process of studying the treatment of the examples in the three sets of textbooks, we find that the US textbooks contain merely "solution", "answer" and "check", and clearly identifies "solution" from "answer"; there is neither "analysis" nor "summaries" in the US textbooks. "Solution" and "answer" in Chinese textbooks are integrated with the "answer" usually led by the conclusive words such that "so" and "therefore"; "check" is also integrated in the solution. There is almost no "check" in the German examples and the "answer" is quite similar to the Chinese examples. The statistical results are shown in Fig. 8-6.

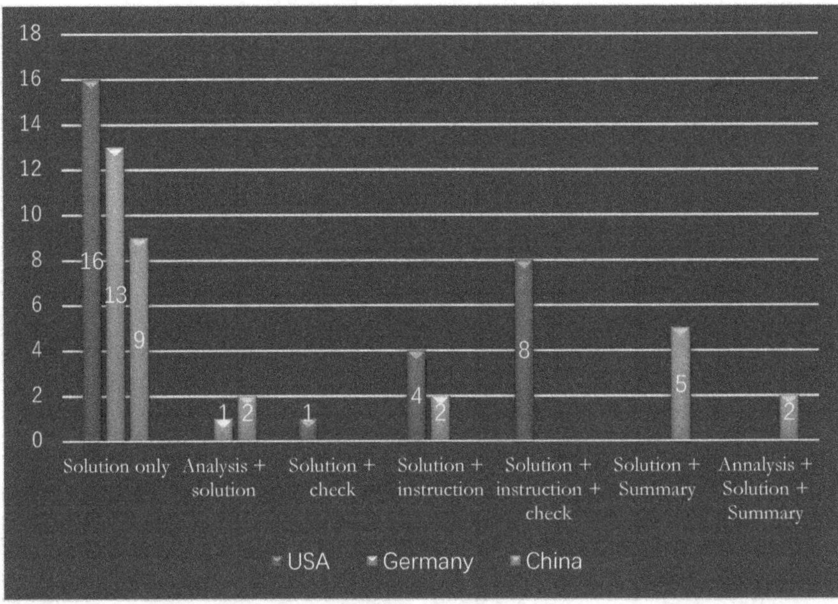

Fig. 8-6 Treatment comparison of example

It can be shown from Fig. 8-6 that the treatment of the examples in the three sets of textbooks is both common and unique. Firstly, the examples presented in the form of "question + solution (answer)" rank at the top and more than other presentation forms to a large extent. There are four types of treatment methods in American examples, namely "question + solution (answer)", "question + solution + check", "question + solution + major step instruction" and "question + solution + major step instruction + check". It can be found that American examples pay much attention in presentation to the four stages as specified in the Polya's problem-solving table: understand the question — give question type or solving tips before the stem, formulate a plan — give major step instructions, implement the plan (sometimes involve multiple steps) and check (sometimes adopt multiple methods for check such as substitution or graphical calculator, see American exemplary Example 3). Compared to the major step instructions and the check stage in American examples, the examples in the Chinese textbook pay attention to the "analysis" and "summary" of questions and its presentation also includes four types, "question + solution", "question + analysis + solution", "question + solution + summary" and "question + analysis + solution + summary". Comparatively, the German textbook adopts relatively less example treatment methods. There are only three presentation forms including "question + solution", "question + analysis + solution" and "question + solution + major step instructions". The last two are rarely used and the major step instructions appear brief. For example:

Solve exponential equations

An equation whose exponent includes a variable is called an exponential equation. For example, $a^x = b$. From this equation we can get $x = \log_a b$, according to the definition of logarithm, or $x = \frac{\lg b}{\lg a}$, using the logarithm to base 10, where $a, b \in \mathbf{R}^+, a \neq 1$. In this solution, we took logarithms at both sides to obtain logarithmic equation. If we use the logarithm to base 10, the approach is as follows:

$$\begin{aligned} a^x &= b \\ \lg a^x &= \lg b \\ x \lg a &= \lg b \\ x &= \frac{\lg b}{\lg a} \end{aligned}$$

We can solve more complex exponential equations by taking logarithm.

EXAMPLE 2 Solve $2^x \cdot 4^{3x+1} = 8^{-x}$.

Approach 1: taking logarithm to base 10	Approach 2: by equaling the exponents
$2^x \cdot 4^{3x+1} = 8^{-x}$	$2^x \cdot 4^{3x+1} = 8^{-x}$
$\lg (2^x \cdot 4^{3x+1}) = \lg 8^{-x}$	$2^x \cdot 2^{2(3x+1)} = 2^{-3x}$ (using a common base)
$\lg 2^x + \lg 4^{3x+1} = -x \lg 8$	$2^{x+2(3x+1)} = 2^{-3x}$ (taiking logarism)
$x \lg 2 + 2(3x+1)\lg 2 = -3x \lg 2$	$x + 2(3x+1) = -3x$
$x + 6x + 2 = -3x$	$x = -0.2$
$x = -0.2$	$L = \{-0.2\}$

EXAMPLE 3 Solve the equation $3^{x+1}+3^{x+2} + 3^{x+3} = 93$.

Solve by separating the lowest power:

$$\begin{aligned} 3^{x+1}(1+3+3^2) &= 93 \\ 3^{x+1} \cdot 31 &= 93 && \text{(divided by 31)} \\ x+1 &= 1 && (-1) \\ x &= 0 \\ L &= \{0\} \end{aligned}$$

3.1.3 Summary

According to the external and internal analysis of the three sets of textbook examples above, the commonality in the design of examples are summarized and their differences are analyzed along with the commonality.

The first is the allocation percentage between examples and knowledge points. From the perspective of the presentation function of examples on knowledge points, the ratio between big questions and knowledge points should be at least 1:1, and the average ratio between the numbers of three sets in terms of sub-questions is around 1.6:1, showing a reasonable allocation should be 1.5 to 2. The German textbook samples are quite deficient in this aspect and their percentage may be increased to some extent.

The second is the diversity of example background material design. The examples for the three sets of textbooks all adopt pure math and background questions. The background questions are either derived from other disciplines such as physics, chemistry and biology in science and geography and politics in the arts or from real life such as shares and loans. It is worthwhile to note that the textbooks of the United States and Germany have obviously fewer background examples than that of China and focus more on pure math operations. If students are good at calculation, the emphasis on the background questions of application type in the examples could strongly stimulate students' comprehensive abilities. The function model is normally abstracted from practical problems in real life, so the design of background examples deserves much attention.

The third is that the example types are designed in an all-round manner, which are exemplary and typical. All the concept- and principle-oriented, approach-oriented and application-oriented examples are designed in the three sets of textbooks, and the application of exponential and logarithmic functions is presented in all the three types. However, the three sets have different focuses in the design of example types. The textbooks of the United States and Germany focus on the design of approach-oriented examples, whereas Chinese textbooks concentrate more on the presentation of application-oriented examples but obviously appear deficient in the design of approach-oriented examples. For example, as for the part of Logarithmic operation, there is a lack of presentation of calculation questions in the deduction and method application of base change formula, and the design of example type would affect the exemplary character of examples to some extent. The American textbooks focus on the demonstration of mathematical methods and processes. The "question type" or "solving tips" would appear before the main question and the explanations to relevant procedures are available right beside the solving procedures. Also, checking is required. The German textbooks focus on the demonstration of mathematical methods and thinking, as well as level and hierarchy. The examples in Chinese textbooks focus on the application demonstration of mathematics modeling.

The above commonalities and differences also affect the basic nature in the example setting of the three sets of textbooks. The examples in the German textbooks are characterized by multiple single knowledge points and examples of single thinking, and the examples involving multiple steps are less than those involving a single step, which shows a highly targeted character of German examples; the American examples are simple and involve multiple steps, which indicate their direct nature; while Chinese examples seek a balance of single and multiple thinking in the amount, especially in the examples involving multiple knowledge points and steps, with even greater comprehensive nature.

The fourth is that the form of "questions + answers" is emphasized in the examples treatment method of all three sets of textbooks. The presentation means of American and Chinese examples tend to be diversified. In fact, the example treatment method cannot always pursue diversity, instead, it depends on the function of examples. For a simple and clear example, it could be understood by students only by means of "questions + answers", hence the completeness of unnecessary ASG method. Approach-oriented or background examples are more complex and comprehensive, so the treatment method of such examples should be more carefully considered. Generally, the design of examples should not only conform to the requirements of the Polya's solution chart but also the cognitive development of students.

3.2 Comparison on the design of exercises

3.2.1 External comparative analysis

According to the basic principles for external design of exercises, the external analysis of exercises is made mainly around three aspects, namely quantity, type and material of exercise.

(1) Exercise quantity comparison

The words concerning exercise type given in the textbook published by the People's Education Press (Edition A) include "练习 (practices)", which are to be done in the classroom, "习题 (exercises)", which may be done after class and "复习题 (review questions)" or "复习参考题 (review questions for reference)". The exercises in the American textbook (Edition ML) include "checkpoint", post-section exercise and chapter testing exercise. Although the German textbook (Edition LS) does not explicitly specify the words for exercise type, we could definitely judge that the questions marked with black sequence numbers (❶❷❸......) are as class practices; those marked with 4, 5, 6... serve as after class exercises and the "Vermischte Aufgaben" in the last section of each chapter are review exercises. It could demonstrate that the arrangement of exercises in the three sets of textbooks is basically identical.

The statistical results of the exercises in the three sets of textbooks according to exercises (which will be referred as big questions in the sequel) and sub-questions are shown in Fig. 8-7.

From Fig. 8-7, we can see that in the US textbook, one exercise contains only one question, similar to the situation of examples in that textbook. And, either in terms of big question or sub-question, the number of exercises in the United States textbook is much larger than that in Chinese textbook and in German textbooks, while the Chinese textbook has the smallest number. When measured by sub-questions, the quantity of in-class exercises in the US and Chinese textbooks is significantly larger than that in German textbook, showing the difference between the time in learning knowledge and completing class exercises under identical teaching conditions. As for after-class exercises, the German textbook has more than twice that of the exercises in the Chinese textbook, and the US

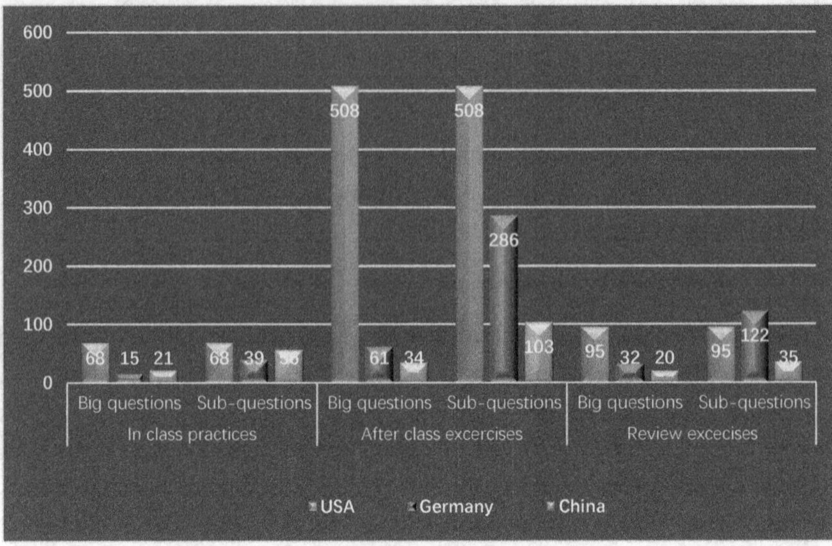

Fig. 8-7 Exercise quantity comparison

textbook has even larger number. Among the review exercises, the German textbook has 3 times that of Chinese textbook and obviously exceeds that of the US textbook. It shows that Germany has relatively high requirements for after-class and review exercises.

(2) Comparison of exercise types

The common exercise types of the three sets of textbooks are shown in Fig. 8-8. From the figure, we can see that both the US and Chinese textbooks are abundant in the type of exercises. The former covers the other six types except blank filling and the latter covers the other six types except true-false. The types of exercises in the German textbook are relatively simple and only include "short-answer, detailed-answer, true-false and construction". It is not difficult to find that the traditional short-answer, detailed-answer and proof questions still occupy a high proportion. The construction question is also essential in combination of the teaching objectives of exponential and logarithmic functions. As students differ in their means of understanding, the proper design of other types of exercises is essential in order to enhance students' understanding of exponential and logarithmic functions.

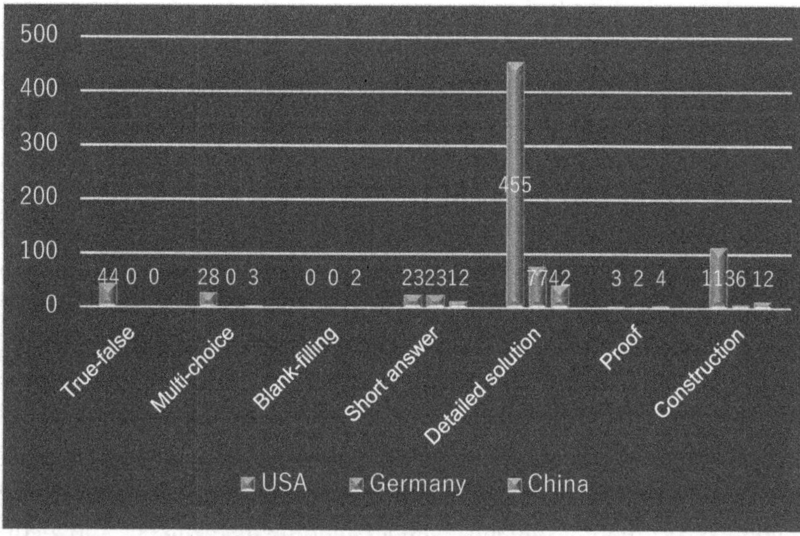

Fig. 8-8 Exercise type comparison

(3) Exercise material comparison

The exercises of the three sets of textbooks can be classified into these with discipline background, with life background and with no background.

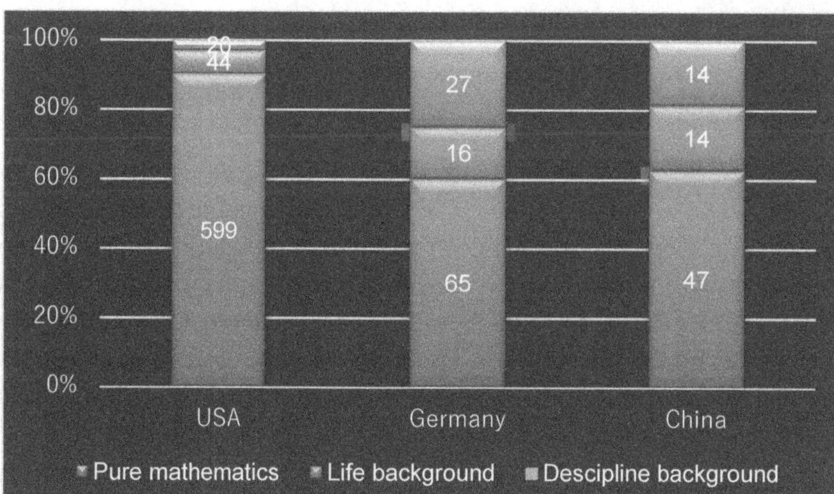

Fig. 8-9 Exercise material comparison

As shown in Fig. 8-9, pure mathematics questions dominate in the exercises of the three sets of textbooks and background questions are provided in all of them. However, there is an obvious difference in the design of background questions. American exercises focus on the design of questions involving life background; German exercises focus on designing background questions of disciplinary knowledge, especially science and engineering; while Chinese background questions are balanced in life and disciplines. As for functions, background questions are essential. On the one hand, many practical issues in life could be abstracted into functional models and the questions derived from life could enable students to better understand the applicability of mathematics; on the other hand, functions play an irreplaceable role in the calculation and problem solving process of physics, chemistry and other disciplines. Material selection and background question design should be based on the principle of facilitating students in understanding mathematics knowledge, acquiring mathematical methods and applying mathematical thoughts.

3.2.2 Internal analysis on the exercise design

According to the basic principles for internal design of exercises, the internal analysis of exercises is mainly conducted from three aspects, namely hierarchy, variation and coverage.

(1) Hierarchy comparison of exercises

The statistical result of the Chinese and German textbooks on the classification of the question hierarchy is shown in Fig. 8-10.

As shown in the figure, both Chinese and German textbook exercises are clear in hierarchy and focus on variation training, whose has a ratio near or equal to 40%. The percentage of open exploration questions is about 8%. However, the exercises in the two editions have an obvious difference in the quantity of imitation exercise and comprehensive application questions. Imitation exercises account for 33% in the German exercises and 19% in the Chinese exercises, with a difference of 14%. Comprehensive application questions account for 22% in the German exercises and 33% in the Chinese exercises, with a difference of 11%. It can be observed that there are more imitation exercises in the German

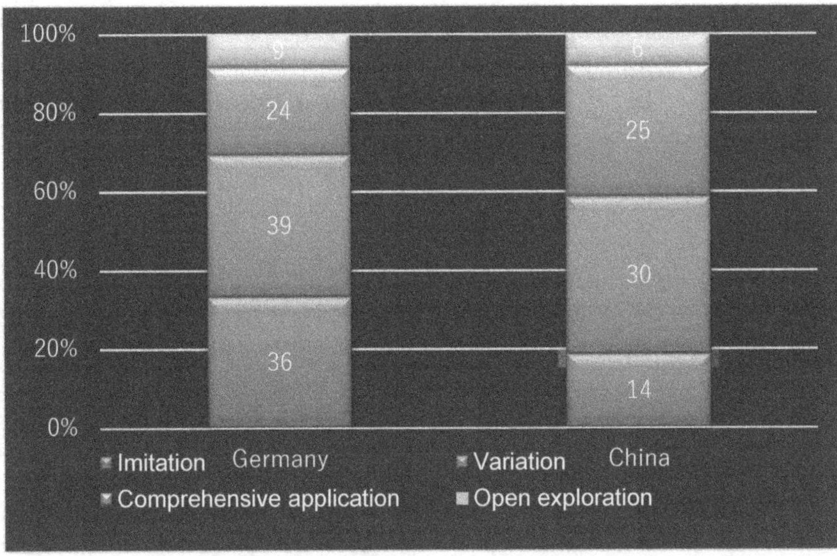

Fig. 8-10 Hierarchy comparison of exercises

textbook, while comprehensive application exercises are emphasized in the Chinese textbook.

It is found according to further qualitative analysis that the exercises in the Chinese, American and German senior high school mathematics textbooks are unique in hierarchy. The consolidation and divergence of imitation are emphasized in the hierarchy design of American textbook exercise. Take the exemplary example 3 above for example. The designed supporting exercises include (pp. 449–452):

Checkpoint (In-class Exercises): Take a common logarithm of each side

Solve the equation:

10. $10^{3x} = 5$ 11. $10^{2x+5} = 6$ 12. $10^{-x} - 4 = 13$

Post-section Exercises — Skills Check:

Solve the equation:

10. $10^{7x+3} = 3$ 11. $10^{-x} + 4 = 15$

Practice and Application — Chapter Exercises

Solve exponential equation

29. $7^{3x} = 5$ 30. $10^{5x} = 27$ 31. $5^{4x} = 100$

32. $10^{2x+1} = 13$ 33. $10^{-x} + 5 = 8$ 34. $10^{4x-5} + 11 = 20$

Those exercises are very similar to Example 3. The solving method and steps as shown in Example 3 could be easily transferred to the solving of the questions. The "valuable information" is enhanced at the same time when the exercises are successfully solved.

Take one more example: Question 54 of Post-section Exercises, Section 8.3 "Modeling with exponential functions" (p.432).

Multi-Step Problem. You have \$3000 to put into an account. A teller at a bank tells you that an account pays 5% annual interest compounded quarterly.

a. Write an exponential model giving the balance A in the account after t years.

b. Does the model show *exponential growth or exponential decay*?

c. Evaluate the model when $t = 0,\ t = 5$, and $t = 10$.

d. An account at another bank yields a balance *A* after *t* years given by the model $A = 3000e^{0.0498t}$. What is the annual interest rate of this account? How is it compounded?

e. Evaluate the model in part (d) when $t = 0,\ t = 5$, and $t = 10$.

f. *Critical Thinking.* Does the account with a greater interest rate or lesser interest rate yield a greater balance? Explain your answer.

Firstly, seek a model in the given context, and then point out the scene of the question according to the new model. Finally, determine the relation between annual rate and the sum of principal and interest through the comparison of the two different models. The convergence and improvement in knowledge and thinking are gained based on examples, from the forward to inverse application of the exponential model and finally to the overview summary of a higher level. Meanwhile, the exercises are highly life styled with the same settings as the examples, namely money saving. It forcefully consolidates the students' learning and promotes divergent thinking as well.

Variations and development are emphasized in the hierarchy design of German textbook exercises. For example, there are two questions on Page 49, whose functions are quite different though the material background is identical.

EXERCISE 13 In exponential growth processes, the time span in which an existing mass doubles is called the doubling time. An algal culture doubles its initial mass of 1 g every four days.

a) Determine the assignment rule which assigns the algae mass to the time (in days) and draw the graph.

b) After how many days has algal culture grown to a mass of 32 g?

c) Another species of algae with a starting mass of 200 g grows daily by 30%, i.e. with a growth factor 1.3. Draw the graph for this growth process and use the graph to determine an approximate value for the doubling time.

EXERCISE 15. An algal culture doubles its initial mass of 1 g every four days. This growth process can therefore be described by reference to the assignment rule$f: x \longmapsto 1 \cdot a^x$. After four days the algae mass has doubled to 2 g. It therefore applies $f(4) = 2$.

a) Use the power calculation to determine the basis a.

b) Determine the assignment rule for the growth of an algae culture, to which its initial mass of 25 g doubles in three days.

Exercise 13 has changed the basic frame from the similar example in the text. However, students may identify the essence of the issues and apply the knowledge acquired for solving and construction. Exercise 15, with a similar background, is a further development. The assignment rule is explicitly in Exercise 15 while the exponential variable is changed from n to x, which differs from the main text and all the previous examples and exercises. Also, the functional relation $f(4) = 2$ is clearly given. All these make a preparation of development for the content of the next section, "Exponential Function".

Likewise, look at the following exercise on Page 59 in the German textbook:

EXERCISE 15. With a pocket calculator you cannot directly determine $\log_2 5$. The following approach is therefore used:

$$x = \log_2 5 \Rightarrow 5 = 2^x = 10^{x \lg x} \Rightarrow x = \lg 5 : \lg 2.$$

Calculate by calculator the following expressions:

a) $\log_7 5$ b) $\log_5 13$ c) $\log_7 20$

d) $\log_{1.1} 2$ e) $\log_{0.96} 0.5$ f) $\log_{1.4} 0.5$

This exercise is not only a variant application of logarithmic identities but also implies the derivation of the base change formula, leading to the study of "Logarithmic Function".

In contrast to the hierarchy of the exercises in the American and German textbooks, comprehensiveness and complication are focused in the design of Chinese textbook exercises.

For example, there is the following example in the text:

EXAMPLE 7 Compare two values:

(1) 1.7^{25} and 1.7^{3}; (2) $0.8^{-0.1}$ and $0.8^{-0.2}$; (3)$1.7^{0.3}$ and $0.9^{3.1}$

The example is companied with two exercises in Group A and Group B respectively.

In Group A

EXERCISE 12. Suppose $y_1 = a^{3x+1}$ and $y_2 = a^{-2x}$, where $a > 0$ and $a \neq 1$. Determine the value(s) of x so that

(1) $y_1 = y_2$; (2) $y_1 > y_2$.

In Group B

EXERCISE 4. When a drug is injected into human body, its residual amount in the blood decreases exponentially.

(1) Suppose that the residual amount of a drug S in the blood is described by exponential function $y = 5e^{-0.2t}$, where t is the time (in h) after the injection and y is the residual amount (in mg) of drug S in the body.

a. What is the initial value of y?
b. What is the value of y when $t = 10$?
c. Draw the graph of the function.

(2) If the residual amount of another drug in the body can be expressed as $y = 5e^{-0.5t}$, then does it decay slowly or quickly in the body compared with the first drug?

Exercise 12 in Group A has a similar question type to Example 7. But the base in Exercise 7 is given while that in Exercise 12 is unknown, which makes Exercise 12 more complex and requires a discussion according to different bases. However, Exercise 4 in Group B has a question type different from that of Example 7. It is an application of the exponential model and confers the model the context for application. The premise for problem solving is to convert the practical issue to a mathematical model, and, by comparing the value of $5e^{-0.2t}$ and $5e^{-0.5t}$, made clear the correspondence of the decay rate with the function expressions.

(2) Comparison of the variation of exercises

According to the basic principles for variation design of exercises, the statistical result for the horizontal variations (BS1) and vertical variations (BS2) of exercises from both the Chinese and German textbooks is shown in Fig. 8-11.

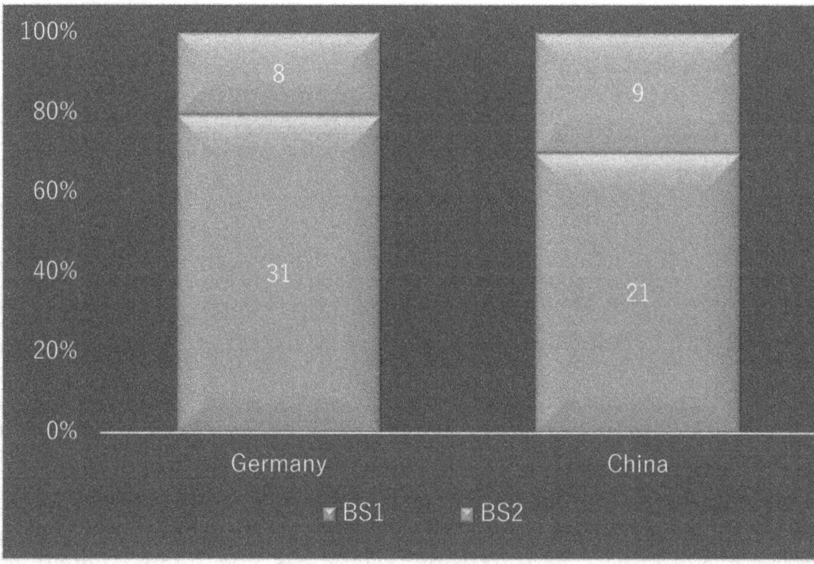

Fig. 8-11 Comparison of variations of exercises

As seen from the statistical result, the ratio of BS1 and BS2 approaches 8:2 in the German textbook and is 7:3 in the Chinese textbook. Both textbooks' exercises have a tendency towards BS1 questions, and the variation is in the surface form characteristics of questions. It can be found that the main functions of the exercises in the textbooks focus on knowledge consolidation, skills acquisition and method mastering, based on which vertical variation questions are added. On the one hand, it can avoid boring students due to repeated learning of simple knowledge; on the other hand, the moderate increase of the difficulty level of exercises could stimulate the desire of students for learning and provide students with more space for learning. It is advised to control the percentage of such questions between 21% and 30%.

As for logarithmic operations, there are two important logarithmic identities, $\log_a a^n = n$ and $a^{\log_a N} = N$, which are not directly given in any of the text or examples or exercises in the Chinese textbook. Instead, teachers should guide students in classroom to deduce according to the definition of logarithm or operation property of logarithm. However, there are two such questions in the Section 5 "Logarithmic Functions" of Chapter III in the German textbook (Pages 58 to 59):

EXERCISE 5. It is $\log_2 32 = \log_2 2^5 = 5$. Determine also:

a) $\log_2 4$	b) $\log_2 16$	c) $\log_2 64$	d) $\log_3 27$	e) $\log_5 125$
f) $\lg 100$	g) $\lg 10000$	h) $\lg 10$	i) $\log_2 2$	j) $\log_3 3$
k) $\log_2 1$	l) $\lg 1$	m) $\log_5 1$	n) $\log_c 1$	o) $\log_u u$
p) $\log_3 \frac{1}{3}$	q) $\log_4 \frac{1}{4}$	r) $\lg 0.1$	s) $\log_3 \frac{1}{9}$	t) $\log_5 \frac{1}{25}$
u) $\log_2 \frac{1}{16}$	v) $\lg 0.001$	w) $\log_5 \frac{1}{625}$	x) $\log_3 \frac{1}{27}$	y) $\log_9 \frac{1}{9}$

EXAMPLE 13. 1) Write numbers 23, 14.2, 63.4, 8.32 and 0.004 in the form of 10^r (r to four decimal places).

2) Determine $10^{\lg 3}$, $10^{\lg 7}$, $10^{\lg 0.5}$, $10^{\lg 10}$, $2^{(\log_2 0.9)}$, $2^{(\log_2 12)}$, and $3^{(\log_3 5)}$.

The two exercises appear in the section of Logarithmic Functions rather than in Section 6 "Logarithmic Operations". Easy operations are used to make students familiar with two basic identities of logarithm, as well as the properties "the logarithm of 1 is 0" and "the logarithm of the base is 1". Positive, zero and negative exponents are all illustrated. Relatively speaking, it is of BS2 type. Exercise 5 is actually the training featuring "inferring other cases from one example". There is also an "Extended Drill" for Exercise 5 —

EXERCISE 7.

a) $\log_2 \sqrt{2}$	b) $\lg \sqrt{10}$	c) $\log_3 \sqrt{3}$	d) $\log_4 \sqrt{4}$	e) $\lg \sqrt{1000}$
f) $\log_2 \frac{1}{\sqrt{2}}$	g) $\log_5 \frac{1}{\sqrt{5}}$	h) $\log_3 \frac{1}{\sqrt{3}}$	i) $\log_8 \frac{1}{\sqrt{8}}$	j) $\log_6 \frac{1}{\sqrt[3]{6}}$
k) $\log_4 0.25$	l) $\log_2 0.25$	m) $\log_2 0.125$	n) $\log_2 0.5$	o) $\lg 0.1$

Six mathematical abilities are proposed in the Standards of German Mathematics Education (CMFRG, 2012). The fifth one is the ability to understand and acquire mathematical symbols, formulas and skills. The ability elements include mathematics definitions, rules, algorithms, awareness and application of formulas. "Level I" to be reached is the ability to use basic solutions and directly apply formulas and symbols. Relevant requirements are shown in Exercise 5 and 7.

The exercises in the American textbook focus on vertical variations and offer life background so as to reach the goal of variations. For example, there is a variation exercise following Example 3, Section 8.6:

EXERCISE 35. (Population) The population of a county can be approximated by the model $P = 3762(0.964)^t$, where t is the number of years since 2000. According to this model, when will the population of the county be about 2,800?

The profile of the problem has changed, but the nature remains the same. This kind of variation questions is also shown in both Chinese and German textbooks (omitted for examples).

(3) Coverage comparison of exercises

The six types of coverage (see Tab. 8-2) in the Chinese and German textbook exercises are counted as follows according to the basic principles of exercise coverage design:

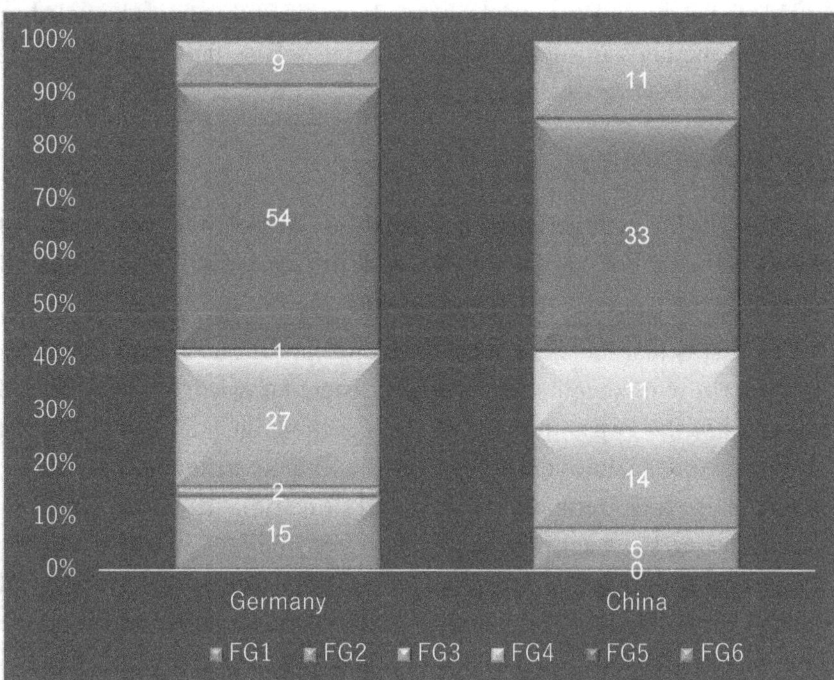

Fig. 8-12 Coverage comparison of exercises

As seen from the statistical result, the two textbooks share the common feature on the coverage: they have the largest proportions of their exercises on the thoughts and methods (FG5 and FG6) appearing in the text and examples, the second one on the skills (FG3 and FG4), and the least one on the knowledge (FG1 and FG2). However, there is a notable difference between the two textbooks: the total percentage of the three "news" — new knowledge (FG2), new skills (FG4) and new methods (FG6) — is 11% in the German textbook, while it is 30% in the Chinese textbook with new skills occupying 15% and methods occupying 14%. It can be seen that German exercises pay more attention to foundation and consolidation, while Chinese textbook exercises are quite flexible and comprehensive. The ability model proposed by the Standard for Mathematics Education of Germany (Cultural Ministry, Federal Republic of Germany 2012) specifies the level of mathematical ability required to be reached by all graduates of Grade 10. Thus, the foundation and consolidation of exercises should be suitable for most students and then a moderate quantity of new knowledge, skills and methods is supplemented so as to satisfy the demands of the students who have spare capacity and desire for further study.

3.2.3 Summary

Through the comparisons and analyses above, the three sets of textbooks have both commonality and differences in the design of exercises, which are discussed and summarized as follows.

The first is that there is a huge difference in the quantity of exercises. On the premise of keeping relatively consistent knowledge points, the US textbook has the largest number of sub-questions in exercises (671); the German textbook comes the second and the Chinese textbook is at the last. Both Chinese and German textbooks have plenty of exercises and their percentages of big questions in revisions are both around 30%. Further comparison shows that US textbook and German textbook has plenty of imitation exercises, while Chinese textbooks pay more attention on variations, with 9% difference between the variation exercises (BS2) in the Chinese textbook and that of the German one. Small proportion of variation exercises may lead to the increase in the quantity of exercises to some extent.

The second difference lies in different choices in the question type design and varying trend in material design. There are only four types concerned in the German textbook exercises and six in both the American and Chinese textbook exercises. Solution-answer questions dominate in the three sets of textbooks; short-answer, construction questions, and proof questions follow. In terms of question material, 60% of the three sets of textbooks are pure mathematics questions and there are different focuses in the background exercises for the three sets of textbooks. With respect to exercise material selection, we should not only emphasize basic knowledge and skills but also pay close attention to the interaction of mathematics and discipline knowledge with actual life to develop students' awareness and ability of application, thus improving their overall understanding towards mathematics.

The third difference is in the unique hierarchies of exercises, and different variations have their own focuses. Consolidation and divergence are emphasized in the hierarchy design of American textbook exercises; imitation is emphasized in the exercise hierarchy design of German textbook exercises, focusing on variations and development; comprehensiveness and complication are focused in the design of Chinese textbook exercises. Meanwhile, in terms of variation drills, the horizontal variation questions are 2 to 3 times of the vertical variation exercises both in the Chinese and German textbook exercises. The vertical variation questions in the Chinese exercises are 9% more than those in the German exercises, which is greatly associated with the educational tradition of both countries. China adopts the college entrance examination system and the importance and selection effort of mathematics in the examination determines the difficulty of mathematics textbook exercises to a large extent. Thus, more comprehensive and vertical variation questions would be integrated into Chinese exercises. Since the German education system is a humanized education system through long-time exploration and practices of governments, enterprises, schools and parents, basic and imitation characters are emphasized in Germany exercises.

The fourth difference is in the varying ratios of exercises covering different aspects. There is a 19% difference in the coverage in terms of new knowledge, new skills and new methods between the exercises in the Chinese and German textbooks. The thoughts and methods appearing in the body and examples have the largest coverage in the two editions of exercises, then skills and finally the knowledge. The coverage of exercise could actually be even broader. Some content that has not appeared in the main text or is not suitable for arrangement may be moderately designed in exercises.

4. Conclusions and Suggestions

4.1 Conclusions

In all the three editions of Chinese, American and German textbooks, the allocation of examples corresponds to the knowledge, with proper percentage between example quantity and quantity of knowledge points. Pure mathematics and background questions are all designed, with diversified background materials involved. The Chinese background examples are obviously more than those in the American and German textbooks. In the three sets of textbooks, concept-principle oriented, approach-oriented and application-oriented examples are designed, with strong exemplary and typical characters. Examples in American textbook more focus on exemplary mathematical methods and processes, the German examples emphasize exemplary mathematical methods and thinking, while Chinese examples pay more attention on exemplary mathematics modeling. "Questions + answers" dominate in the presentation of the examples in all the textbooks, but the presentation of examples in the American and Chinese textbooks tends to be more diversified.

There is a large difference in the allocation of exercises in the textbooks. Relatively speaking, the United States has the most exercises, German comes the second, and China has the least. In terms of exercise type design, there are only four types in the German textbook and there are six in both the American and Chinese textbook. Solution-answer questions dominate in the three sets of textbooks, followed by short-answer questions, construction questions, and proof questions. Pure mathematics questions are the majority in all textbooks, and background questions have different focuses in different textbooks. Each set of textbook has its own characteristics in terms of the hierarchy of exercises, and has the different focuses on the variation drills. For the hierarchy design, consolidation and divergence are emphasized in American textbook, imitation with variations in German textbook, while comprehensiveness and complication in Chinese textbook. The coverage rate of exercises for knowledge, skills, and thought and methods has a common feature in the three set of textbooks, with great difference in details. In particular, there is a big gap in the coverage of new knowledge, new skills and new methods in German behind Chinese textbook.

4.2 Suggestions on the design of examples and exercises

4.2.1 Suggestions on optimizing example design

(1) Enhance targeted exercise design

Considering the effect of different examples as required by different knowledge points, three types of examples, namely concept and principle-oriented, approach-oriented and application-oriented are recommended to be adequately selected and designed. In order to enhance the functions of examples in teaching, we must clarify our targets in designing examples. As for the examples featuring concepts and principles, we recommend the illustrated examples in the German textbook which can facilitate our understanding towards connotation and extension of mathematics concepts. With regard to the approach-oriented examples, the question type or solving tips may be provided before the main questions of the examples just like the American textbook in order to present the application of theorems and formulas. As for the application-oriented examples, in order to present the solving steps, margin notes may be properly designed in the solving process to explain major steps. Adequate enrichments of example design could make the purposes of examples more specific and clear. This not only enables students to grasp basic concepts but also cultivates their ability to formulate problem-solving plans and ultimately develops their thinking ability.

(2) Strengthen basic and enlightening example design

Receiving new knowledge is always on the premise of the knowledge previously obtained and it also provides a foundation for getting newer knowledge. Examples are the "anchoring point" for students to learn new knowledge and represent the initial point for further exercise and extension. Thus, basic and enlightening exercise design must be enhanced. According to targeted example design, for new knowledge and skills, the examples involving a single knowledge point and straightforward thinking should be focused on in basic design, and examples with multiple knowledge points, multiple steps, and multiple thinking dimensions serve as the supplement and support. Expanded and complex examples within the acceptable range of students can also be provided for the comprehensive demonstration of knowledge application, but they should be accompanied with guidance language in line with students' cognitive law to lead students to analyzing problem-solving strategies, associating knowledge and principles to solving procedure, so that the students can summarize and refine the rule by themselves, and finally their self-learning ability can be improved.

(3) Enhance typical and exemplary example design

In order to improve the transference in study, the design of the typical and exemplary examples must be enhanced. The knowledge and principles reflected in examples, methods used in problem-solving and background of questions presented should serve as a sufficient demonstration for subsequent exercises. In consideration of limited teaching hours, limited space for textbook arrangement and other practical problems, examples must be refined and typical in design. Both pure mathematics questions and questions with backgrounds need to be meticulously prepared. The pure mathematics examples must be refined from the diversity and essence of the mathematics. Coupled with diversified life and disciplinary backgrounds, the questions with backgrounds selected should be able to contain typical mathematics models and enable students to feel and experience the application of mathematics in the real world. Examples should accompany with complete solving/answering process to necessary details. The mode of "questions + answers + analysis + major step instructions + summary" is not compulsory, but some parts therein to be cut off only if the exemplary function of examples is guaranteed as practically as possible.

(4) Enhance systematic and progressive example design

Systematicness is the most important characteristic of mathematics, which emphasizes the completeness of knowledge network and realization of basic functions. For some knowledge that is simple and easy to understand, such as familiarity with formulas and interpretation of theorems, students may understand through their self-learning, together with the help of teachers by compiling relevant examples according to the circumstances of classes. For some knowledge that is crucial but not easy to understand, examples or families of examples are essential for students learning, and the examples may be accompanied with some aided questions such as "analysis", "summary" and "think". Systemticness is never accomplished at once but requires subsequent continuous accumulation. Thus, in addition to the examples in the text, the "examples in the exercises" may be designed in a progressive manner. For example, in the summary and review at the end of each chapter of the American textbook, examples are designed before exercises to perfect the systemticness. In German textbook, "examples in the exercises" are often demonstrated in the statements of exercises. This not only enables students to digest relevant knowledge but also to discover the laws therein.

(5) Properly design the variations and summary of examples

Examples should be provided with some variations on a refined and typical basis. "One question, multiple solutions" and "one question, multiple variations" may be adopted in the designs. Extended considerations may also be made after answering so as to reserve more space of thinking for students. The examples should be flexible and varying in question type and answering. For example, several examples may be grouped into a family or a block to reflect a key point or a key concept, or to train a basic skill from different perspectives. Also, summaries should be properly designed in the presence of examples so as to further sublimate new knowledge, new methods and new thoughts to the mathematical system theory and perfect the mathematics mind map of students. A summary may be given in text or be combined with the use of graph calculator, infiltrating the application of IT to help students' learn.

4.2.2 Suggestions on the optimization of exercise design

(1) Enhance targeted and comprehensive exercise design

The exercise design in the textbook should interact with the examples and knowledge and especially facilitate the transfer of knowledge, methods and thoughts. The association between exercises and examples include question background, type, solving methods, skills, knowledge and principles, etc. Questions should be designed in consistency with examples. Some questions of a higher difficulty level should also be designed based on examples or new knowledge so as to consolidate the new knowledge acquired and promote students to gain further development. For the systematicness of knowledge and examples, the comprehensiveness of exercise design should be paid more attention to. Based on the knowledge, skills and methods covered in the text and examples, we should also supplement the exercises with more knowledge and challenges so as to satisfy the demands of different students.

(2) Enhance basic and hierarchical exercise design

Exercise design should not only be complete, systematic and targeted, but also fundamental and hierarchical. Due to differences in the initial cognition and learning ability of students, it is impossible for all students to reach the same level of problem solving. Thus, hierarchical design of exercises must be conducted on the basis of guaranteeing that most students can do most of the exercises. It is found through comparison that the examples and exercises in the American and German textbooks have a higher degree of matching, though in German textbook certain question type in exercises not appeared in text or as examples, but tips would always be given in the these questions. However, the matching degree in Chinese textbook is not so ideal. Despite obvious hierarchy, most question types never appeared in the text or examples, paying more attention on comprehensiveness and difficulty. Exercise design is advised to move from the mostly fundamental to partly medium and then moderately comprehensive variation questions in a gradual and subtle manner.

(3) Properly design the variations and comprehensiveness of exercises

On the premise of ensuring fundamentality and hierarchy, exercises should be of moderate variations and comprehensiveness in design. While keeping the basic horizontal variations, the vertical variations should be developed in a moderate manner so as to provide students with different space for improvement. Variations could start either from question types or materials. Questions can be designed with the same nature yet with varying types and materials. Life settings and disciplinary styles should be adequately designed so as to spur students' interest in learning. Alternatively, we could start from question structure variations to interpret the connotation and extension of questions in an all-round manner so as to set up a platform for students to march towards a higher level.

(4) Increase the interest and applicability of exercise design

The mathematics of life style could arouse students' learning interest and stimulate their desire for exploration. Question types may moderately transit from traditional exercises to new ones such as exploration, think, investigation and simulation. A commonplace example/exercise with no life settings or background couldn't tell students the reason they need to learn mathematics other than being a course taught at school. We should put the mathematics knowledge learnt into the settings of our daily life that are familiar to students. This would help to make students aware that studying mathematics could better solve the problems in the real world and encourage them to apply their mathematics knowledge, hence remarkably increase the efficiency of teaching and learning.

References

[1] Atklinson, R. K., Derry, S. J., Renkl, A. and Wortham, D. (2000). Learning from examples: instructional principles from the worked examples research. *Review of Educational Research*, 70(2): 181–214.

[2] Bao, Jiansheng (2002). Comparative study on composite difficulty of Chinese and British school mathematics Curricula [in Chinese]. *Global Education*, 31(9): 48–52.

[3] Cai, Shanghe (1990). Talk about the exercises in mathematics textbook in junior middle school [In Chinese]. *Middle School Mathematics*, (2): 30–32.

[4] CMFRG (Cultural Ministry, Federal Republic of Germany) (2012). *Bildungsstandards im Fach Mathematik für die Allgemeine Hochschulreife*, Beschluss der Kultusministerkonferenz vom 18. 10. 2012.

[5] Dai, Zaiping (1999). *Theory of Mathematics Exercise* [In Chinese]. Shanghai: Shanghai Education Press.

[6] Halmos, P. R. (1982). The heart of mathematics [translation into Chinese]. *Bulletin des Sciences Mathematics* (*Shuxue Tongbao*), 21(4): 27–31.

[7] Howson, G. (1995). *Mathematics Textbooks: A Comparative Study of Grade-8 Texts.* Vancouver: Pacific Education Press.

[8] Kalyuga, S., Chandler, P., Tuovinven, J. and Sweller, J. (2001). When problem solving is superior to studying worked examples. *Journal of Educational Psychology*, 93(3), 597–588.

[9] Lai, Risheng (2005). Research on effects of worked-example presentation form and cognitive style on mathematics learning performance of senior high school students [In Chinese]. Jiangxi Normal University.

[10] Luo, Zengru (2001). *Introduction to Mathematical Problem Solving Theory* [In Chinese]. Xian: Shaanxi Normal University Press.

[11] V. A. Oganesian (В. А. Оганесян) *et al.* (eds) (Translated by Liu Yuantu et al) (1983). *Pedagogy of School Mathematics* [In Chinese]. Beijing: SinoMaps Publishing House.

[12] Ren, Jinjie and Lu, Haidong (2009). Effects of self-explanation and learning from worked examples on university students' transfer of mathematics learning [In Chinese]. *Heilongjiang Researches on Higher Education*, (9): 188–190.

[13] Ross, B. H. and Kilbane, M. C. (1997). Effects of principle explanation and superficial similarity on analogical mapping in problem solving. *Journal of Experimental Psychology: Learning, Memory, and Cognition*, 23(2), 427–440.

[14] Shao, Guanghua (1997). Preparation requirements of "four exercises" in high school mathematics textbooks [In Chinese]. *Bulletin des Sciences Mathematics* (*Shuxue Tongbao*), 36(11): 10–11.

[15] Shao, Guanghua (2003). Theoretical and experimental study on learning from mathematical example [In Chinese]. Doctoral Dissertation. Shanghai: East China Normal University.

[16] Sun, Xuhua, Huang,Yiying, Lin, Zhizhong and Zhang, Dianzhou (2006). The unification of structure and function in mathematics problems variations [In Chinese]. *Curriculum, Teaching Material and Method*, 26(5): 25–31.

[17] Wang, Xiaoming (2007). Research on Worked-out Examples Study Abroad [In Chinese]. *Elementary and Secondary Schooling Abroad*, (1): 48–51.

[18] Xu, Binyan (2007). On mathematical competence model in mathematics education standards in Germany [In Chinese]. *Curriculum, Teaching Material and Method*, 27(9): 84-87.

[19] Zhu, Xinming, Qin, Yulin, Shi, Tieru and Si, Mahe (1987). Production establishment through examples and problem solving [In Chinese]. *Acta Psychologica Sinica*, (2): 176–183.

https://doi.org/10.1142/9789814713955_0009

Chapter 9

International Comparison on the Application of Information Technology in High School Mathematics Textbooks

Chao Zhou

School of Mathematical Sciences, Soochow University
Suzhou, Jiangsu 215006, China
zhouchao@suda.edu.cn

The application of information technology has become a major area of study for the international mathematical education field; however, as for the integration of information technology and mathematics curriculums, we should also consider how to make information technology an effective classroom teaching means. Our research idea is to compare both similarities and differences in information technology handling of China and other countries under a rational framework and then to explore how to adjust the information technology in the high school mathematics textbooks of China.

1. Introduction

For long, textbook has been regarded as one of the comparative objects in international comparative studies of mathematical education. The fact that mathematics textbooks are considered as one factor influencing students' mathematical accomplishments in large-scale international comparative studies of mathematical education can be dated back to the Second International Mathematics Study (SIMS, 1981−1982) conducted in the 20th century, after which an international mathematics textbook resource database was established in Michigan State University. In the Third International Mathematics and Science Study (TIMSS, 1955, 1999) and Trends in International Mathematics and Science Study (TIMSS, 2003, 2006, 2009), the mathematics textbooks of participating countries were regarded as an important background factor of study. Also, a series of research frameworks, methods and tools were proposed. For example, while studying textbooks, TIMSS focuses on teaching content, expectation on students, points of view for training, teaching forms, teaching methods, applications of practical activities/experiments and other aspects. Let us take one more example. Since 1999, the KASSEL project adopted in 18 countries has made comparisons on textbooks from the perspectives of introduction of new concepts and methods, importance of mathematical theories and rules, organization of disciplinary structures, role of proof and proving, emphasis on laws and examples, etc. Many small-scale multinational research projects also compared textbooks.

As for the textbook comparisons in the existing studies of China, some use various difficulty models to compare the difficulty levels of textbooks in two countries (see Jiansheng Bao, 2009; Lili Hu, 2008); some compare the similarities and differences of the textbooks in two countries in terms of introduction modes, teaching objectives, presentation means, knowledge scope, knowledge structures, examples and exercises by focusing on the content of specific mathematical topics (e.g., Fenping Wang, 2011); still others compare the existing Chinese textbooks of different versions in terms of system arrangement, material selection, content presentation (e.g., Lili Song, 2010).

The study of the international mathematical education field on technologies originates from the 3rd International Congress on Mathematical Education (ICME-3, 1976), at which the "Application of Computers and New Technology in Classroom Teaching" is one topic of the congress. Since then, each session of ICME regards technologies and mathematical education as important topics. Technologies have become a hot topic that is continuously paid attention to: from the first study series "The Influence of Computers and Informatics on Mathematics and Its Teaching" conducted by the International Commission on Mathematical Instruction (ICMI) in 1985 to the 17th study series "Digital Technologies and Mathematics Teaching and Learning-Rethinking the Terrain" by ICMI in 2006 (Jiahong Xu, 2011).

Seen from literatures published in English, most of them focus on technological tools. For example, F. K. S. Leung *et al.* (2008) made a detailed classification of the technological tools in mathematical education based on whether they are software or hardware at first (Fig. 9-1).

Hardware
- Processors
  - Computers
  - Calculators
  - Other hand-held devices
- Input/Output Devices
  - Input
    - Data loggers
    - Internet
    - Digital cameras etc.
  - Output (whole-class display)
    - Interactive white boards
    - OHP panels etc.

Software
- General purpose
  - Spreadsheet
- Programming
  - LOGO
  - Calculator (BASIC)
- Mathematics
  - Computer Algebra Systems (CAS)
  - Dynamic Geometry Software (DGS)
  - Graph Plotting Software (GPS) (which run on computers or graphing calculators)
  - Data Handling Software (DHS)
- Other specific programs

Fig. 9-1 Classification of technological tools in mathematics education

Then, the author summarized the existing studies and findings of the potential role of five technological tools, (namely interactive white boards, graphing calculators, Spreadsheet, Computer Algebra System and Dynamic Geometry Software) in mathematics teaching and learning by means of literature review. He also summarized by means of cases the official policies concerning technological tools of mathematics teaching and learning adopted in different countries and introduced the results of actual use of technical tools in class obtained from the TIMSS1999 teachers and students' questionnaires.

Another typical characteristic of literatures written in English is to focus on students' ways of using technological tools during the process of learning. For example, Doerr and Zangor (2000) paid attention to the role of graphing calculator and its application forms. They made a classification from the perspective of student activities and pointed out students' orientation on the role of graphing calculator and their activities respectively (see Tab. 9-1). The paper also mentioned the two limitations of the graphing calculator: (1) Students' attempted uses of the device as a

Tab. 9-1 Using modes and methods of graphing calculator

Role of the graphing calculator	Description of student activity
Computational tool	Evaluating numerical expressions, estimating and rounding
Transformational tool	Changing the nature of the task
Data collection and analysis tool	Gathering data, controlling phenomena, finding patterns
Visualizing tool	Finding symbolic functions, display data, interpreting data, solving equations
Checking tool	Confirming conjectures, understanding multiple symbolic forms

"black box" without attending to meaningful interpretations of the problem situation; and (2) the personal (or private) use of the tool.

Some other literatures discuss how to use technologies to facilitate mathematics teaching and learning. For example, Hans-Georg Weigand (2010) conducted a pre- and a post-test of teaching by calculator through teaching experiments and constructed an ability model to describe, analyze and evaluate students' ability to use Symbolic Calculator (SC) during the process of solving function problems. The ability model is shown in Tab. 9-2.

Tab. 9-2 Ability to use tools

Levels	Meanings	Examples
1（Static Mode）	Using the SC as a tool which produces static representation	Graphing the function f with $f(x) = x^2 + 1$
2（Dynamic Mode）	Creating dynamic representations	Graphing a family of functions f_a with $f_a(x) = ax^2 + c$, $a, c \in \mathbb{R}$, while changing the parameters via sliders, dragging a graph, tracing a graph or scrolling a table
3（Symbolic Mode）	Working with the SC on the symbolic level	Doing term manipulations by special commands (factor, expand) or solving equations with the "solve-command

The author also associated understanding of functions with tool competence (symbolic calculators) and constructed a relational table between the two (see Tab. 9-3), which provided a good example for us to analyze the influence, role and expression forms of technological tools in the learning of specific mathematical content.

Many studies paid attention to using specific technology in the teaching of the specific mathematical content. However, as a whole, there is a lack of research concerning comparative study of technological usage in textbooks.

As for the existing studies concerning the combination of information technology and textbooks in China, some regard technical comparison as a small proportion of textbook comparison, such as studies conducted by Weizhong Zhang and Fangqi Li (2009), Xiaoping Zhao and Xue Yao (2010) and Tianxiao Zhang and Caibin Tang (2005) and quite a few master's theses on textbook comparison; some make a special comparison

Tab. 9-3 The UF (understanding of functions) and TC (tool competence) relationship

Mode	Intuitive understanding	Conceptual understanding	Relational understanding	Structural understanding
Symb		Simplification of — simple — terms as a substitute for hand calculations	Seeing relations between properties of functions in different representations	Simplification of complex terms
Dyna	Plotting the given values e.g. of a table dynamically	Plotting representations of families of functions. Using sliders for dynamic representations.	Seeing relations between families of functions and their representations.	Explaining composition of functions by dynamically interpreting operations.
Static	Plotting graphs and diagrams of given data	Plotting representations of functions (GC)	Seeing the relation between functions and their representations.	Explaining solutions of equations from a functional point of view

on the application of technologies in mathematics textbooks, such as comparisons in terms of types of information technology adopted and integration models (see Zhiqiang Yuan, 2008), models on integration of information technology and textbooks, hardware and software adopted and their functions, content in which information technology and textbook integrated, etc. (Chao Liu and Zhijun Wang, 2011); and two master theses (see Yuanyuan Zhao, 2009 and Xue Yao, 2010) make a special comparative study of the application of technologies in textbooks. In addition, the application of technologies in textbooks is mentioned in a lot of papers, such as study made by Yingmei Cui (2006). However, throughout those studies, there is a lack of relatively systematic and standard research framework and comparison in the cross-cultural field.

In conclusion, the application of information technology has become a major area of study for the international mathematical education field; however, as for the integration of information technology and mathematics curriculum, we should also consider how to make information technology an effective classroom teaching means. Our research idea is to compare both similarities and differences in information technology handling of China and other countries under a rational framework and then to explore how to adjust the information technology in the high school mathematics textbooks of China.

2. Research Object

In addition to the high school mathematics textbooks (Edition A) published by the People's Education Press, which are widely used in China, the other textbooks selected in this study are from Singapore, USA, Australia and Japan.

The main reasons for selecting the textbooks from those countries for comparison include the following aspects:

(1) Although the mathematics curriculums in the countries above differ in the background of development, they are amazingly identical in the basic concept of technological application. They pay attention to the integration of information technology and mathematics curriculums and identify the infiltration of information technology in mathematics curriculums. However, there is a slight difference in the assessment of students' application of information technology.

For example, the teaching syllabus, which started to be implemented in Singapore, 2013, identifies "teaching should connect learning to the real world, harness ICT tools and emphasize 21st century competencies" (Singapore Ministry of Education, 2012; p. 21) as one of the teaching principles and "effective application of mathematical tools (including information technology tools) during the process of learning and applying mathematics" (ibid, p. 33) as target.

In the USA, the "technology principle" is regarded as one of the six major principles in the curriculum standards issued by the well-known NCTM (abbreviation for the National Council of Teachers of Mathematics) in 2000. In the *Common Core State Standards for Mathematics* (CCSSM) finalized and announced in the early June, 2010, the application of mathematical tools has been listed as the core mathematical thinking ability and an important part of mathematical competence or practice. "The use of mathematical tools (or technologies) is never new ... however, it is undoubtedly a breakthrough by improving it to the core mathematical competence dimension, which is no doubt an important influence of modern mathematical technology development (especially the support from computer technology development) over the mathematicians' working modes and ways of mathematical thinking. Also, such influence is reflected in mathematical education." (Yiming Cao *et al.*, 2010, p. 10)

The Australian Curriculum (AC), the first unified curriculum standard in Australia ever issued in history, lays special emphasis on students' mathematical competence, in which communication information technology ability is listed as one element (Yueyuan Kang, 2011). Although no universal high school graduation examination is offered in the country, it is strictly specified in some provincial examinations that the graphing calculator should be used. For example, the graphing calculator is allowed to be used in the high school mathematics teaching and examinations of Victoria and South Australia, (Yaohua Hu, 2009). In Victoria, both teachers and students are encouraged to use the technology [including graphing calculators, Computer Algebra System (CAS), calculators, Spreadsheet, graph and digital analysis software package, Dynamic Geometry System (DGS)] as an appropriate tool for new knowledge learning, skills training, standard application and study (Leigh-Lancaster, 2004).

The concept of computer has been introduced into mathematics curriculums in the *Guiding Essentials for Learning of Higher Institutions* revised by Japan in 1978 (Jiahong Xu, 2012).

The prevailing *High School Mathematics Curriculum Standards (Experimental)* of China (MOE of PRC, 2003; hereinafter referred to as "the Standards") also regards the emphasis on the integration of information technology and mathematics curriculum as one of the ten basic concepts. It is specified in the Standards (p. 6) that, "the high school mathematics curriculums should encourage the presentation of curriculum content that is hard to be presented in the previous teaching by information technology. While guaranteeing manual computation practice, we should apply scientific calculators and mathematical education technology platforms, enhance the combination of mathematical education and information technology and encourage students to apply computers and calculators for exploration and discovery."

(2) The educational result of each country has its own characteristics.

First, the excellent mathematical education of Singapore makes a remarkable contribution to the economic success of the country. In the TIMSS (1995, 1999, 2003), Singaporean students achieved the highest scores on average in the mathematical achievement testing for three consecutive sessions. The proud achievement made by Singapore in mathematics attracted the attention from many other countries. Those counties tried to find out the reasons of the country's success in the discipline. The US Department of Education also provided special funds and organized experts to conduct studies (Hongyu Su, 2007).

Second, since the 1970s, the USA has dedicated itself to the integration of information technology and curriculums encouraged "student-centered teaching", held the idea that students should be capable of self-management and emphasized the cultivation of students' self-learning ability and provided sufficient autonomously allocated time. Most of the assignments for American students require manual operation and most of them need a computer to acquire information. Thus, American students have a potential ability of operation.

In the TIMSS study, the students of lower grades in Australian high schools are high in performance level (Lokan *et al.*, 1996), so are the pupils at primary schools (Lokan *et al.*, 1997). In addition, in terms of expression ability, Australian students are obviously better than Chinese students when expressing their ideas or raising questions in public. They speak frankly with assurance, never act shy, have a broad knowledge scope and are quite familiar with the topics concerning politics, economy, history and culture. They are skillful at manual operation, initiative and creativity, quite independent and confident. All these distinguished advantages of Australian students do not only relate to cultural heritage and family education but also be associated with the education in secondary and primary schools of Australia (Xiangyang Chen *et al.*, 1997).

The TIMSS video tape study in Japan has left us a quite deep impression: Teachers almost make all their efforts in the skills for solving special problems, mathematical thinking and understanding; the routine problems and the problems to be thoroughly considered are almost identical in quantity; there is a high frequency for occurrence of multiple solutions corresponding to one problem; the mathematical level in classroom is relatively high (the proportion of low, medium and high levels is 13:57:30) and the requirements on classroom assignments is high as well (the proportion of procedure utilization, application of concept and creative thinking is near 3:1:3) (National Center for Education Statistics, 1996), which matches the ideal mathematics classroom in China.

(3) The mathematical education in the four countries has a close connection with that in China. China, Japan, USA, Australia and Singapore differ in politics, economy, culture, etc., but they have a long history in the communication of mathematics and mathematical education. A lot of researchers have paid great attention to the mathematical education of those countries (Daiqin, 2011; Jinfa Cai, 2007; Han Fang, 2010; Shenglan Li, 1995; Weiping Zhang, 2012).

In Singapore, the high school mathematics curriculums are divided into three types, H1, H2 and H3. H2 (designed for the students who plan to study mathematics, science and engineering) includes the content of H1 (designed for the students who further study business and social science in the future) and H3 aims to help the students who have special emotions towards mathematics to deepen and develop advanced mathematics concepts and skills (Weiling Lu, 2012). Thus, this study chooses the H2 (Mathematics Volume 1, 2) (Federick Ho *et al.*) and published by Panpac Education as the objects of study for high school mathematics textbooks in Singapore.

For Australia, the high school mathematics textbooks chosen is the Mathematics Higher Level (core) (The Third Edition) (hereinafter referred to as AU-MHL textbooks). The AU-MHL textbooks are subject to the review by the Victorian Curriculum and Assessment Authority (VCAA) and makes refer to the October Examination Papers of the Incorporated Association of Registered Teachers of Victoria. Thus, they do not only reflect the current status of mathematics textbooks in Australia but also keep an international level and consistent style of international mathematics textbooks (Didi Lin, 2012).

Two editions of high school mathematics textbooks are selected in the USA: one is the Prentice Hall (hereinafter referred to as "US-PHM") for high schools issued in 2009 by the Pearson Education Press in accordance with the NCTM curriculum standards, including PreAlgebra, Algebra 1, Algebra 2 and Geometry, which are typical American curriculums (Shushu Zhou, 2012). The textbooks for primary and high schools published by Pearson Education, one of the global renowned education publishing companies, occupy about 24% of the entire American market (Wangquan Zheng, 2012). The US-PHM (2009 Edition) also represents the latest educational concepts and thoughts of the USA. Another edition is the Algebra and Geometry in the UCSMP (abbreviation for The University of Chicago School Mathematics Project) textbooks complied by the University of Chicago and it is one of the textbooks with the widest application scope. One of its compiling requirements is that the application of calculators and computers should run through the whole mathematics curriculums (Chao Zhou, 2013).

The Japanese textbooks include 7 volumes including *Mathematics Fundamentals, Mathematics I, Mathematics II, Mathematics III, Mathematics A, Mathematics B* and *Mathematics C, which are* published by So-In Education. The reason for selecting those textbooks is that the So-In publishes many types of textbooks, which can satisfy the demands of different schools (Yuelan Chen, 2010). Thus, more schools opt to select the press.

The Chinese textbooks are the High School Curriculum Standard Experimental Textbooks — *Mathematics (Compulsory 1–5)* and *Mathematics Optional (1–1, 1–2)* (Edition 2007) published by the People's Education Press. The main reason for selection of such textbooks is that they have the widest application scope, the most users and the largest influence throughout the country.

3. Research Framework

The analysis index system for information technology in the textbooks of this study is based on four dimensions including technological tools, presentation means, content links (where they appear) and application forms (Jiahong Xu, 2011). The secondary indexes under those dimensions are shown as Tab. 9-4.

Tab. 9-4 Analysis framework and index system in this study

Technological Tools	Calculators Computers Internet Others
Presentation Means	Integration Topics Asides
Content Links	New knowledge introduction New knowledge exploration Examples Exercises Extension
Application Forms	Tools Problem solving Programming

Based on the existing comparative studies, we established an index system upon gradual perfection and segmentation, chose the textbooks from 1 to 2 countries for pilot study, and further perfected the index system. While comparing the information technology in different textbooks, this study mainly counts the frequency of the appearance of each secondary index and then transfers to the percentage of the secondary index so that a full understanding of each primary index for the textbooks in different countries would be obtained.

3.1 Instructions to the dimension of "technological tools"

According to the types of technological tools as shown in Tab. 9-4, the "calculators" include scientific and graphing calculators; "computers" refer to the computers not linked to the internet and only rely on the in-built functions of computers such as PPT, Geometric Sketchpad and Excel.

3.2 Instructions to the dimension of "presentation means"

According to the presentation means shown in Tab. 9-4, the meanings of the three secondary indexes are shown as Tab. 9-5.

Tab. 9-5 Meanings of indexes in presentation means

Integration	The use of technologies appears in the body of a textbook (introduction, new knowledge and examples) and constitutes a part of the textbook (If such part is deleted, the context will be short of continuity and completeness). It may appear in the places such as new knowledge introduction or background presentation, new knowledge exploration or law discovery (properties and conclusions or theorems) as well as knowledge application (such as examples and conclusions verification).
Topics	The centralized collection, special study or discussion of the content concerning information technology appear in textbooks in a relatively independent manner, which introduces the basic operation and application of information technology, presents the information in relatively independent forms or shows users how to use technology (including hardware (such as graphing calculators) and software (such as CAS)) and apply information technology for problem solving or mathematical exploration activities.
Asides	Supplements or instructions to the body of a textbook such as giving tips on the body or enlightening learners on consideration. The aside mentioned in this study particularly refers to the aside concerning technologies.

3.3 Instructions to the dimension of "content links"

According to the content links in Tab. 9-4, which the links appear in the process of teaching is analyzed. "New knowledge introduction" means information technology appears in new knowledge introduction or background presentation. For example, in Section 3.1.1 of Compulsory 3, Edition A, People's Education Press, the concept of probability is introduced (see Fig. 9-2).

We can use computers to run simulation tests of coin toss. The figure below shows the result of a simulation test of coin toss.

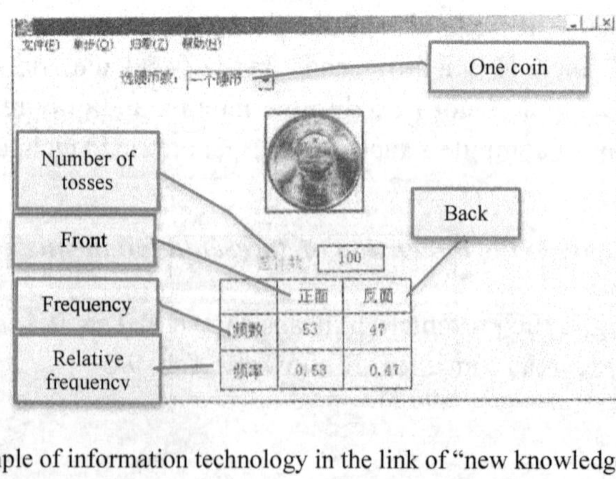

Fig. 9-2 Example of information technology in the link of "new knowledge introduction"

The "new knowledge exploration" in the content link means that information technology is used as an exploring tool to discover laws, properties or theorems of mathematics. For example, in Section 3.1.2, Compulsory 1, Edition A, People's Education Press ("Solving the Approximate Solutions of Equations by Dichotomy"), the exploration on the mathematical laws contained in the dichotomy is carried out by computer (see Fig. 9-3).

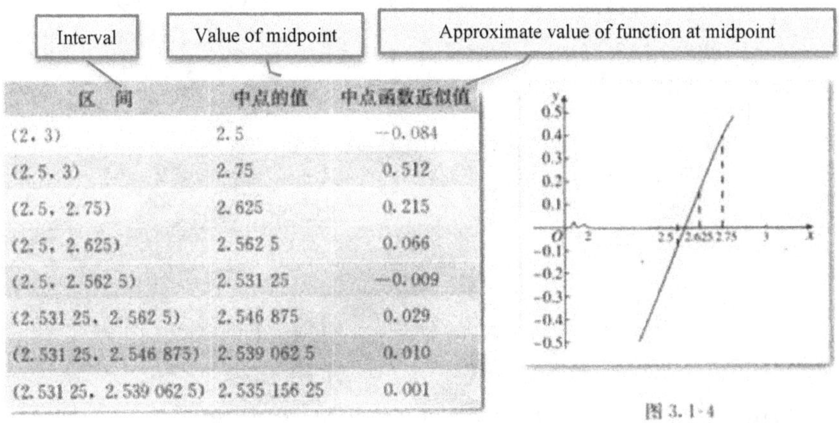

区 间	中点的值	中点函数近似值
(2, 3)	2.5	−0.084
(2.5, 3)	2.75	0.512
(2.5, 2.75)	2.625	0.215
(2.5, 2.625)	2.562 5	0.066
(2.5, 2.562 5)	2.531 25	−0.009
(2.531 25, 2.562 5)	2.546 875	0.029
(2.531 25, 2.546 875)	2.539 062 5	0.010
(2.531 25, 2.539 062 5)	2.535 156 25	0.001

Fig. 9-3 Example of information technology in the link of "new knowledge exploration"

The "examples" and "exercises" in the content links have explicit meanings, which refer to the application of information technology in the examples and exercises in class. There is no need for further elaboration here.

The "extension" in the content link refers to the parts beyond teaching links for teachers' selection at their discretion, aiming to broaden students' vision and help them understand the history of mathematics and the applicability of mathematics. For example, the extended program in Chapter 3, UCSMP Algebra, USA advises students to explore the integer solutions of linear equations by calculator or computer, as shown in Fig. 9-4.

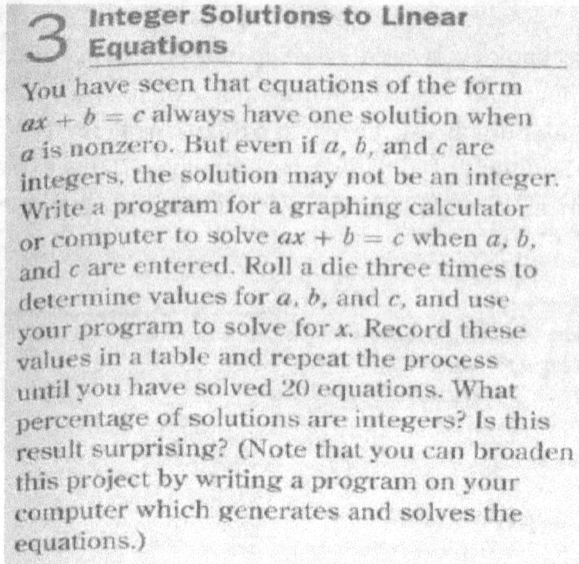

3 Integer Solutions to Linear Equations

You have seen that equations of the form $ax + b = c$ always have one solution when a is nonzero. But even if a, b, and c are integers, the solution may not be an integer. Write a program for a graphing calculator or computer to solve $ax + b = c$ when a, b, and c are entered. Roll a die three times to determine values for a, b, and c, and use your program to solve for x. Record these values in a table and repeat the process until you have solved 20 equations. What percentage of solutions are integers? Is this result surprising? (Note that you can broaden this project by writing a program on your computer which generates and solves the equations.)

Fig. 9-4 Example of information technology in the link of "extension"

3.4 Instructions to the dimension of "application forms"

The meanings of the secondary indexes with respect to application forms are interpreted in Tab. 9-6.

Tab. 9-6 Meanings of indexes with respect to application forms

Tools	Methods to solve problems by technology, which lays a direct emphasis on the mechanical operation by the application of information technology, without or almost without the involvement of mental activities, such as four arithmetic operations, extraction of square root and calculation of trigonometric function values, etc. by using calculators.
Problem solving	The process of applying technologies to problem solving, which emphasizes the involvement of a lot of mental activities, possibly accompanied by some mechanical operations. For example, in the content of function, a graphing calculator is used to draw the graphs of two functions and calculate the intersection point of two graphs.
Programming	A form to use computer programming languages (e.g. VB) to solve mathematical problems, such as method of successive division in the algorithms.

To be clear, we make a slight revision of the index system when encoding the first set of textbooks of the USA (US-PHM), changing the "programming" in the "application forms" to be "extracurricular tutoring". Some online courseware, resources or technological application guidance are given in those textbooks for students to select at their discretion in their spare time.

3.5 Internal consistency test of encoders

In order to guarantee the consistency of encoding by different researchers on the same research object, upon the determination of the analytical framework and index system, we selected three persons in the research group (marked as Groups A, B and C), conducted a pilot study of Chapter 3, Algebra 2 of the textbook Prentice Hall in the USA (Bellman *et al.*, 2003) and ran a Kappa consistency test on the study result using the SPSS software.

In order to conduct Kappa consistency test, we encode the index system, which is shown as Tab. 9-7.

Tab. 9-7 Index encoding

Index System		Code
Technical tools	Calculators	1
	Computers	2
	Internet	3
	Others	4
Presentation means	Integration	5
	Topics	6
	Asides	7
Content links	New knowledge introduction	8
	New knowledge exploration	9
	Examples	10
	Exercises	11
	Extension	12
Application forms	Tool-oriented applications	13
	Problem solving	14
	Extracurricular tutoring	15

Convert the codes of statistical data. Input the code in the data editor of PASW Statistics. Choose Analyze→Descriptive Statistics→Crosstabs by using menu and select KAPPA on the statistics button. The final result is shown as follows (Tab. 9-8–Tab. 9-10).

Tab. 9-8 Consistency test result of Groups A and B

		Value	Asymptotic standard error[a]	Approximate value (T[b])	Approximate value (Sig.)
Consistency measurement	Kappa	.792	.039	24.496	.000
N in valid cases		120			

a. Zero consumption is not supposed.
b. Zero consumption is supposed using asymptotic standard error.

Tab. 9-9 Consistency test result of Groups B and C

		Value	Asymptotic standard error[a]	Approximate value (T[b])	Approximate value (Sig.)
Consistency measurement	Kappa	.810	.038	24.481	.000
N in valid cases		120 120			

a. Zero consumption is not supposed.
b. Zero consumption is supposed using asymptotic standard error.

Tab. 9-10 Consistency test result of Groups A and C

		Value	Asymptotic standard error[a]	Approximate value (T[b])	Approximate value (Sig.)
Consistency measurement	Kappa	.867	.033	26.059	.000
N in valid cases		120			

a. Zero consumption is not supposed.
b. Zero consumption is supposed using asymptotic standard error.

It can be known from the tables above that the Kappa between any two is larger than 0.75, with good consistency effect, which can guarantee the feasibility of study and consistency of study result.

4. Results of the Comparative Study

Our team sorted out tool types, presentation means, content links, application forms, etc. of high school mathematics textbooks in different countries under the above index system and got both advantages and disadvantages of information technology in high school textbooks of China.

4.1 Comparison on tool types of information technology in the textbooks of different countries

First, from the perspective of the variety of tool types, except the Chinese and American textbooks, the textbooks of the other three countries mentioned only one type of technological tools: (1) the only technological tool used in the high school mathematics textbooks of Singapore is the graphing calculator TI-84 plus; (2) only the graph calculator TI-83 is used as technological tool in the Australian textbooks; (3) the information technology in the Japanese textbooks So-In edition is only computer. There are two versions of software: one is universal software — Spreadsheet (22.89%) and the other is programming software — BASIC (77.11%).

Both Chinese and American textbooks use wide ranges of technological tools. The technological tools adopted by the teachers for high school mathematics textbook (Edition A) published by the People's Education Press include calculators, computers and internet and the frequently used software are Excel, Word and Sketchpad. PowerPoint is also popular in the actual application. The technologies adopted in the curriculums for primary and high schools of America are in a certain order (Chao Zhou, 2013): In the lower grades of primary school, simple calculators are used; in the higher grades of primary school, scientific calculators are used to deal with fractions; in the early stage of junior high school, graphing and geometric calculators are used; in the late stage of junior high school, CAS is used. In comparisons, the American textbooks use graph calculators, Spreadsheet and CAS. Also, the USA is the only country to use internet in textbooks. Each volume of the first American textbooks (US-PHM) of the USA is provided with a code for logging after class to find out the content of learning, including video tutoring of key concepts and other support.

Second, from the perspective of "the role of technological tools during the process of learning and teaching of mathematics", the tools in the textbooks of the countries above are rich in function: (1) The TI-84 plus of Singapore could play various roles during the process of mathematics teaching and learning such as calculation, transformation, data acquisition and analysis, visualization and mathematical demonstration (Doerr and Zangor, 2000) and it is one of the technological tools designated in the GCE A-Level test of Singapore; (2) In addition to basic functions as a scientific calculator, the TI-83 of Australia has many other functions including function chart making, polar coordinate charting, parameter charting, sequence charting, statistical analysis, triangular and algebraic functions. (3) The graphing calculators prevailing among American students have included TI-Nspire and TI-Nspire CAS (launched in September 2007). They can not only calculate numbers and display function graphs but also play the functions for solving equations (including differential equations), factorization, derivation, integration, matrix manipulation, data processing, solving regression curves, etc. By using the geometric functions of calculator, students are able to carry out compass and straightedge constructions, draw point paths and curve envelopes, calculate length and angle, measure area, conduct translation, rotation and turnover of graphs. Also, functions like animation and automatic data acquisition as well as program functions of many complicated tasks can be finished. (4) The calculators used in the Chinese high school textbooks have plenty of functions including four arithmetic operations, power, root and reciprocal operations, trigonometric functions and logarithm operations, angle-radian conversion, Cartesian-polar coordinate conversion, permutation and combination, factorial operation, computational programming, statistical operation, etc.

There are obvious differences in the types of technological tools used in the Chinese and foreign high school mathematics textbooks. Comparatively speaking, hand-held technologies including graphing calculators are used in the textbooks of other countries except China. Although the calculators used in the high school textbooks of China have diversified functions, the tools used by Chinese students are more backward than other countries. In China, students use common calculators, scientific calculators or four arithmetic operation function calculators.

Only a few high schools in some major cities have tried to use leading TI graphing calculators that have the functions of data processing, functions, graphing, simple programming, mathematical and physical test and can visually draw graphs, carry out dynamic demonstration and locus tracking, etc. (Meiying Chen, 2008).

4.2 Comparison on information technology presentation means

The presentation means of information technology in the textbooks of the countries above are shown in Tab. 9-11.

Tab. 9-11 Distribution of information technology presentation means in the textbooks of the countries

Country		Types of Presentation Means		
		Integration	Topics	Asides
Singapore		87.3%	2.4%	10.3%
Japan		88.4%	5.8%	5.8%
Australia		87.9%	8.1%	4.0%
USA	1 (PHM)	16.5%	3.6%	79.9%
	2 (UCSMP)	66.0%	30.9%	3.1%
China		50.7%	28.9%	20.4%

Tab. 9-11 informs us that, except the first textbooks in the USA, the high school textbooks in the five countries above have commonality in terms of information technology presentation means: most technology is reflected by means of integration; topics and asides only occupy a small proportion and there is an obvious difference in the ratio of presentation means.

However, when studying the means with technologies and without technologies in the handling of high school textbooks, we found a clear difference between China and any other country.

On the one hand, technological tools are used in the counties except China in the teaching and learning of the identical mathematical content. For example, as for the content of "solving inequalities", both manual and calculator solutions are presented in Singapore (see Fig. 9-5) but there are only manual solutions in China. "Different features would lead to different ways of thinking". The actions of Singapore's textbooks could facilitate students to understand in multiple aspects, improve students' ways of thinking, enable students to select, use and transform various mathematical representations and establish extensive association among different representations. Thus, diversified representations should be provided in the preparation of textbooks.

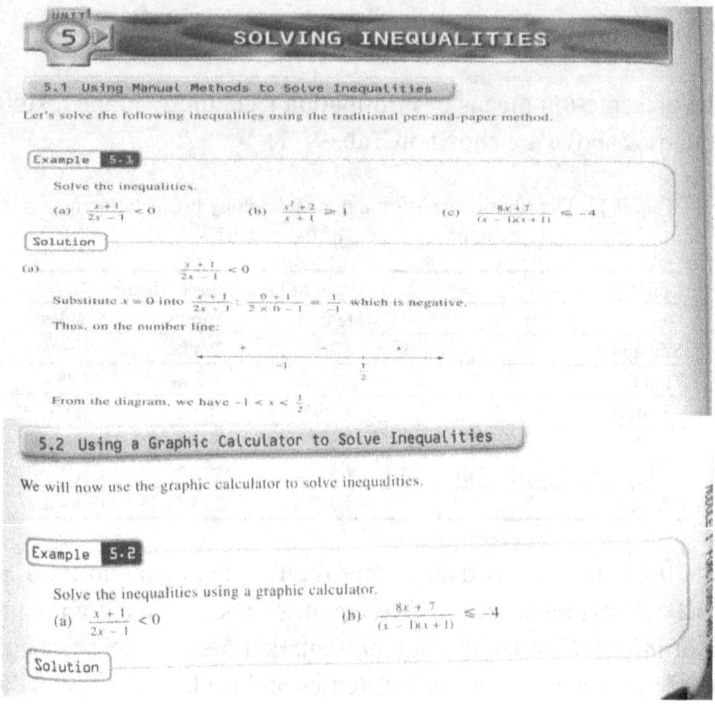

UNIT 5 SOLVING INEQUALITIES

5.1 Using Manual Methods to Solve Inequalities

Let's solve the following inequalities using the traditional pen-and-paper method.

Example 5-1

Solve the inequalities.

(a) $\frac{x+1}{2x-1} < 0$ (b) $\frac{x^2+2}{x+1} \geq 1$ (c) $\frac{8x+7}{(x-1)(x+1)} \leq -4$

Solution

(a) $\frac{x+1}{2x-1} < 0$

Substitute $x = 0$ into $\frac{x+1}{2x-1}$: $\frac{0+1}{2\times 0-1} = \frac{1}{-1}$ which is negative.

Thus, on the number line:

From the diagram, we have $-1 < x < \frac{1}{2}$.

5.2 Using a Graphic Calculator to Solve Inequalities

We will now use the graphic calculator to solve inequalities.

Example 5-2

Solve the inequalities using a graphic calculator.

(a) $\frac{x+1}{2x-1} < 0$ (b) $\frac{8x+7}{(x-1)(x+1)} \leq -4$

Solution

Fig. 9-5 Two solutions for "solving inequalities" in Singapore's textbooks

On the other hand, the presentation means are identical but their positions and roles are different. For example, the topics in Edition A published by the People's Education Press are presented in the column of "information technology application". They usually appear in the end of each chapter or section and are mainly introductory. And all asides are the tips in the places where it is appropriate to use information technology or use the tips of "calculators or computers may be used" for reminding the adoption of information technology … (see Fig. 9-6). However, in the AU-MHL textbooks of Australia, the information technology by means of asides is not used to remind the use of technologies but to introduce other operating procedures of TI-83.

> *You can also use calculator to draw the graphs of the four functions in Example 2 and make decisions based on the graphs.*

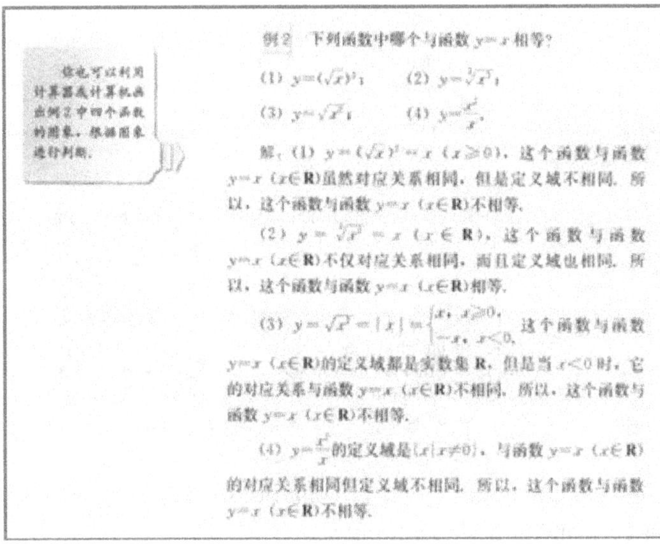

你也可以利用计算器或计算机画出例2中四个函数的图象，根据图象进行判断。

例2 下列函数中哪个与函数 $y=x$ 相等?

(1) $y=(\sqrt{x})^2$; (2) $y=\sqrt[3]{x^3}$;

(3) $y=\sqrt{x^2}$; (4) $y=\frac{x^2}{x}$.

解：(1) $y=(\sqrt{x})^2=x\ (x\geqslant 0)$，这个函数与函数 $y=x\ (x\in \mathbf{R})$ 虽然对应关系相同，但是定义域不相同。所以，这个函数与函数 $y=x\ (x\in \mathbf{R})$ 不相等。

(2) $y=\sqrt[3]{x^3}=x\ (x\in \mathbf{R})$，这个函数与函数 $y=x\ (x\in \mathbf{R})$ 不仅对应关系相同，而且定义域也相同。所以，这个函数与函数 $y=x\ (x\in \mathbf{R})$ 相等。

(3) $y=\sqrt{x^2}=|x|=\begin{cases} x, & x\geqslant 0, \\ -x, & x<0, \end{cases}$ 这个函数与函数 $y=x\ (x\in \mathbf{R})$ 的定义域都是实数集 $\mathbf{R}$，但是当 $x<0$ 时，它的对应关系与函数 $y=x\ (x\in \mathbf{R})$ 不相同。所以，这个函数与函数 $y=x\ (x\in \mathbf{R})$ 不相等。

(4) $y=\frac{x^2}{x}$ 的定义域是 $\{x|x\neq 0\}$，与函数 $y=x\ (x\in \mathbf{R})$ 的对应关系相同但定义域不相同。所以，这个函数与函数 $y=x\ (x\in \mathbf{R})$ 不相等。

Fig. 9-6 Example of Asides in the Edition A Textbooks Published by the People's Education Press

In terms of presentation means of technology, the high school mathematics textbooks published by the People's Education Press of China adopt a balanced technology presentation means. However, the information technology has not been fully integrated with teaching content. The other countries in Asia except China, which enjoy excellent high school mathematics education, present most information technology by means of integration, thus contributing to the integration of information technology and mathematics curriculums. A lot of information technology in Chinese textbooks are centralized (topics + asides), which deviates from teaching content. Although the role of technology on mathematics is one for one, such approach is not helpful for students to learn mathematics and technological application in a clear and orderly manner but would segment mathematics and technology and will not help students to form a correct view of mathematics. Also, students could not be able to acquire

Translation of Fig. 9-6

You may draw the graphs of the four functions in Example 2 by a calculator or a computer first, and then make your judgements.

Example 2 Among the following functions, which one identifies with the function = x ?

(1) $y = (\sqrt{x})^2$; (2) $y = \sqrt[3]{x^3}$;

(3) $y = \sqrt{x^2}$; (4) $y = \frac{x^2}{x}$.

Solution: (1) We have $y = (\sqrt{x})^2 = x\ (x > 0)$. This function obeys the same correspondence relation with the function $y = x\ (x \in \mathbf{R})$, but have different domains. Thus, the function does not identify with the function $y = x\ (x \in \mathbf{R})$.

(2) We have $y = \sqrt[3]{x^3} = x\ (x \in \mathbf{R})$. This function not only obeys the same correspondence relation, but also has the same domain with the function $y = x\ (x \in \mathbf{R})$. Thus, the function identifies with the function $y = x\ (x \in \mathbf{R})$.

(3) We have $y = \sqrt{x^2} = |x| = \begin{cases} x, & x \geq 0, \\ -x, & x < 0. \end{cases}$ Although both of this function and the function $y = x$ have the real numbers $\mathbf{R}$ as their domains, but the function obeys a correspondence relation different from that of the function $y = x\ (x \in \mathbf{R})$ when $x < 0$. Thus, the function does not identify with the function $y = x\ (x \in \mathbf{R})$.

(4) The domain of the function $y = \frac{x^2}{x}$ is $\{x | x \neq 0\}$. So the function has the same correspondence relation but with different domains with the function $y = x\ (x \in \mathbf{R})$. Thus, the function does not identify with the function $y = x\ (x \in \mathbf{R})$ too.

comprehensive understanding of technology and only learn scattered mathematical content and technology.

4.3 Comparison on the content links where information technology is used

The distribution of information technology in terms of content links in the textbooks of all the countries is counted in Tab. 9-12 below.

Tab. 9-12 Distribution of Information Technology of Textbooks in the Countries in Terms of Content Links

Country		Types of Content Links				
		New knowledge introduction	New knowledge exploration	Examples	Exercises	Extension
Japan		1.0%	16.5%	25.2%	48.5%	12.6%
Australia		0.6%	32.2%	63.7%	3.5%	0.0%
USA	1 (PHM)	1.0%	1.2%	25.9%	62.9%	9.0%
	2 (UCSMP)	25.7%	29.7%	19.6%	15.5%	9.5%
China		13.3%	17.8%	26.7%	11.1%	31.0%

It is shown from the data of Tab. 9-12 that:

(1) China is the only country arranging more than 30% of technology in the link of extension; however, such link is usually ignored by both teachers and students in practice.

(2) According to the role played by teachers and students in traditional mathematics teaching, teachers guide students to learn new knowledge, show how to use old knowledge or transform thoughts to acquire new knowledge and demonstrate in person how to use new knowledge to solve typical examples. What students should do is to understand and accept the exploration process shown by teachers and the mathematical results, simulate teachers' thoughts and specifications in problem solving for attempts in exercises and further exploration after class. Thus, except the first textbooks of America and the textbooks of Japan, the other countries use more technologies in the teachers' parts (first three links); the parts of teachers and students (the last two links) in China are almost equal in the percentage of comparison, about 6:4. When the general ignorance of teachers and students on the link of extension is considered, the Chinese textbooks' disadvantages of information technology use are the heavy attention on the "in-class integration", with the percentage of teacher and student parts being 6:1.

According to the approaches on the actual treatment of textbooks, we also get the following conclusions:

(3) Just like the first American textbooks, the maximum proportion of technology of Japanese textbook is used in the parts of exercises to encourage students to solve problems using calculators. If we carefully study the information technology in the part of exercises of Japanese textbooks, the technology used in many exercises matches the technology in examples, such as Example 19, Page 154 of Mathematics B in the Sukan Shupan Edition (see Fig. 9-7): Give the algorithm of "inputting two natural numbers and calculate the greatest common divisor of the two numbers by Euclidean division algorithm". The algorithm is just used in the following exercises to calculate the greatest common divisor of the two groups of numbers, which could not only reduce the difficulty of workload but also extent technological applications from class to the outside of the class.

例 19 2つの自然数 A，B の最大公約数を，ユークリッドの互除法で求めるプログラムは，たとえば次のようになる。

```
100 INPUT PROMPT "A, Bは": A, B
110 LET R=A-INT(A/B)*B
120 IF R=0 THEN 160
130 LET A=B
140 LET B=R
150 GOTO 110
160 PRINT "最大公約数は"; B
170 END
```

練習 22 例 19 のプログラムを実行し，次の 2 つの自然数の最大公約数を求めてみよう。

(1) 437，943　　(2) 21318，29716

Fig. 9-7 Example 19, page 154, Mathematics B by Sukan Shupan of Japan

(4) There is a great disparity in the content links of information technology of Australian textbooks and almost no aspects of extension and new knowledge introduction. They mainly concentrate on the exploration of new knowledge solution and in-class examples by using technology. There are different features in the specific handling. The process and result of technological application are subject to screenshot in the examples of textbooks. For example, in Page 38, AU-MHL, when TI-83 is used to solve the simultaneous linear equations with two unknowns $\begin{cases} y = -x + 7 \\ y = 2x + 1, \end{cases}$ the process and operating procedures are shown as Fig. 9-8.

(5) There is clear text in the American UCSMP textbooks showing that technology is essential for students to use while it is not such case in the Edition A textbooks of the People's Education Press of China. Although technology is reflected in the application of mathematics, the practice in classroom teaching is topic proposing, orientation reminding and result demonstration by teachers using information technology. Students could

Step 1: Enter both equations in the required form, i.e., $y = \ldots$

Step 2: Choose an appropriate window setting, in this case we have [−2,8] by [−1,8].

Step 3: Sketch the straight lines using the **GRAPH** key.

Step 4: Call up the **CALC** menu (i.e., press **2nd TRACE**) and choose option **5: intersect**.

Step 5:
- 5.1 Move the cursor to where the lines intersect and press **ENTER** – this confirms that you have selected your first equation.
- 5.2 Press **ENTER** again, this confirms that you are using the second equation.
- 5.3 Because you have already placed your cursor near the point of intersection, when prompted to Guess? simple press **ENTER**.

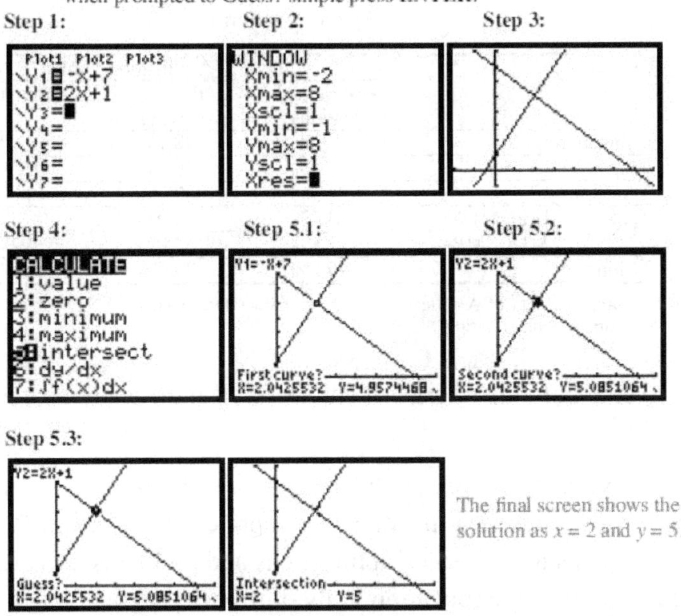

Fig. 9-8 Examples of TI-83 Application in the Form of "Problem Solving" in the AU-MHL Textbooks of Australia

only understand the role of information technology in mathematical content learning through the secondary processing of teachers on textbooks in class. There is almost no opportunity for students to practice in class and there are less after-class exercises, which would deprive students of independent exploration opportunity. Then, it is impossible to cultivate their interests and ability in exploration and unfavorable for the cultivation of students' innovation ability and practice spirits.

4.4 Comparison on the application forms of information technology

The distribution of information technology of all the countries in terms of application forms is counted in Tab. 9-13.

Tab. 9-13 Distribution of Information Technology of the Countries in terms of Application Forms

Country		Type of Application Forms		
		Tool-oriented applications	Problem solving	Programming
Singapore		1.9%	98.1%	0.0%
Japan		38.6%	46.4%	15.0%
Australia		33.3%	67.7%	0.0%
USA	1 (PHM)	53.7%	44.0%	2.3%*
	2 (UCSMP)	42.8%	53.0%	4.2%
China		45.1%	29.6%	25.3%

Note: The data of American US-PHM textbooks marked with "*" in Table 9-13 is "extracurricular tutoring".

By summarizing a series of studies, we get the common conclusions as follows:

(1) Except the Chinese and Japanese textbooks, the textbooks of the other three countries position the application forms of information technology as tool-oriented applications and problems-solving means and there is almost no application in the form of "programming". The reason is that algorithms are not included.

For example, "problem-solving" is the major application form of technology in the high school mathematics textbooks of Singapore, only a small percentage for tool-oriented applications and zero for programming. Also, there are mathematical exploration topics related to the application of technology in problem solving in its appendix.

There is also no application in the form of "programming" in the textbooks of Australia and two thirds of the technological applications are in the form of problem solving. While in the application of the technological tool TI-83, operating procedures and operating results are presented in textbooks by means of screenshot.

As for Chinese and Japanese textbooks, if the content of statistics and algorithms in the Compulsory 3 of the People's Education Press and the So-In Edition Mathematics B are not considered, there are few programs. Tools and problem solving dominate in the application forms. The application of calculators as a tool is mostly used as a solution to complex calculating problems.

(2) In terms of application forms of information technology, the two sets of American textbooks have less difference compared with the presentation means of information technology and the distribution of content links. Almost half of the technology application forms are tool-oriented and the other part is "problem-solving" type.

(3) Similar to the USA, the tool-oriented applications in the Chinese textbooks under comparison are more than half of the total number. The difference is that nearly one fourth of information technology application forms is "programming", especially in the "algorithm fundamentals" of Chapter I, Compulsory 3 (see Fig. 9-9).

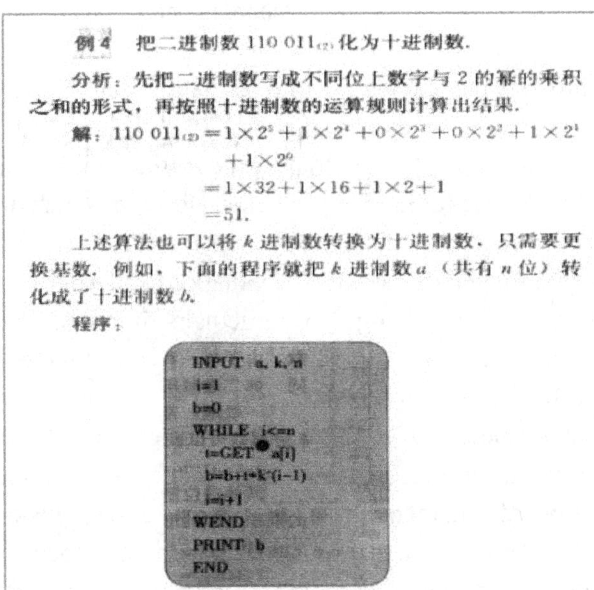

例 4 把二进制数 $110\,011_{(2)}$ 化为十进制数.

分析：先把二进制数写成不同位上数字与 2 的幂的乘积之和的形式，再按照十进制数的运算规则计算出结果.

解：$110\,011_{(2)} = 1\times2^5+1\times2^4+0\times2^3+0\times2^2+1\times2^1+1\times2^0$

$= 1\times32+1\times16+1\times2+1$

$= 51.$

上述算法也可以将 k 进制数转换为十进制数，只需要更换基数. 例如，下面的程序就把 k 进制数 a（共有 n 位）转化成了十进制数 b.

程序：

```
INPUT a, k, n
i=1
b=0
WHILE i<=n
  t=GET a[i]
  b=b+t*k^(i-1)
  i=i+1
WEND
PRINT b
END
```

Fig. 9-9 Example of the Application of Programming in compulsory 3, the Edition A Textbook of the People's Education Press of China

Translation of Fig. 9-9

Example 4 Convert the binary number $110\,011_{(2)}$ into decimal.

Analysis: We write the binary number as a sum of the products of the digits in different places with corresponding powers of 2 first, then calculate the result in decimal system.

Solution:
$$\begin{aligned}110\,011_{(2)} &= 1 \times 2^5 + 1 \times 2^4 + 0 \times 2^3 + 0 \times 2^2 + 1 \times 2^1 + 1 \times 2^0 \\ &= 1 \times 32 + 1 \times 16 + 1 \times 2 + 1 \\ &= 51.\end{aligned}$$

The above method can be used to convert a number in base-k system into a decimal number, provided we change the base. For example, the following program converts a number a in base-k system (with n digits) into a decimal number.

Program: (omitted.)

Only from the point of "application forms", the advantage is there are various forms in the Chinese textbooks, which are relatively balanced. However, compared with the textbooks in other countries, the Chinese textbooks have the following disadvantages: (1) There is a lack of concreteness in handling and the operating process is not fully presented. Although page layout is saved, teachers need to make additional preparations and it is not favorable for students to learn, operate and explore; (2) As for the application of information technology for problem solving, the Chinese textbooks mainly uses its powerful calculating function, mostly apply directly to the pure mathematical problems, with little change in the background of problems. However, the textbooks in the other countries frequently change the background of the problems and point out for their students the possible situations and handling methods in the process of problem solving; (3) While using information technology, the foreign textbooks will also use technologies to test the logic thinking result or use ICT in testing the accuracy of the conclusions, which does not only enable students to use information technology but also increase their application awareness and skills of mathematics and improve their perception towards mathematics. The testing of conclusions in most Chinese textbooks is conducted by logic thinking and focus more on the result inferred by the technologies according to logic thinking test.

5. Implications and Suggestions

As the application of information technology in textbooks not only relates to curriculum standards but also has a close relationship with teachers' teaching and assessment of mathematical learning, this study proposes suggestions from the following four aspects based on the above-mentioned results.

5.1 Suggestions on high school mathematics curriculum standards of China

Although there is explicit text in the high school mathematics curriculum standards of China which points out that there is requirement on information technology, the wording, in the opinion of the author, is not specific and deficient in operability. High school mathematics curriculum should encourage the organic integration of information technology and curriculum content (for example, integrating algorithms into all relevant parts of mathematics curriculum). The basic principle of integration is to favor students to understand the essence of mathematics. It should encourage the application of information technology to present contents that are hard to be presented in the previous teaching, use scientific calculators and various mathematics education technology platforms as practically as possible while guaranteeing manual trainings, enhance the combination of mathematics teaching and information technology and encourage students to make exploration and discovery using computers and calculators (Ministry of Education of the people's Republic of China, 2003). If calculators are forbidden in the recruitment exams or the use of computer is under restricted in the recruitment exams, teachers and schools will choose not to use information technology for teaching and homework assignment.

Thus, the author suggests that the curriculum standards should specify the fundamental requirements on the use of information technology under the consideration of the imbalance of educational resources among different regions. For example, it should point out that, for which contents, the teaching and learning must use information technology, specify the basic times of information technology use, the application forms of information technology or provide specific examples given according to different content, students' and teachers' conditions and teaching environments. Despite of inflexibility, the due role of information technology in the mathematics curriculums could be guaranteed at least.

5.2 Suggestions on textbook writing

The High School Mathematics Curriculum Standards (Experimental) of China (MOE of PRC, 2003) explicitly identify the value of modern information technology and make suggestions on the textbook writing, "…in the handling of some textbook content, calculators or computers are encouraged … Also, students should be encouraged to use modern technological means to deal with complicated calculation and solve actual problems so as to explore and discover the laws of mathematics and cultivate innovation spirit and practical ability. On the other hand, modern information technology could not only exhibit huge potential in improving the learning methods of students but also infiltrate into mathematics curriculum content. Textbook writing should pay attention to the integration of these resources." (p. 119–120)

Thus, explicit requirements should be proposed for the part using information technology in textbooks:

(1) Sequence of information technology: In order to prevent students from learning scattered mathematics and technology, the consistency and systematicness of the context should be paid attention to when information technology is mentioned. Thus, mathematics should be regarded as the core. In addition to the problems that should be solved (such as statistics) or optimized (such as algorithms) by using information technology, the technological tools should also be introduced gradually. Textbooks should start from the initial functions of various technological tools (such as calculating tools) and simple mathematical problems (such as solving equations), then, use the further functions of various technological tools (such as construction), make hypotheses and explore the laws of mathematics (such as graphs and properties of functions and graph

transformation). Finally, textbooks can use the senior programming function of technological tools to solve the problems concerning statistics and algorithms. Thus to provide students with a continuous opportunity which increases skills to use technology and develop information awareness, and enable them to feel the revolutionary changes to learning modes of mathematics with the integration of technological tools.

(2) Type of information technology. According to the result of comparisons in 3.1, the technical tools used in the Edition A textbooks published by the People's Education Press are more diversified than the textbooks of Singapore, Australia and Japan, but the handheld tools are more backward. In addition, the internet, the powerful technological tool has never occurred. Thus, it is suggested that textbooks can present example that uses internet for mathematics learning and be equipped with online study tutoring.

(3) Presentation means of information technology. According to the result of comparisons in 3.2 that, compared with the textbooks of Singapore, Japan, Australia and the UCSMP textbooks of USA, the Edition A textbooks of the People's Education Press use more "topics" and "asides" to present information technology. However, they count for little in the actual teaching and has not attracted the attention of teachers and students. Thus, it is advised to learn from the compiling style of Singapore's textbooks and regard the treatment of mathematical content by information technology as understanding of the content in another perspective, or formulate the diversified representations of mathematical content so as to enlarge students' knowledge scope and help them to formulate more diversified and better knowledge structure. That is, different problems should be frequently solved in the form of "integration" to reflect the use of information technology as well as the functions of information technology in laws exploration and inspection. For example, verbal descriptions need to be added to the textbooks when information technology is used to show what kind of tool is introduced, what requirements of processing skills are raised for students and what requirements of cognition and thinking are. Take a calculator as an example when it is used as a calculating tool, students should estimate and make rounding; when it is used as an analytical tool of data, students

should try to seek for models; when it is used as an exhibiting tool, students should try to explain data and solve equations, etc.; when it is used as a testing tool, students should understand multiple symbolic forms. It can help students to better understand information technology, better learn mathematics and mathematical thinking.

(4) Distribution of information technology in content links. According to the comparisons in 3.3, the information technology in the Chinese textbooks rarely appears in the after-class examples and exercises. Thus, teachers have other considerations when selecting the mathematics conclusions for information technology introduction and exploration. This also makes students to focus on one side when interpreting teachers' teaching that "information technology means nothing for students". As a result, it would become impossible for students to pay attention to the information technology in the extension content that is not taught in class, not assigned in after-class exercises and not tested. Therefore, in terms of the link of information technology application, we should refer to the Japanese textbooks So-In Edition, increase the application of information technology in the after-class examples and exercises (instead of extension) and match it with the content in the part of exercises of textbooks so that students could observe teachers' demonstration in class and apply the information technology in practice in the learning enhancement after class.

(5) Application forms of information technology. It can be known from Tab. 9-13 that the Edition A textbooks published by the People's Education Press use almost a quarter of information technology in the form of programming in Compulsory 3, which poses a severe challenge on teachers' controlling ability. Almost sixty percent of the other three fourths are tool-oriented applications, which greatly differ from the distribution of the textbooks in the other four countries in the study. Thus, it is suggested that the applications in the form of problem solving should be increased so as to enable students and teachers realize the role of information technology better, enhance the awareness and ability of using information technology as tool and approach for mathematical exploration and facilitate students to obtain more profound understanding on the process of mathematical conclusion formulation.

(6) Descriptions of information technology. As most places in the Chinese textbooks use text to describe the application of information technology, students' visual feeling and understanding of the operation and results may be limited. Therefore, it is suggested that we should make reference to the handling method of Australian textbooks and present the process and result of information technology application by means of screenshot, which could make the textbooks closer to students' cognitive law in content presentation and facilitate their cognitive development. However, the issue of page layout should also be taken into consideration, and it should be specified in detail when a certain technology is used or a certain type of problem is solved for the first time.

5.3 Suggestions on teaching

It is explicitly pointed out in the high school mathematics curriculum standards of China that, "Teachers should use information technology properly, improve students' style, guide them to use information technology to learn relevant content of mathematics, explore and study some meaningful and valuable mathematical problems." (MOE of PRC, 2003, p. 112)

It can be inferred that teachers should not only regard information technology as a calculation or representation tool but also consider it as a cognitive tool which helps students to experience various thinking activities including visual perception, observation, analogical induction, space imagination, and abstract generalization so as to obtain the understanding of the essence of mathematics. Meanwhile, more space should be reserved for the students' exploration. We should not only use technology in example explanation but also use information technology in exercises.

5.4 Suggestions on assessment

As for the application of information technology, there is a fear in the international mathematical education field: technology would become a "black box" and lower the level of students' thinking. The reasons for forbidding the use of calculators in some high schools and college entrance examinations lie in the storage and transmission functions of calculators, which has become a potential hazard to examination disciplines and would affect the organization and management of examinations. In addition, most front-line teachers do not agree on the use of calculators. The reason is that students would depend too much on calculators, which is not favorable for the acquisition of "basic knowledge and skills". However, calculators are allowed to be used in examinations in many other countries. Therefore, if the model of the calculator is specified, students would not excessively rely on graphic calculators in their regular study. It is believed that the fear could be done away with to a large extent.

References

[1] Bao, Jiansheng (2009). *Comparative Study of Comprehensive Difficulty for Junior High School Mathematics Curriculums in the UK and China* [in Chinese]. Guilin: Guangxi Education Press.

[2] Bellman, A. E. *et al.* (2003). *Algebra 2*. New York: Prentice Hall.

[3] Cai, Jinfa (2007). *Empirical Study Series for Mathematics Learning by American and Chinese Students: How to Learn from Others* [in Chinese]. Beijing: Educational Science Publishing House.

[4] Cao, Yiming, Wang, Lidong and Cobb, P. (2010). Review on the high school mathematics part in the Common Core State Standards of the USA [in Chinese]. *Journal of Mathematics Education*, 19(5): 8–11.

[5] Chen, Meiying (2008). Comparison on calculator application in mathematical teaching in America and China [in Chinese]. *Journal of Tianjin Normal University (Fundamental Education)*, 9(3): 70–72.

[6] Chen, Xiangyang, Jia, Hongli and Chen, Xi (1997). Australian education in our eyes [in Chinese]. *People's Education*, (5): 43.

[7] Chen, Yuelan (2010). A survey of Japanese mathematics education [in Chinese]. *Monthly of High School Mathematics*, (5): 1–4.

[8] Cui, Yingmei (2006). Study on presentation of information technology in middle school mathematics textbooks and its effects recognized by students [in Chinese]. Master thesis. Changchun: Northeast China Normal University.

[9] Daiqin (2011). Review and prospects of Chinese and Japanese education research exchanges [in Chinese]. *Shuxue Tongbao (Bulletin des Sciences Mathematics)*, 50(11): 58–59.

[10] Doerr, H. M. and Zangor, R. (2000). Creating meaning for and with the graphing calculator. *Educational Studies in Mathematics*, (41): 143–163.

[11] Fang, Han (2010). Comparative study of college entrance examination systems in China and Australia [in Chinese]. Master thesis. Shanghai: East China Normal University.

[12] Hu, Lili (2008). comparative study of difficulty for middle school mathematics textbooks in the USA and China, with the People's Education Press Edition and Prentice Hall Edition as examples [in Chinese]. Master thesis. Shanghai: East China Normal University.

[13] Hu, Yaohua (2009). Mathematics assessment under the information technology environment: the use of graphing calculator in college entrance examination of foreign countries and its implications [in Chinese]. *Modern Education Technology*, 19(5): 37–40.

[14] Kang, Yueyuan (2011). Review and analysis on the national unified mathematics curriculum standards of Australia [in Chinese]. *Journal of Mathematics Education*, (5): 81–85.

[15] Leigh-Lancaster, D. (2004). Mathematics, technology and mathematics education [in Chinese]. *Journal of Mathematics Education*, 13(4): 62–66.

[16] Leung, Frederick K. S. (2008). Information and communication technology in mathematics education. In N. Mogens (Ed.), *ICME-10 Proceedings* (pp. 228–243), Kailow Graphic, Denmark.

[17] Li, Shenglan (1995). Comparison on children family education in China and Australia [in Chinese]. *Shanghai Education Scientific Research*, (3): 37–38.

[18] Lin, Didi (2012). Comparative study on the technological application in high school textbooks of China and Australia [in Chinese]. Master thesis. Suzhou: Soochow University.

[19] Liu, Chao and Wang, Zhijun (2011). Comparative study on the integration of information technology and high school mathematics textbooks, with PEP Edition A, Beijing Normal University Edition and Jiangsu Education Press Edition as examples [in Chinese]. *Mathematics Education in China*, (10): 10–12.

[20] Lokan, J., Ford, P. and Greenwood, L. (1996). *Maths & Science on the Line: Australian Junior Secondary Student's Performance in the Third International Mathematics and Science Study*. Melbourne: Australian Council for Educational Research.

[21] Lokan, J., Ford, P. and Greenwood, L. (1997). *Maths & Science on the Line: Australian Middle Primary Student's Performance in the Third International Mathematics and Science Study*. Melbourne: Australian Council for Educational Research.

[22] Lu, Weiling (2012). Comparative study on the technological application in high school textbooks of China and Singapore [in Chinese]. Master thesis. Suzhou: Soochow University.

[23] MOE of PRC (Ministry of Education, the People's Republic of China) (2003). *High School Mathematics Curriculum Standards (Experimental)* [in Chinese]. Beijing: People's Education Press.

[24] National Center for Education Statistics (1996). *Pursuing Excellence.* Washington, DC: U.S. Government Printing Office.

[25] NCTM (2000). *Principles and Standards for School Mathematics.* Reston, VA: National Council of Teachers of Mathematics.

[26] Singapore Ministry of Education (2012). *O- & N(A)-Level Mathematics Teaching and Learning Syllabus*, First edition.

[27] Song, Lili (2010). Comparison and analysis of the content of algorithm in five versions of high school mathematics textbooks as per the national curriculum standards [in Chinese]. *Shuxue Tongbao (Bulletin des Sciences Mathematics)*, 49(12): 5–10.

[28] Su, Hongyu (2007). Comparison on the Level of Questions in the Primary School Mathematics Textbooks in China and Singapore [in Chinese]. *Modern Teaching*, (4): 56–58.

[29] Usiskin, Z. (2010, March). Grades 7−12 learning progressions in mathematics content. Paper presented at APEC Conference on replicating exemplary practices in mathematics education. Koh Samui, Thailand. Chinese translation Xiaoli Lu, *Mathematical Teaching*, 2010(9): 19–21.

[30] Wang, Fenping (2011). Comparative study of the contents of complex numbers in high school mathematics textbooks of China and Britain: the case of the comparison between AQA textbook and the version A textbook of People's Education Press [in Chinese]. *Journal of Mathematics Education*, 20(3): 83–86.

[31] Weigand, H. (2010). Application of symbol calculators in mathematics classrooms (Translated into Chinese by Jiansheng Bao). *Mathematical Teaching*, (12): 1–3.

[32] Xu, Jiahong (2011). Comparative study of technological application in the high school mathematics textbooks of certain countries [in Chinese]. *The Monthly Journal of High School Mathematics*, (5): 27.

[33] Xu, Jiahong (2012). Comparative study of technological application in the high school mathematics textbooks of Japan and China [in Chinese]. *The Monthly Journal of High School Mathematics*, (7): 53–55.

[34] Yao, Xue (2010). Comparative study on the high school mathematics textbooks in Shanghai, Hong Kong and Singapore [in Chinese]. Master thesis. Shanghai: East China Normal University.

[35] Yuan, Zhiqiang (2008). Comparative study on the integration of information technology and high school mathematics curriculums of the new standards: with Math 1 as example [in Chinese]. *Journal of Mathematics Education*, 17(3): 88−90.

[36] Zhang, Tianxiao and Tang, Caibin (2005). Common features of primary school textbooks in the USA, Japan and Germany and enlightenments [in Chinese]. *Comparative Education Review*, 26(1): 78−81.

[37] Zhang, Weiping (2012). Presentation means of Mathsquest Textbooks in Australia: perspective of metaphor of experiencing type [in Chinese]. *Global Education*, 41(1): 72–76.

[38] Zhang, Weizhong and Li, Fangqi (2009). Comparison on the characteristics of the mathematics textbooks of Singapore and China [in Chinese]. *Foreign School Education*, (2): 32–36.

[39] Zhao, Xiaoping and Yao, Xue (2010). A comparative study of statistics contents in textbooks used in Shanghai and Singapore high schools [in Chinese]. *Mathematical Teaching*, (2): 9–11.

[40] Zhao, Yuanyuan (2009). Study on the integration of information technology and high School mathematics curriculums, with Math 1 as example [in Chinese]. Master thesis. Shenyang: Liaoning Normal University.

[41] Zheng, Wanquan (2012). Overview to the school textbooks in the USA [in Chinese]. [2012-03-12]. Retrieved from https://old.pep.com.cn/czyw/jszx/jxyj/yycf/201008/t20100825_731683.htm.

[42] Zhou, Chao (2013). Status and role of information technology in high school mathematics textbooks: comparative perspectives of China and America [in Chinese]. *The Monthly Journal of High School Mathematics*, (2): 34–37.

[43] Zhou, Shushu (2012). Comparative study on the technological application in the high school textbooks of China and Singapore [in Chinese]. Master thesis. Suzhou: Soochow University.

https://doi.org/10.1142/9789814713955_0010

Chapter 10

An International Comparative Study on Exercise Questions in Middle School Mathematics Textbooks

Yiming Cao* and Libao Wu

School of Mathematical Sciences, Beijing Normal University
Beijing 100875, China
**caoym@bnu.edu.cn*

This chapter presents a comparative study on middle school mathematics textbooks selected from 10 countries including China, Australia, the United States, Britain, France, Germany, South Korea, Singapore, Japan and Russia. Using both quantitative and qualitative approaches, the comparison focuses, on the quantity, type, openness and difficulty level of all the exercise questions in the topics of "numbers and operations", "equations", "triangles", "solid figures" and "statistics". The results reveal that the exercise questions in these textbooks have their own distinctive features. For instance, the US and Australian textbooks offer simpler exercises, while the Russian textbooks contain the most difficult ones, and the exercises in the Chinese textbooks are above the average level in terms of difficulty. Further analysis shows the selected textbooks bear similarities as well, such as regional characteristics, prioritized exercise allocation, hierarchy of exercise design and the emphasis on the association of mathematics, real life and other disciplines. These findings shed some light on developing Chinese mathematics textbooks: increasing to some extent the amount of exercises, enhancing the openness of exercise conclusions and maintaining a reasonably difficulty level of exercises.

1. Introduction

Textbooks are the main materials for teachers and for students to learn subject knowledge. In learning activities, textbooks are the object one of the three elements (i.e., subject element, leading element, and object element). It is also the physical carrier of information in the process of imparting social and life experience, or in other words, in educational activities (Yongchun Zhang, 1996). Earlier research shows that textbooks directly affect teachers on what to teach and how to teach (Robitaille and Travers, 1992). The Third International Mathematics and Science Study reveals that textbooks are the main text resource for teachers' instructional design and for guiding their classroom practice (Beaton *et al*., 1996). Researchers generally believe that teachers are largely dependent on textbooks in teaching activities and they are inclined to decide what to teach and how to teach, as well as what exercises students should do after class, based on the textbooks they use (Meiling Jiang, 2008). There is also evidence that the reliance on textbooks is an important feature of classroom teaching (Macintyre and Hamilton, 2010). Traditionally, textbooks were used mainly to impart knowledge and to serve as a reservoir of exercises. They also have an implicit function to bear the social and cultural values (Gerard and Roegiers, 2003). In the past 20 years, the role textbooks played in mathematics study for both teachers and students has attracted increasing attention from researchers (Ball and Cohen, 1996). The above all point to the value of textbooks in teaching and learning. In suggestions for textbook compilation provided in the Chinese *Mathematics Curriculum Standards for Compulsory Education (2011 edition)* (MOE of PRC, 2012), it explicitly points out that "mathematics textbooks shall provide learning themes, basic guidelines and knowledge structures for students' mathematics learning activities, and they are important resources for realizing mathematical curriculum objectives and implementing mathematics teaching". Teaching materials typically refer to textbooks. Though the structure of mathematics textbooks varies across different countries, mathematics textbooks typically consist of three basic components, namely, the main text, examples and exercises (Curriculum Research Institute, 1993), and the three components are included in mathematics textbooks in all the countries.

As an important component of mathematics textbooks, exercise questions (or simply exercises) and their differences in quantity, difficulty level, style and focuses not only reflect different countries' overall levels, styles and inclinations in mathematics textbook compilation, but also have a remarkable influence upon mathematics classroom teaching. If the concepts and laws of propositions in a textbook are considered as a point, then exercise questions are a bundle of rays starting from this point of fundamental principles to help students consolidate, extend and use new knowledge; they also help the training and infiltration of mathematical thinking and provide the diagnosis, feedback and remedy (Libao Wu, 2013).

The majority of international comparative studies on the exercises of mathematics textbooks have been done for only two countries. For example, Libao Wu (2013) conducted a comparative study on the exercise questions in Grade 8 textbooks between China and Australia, and Libao Wu and Yiming Cao (2013) conducted a comparative study at the middle school level between China and the UK. Comparative studies focusing on three or more countries are not commonly seen. Moreover, in those studies, textbook samples selected are also varied in different studies. Such comparisons of two countries provided us with a base for research on the comparison of exercise questions, from which we can identify the dimensions of international comparison of exercises and establish an indicator system for exercise comparison, based on which we can conduct further comparative study on exercise questions concerning multiple countries to obtain more valuable conclusions. On the other hand, we carried out a comparison and analysis on the distributions of mathematics contents in middle school (Grades 7–9) mathematics textbooks from ten countries including China, Australia, the USA, the UK and France in three mathematics topic areas: number and algebra, measurement and geometry, and statistics and probability. Based on this, we are conducting a comparative study on exercise questions in these textbooks using a unified indicator system. Detailed design is explained below.

2. Research Design

2.1 Country selection: ten countries from four continents

In this study, the countries selected including four Asian countries — China, Japan, South Korea and Singapore, four European countries — UK, France, Germany and Russia, one American country — the USA, and one Oceania country — Australia, a total of ten countries from four continents.

These ten countries have different cultural backgrounds, but all of them have a high level of education and economic development; in terms of geography, they are located in four continents and to some extent are typical and representative, with all the five permanent members of the United Nations Security Council being included. In addition, these countries ranked high in important international comparison projects like TIMSS and PISA.

2.2 Grade selection: Grades 7–9

The educational system for middle school differs in these ten countries, among which China, Japan, South Korea and the UK have a three-year middle school system, corresponding with Grades 7–9 while Singapore, Australia, Russia, France and Germany have four or five years with all including Grades 7–9. In comparison, textbooks selected from the USA correspond with Grades 6–8, just at its middle school level. Though there is one year lagged behind, it is a complete school phase. For all the remaining countries, this study selected mathematics textbooks for students of Grades 7–9 to compare.

2.3 Subject of study: textbooks

All the textbooks selected in the study are released or recommended textbooks by the concerned education administrative authorities in these countries, rather than informal school-based teaching materials (resources or handouts prepared and developed by the teachers) or reference books. They are all mainstream textbooks in these ten countries and are used in certain regions; in addition, they are all well-received or recommended by experts in the countries. The detailed information is described as follows:

China: The *Mathematics* textbooks (2008 edition) for middle schools published by the China People's Education Press are selected, with Academician Qun Lin as the chief editor. This series of textbooks consist of six volumes, two for each academic year. There are in total 942 pages and 95 sections allocated in 29 chapters.

Japan: *The Mathematics Course books (revised in 2005, the 5th edition in 2010)* published by the Dai Nippon Printing Co., Ltd are selected, consisting of three volumes.

South Korea: Since the reform on educational curriculum in 2007, the mathematics textbooks for middle school are of a great variety in South Korea. There are altogether 14 types of textbooks for middle school students. The South Korea textbooks have been following the national standards with virtually the same pattern. We select the *Mathematics Textbooks for Middle School* edited by Wongyeong Kim *et al.* and published by VISANG Education INC in 2009 with three books in total.

Singapore: The *Math Insights Secondary 1A, 1B, 2A, 2B, 3A, 3B, 4, 5 for Normal (Academic)* published by the Pearson Publishing Group are selected, totally eight books included, among which Book 1A, 1B, 2A, 2B, 3A and 3B correspond with students of Grades 7–9 and are developed for the use of the average to lower-than-average middle school students of Singapore.

UK: There is no unified education system in the UK. England, Scotland, Northern Ireland and Wales have their own Ministries (Departments) of Education and education systems, and the school systems and curriculum arrangement also differ from each other. *Math Links* was designed in accordance with the UK mathematics curriculum standards issued in 2007 and published by the Oxford University Press in 2009. According to the varied levels of students, this series of textbooks is divided into series A, B and C. Series A is relatively easier and mainly designed for average students. Grades 7–9 are supplied with three books, each for one grade; they are mainly used in some middle schools of England.

France: With no unified national mathematics textbooks in France, we have selected the *SéSAMATH LE Manual 3e (4e、5e、6e)* published by the SéSAMATH Company, among which the 6e, 5e, 4e and 3e correspond with the Grades 1, 2, 3, and 4 in middle school, one book for each academic year; and Book 5e, 4e and 3e correspond with textbooks for students of Grades 7, 8 and 9 in China, published in 2010, 2011 and 2012, respectively.

Germany: There are no unified textbooks in Germany. A series of textbooks used in comprehensive secondary schools (Gymnasium) of the Bavaria State is selected; they were published in 2005 by the Ernst Klett Publishing House, consisting of eight books for students of Grades 5–12. Bavaria State is the largest and oldest state in the Federal Republic of Germany. In the recent PISA (Programme for International Student Assessment) test, students from Bavaria perform better in all the tested subjects than their peers from all the other states in Germany, ranking among the best in the world.

Russia: The selected series of textbooks are published by the Russia Education Press with six volumes in total. Ю. Н. Макарычев is the overall editor-in-chief for five of the six volumes, *Mathematics*, with Н. Г. Миндюк, К. И. Нешков and С. Б. Суворова as editors-in-chief for different volumes corresponding with Grades 5–9, among which the textbooks for Grades 7–9 are in their 20th edition, 18th edition and 18th edition, respectively; the other volume, in its 21st edition, is *Geometry,* whose editor-in-chief is Л. С. Атанасян and the book is used throughout the Grades 7–9.

USA: The selected textbooks are *IMPACT Mathematics* published by the McGraw-Hill Companies in 2009, including three books: the *IMPACT Mathematics (Course 1)*, *IMPACT Mathematics (Course 2)* and *IMPACT Mathematics (Course 3)*, designed for Grades 6–8.

Australia: There are no nationwide unified mathematics textbooks in Australia. Victoria is the most populated and industrialized state in Australia and the most developed area in education; correspondingly, its textbooks are fairly representative. The study selected the *Heinemann Maths Zone 7–10 VELS Enhanced* textbooks, published by the Pearson Publishing Group, which consist of 4 volumes in total.

2.4 Research contents

For these ten countries, the mathematics textbooks for middle schools mainly consist of three parts — "number and algebra", "figures and geometry (including measurement)" and "statistics and probability". In terms of the number of pages for the contents, the "number and algebra" part occupies the highest proportion, 56% on average, among which the mathematics textbooks in seven countries have surpassed 50% and the USA textbooks reach a 75%, ranking highest. The "figures and geometry" part occupies the second place, 32% on average, among which the Chinese textbooks reach a proportion of 41%, ranking highest; while the "statistics and probability" part only takes an average proportion of 12% (Libao Wu and Yiming Cao, 2013).

Since the contents of mathematics textbooks for middle school are of great loads, this study only selected five common themes for core knowledge to have a better comparability; the five common themes are "numbers and operations" and "equations" in "number and algebra", "triangles" and "solid figures" in "figures and geometry", and "statistics" in "statistics and probability". These five core themes take different proportions, among other differences, in the textbooks of the respective countries. For instance, in Japan, there is no statistics in their textbooks at the middle school level, and hence we can only analyze the other four themes in this study. In addition, as the French textbooks are hard to translate, here we only selected "equations", "triangles" and "statistics" for the analysis, whose contents cover 36% of the textbooks for the three grades of middle school in France. To sum up, the core knowledge themes take more than 32% of the proportion on average in the selected mathematics textbooks for the whole 3-year middle school stage in all the ten countries. The contents were studied for more than one year and the UK takes the highest proportion in covering these contents, at a striking 62%. The detailed information is shown in Fig. 10-1.

2.5 Research methodology

Both the quantitative analysis and qualitative description methods are adopted to carry out the comparative study on the exercises of the five knowledge themes: "numbers and operations", "equations", "triangles", "solid figures" and "statistics".

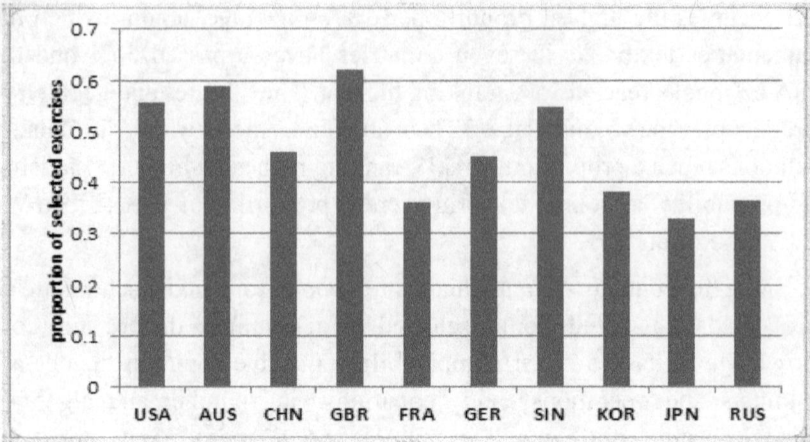

Fig. 10-1 Proportion of exercises in the five knowledge themes among all the exercises in textbooks

2.6 Quantitative indicators

2.6.1 Quantity of exercise questions

The number of exercises can reflect the quantitative level of the exercise systems in each series of textbooks in different countries. In the study, in counting the number of exercise questions, we refer to the big questions, that is, no matter how many sub-questions or sub-items a big question has, it is counted as one exercise question only.

2.6.2 Openness of conclusions

By whether there is only one conclusion, the exercises are divided into "closed exercises" and "open-ended exercises". The open-ended exercises, compared to those closed exercises with definite conditions and conclusions, consist of both exercises with inadequate conditions and exercises with indefinite conclusions. Rather than seeking one definite answer, an open-ended question puts emphasis on the variety of methods for problem solving, and they provide students with a vast knowledge seeking space, e.g., the item "write three expressions that will expand to give the first term as $12e^3f$" is a typical exercise with inadequate conditions.

2.6.3 Exercise types

Exercise types are generally divided into objective items and subjective items. Objective items consist of multiple-choice questions, true or false questions, fill-in-blank questions etc., while the subjective items consist of calculation questions, solution questions (i.e., questions that require solutions), proof questions, etc. Upon analysis over the given textbooks, it is found the exercises mainly consist of six types of questions: multiple choice questions, true or false questions, fill-in-blank questions, calculation questions, proof questions and solution questions. In the case an exercise question cannot be classified to one of the first five types, it is coded as a solution question.

2.6.4 Difficulty level of exercises

The difficulty level of exercise questions can be used to depict the characteristics of the quality of the exercise system concerned. There are many perspectives to describe the difficulties of exercises, such as level of exercise requirement, number of knowledge points, background of exercises, problem-solving steps or reasoning steps, closeness and openness of conclusions, amount of time for solving problem solving, and the implicitness of the conditions. As to the difficulty level of the static exercises in textbooks, how to define an operational and scientific indicator? By surveying and interviewing 13 experts as well as teachers, we conducted factor analysis to identify the difficulty level of exercises (*N*), which is described through three aspects, namely the requirement level (*YQ*), number of knowledge points (*ZS*) and background (*BJ*). The computation formula is shown as follows:

$$N_i = \alpha * YQ_i + \beta * ZS_i + \gamma * BJ_i \quad (i = 1, 2, 3, 4, 5),$$

amongst which α, β, γ are the weights for *YQ*, *ZS* and *BJ*, respectively, and $i = 1,2,3,4,5$ refers to the five knowledge themes, namely, the "numbers and operations", "equations", "triangles", "solid figures" and "statistics".

The values of α, β, γ were obtained via questionnaire surveys; to be specific, 23 experts from three provinces (Jilin, Jiangsu and Sichuan), two province-level cities (Beijing and Tianjin) and Inner Mongolia Autonomous Region were interviewed, and they were asked to grade the weights, and the mean value was calculated; moreover, 143 middle school mathematics teachers from Shandong and Sichuan Provinces and 18 masters students majored in education from Beijing Normal University (most of whom are mathematics teachers for middle schools) were investigated and required to grade the weights. The results were also averaged and the detailed values are shown in Tab. 10-1. Upon the discussion of the research group, the grades were rounded correct to the second decimal place as 0.38, 0.36 and 0.26, respectively (Yiming Cao and Zhuting Wang, 2007).

Tab. 10-1 Weighting indicators on exercise difficulty level from the expert/teacher investigation

	Required level	Knowledge points	Background
Expert evaluation	0.3757	0.3630	0.2613
Teacher questionnaire	0.3643	0.3597	0.2759

(1) Requirement level

The requirement levels are divided into four grades and they are imitation, comprehension, application and investigation, respectively, assigned with values 1, 2, 3 and 4.

Imitation: knowing from the examples or give examples to describe the relevant characteristics of learning targets; identifying the targets from specific situations or explaining the targets with examples; and solving problems by simulating the approaches given in the textbooks directly.

Comprehension: based on the given example, being able to elaborate the differences and relationships between the target and the given other related targets.

Application: based on the comprehensive use of knowledge, solving new problems with such knowledge in new situations, specially referred to application problems.

Investigation: it requires a complex and non-algorithmic thinking. There are no fixed ways, or in other words, there is no obviously predictable and fixed route or approach to follow from the given tasks, explanations or solved examples.

The average requirement level (YQ_i) of exercises for the ith knowledge theme is calculated as follows:

$$YQ_i = \frac{1 * A + 2 * B + 3 * C + 4 * D}{A + B + C + D} \quad (i = 1, 2, 3, 4, 5),$$

where A (or B or C or D) refers to the number of the exercise questions using "simulation (or comprehension or application or investigation)" in the ith knowledge theme.

(2) Number of knowledge points

The number of knowledge points is reckoned by the above-mentioned agreement; according to the statistics, the maximal number is no more than four; the average knowledge points (ZS_i) of the ith knowledge theme are calculated as follows:

$$ZS_i = \frac{1 * A + 2 * B + 3 * C + 4 * D}{A + B + C + D} \quad (i = 1, 2, 3, 4, 5),$$

where A (or B or C or D) refers to the number of exercise questions containing one (or two or three or four) knowledge point" in the ith knowledge theme.

(3) Background

Background mainly consists of four categories: no background (pure mathematics background is also deemed as no background), personal life background (specifically for students), public life background (for adult world) and scientific background; the average exercise background level (BJ_i) of the ith knowledge theme is calculated as follows:

$$BJ_i = \frac{1 * A + 2 * B + 3 * C + 4 * D}{A + B + C + D} \quad (i = 1, 2, 3, 4, 5)$$

where A (or B or C or D) refers to the number of the exercise questions with "no background (or personal life background or public life background or scientific background)" in the ith knowledge theme.

The ratios of the page numbers for "numbers and operations", "equations", "triangles", "solid figures" and "statistics" against the total number of pages for all the five knowledge themes in every surveyed country are taken as the weight of the respective knowledge theme, denoted as A_1, A_2, A_3, A_4, and A_5, which are different for different countries. For instance, there is no statistics content in Japanese middle school mathematics textbooks, and the French mathematics textbooks contain only three knowledge themes (equations, triangles and statistics) due to the translation difficulties. Below is the computation formula for the exercise difficulty in each country:

$$N = A_1N_1 + A_2N_2 + A_3N_3 + A_4N_4 + A_5N_5.$$

Tab. 10-2 Proportion distribution of the five knowledge themes

Knowledge theme	USA	AUS	CHN	GBR	FRA	GER	SIN	KOR	JPN	RUS
Numbers and Operations	0.605	0.462	0.219	0.558	—	0.149	0.392	0.387	0.332	0.172
Equations	0.198	0.145	0.245	0.144	0.286	0.305	0.159	0.188	0.337	0.415
Triangles	0.05	0.131	0.299	0.05	0.519	0.383	0.148	0.175	0.181	0.259
Solid figures	0.056	0.058	0.106	0.074	—	0.109	0.093	0.106	0.151	0.073
Statistics	0.091	0.204	0.132	0.174	0.195	0.055	0.208	0.144	—	0.081

Note: only the knowledge themes "equations", "triangles" and "statistics" are examined in the French textbooks and the "statistics" theme is not introduced in the Japanese books.

3. Research Outcomes and Analysis

3.1 Number of exercise questions

The overall exercise quantity of the five knowledge themes of mathematics textbooks for middle schools in the ten countries is shown in Fig. 10-2. American mathematics textbooks provide 4,192 questions in the exercise part, ranking No. 1, which is 2.29 times that of Australian textbooks, which rank No. 2 among all the ten countries. One striking feature of the exercises in Australian textbooks are that they contain many sub-questions, with the most contains 26 sub-questions. So if the sub-questions are counted, the Australian textbooks for Grade 8 provides 4,725 questions, 4.7 times the number offered in the Chinese counterpart (Libao Wu, 2013), and no mathematics textbooks from other countries could

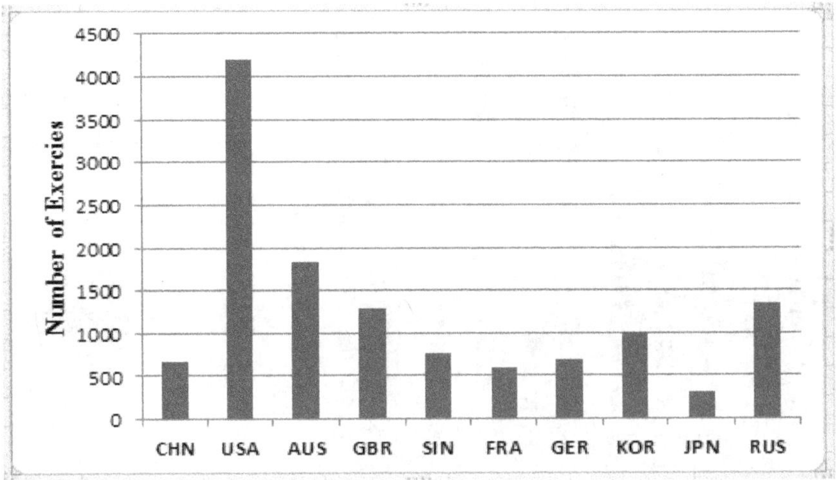

Fig. 10-2 The number of exercises in the five knowledge themes

match this. The Russian and British mathematics textbooks present 1,346 and 1,295 exercise questions, ranking No. 3 and No. 4 respectively, and the difference in the numbers of exercise questions between the two countries is not striking. South Korea ranks No. 5 with 999 exercise questions in total and followed by Singapore, Germany and China, ranking No. 6, 7 and 8, respectively. As to the exercise of the French textbooks, we found that the number of the exercises in the three themes has surpassed that of Japan, about 2.11 times the number in the Japanese counterparts. Due to translation difficulties, we estimated that the number of exercise questions provided in the French textbooks is between that of South Korea and Singapore, probably a little more than Singapore. Japanese textbooks present the least number of exercises; in addition, they do not cover statistics content.

Overall, the number of exercises provided in American textbooks is 15 times the number provided in the Japanese textbooks; while the Australian textbooks offer 6.59 times as many, and Chinese textbooks offer 2.41 times as many as the Japan textbooks.

The quantitative distribution of the exercise questions about the five knowledge themes in the mathematics textbooks in the ten countries is shown in Fig. 10-3. In the textbooks of the USA, Australia, the UK, Singapore, South Korea and Japan, the number of exercises in "numbers

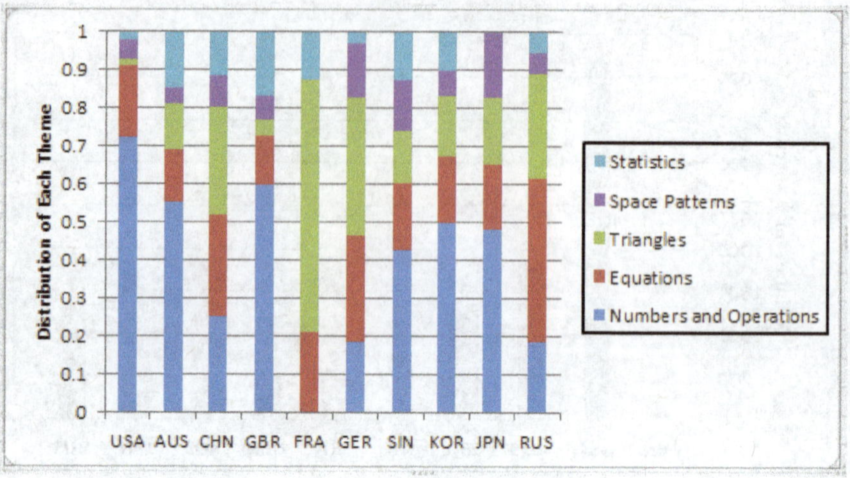

Fig. 10-3 The distribution of exercise questions for the five knowledge themes

and operations" is the most among the five knowledge themes. The USA, Australia and the UK rank to p three, with the exercises in this theme occupy 72.3%, 55.1% and 59.8% of the total exercises in the five knowledge themes, respectively. American textbooks present the greatest number of exercises in both "numbers and operations" and "equations", which might be related to the fact that the US education put much emphasis on "the status and role of algebraic thinking in mathematics education in primary and secondary schools" (Libao Wu and Silin Zhao, 2013). Also, it could also be related to the fact that the middle school phase starts from Grade 6 in the USA. By comparison, the proportion of "numbers and operations" is relatively low in Chinese mathematics textbooks, only a little higher than that of Russia; the learning of this theme completed mainly in the upper primary school stage in China. In the Russian mathematics textbooks, "equations" takes the highest proportion in exercises, 574 questions in total, which is about 42.64% of all the exercises in the five knowledge themes. In addition, Russian textbooks covers the broadest range in the "equations" part, including quadratic equations with two unknowns, linear equations, equations of circles, parametric equations and polynomial equations, which are typically for in senior high schools in many other countries. This is a unique feature of Russian mathematics textbooks and thus the number of the corresponding exercise questions becomes more. The highest proportion of exercises in China, France and Germany is the "triangles" part, among which Germany presents a 37.19% in all the exercises in its textbooks, which is higher than the 28.21% in the Chinese textbooks, while the Russia textbooks present the largest number in this part. In contrast, the UK textbooks only provide 54 questions, ranking second lowest among all the ten countries surveyed; on the whole, the exercises for the geometry part in the UK only take about 10%, which is relatively low. As for the knowledge theme "statistics", the Australia textbooks rank No. 1 with 266 exercise questions, covering a wide range. In terms of the quantity, we can see that the five knowledge themes in the textbooks are given varied priority in those ten countries.

3.2 Exercise types

The numbers of different types of exercise questions in the five knowledge themes in the mathematics textbooks for middle schools in the ten countries are shown in Tab. 10-3. The study reveals that the textbooks in seven countries each contain six exercise types; they are China, the USA, Australia, the UK, Singapore, South Korea and Russia. However, the numbers of questions in these types differ drastically across the seven countries; for certain exercise type, there is only one question throughout the whole series of textbooks in some countries, which is practically negligible in the study. In comparison, the exercise types are most evenly

Tab. 10-3 Type distribution of exercise questions in the five knowledge themes

Country	Fill-in-blank	Multiple choice	True or false	Calculation	Proof	Solution questions
CHN	22	23	7	62	28	528
USA	253	291	12	1269	11	2356
AUS	50	177	21	248	5	1341
GBR	67	4	13	225	3	983
SIN	15	58	1	45	1	648
FRA				96	11	480
GER	0	5	0	283	74	329
KOR	37	86	3	151	4	718
JPN	1	1	4	145	0	168
RUS	7	10	12	202	158	664

distributed in the US mathematics textbooks; in particular, there are also test-oriented exercises for students to practice (test-taking practice). The exercise types are relatively monotonous in the French and German textbooks, mainly focusing on calculation and solution questions, and then mathematical proof questions.

Though one cannot identify the textbook features of a country only through the exercise types, exercise types as part of the presentation methods of exercise can help discover some characteristics in the respective textbooks. The common characteristic of the mathematics textbooks for middle school in the ten countries is that the solution questions rank top across all the exercise types, far more than the other types. In addition, the calculation exercises come second highest, which is possibly related to the nature of knowledge themes "numbers and operations" and "equations", and in particular the "numbers and operations" theme occupies a relatively high proportion in a few countries. On the other hand, the Russian textbooks pay most attention, in terms of the quantity, to proof questions with a total number of 171, accounting for 19% of the total number of all the exercise questions; such a large number and a higher proportion indicate that Russia attaches more importance to the cultivation of students' deduction and proof competence, especially the deductive reasoning competence. The exercises of proof in Russian textbooks concentrate mainly on the geometry section, especially in the triangles part with a total of 126 exercises. Coming next is the German middle school mathematics textbooks, whose proof exercises reach 74 altogether, about 22.5% of the total number. Though the total number in Germany is smaller than Russia, the proportion is higher than Russia. The Chinese textbooks present 28 proof exercises, taking a proportion of only 5.3% and mainly in the "triangles" part. American and French mathematics textbooks for middle schools both have 11 exercises of proof and textbooks from the other countries such as Australia, the UK, Singapore and South Korea present few proof type exercises so as to be practically negligible. The detailed data are shown in Fig. 10-4.

From the number of proof questions presented, the inclinations for mathematics in some countries could be seen, and they also reflect, in

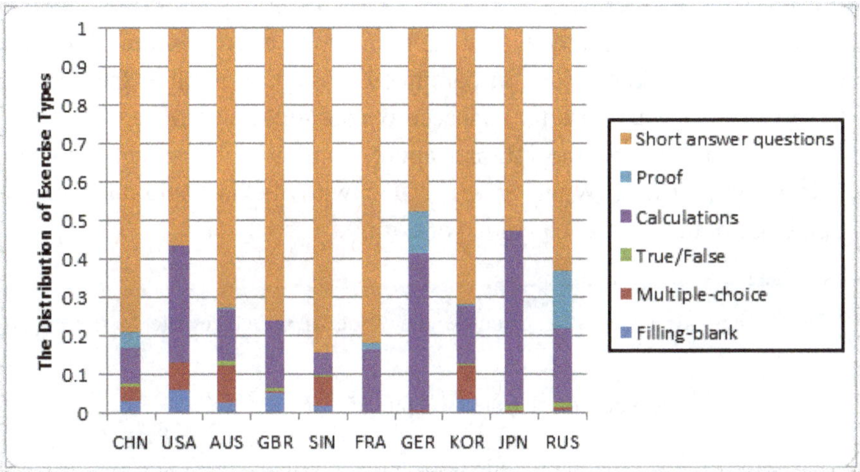

Fig. 10-4 Statistical graph for the exercise types in the five knowledge themes

some sense, their values about mathematics education. The US mathematics textbooks also present a large number of fill-in-blank and multiple-choice questions, each with 253 and 291 questions, respectively. The number of fill-in-blank questions in the US textbooks is more than 5 times that in the Australia textbooks, which rank second most, and the number of fill-in-blank questions is approximately 2 times the number in the Australia textbooks.

3.3 Openness of conclusions

The openness of exercise conclusions in mathematics textbooks for middle school could be divided into two categories: some questions require students to give an example on their own. For instance, the American textbooks have an exercise type entitled "*in your own words*", and in some cases, students are required to give counterexamples; both could be categorized into the "condition-open types", which are particularly prominent in the Australian and American textbooks. The other is extension exercises. In Chinese textbooks these are usually analytical questions with no fixed or standard answers, and some of them ask students to discover patterns. The descriptions of the discovery process provided by the students are not necessarily the same, nor are their results.

For instance, in Section 5, Chapter II "digital technique" of the Grade 8 Australian textbook, the analytical homework exercises are designed according to the knowledge of continued fraction in number theory (see Case I below). There are five consecutive sub-questions raised one upon the other to describe the rational approximation of irrational numbers (Libao Wu, Fuying Wang *et al.*, 2013) with the last sub-question in particular not easy to get a unique or definite answer for it.

Case I

(1) Using a calculator to calculate the value of $\sqrt{13}$ (accurate to the eighth decimal);

(2) Using a calculator to calculate the value of $3+\frac{4}{6+\frac{4}{6}}$;

(3) Using a calculator to calculate the value of $3+\frac{4}{6+\frac{4}{6+\frac{4}{6}}}$;

(4) Using a calculator to calculate the value of $3+\frac{4}{6+\frac{4}{6+\frac{4}{6+\frac{4}{6}}}}$;

(5) For the calculated results, what did you find out by observation? If we continue such calculation in this manner, what pattern do you think you could get?

On the whole, the majority of the conclusions of exercises for mathematics textbooks for middle school in the ten countries are close-ended, accounting for more than 90% of all the questions. In other words, open-ended exercises occupy a relatively low proportion. In addition, there exist relatively large differences in the proportions of the open-ended questions in the ten countries. The specific proportions are as follows: the German mathematics textbooks take the highest proportion, approximately 9.12% of all the exercises provided; the Australian textbooks rank No. 2 with 141 such questions, approximately 7.69% in proportion; the next are the USA, South Korea and the UK ; the Japanese textbooks present only 14 such questions, accounting for 4.44% due to the relative fewer exercises in total; then are the Singaporean textbooks; the Chinese textbooks only present 15 questions of this type, accounting for 2.24% of the total exercises and rank third lowest; the Russian textbooks offer 4 questions in this type, approximately 0.3% in proportion; and the lowest ranked textbooks are from France, which actually contain no open-ended exercises.

3.4 Difficulty level of exercise questions

The exercise difficulty, overall difficulty and ranking of overall difficulty in the exercises of the five knowledge themes in the mathematics textbooks of middle school in the ten countries are presented in Tab. 10-4, and their average number of knowledge points, requirement level and background are given in Fig. 10-5.

Tab. 10-4 Statistical table for exercise difficulty for the five knowledge themes

Country	Numbers & operations	Equations	Triangles	Solid figures	Statistics	Overall difficulty	Ranking
USA	1.173	1.31	1.329	1.316	2.215	1.311	10
AUS	1.619	1.577	1.543	1.705	1.959	1.677	8
CHN	1.586 7	2.071	1.926	1.697	2.142	1.892	2
GBR	1.248	1.254	1.38	1.384	1.718	1.347	9
FRA	—	1.924	1.546	—	1.926	1.728*	5
GER	1.638	1.639	1.753	1.867	1.796	1.716	6
SIN	1.776	1.817	1.644	1.907	1.770	1.774	3
KOR	1.567	1.647	1.699	1.681	1.987	1.677	7
JPN	1.712	1.601	1.594	2.506	-	1.774	4
RUS	2.004	2.469	2.646	2.843	2.885	2.496	1

*The average difficulty is adopted; obviously it will not affect the ranking.

The Russian textbooks rank No. 1 in exercise difficulty across all the five knowledge themes, and the degree of difficulty is much higher than the next top country — China. In the statistics graph for the three dimensions of exercises, the exercises of the Russian textbooks outshine others and rank No. 1 in every dimension. In particular, the requirement level is about 2.87, knowledge point quantity 2.68 and background level 1.98, and all of these are far more difficult than the textbooks from the other countries. The Chinese mathematics textbooks rank No. 2 in "overall difficulty", with the difficulty level in the "numbers and operations", "equations", "triangles", "solid figures" and "statistics" ranked No. 6, No. 2, No. 2, No. 6 and No. 3, respectively; among these ranks, the "numbers and operations" and "solid figures" rank in the middle, while both the difficulty level of the exercises in "equations" and "triangles" rank No. 2. Judging from the three different dimensions, the exercises of the Chinese mathematics textbooks rank No. 2 in the requirement level, No. 3 in the background level, and No. 6 in the knowledge point quantity, which indicates that the degree of the integration of different knowledge in the exercises of the Chinese textbooks is not very high.

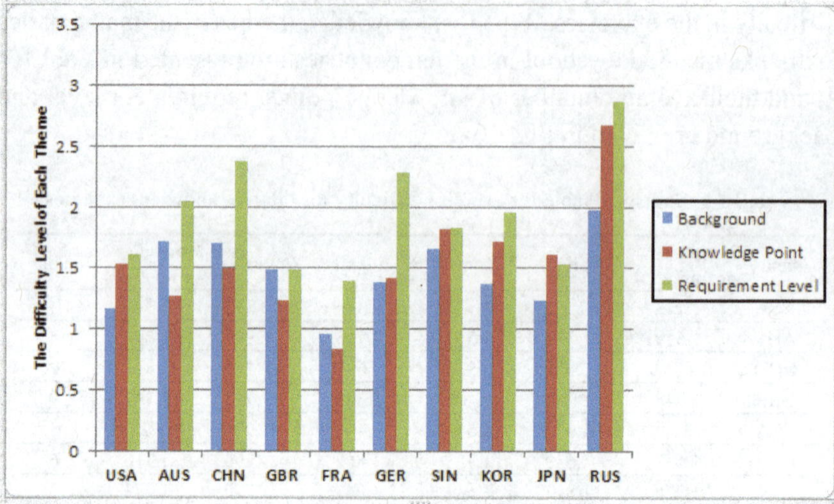

Fig. 10-5 Statistical graph for the three dimensions of exercises in the five knowledge themes

The difficulty level of the exercises in the mathematics textbooks in Japan and Singapore is 1.7741 and 1.7736, respectively, ranking No. 3 and No. 4, with a very small difference. The number of knowledge points found in the exercises in the Singaporean textbooks is 1.82 on average, ranking No. 3 and the background level ranks No. 4; by comparison, the number of knowledge points in the Japanese mathematics textbooks rank No. 2 and their background level and requirement level rank No. 6 and No. 7, respectively. The mathematics textbooks of France and Germany score 1.7279 and 1.716 respectively, ranking No. 5 and No. 6, and the difference between them is very small; both Germany and France have high requirements in their exercises, yet the number of knowledge points involved is relatively few, similar to Chinese textbooks.

The mathematics textbooks of South Korea and Australia score 1.6774 and 1.6773, ranking No. 7 and No. 8, respectively, among which the background level of South Korea exercises is relatively low with a slightly higher level of the number of knowledge points, while the background level of the Australian mathematics textbooks ranks No. 2, next to Russia, and its requirement level is above average, although the average number of knowledge points is relatively low.

The difficulty levels of the US and UK mathematics textbooks are in the bottom of the list. In particular, both the requirement level and the number of knowledge points in the exercises in the UK textbooks come last, and this has been confirmed in another analysis on the five knowledge themes in another series of the UK textbooks.

4. Further Conclusions

4.1 Regional characteristics

Upon further analysis, the study finds that the mathematics textbooks of the ten countries demonstrate striking regional characteristics. The Russian mathematics textbooks are, on the one hand, of great difficulty, far more difficult than the other countries, and on the other hand, they attach more importance to deductive reasoning; and Russia is the only country among the ten that devotes a whole volume of textbook to the topic of geometry, and the number of proof questions turned out to be the most. The difficulty level of exercise questions in the four Asian countries, i.e., China, Singapore, Japan and South Korea, and the two European countries, Germany and France, is comparatively closer to each other, with China being the highest and South Korea the lowest. Under the influence of Confucian culture, the four Asian countries form a unique mathematics education pattern. In numerous series of large-scale international tests for primary and secondary school students, the East Asian students outperform their western peers in mathematics. Both Germany and France have good mathematics tradition in the European Continent; the overall difficulty levels in the two countries are slightly lower than the four Asian countries, which are about 1.728 and 1.716, ranking No. 5 and No. 6, respectively. The mathematics textbooks in the USA, Australia and the UK give priority to "numbers and operations", all accounting for a proportion of over 50% of all the exercises in the five knowledge themes; in addition, the US and Australian mathematics textbooks present a large number of exercises, and on the other hand, the exercises are of low difficulty, in particular, the case in the mathematics textbooks for middle school in the USA and the UK.

4.2 Focusing on different themes

Judging from the knowledge themes, "numbers and operations" takes the highest proportion in the textbooks from the USA, Australia, the UK, Singapore, South Korea and Japan, among which the USA, Australia and UK present exercises in this theme with a proportion of 72.3%, 55.1% and 59.8%, respectively, all exceeding 50%. In the mathematics textbooks of China, France and Germany, the "triangles" part contains the most exercises, among which Germany presents exercises for this part with a proportion of 37.19% of all the exercises for the five knowledge themes, higher than the 28.21% in the Chinese textbooks; it helps students form a rigorous and practical learning style. The Russian textbooks present the most exercises in the "equations" part and the overall number is 574 questions, taking a proportion of 42.64% of all the exercises for the five knowledge themes. By and large, one can see that the quantity distribution of the exercise questions in the textbooks shows varied priority in these ten countries. Judging from exercise types, on the other hand, Russia, China and Germany pay more attention to proof questions, emphasizing the development of students' deductive reasoning competence. In particular, Russia has compiled geometry into a single volume of textbook.

4.3 Hierarchical design in exercises

In terms of exercise design and the form of presentation, many countries adopt a hierarchical design in their textbooks according to the requirement level for knowledge. Taking the Australian textbooks for example, they divide all the exercises, be it homework exercises or review questions, into three categories, namely, "skills, applications and analysis"; the US textbooks divide exercises into three modules, namely, the "practice and apply", "connect and extend" and "mixed review"; the exercises in the Chinese textbooks consist of three types as well — "revision", "comprehensive application" and "analysis and questioning"; the UK exercises consist of the types of revision and quiz, practice (challenge), consolidation, etc.; the Singaporean textbooks are also of three types — practice, exercises and review questions; and two types in France textbooks — in-class exercises and out-class exercises. In the initial section, part of such exercises in these countries help students prepare for the new lessons; and the other parts are designed in correspondence with the examples given in textbooks, all following the model of examples to help students consolidate the new concepts, proposition, formula or theorems; questions of these parts could be answered in the ways explained in examples which reveal the exemplary function of examples, such as the "skills" questions in Australian textbooks, the revision part in Chinese mathematics textbooks, the "revision and quiz" part in the UK textbooks; while in the later comprehensive application and analysis parts, the exercises are mainly designed to apply the newly-gained knowledge and in the extension part for new knowledge, and on the other hand, the requirement for students are relatively higher.

4.4 Diversified exercise background

The mathematics learning of students aims to cope with the challenges from the complex world in future, so mathematics education shall lay emphasis on the cultivation of students to establish linkage between mathematics and reality, and solve varied and changing problems in the real world. Therefore, the exercise background of mathematics textbooks from various countries needs to be diversified, and the realness of exercise background and universality of relation shall be emphasized. The textbooks in the UK, Australia and the USA are the most representative ones in this aspect. In terms of exercise design, the textbooks shall, on the one hand, rely on the current social production and daily life to establish linkage, thus making the information closer to reality; on the other hand, textbooks shall also consider design of different types for exercise questions. The textbooks in China, Australia and the USA explicitly contain application parts. In particular, the connection and extension module in the US mathematics textbooks, provides not only the regular exercise types, but also open-ended questions like "pattern searching" and "in your own words", with the purpose to help enhance students' communication ability. In terms of content, they involve the relevant knowledge in disciplines like science, economics, physics, statistics and dietetics. In addition, there is background knowledge from many other aspects in the exercise questions.

5. Inspirations for Compiling Exercises in Chinese Mathematics Textbooks

5.1 The exercise quantity shall be increased appropriately

In terms of exercise quantity, there are altogether 12700 exercise questions for the five knowledge themes in the mathematics textbooks in the ten surveyed countries with an average of 1270 for each country; should the "numbers and operations" and "solid figures" be included in the French textbooks, the average amount would be over 1300. In comparison, the Chinese mathematics textbooks for middle school have 670 exercises in total, ranking No. 8 or the third lowest, far below the US and Australian mathematics textbooks, which is only slightly higher than the Japanese textbooks; it seems that there are too few exercises presented. For one thing, the Chinese textbooks are written in accordance with the mathematics curriculum standards, in which even the book size and sheet numbers are specified. When compared with the US and Australian textbooks, China counterpart is dwarfed by both the number of sheets and the book size. From a comparative perspective, the Chinese mathematics textbooks for middle school shall provide more exercises as appropriate and further optimize the exercise types of textbooks. Instead of flooding supplemental materials (mostly in the form of exercises, or problem sets) with uncertain qualities in the Chinese mainland, it would be better to give more chances to the professional textbook writers to add more exercises to the textbooks so as to ensure the quality of exercises.

5.2 The openness of exercise conclusions shall be enhanced

Regarding the openness of exercise conclusions, China strikes the bottom three among the surveyed countries. Though open-ended exercises are not necessarily better than closed exercises, students could practice their open thought with more open-ended exercises. Therefore, we need to raise the proportion of open-ended questions in the textbooks, especially the exercises with open conditions or conclusions. The increment of open-ended exercises shall be carried on from two perspectives: on the one hand, the textbooks shall give students more chances to think about the examples by themselves, including positive examples and counterexamples, so as to enhance the discernment of students; on the other hand, more open-ended exercises should be further designed in the textbooks on an optimal basis to nurture the students' self-study competence, help them discover, raise, analyze and solve problems on their own. For example, the open-ended exercises in Australian textbooks could serve as reference with its features like low starting point, gentle slope, and step-by-step extension, which actually give the textbooks, as a teaching medium, the quality to attract students to explore and discover proactively, thus leading them to efficient study.

5.3 The difficulty level of exercises shall be maintained

According to the statistical analysis on the difficulty indicator system for exercises, the difficulty level of the Chinese mathematics textbooks for middle school is among the top few. Actually, both the teachers and students surveyed still believe the exercises of textbooks are relatively simple, so it will not be necessary to reduce the difficulty level; instead, they could be increased as appropriate on the current basis. To maintain or further increase the difficulty level of exercises, we should make endeavors in the following aspects. Firstly, the background of exercise design shall be further improved to make it more multifarious and interdisciplinary, and help students enhance their mathematical application ability; for instance, the exercises of the US and Australian mathematics textbooks of middle school are designed more interrelated to the subjects of chemistry, biology, geography, economics, physics and so on; and in the US textbooks, there is always a background knowledge heading explicitly listed before an exercise. Secondly, we should make exercises more comprehensive on an appropriate basis and pay attention to the context and linkage of knowledge, especially the permeation of mathematical thinking and methods. Thirdly, the hierarchy in exercise design shall be given priority and the exercises with higher difficulty level shall be given adequate groundwork to make them more accessible by students thus help them meet or exceed the expected requirements through this inch-by-inch and step-by-step progression.

Through comparison, we know more clearly about the features of our own textbooks. Moreover, it broadens our vision and will serve as good references for our textbook writing and mathematical class teaching in middle schools.

Acknowledgement: We would like to express our heartfelt appreciation to Professor Daiqin of the Inner Mongolia Normal University, Professor Guangming Wang of Tianjin Normal University, Professor Lianhua Ning of Nanjing Normal University, Associate Professor Kangbiao Jin of Yanbian University, Dr. Yuhuan Zhang of Henan University, Dr. Xiaoxu Zhang of Beijing Normal University and other scholars for their support in data collection! We sincerely appreciate the generous support from the School of Mathematics Division of Beijing Normal University Library.

References

[1] Ball, D. L. and Cohen, D. K. (1996). Reform by the book: What is — or might be — the role of curriculum materials in teacher learning and instructional reform? *Educational Researcher*, 25(9): 6–8; 14.

[2] Bao, Jiansheng (2008). *A Comparative Study of the Difficulty of Middle School Mathematics Curriculum of China and UK* [in Chinese]. Nanning: Guangxi Education Press.

[3] Beaton, A., Martin, M., Mullis, I., Gonzales, E., Smith, T. and Kelly, D. (1996). *Mathematics Achievement in the Middle School Years: IEA's Third International Mathematics and Science Study (TIMMS).* Boston: Center for the Study of Testing, Evaluation and Educational Policy, Boston College.

[4] Cao, Yiming and Wang, Zhuting (2007). Research about teaching algebraic thinking and big ideas [in Chinese]. *Journal of Mathematics Education*, 16(1): 8−11.

[5] Curriculum Research Institute (1993). *Research in Curriculum: a Review of Ten Years' Work* [in Chinese]. Beijing: People's Education Press.

[6] Gerard, F.-M. and Roegiers X. (2003). *Des Manuels Scolaires pour Apprendre: Concevoir, Évaluer, Utilizer.* Bruxelles: De Boeck & Larcier s.a.. Chinese Translation by Ling Wang and Zhenping Zhou. Shanghai: East China Normal University Press.

[7] Jiang, Meiling (2008). *The Study of Teachers' Practical Knowledge* [in Chinese]. Shanghai: East China Normal University Press.

[8] Macintyre, T. and Hamilton, S. (2010). Mathematics learners and mathematics textbooks: A question of identity? Whose curriculum? Whose mathematics? *The Curriculum Journal*, 21(1): 3–23.

[9] MOE of PRC (Ministry of Education of the People's Republic of China) (2012). *Mathematics Curriculum Standards for Compulsory Education (2011 Edition)* [in Chinese]. Beijing: Beijing Normal University Press.

[10] Robitaille, D. F. and Travers, K. J. (1992). International studies of achievement in mathematics. In D. A. Grouws (Ed.), *Handbook of Research on Mathematics Teaching and Learning* (pp. 687−709). New York: Macmillan Publishers.

[11] Wu, Libao (2013). A comparative study of mathematics textbook exercises between China and Australia — a case study of the Grade 8 textbooks published by HMZ and People's Education Press [in Chinese]. *Journal of Mathematics Education*, 22(2): 58–61.

[12] Wu, Libao and Cao, Yiming (2013). An international comparative study of the content allocation of mathematics curriculums in middle school [in Chinese]. *Journal of Educational Studies*, 9(2): 29–36.

[13] Wu, Libao and Zhao, Silin (2013). Inquiry learning of $\sqrt{13}$ [in Chinese]. *Mathematics Teaching in Secondary Schools (Middle School Edition)*, (4): 67–69.

[14] Wu, Libao, Wang, Fuying and Qin, Hua (2013). The role of examples in mathematics textbooks [in Chinese]. *Bulletin des Sciences Mathematics (Shuxue Tongbao)*, 52(3): 18–20; 23.

[15] Wu, Libao, Wang, Jianbo and Cao, Yiming (2014). An international comparative study of mathematics textbook exercises [in Chinese]. *Curriculum, Teaching Material and Method*, 34(2): 112–117.

[16] Zhang, Yongchun (1996). *The Theory of Mathematics Curriculum* [in Chinese]. Nanning: Guangxi Education Press.

https://doi.org/10.1142/9789814713955_0011

Chapter 11

Survey and Analysis on the Usage of the New Primary School Mathematics Textbooks

Qiping Kong*, Yeqing Liu and Li Yang

School of Educational Sciences, East China Normal University
Shanghai 200062, China
**qpkong@kcx.ecnu.edu.cn*

In the turn of the century China launched a school curriculum reform, and the new primary mathematics textbooks as per the National *Mathematics Curriculum Standards for Compulsory Education (Experimental)* (MOE of PRC, 2001) have been put into use in 2001, starting from certain experimental districts and then popularized to the whole country. In 2011, the Ministry of Education promulgated a revised version of the standards, the *National Mathematics Curriculum Standards for Compulsory Education (2011 Edition)* (MOE of PRC, 2012), and then the textbooks revised according to the 2011 standards have entered the school classrooms since 2013.

What happens to students since this round of curriculum reform? This question is concerned not only by the policy makers, reform designers, and school teachers, but also by the parents, and even by the whole society.

This chapter reports an investigation we carried out in 2011 on pupils' academic ability in mathematics in the certain experimental districts. Compared with the West, Japan shares more similarities with China in mathematics education, such as the school system, curriculum contents, etc. Therefore, we selected the mathematical test papers for grade 6 of the "2010 National Survey of Academic Level in Japan", in which the examined contents and questions are more in line with the learning process and cognitive development levels of Chinese pupils. It is hoped that the assessment of the pupils' mathematics learning quality could reflect the effectiveness of the new textbooks used in the experimental districts.

1. Overview of the Survey

This survey was conducted mainly through two paper tests to the pupils and supplemented by teacher interviews. We use the mathematical test papers for Grade 6 (NIEPR, 2010a) in the "2010 National Survey of Academic Level in Japan" (NSALJ-2010, hereafter). The content structure of the test papers matches the content system of the primary mathematics textbooks compiled under the experimental edition Standards (experimental textbooks, hereafter) in China. For instance, two categories "number and calculation" and "quantitative relation" in NSALJ-2010 correspond to the learning domain "number and algebra" in the experimental textbooks; and the category "quantity and measure" in NSALJ-2010 includes both the content of the domain "space and figures" and part of the domain "statistics and probability"; while the category "figures" in NSALJ-2010 corresponds to part of the content in domain "space and figures" of the experimental textbooks. In addition, the writing process of the experimental textbooks was guided by basic mathematical thinking in both content selection and arrangement, while in the Japan academic test papers, the "mathematical thinking" is deemed an important academic ability indicator. To some extent, the testing questions in the papers could reflect the important mathematical thinking and methods such as the sense of numbers and symbols, spatial concept, geometric intuition, perception of data analysis, capability of operations and inference, as well as modeling thought.

1.1 Objectives of the survey

The objectives of the survey include:

— To investigate the status quo of mathematical ability of the pupils in the background of adoption of experimental textbooks.

— To compare the performance of Chinese pupils who used the experimental textbooks with the Japanese pupils in the same grade level.

— Through the above analysis and discussion, to summarize the effect of the experimental textbook on the academic performance of pupils, and then explore the directions for further development of mathematics textbooks.

1.2 The sampling areas

— Zhengzhou City, Henan Province:
The Yousheng Road Primary School, 74 pupils;
The Second Primary School on Huanghe Road, 62 pupils.

— Bozhou City, Anhui Province:
Primary School Attached to Lixin Normal School, 60 pupils;
The First Primary School of Kantuan Town, 52 pupils.

— Quzhou City, Zhejiang Province:
Quzhou City Experimental School, 56 pupils;
The Fourth Primary School of Qujiang, 38 pupils.

— Yuhang City, Zhejiang province:
The First Primary School of Linping, 51 pupils;
Tingzhi Experimental School, Yunhe Town, 44 pupils.

Total sample size: 437 (pupils).

1.3 Abilities and skills to investigate

— *Test Paper A (Arithmetic A)*

It mainly tests the "basic knowledge and basic skills" of the pupils. Here the term "basic knowledge and basic skills" refers to the knowledge and skills which "may hinder the normal learning activities in the future if not grasped" and "are expected to be used frequently and flexibly". In other words, they are the "basic requirements corresponding to each academic year" and the "practical knowledge and skills necessary to pupils' viability".

— *Test Paper B (Arithmetic B)*

It mainly tests the "flexible application" of the pupils. Here the term "flexible application" refers to the "ability to use the acquired knowledge and skills flexibly on the various occasions in practical life" as well as "the ability to devise, execute, evaluate and improve a scheme for solving varied problems". It reflects not only the acquisition of knowledge, but also an effective application of knowledge as well.

1.4 The mathematical content covering of the tests

— *Test Paper A (Arithmetic A)*

Test Paper A in NSALJ-2010 includes 9 questions with 19 items and these items can be classified into the following categories: "number and calculation" (8 items), "quantity and measure" (5 items), "figures" (3 items), and "quantitative relation" (3 items). From the question distribution one can see that Test Paper A focuses on "number and calculation" and "quantity and measure". The objectives of procedure and methodology are embodied in the representing and handling of quantity and figures (9 items) as well as the acquiring and understanding of the knowledge in quantity and figures (10 items). These items are mainly multiple choices and short answer questions.

— *Test Paper B (Arithmetic B)*

Test Paper B in NSALJ-2010 includes 6 questions with 12 items and the items can be classified in the following categories: "number and calculation" (1 item), "quantity and measure" (1 item), "figures" (5 items) and "quantitative relation" (6 items). One of the items (Question 1.1) tested the knowledge of both "number and calculation" and "figures". The distribution shows that Test Paper B focuses on "figures" and "quantitative relation". The objective of process and methodology are embodied in the process of mathematical thinking (8 items), the representing and handling of quantity and figures (2 items) as well as the acquiring and understanding of the knowledge in quantity and figures (2 items). This indicates that the process of mathematical thinking is highly emphasized. The items are mainly multiple choices, short answer questions and essay-type questions.

2. The 2011 Survey Results

2.1 Analytical result of Test Paper A

As shown in Tab. 11-1, Tab. 11-2, and Fig. 11-1, in Test Paper A, the pupils from Yuhang, Zhejiang Province performed best on the whole. All items were answered with similar correctness percentages, without extreme cases, though item 7 is also not desirably answered by these pupils,

Tab. 11.1 The statistics of the correctly answered items in Test Paper A on person basis

	Zhengzhou	Bozhou	Quzhou	Yuhang
Average	17.09	17.50	17.10	18.09
Minimum	9.00	9.00	11.00	11.00
Maximum	19.00	19.00	19.00	19.00
Median	18.00	18.00	18.00	19.00
Mode	18.00	19.00	18.00	19.00
Standard deviation	2.01	1.82	1.98	1.41
Deviation	-1.59	-2.16	-1.45	-2.50
Number of samples	136	112	94	95

Tab. 11.2 The statistics of each question in Test Paper A

Item	Zhengzhou			Bozhou			Quzhou			Yuhang			*Overall correctness rate (%)*
	Correctness rate (%)	Error rate (%)	Unanswered rate (%)	Correctness rate (%)	Error rate (%)	Unanswered rate (%)	Correctness rate (%)	Error rate (%)	Unanswered rate (%)	Correctness rate (%)	Error rate (%)	Unanswered rate (%)	
1.1	94.9	5.1	-	96.4	3.6	-	91.5	8.5	-	95.8	4.2	-	*94.7*
1.2	81.6	18.4	-	88.4	11.6	-	83.0	17.0	-	95.8	4.2	-	*86.7*
1.3	89.7	10.3	-	94.6	5.4	-	89.4	10.6	-	95.8	4.2	-	*92.2*
1.4	100.0	0.0	-	98.2	1.8	-	98.9	1.1	-	98.9	1.1	-	*99.1*
1.5	97.8	2.2	-	98.2	0.9	0.9	100.0	0.0	-	96.8	2.1	1.1	*98.2*
1.6	87.5	12.5	-	94.6	5.4	-	96.8	3.2	-	100.0	0.0	-	*94.1*
2.1	82.4	17.6	-	98.2	1.8	-	88.3	11.7	-	89.5	10.5	-	*89.2*
2.2	86.8	13.2	-	88.4	11.6	-	79.8	20.2	-	93.7	6.3	-	*87.2*
3	99.3	0.7	-	97.3	0.9	1.8	97.9	2.1	-	100.0	0.0	-	*98.6*
4.1	91.2	8.8	-	92.0	8.0	-	92.6	7.4	-	95.8	3.1	1.1	*92.7*
4.2	82.4	17.6	-	92.0	8.0	-	75.5	24.5	-	95.8	3.1	1.1	*86.3*
5.1	89.0	11.0	-	86.6	12.5	0.9	78.7	21.3	-	93.7	6.3	-	*87.2*
5.2	87.5	12.5	-	95.5	4.5	-	94.7	5.3	-	95.8	3.1	1.1	*92.9*
6	95.6	4..4	-	95.5	4.5	-	94.7	4.2	1.1	94.7	5.3	-	*95.2*
7	83.8	16.2	-	71.4	26.8	1.8	78.7	21.3	-	87.3	11.6	1.1	*80.3*
8.1	92.6	7.4	-	90.2	9.8	-	90.4	9.6	-	94.7	5.3	-	*92.0*
8.2	95.6	4.4	-	98.2	1.8	-	94.7	5.3	-	90.5	9.5	-	*95.0*
9.1	86.0	14.0	-	81.3	17.8	0.9	93.6	6.4	-	95.8	3.1	1.1	*88.6*
9.2	86.8	13.2	-	92.0	8.0	-	90.4	9.6	-	93.7	5.2	1.1	*90.4*

compared with other items. The pupils from Zhengzhou, Henan Province also did well, but the correctness percentage of few items drop to the low eighties. The distribution of the correctness rate of pupils from Bozhou, Anhui Province is rather divergent, with items 7 and 9.1 merely at 71.4%

and 81.3% respectively. Compared with this case, the correctness rate is more evenly distributed across the items for pupils from Quzhou, Zhejiang Province, but the data also show that the pupils from Quzhou had relatively lower performance on items 1.2, 2.2, 4.2, 5.1 and 7.

The data in the last column (overall correctness rate) of Tab. 11-2 do not used in Fig. 11-1. They will be used in Fig. 11-3.

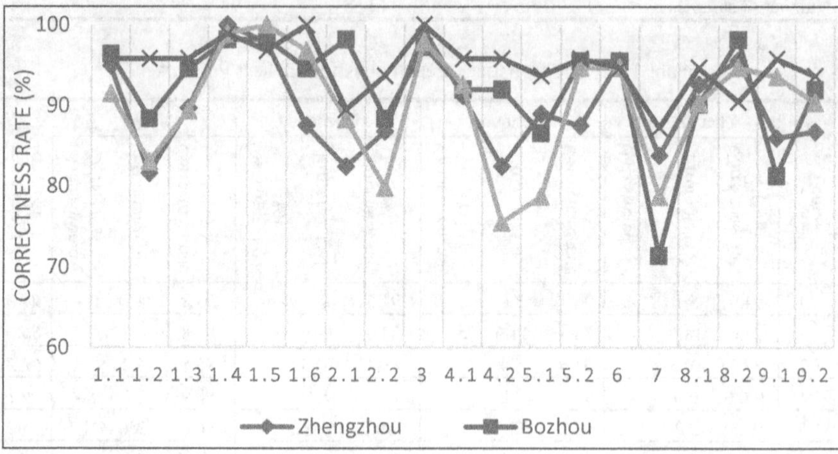

Fig. 11-1 The correctness rate distribution in Test Paper A

2.2 Analytical results of Test Paper B

Tab. 11-3 The statistics of the correctly answered items in Test Paper B on person basis

	Zhengzhou	Bozhou	Quzhou	Yuhang
Average	8.90	7.71	8.20	8.92
Minimum	2.00	2.50	3.50	4.00
Maximum	12.00	11.00	12.00	12.00
Median	9.25	8.00	8.00	9.00
Mode	11.00	9.00	10.00	8.00
Standard deviation	2.46	1.83	2.36	2.08
Deviation	-0.60	-0.339	-0.068	-0.53
Number of samples	136	112	94	95

Tab. 11-4 The statistics of each question in Test Paper B

Item	Zhengzhou			Bozhou			Quzhou			Yuhang			*Overall correctness rate (%)*
	Correctness rate (%)	Error rate (%)	Unanswered rate (%)	Correctness rate (%)	Error rate (%)	Unanswered rate (%)	Correctness rate (%)	Error rate (%)	Unanswered rate (%)	Correctness rate (%)	Error rate (%)	Unanswered rate (%)	
1.1	67.6	32.4	-	45.5	54.5	-	58.5	41.5	-	80.0	18.9	1.1	*62.7*
1.2	74.2	24.3	1.5	89.3	9.8	0.9	81.9	14.9	3.2	93.7	4.2	2.1	*84.0*
2.1	56.7	42.6	0.7	41.1	52.7	6.2	46.8	51.1	2.1	52.6	46.3	1.1	*49.7*
2.2	57.3	41.2	1.5	56.3	43.7	-	62.8	36.1	1.1	69.5	29.4	1.1	*60.9*
3.1	94.9	4.4	0.7	87.5	12.5	-	92.6	5.3	2.1	96.8	2.1	1.1	*92.9*
3.2	91.9	7.4	0.7	67.0	32.1	0.9	80.9	17.0	2.1	71.6	26.3	2.1	*78.7*
3.3	65.5	33.8	0.7	27.7	69.6	2.7	69.1	30.9	-	35.8	64.2	-	*50.1*
4	55.1	34.6	10.3	50.0	42.9	7.1	50.0	42.6	7.4	54.7	44.2	1.1	*52.6*
5.1	87.5	10.3	2.2	84.8	9.8	5.4	85.1	9.6	5.3	82.1	15.8	2.1	*85.1*
5.2	82.3	17.7	-	84.8	13.4	1.8	84.1	13.8	2.1	89.5	10.5	-	*84.9*
6.1	88.2	11.8	-	83.0	17.0	-	83.0	17.0	-	89.5	10.5	-	*86.0*
6.2	72.8	27.2	-	58.9	40.2	0.9	62.7	36.2	1.1	78.9	21.1	-	*68.4*

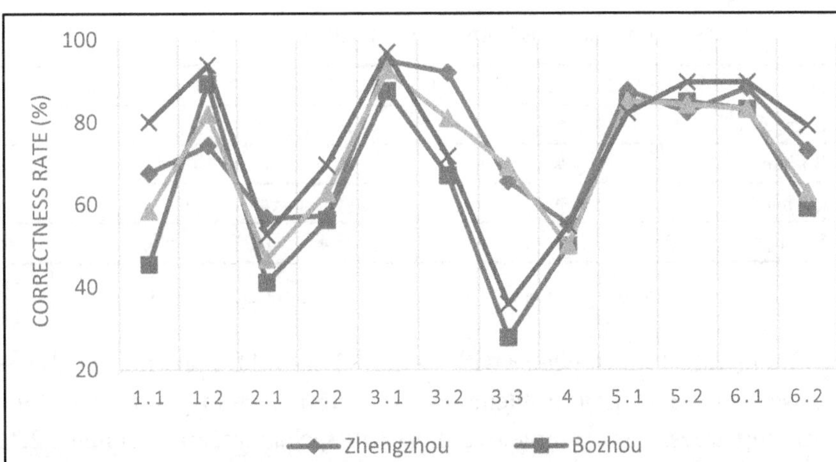

Fig. 11-2 The correctness rate distribution for each question in Test Paper B

As shown in Tab. 11-3, Tab. 11-4 and Fig. 11-2, in all sampling areas, the scores of Test Paper B are much lower than those of Test Paper A, so it can be considered that for the pupils, the difficulty of Test Paper B is much higher than that of Test Paper A. However, the score distribution of Test Paper B is more balanced than that of Test Paper A, and the former is close to a normal distribution.

It seems difficult to compare the overall performances of different sampling areas, but we can say something in this direction. (1) Pupils from Yuhang, Zhejiang Province win the highest correct ness rate in half of the testing items. They are item 1.1 (80%), item 2.2 (although only 69.5%), item 3.1 (96.8%), item 5.2 (89.5%), item 6.1 (89.5%) and item 6.2 (78.9%), showing good strength of pupils from this area. But it is a pity that for item 3.3 these pupils have only 35.8% as the correctness rate, next to the lowest one in the four sampling areas. (2) Pupils from Bozhou, Anhui Province have the most volatile correct answer rates. In some items they have no obvious differences from pupils from the other areas, but in some items (such as item 1.1, item 2.1, and item 3.3), there is a relatively big gap between them and the other pupils, especially in the item 3.3, their correctness rate (27.7%) is lowest of all data.

The items that pupils from each area have low correctness rates (≤ 60% and ≤ 50%) are shown in Tab. 11-5.

Tab. 11-5 The items in Test Paper B with low correctness rates

Items	1.1	1.2	2.1	2.2	3.1	3.2	3.3	4	5.1	5.2	6.1	6.2
Zhengzhou			○	○				○				
Bozhou	●		●	○			●	●				○
Quzhou	○		●					●				
Yuhang			○				●	○				

○ 50 < correctness rate ≤ 60 ● correctness rate ≤ 50

From Tab. 11-5 (together with Tab. 11-3, Tab. 11-4 and Fig. 11-2) we can see some common problems in the performance of pupils from four sampling areas. For example, we must pay special attention to items 2.1, 3.3 and question 4, whose correctness rate is generally lower than the other items. Item 2.1 requires pupils to find the length and width of a rectangle in real situation and to give an explanation in mathematical language, while item 3.3 require students to understand the meaning of data in a two-dimensional table and to relate the data with a pie chart. These two items involve the ability requirement on "mathematical thinking and methods". The low performance shows that pupils' cognitive obstacles to certain degree do exist when mathematics questions are tested in real situations. Question 4 gives an explanation of two triangles with different shapes in a parallelogram to have the equal area, then pupils are required to explain why two triangles with different shapes in a trapezoid have the equal area. This question mainly inspects the student's logical reasoning and the language expression ability. The survey results reflect that pupils in the new curriculum experimentation area still have some deficiencies in the above-mentioned respects.

3. Comparison of the Results in 2011 and 2008 between China and Japan

In this section we compare our survey in 2011 with the results of Japanese survey conducted in 2010 using the same test papers, and also compare it with the survey in 2008 (together with Japanese survey in 2007). Our survey in 2008 using "2007 National Survey of Academic Level in Japan" (NSALJ-2007, hereafter) (NIEPR, 2007a) involving five sampling areas with 270 samples in total (See Qiping Kong & Li Yang, 2012 for more details). In Subsection 3.1 we consider the results on Test Papers A of NSALJ-2010 and NSALJ-2007, while subsection 3.2 is devoted to the comparison of survey results on Test Papers B in both countries and both years.

3.1 Comparison of test results for Test Papers A

We put the overall correctness rates of the two countries together in Fig. 11-3, where the data on China come from the last column of Tab. 11-2, while the data on Japan is quoted from the Japanese report on the survey (NIEPR, 2010b, 2010c). We find that, on the whole, the Japanese pupils and the Chinese pupils from the sampling areas share certain similarities in their mathematical achievement. The Chinese pupils have higher correctness rates and the rates have an evener distribution.

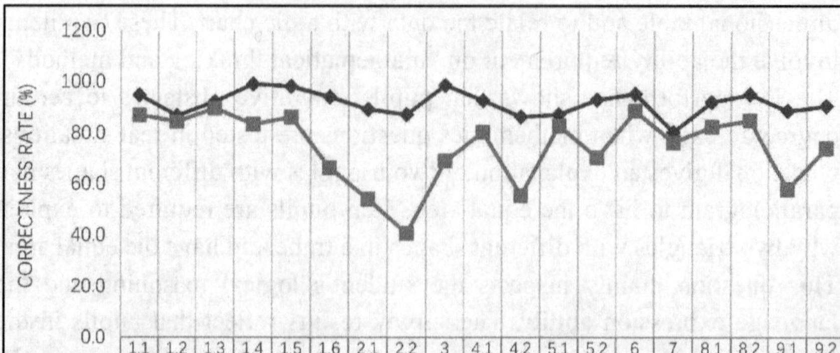

	1.1	1.2	1.3	1.4	1.5	1.6	2.1	2.2	3	4.1	4.2	5.1	5.2	6	7	8.1	8.2	9.1	9.2
China	94.7	86.7	92.2	99.1	98.2	94.1	89.2	87.2	98.6	92.7	86.3	87.2	92.9	95.2	80.3	92.0	95.0	88.6	90.4
Japan	87.0	84.4	89.7	83.4	86.2	66.3	54.1	40.6	69.0	80.3	55.5	82.9	70.4	88.4	76.3	82.3	84.9	57.8	74.0

Fig. 11-3 Comparison between China's 2011 survey and Japan's 2010 survey using NSALJ-2010's Test Paper A

The data and the line chat of the survey in 2008 and its comparison with Japanese survey in 2007 are reproduced in Fig. 11-4. The data of both countries can be found in the 2008 survey report (Qiping Kong and Li Yang, 2012), and the original data of Japan are from Japanese report on the survey (NIEPR, 2007b, 2007c).

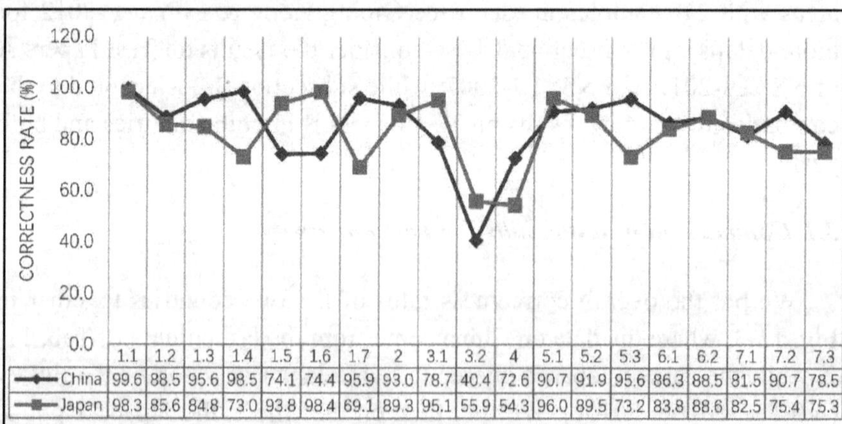

	1.1	1.2	1.3	1.4	1.5	1.6	1.7	2	3.1	3.2	4	5.1	5.2	5.3	6.1	6.2	7.1	7.2	7.3
China	99.6	88.5	95.6	98.5	74.1	74.4	95.9	93.0	78.7	40.4	72.6	90.7	91.9	95.6	86.3	88.5	81.5	90.7	78.5
Japan	98.3	85.6	84.8	73.0	93.8	98.4	69.1	89.3	95.1	55.9	54.3	96.0	89.5	73.2	83.8	88.6	82.5	75.4	75.3

Fig. 11-4 Comparison between China's 2008 survey and Japan's 2007 survey with NSALJ-2007's Test Paper A

Through the consecutive surveys and research before and after the year 2008, we found the academic performance of Chinese pupils using experimental textbooks took on a rising trend compared with the Japanese counterparts. The correctness rates in 2011 are entirely higher than those of Japan, rather than the tangled data lines of the two countries in 2008. This result indicates that the continuous use of the textbooks based on the new curriculum standards could produce more desirable effect, which is a prominent advantage over the other previous domestic textbooks.

Japan has a well-developed education system and Japanese students have outstanding performance in international mathematical competency tests (such as PISA and TIMSS). Therefore, via the horizontal comparison with their Japanese counterparts, we could, to some degree, better reflect the mathematics academic levels of the pupils using experimental textbooks, and further provide objective reflections on the use and effectiveness of the textbooks on the new curriculum standards.

3.2 Comparison of test results for Test Paper B

We put the data of the surveys in 2011 and 2008 on Test Papers B together with the results of Japanese surveys 2010 and 2007 in two figures (Fig. 11-5 and Fig. 11-6).

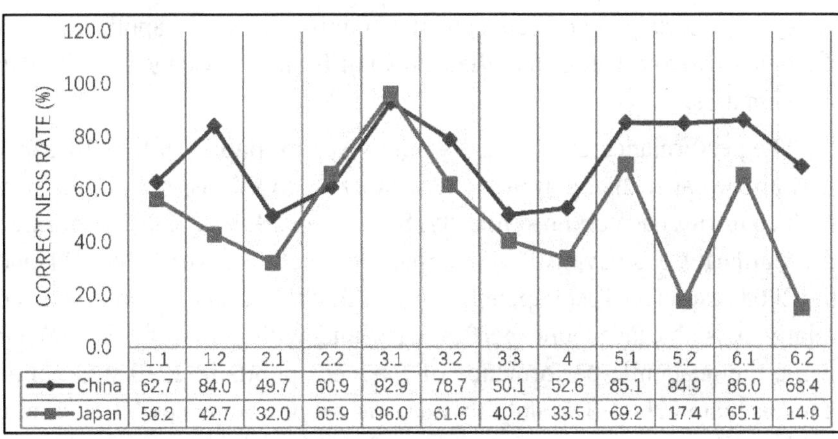

	1.1	1.2	2.1	2.2	3.1	3.2	3.3	4	5.1	5.2	6.1	6.2
China	62.7	84.0	49.7	60.9	92.9	78.7	50.1	52.6	85.1	84.9	86.0	68.4
Japan	56.2	42.7	32.0	65.9	96.0	61.6	40.2	33.5	69.2	17.4	65.1	14.9

Fig. 11-5 Comparison between China's 2011 survey and Japan's 2010 survey with NSALJ-2010's Test Paper B

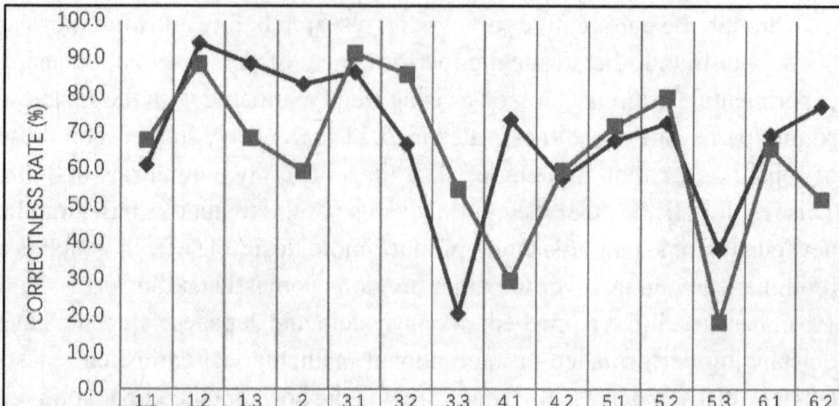

	1.1	1.2	1.3	2	3.1	3.2	3.3	4.1	4.2	5.1	5.2	5.3	6.1	6.2
China	60.7	93.7	88.2	82.6	85.9	66.3	20.4	73.0	57.0	67.4	72.2	37.8	68.9	76.7
Japan	67.5	88.2	68.1	59.0	91.0	85.3	54.1	29.5	59.3	71.4	79.2	18.2	65.1	51.4

Fig. 11-6 Comparison between China's 2008 survey and Japan's 2007 survey with NSALJ-2007's Test Paper B

Test Paper B was designed mainly to test the pupils' skills to flexibly apply mathematical knowledge. In general, the average score of Test Paper B are obviously lower than those of Test Paper A. But it has a higher degree of differentiation. Thus, it can better reflect the real mathematical achievements of the pupils.

Pupils' different performance in Test B and Test A suggests that there is still a certain gap between the pupils' flexible application of mathematical knowledge and their grasp of basic knowledge and skills in mathematics.

The performance of Japanese primary school pupils on the test paper B is also weaker than that of the test paper A. In this aspect, pupils from both countries had certain similarity. Surveys in 2008 and 2007 confirmed the similarity. However, if we compare vertically between the 2011 and the 2008 results on Test Papers B, we will find that Chinese pupils showed relatively stable academic performance and with certain enhancement; while the academic performance of Japanese pupils in 2010 was not as good as that in 2007. The stability in academic performance of the Chinese pupils indicates that this series of experimental textbooks adapt quite well to the pupils and can improve their ability to apply the mathematical knowledge on a flexible basis.

4. Conclusions

Through the surveys and test paper analysis in 2008 and 2011, we can draw the following conclusions:

First, in general, Test Paper A (basic knowledge and ability) for the pupils of Grade 6 did not reveal much difference between the urban and rural pupils while Test Paper B (comprehensive application competence) showed bigger differences.

Second, according to the academic performance of pupils in both Test Papers A and B for all the sampling areas, the comprehensive application competence is positively correlated with the basic knowledge and skills. Thus, they are mutually productive instead of conflicting with each other. The consolidation of basic knowledge and skills is a prerequisite for the improvement of students' comprehensive competence while the improved comprehensive competence may, in return, strengthen the mastery of basic knowledge and skills.

Third, overall, the academic performance in 2011 is higher than that in 2008 and its distribution is also more balanced. In 2008, the average scores of Chinese pupils taking Test Paper A and B are 84.67 and 67.21 respectively; In 2011, the average scores are 91.63 and 70.42 respectively. The grades of Test Paper A in 2011 are clearly higher than those in 2008, while the improvement in Test Paper B from 2008 to 2011 is not as big as in Test Paper A. This survey result may reveal the issue of tendency which merits attention for the research of teaching and learning.

Fourth, in calculating skills, Chinese pupils showed much better performance than their Japanese counterparts, which, on the one hand, reveals the reasonable content arrangement in Chinese experimental textbooks, as well as the mathematics tradition and advantages in China. However, on the other hand, it may also reveal the fact that too much emphasis is laid on the calculating capabilities in the teaching process.

Fifth, the contrast of the test results in 2008 and 2011 suggests that the mathematical reasoning ability (the graphical and numerical reasoning ability) of the Chinese pupils has been rising, which reflects the fact that the experimental textbooks have taken the mathematical reasoning as a priority in the mathematics class. The students' problem-solving ability in mathematics also demonstrates the basic strengths after this round of mathematics reform.

Sixth, some mathematical concepts were not well grasped by the pupils. For example, in Question 3 in NSALJ-2007's Test Paper A, the pupils showed clear lag behind the Japanese pupils in understanding and using the concepts of number axis and scale. Another example is Question 7 in NSALJ-2010's Test A, for which Chinese pupils did not show much difference from Japanese pupils. The question asks to choose the fourth vertex of a parallelogram from several options when the other three vertices and the two connecting sides of the parallelogram are given.

Seventh, the mathematics reading and data processing abilities are yet to be improved. For instance, the pupils gained relatively low scores in items 3.2 in NSALJ-2007's Test Paper B, which requires pupils to interpret the changes in a population engaging in fishery from a bar graph and write down the changes in the total population engaging in fishery. Pupils' performance in item 3.3 is even worse — the correctness rate of the item is only around 20%! In this item pupils are required to read the percentage graph before answering multiple-choice questions and "true or false" questions. Pupils also had low scores in Question 3 in NSALJ-2010's Test Paper B, which requires reading information from a pie graph.

Eighth, there is slight insufficiency in the cultivation of the pupils' space concept and spatial imagination ability. The pupils had difficulties in making transition between figures, especially between 2-dimensional and 3-dimensional figures, which is evidently revealed from the survey in 2011, such as the items 4.1 and 4.2 in NSALJ-2010's Test Paper A and items 2.1 and 2.2 in NSALJ-2010's Test Paper B.

In addition, in the 2008/2007 surveys, both the Chinese pupils and the Japanese counterparts failed to clearly interpret the application context. Hence pupils from both countries did not perform well in Test Paper B. A typical example is Question 4 in NSALJ-2007's Test Paper B (a complex word problem). The pupils cannot apply mathematical concepts and formulas to a specific context in an accurate and reasonable manner, which reflects an inadequacy in the mathematical application ability. Another example is Item 5.3 in Test Paper B, a calculating question about area integrated in a real context. Compared with the Items 5.1 and 5.2 in Test Paper A, which are also area calculations, but without context. The performance of pupils is very different — in Test Paper A, the majority of pupils could obtain correct answers while in Test Paper B, most of them gave wrong answers.

According to the conclusions, the following proposals are recommended for the practice of teaching:

(1) In classroom teaching, the teachers shall, on the one hand, help the pupils consolidate the basic knowledge and skills they have learned, and on the other hand, pay more attention to the training of the pupils' comprehensive application abilities. By searching or unveiling the issues impeding the comprehensive application ability, more innovative classroom teachings could be carried on. Hence the teachers can provide a much more improved learning environment and more opportunities to help develop the pupils' comprehensive application ability.

(2) In teaching practice, appropriate attention shall be paid to improving the pupils' problem-solving ability in real contexts. Teachers are advised to guide the students to understand the context of problems and help them build reasonable mathematical models. In the real world, problem situations could be manifold, such as the individual context which is directly related with the pupils' daily life, the educational contexts which are associated with the pupils' school life or work environment. Thus the students shall be encouraged to observe the more extensive environment around them.

5. Discussion

Through the above-mentioned analysis, we find that by using the experimental textbooks, the mathematics academic performance of pupils in 2011 is clearly better than that of 2008, which is revealed in both Test Papers A and B. When explaining mathematical knowledge for students and assigning exercises for them to practice, mathematics textbooks play an irreplaceable role, whether the knowledge of a mathematics textbook is reasonably arranged and whether such knowledge is closely related to real-life situations will directly influence how well the pupils can grasp such mathematical knowledge and how flexibly they could use such knowledge. This investigation adopted the NSALJ-2010 Test Papers to test how the pupils mastered the basic mathematical knowledge and how flexible they could use such knowledge. Through the study, we concluded that the pupils have a more solid grasp of the basic mathematical knowledge; compared with the 2008 results, their flexible application ability is also improved, steadily though not dramatically (compared with the Japanese counterparts). The investigation about the mathematics academic abilities of pupils indirectly reflects the desirability of experimental textbooks, since they help pupils consolidate and promote the grasp and the application of mathematical knowledge.

In addition, we also interviewed some classroom teachers in our research and communicated with them for long hours. Their talks and reflections provided precious proposals for us, such as they are convinced that the experimental textbooks offered ideal content arrangement with rich mathematical contexts, and the introduction part for new knowledge was well designed. The teachers also mentioned that the pupils were more than willing to learn mathematics now than they used to be and that the experimental textbooks positively guided the pupils' emotional attitude and values in learning mathematics. But they also pointed out that the after-class exercises were relatively too few, and there were actually no exercises in some chapters or sections. Therefore, many teachers had to provide relevant exercises for pupils to practice by themselves. Moreover, there were disagreements over the designed contexts by the rural and urban teachers. The urban teachers held it that the contexts were somewhat over-ruralized, while teachers from rural areas thought the context were, instead, too urbanized. Therefore, the room for further reform of mathematics education in primary school is very large in the future and the revision of mathematics textbooks has significant importance and influence in mathematics education.

There is a common saying that every skyscraper is built from the ground. Primary school mathematics education lays down a foundation for the development of a person, in which primary school mathematics textbooks plays a key role. Looking back at the development of China's primary mathematics textbooks, as Feiyu Cao and Runquan Li (1989) said: "the primary mathematics textbook for general education started from scratch, and grew up from low to higher quality. It has gone through a reform and development process full of twists and turns." This curriculum reform from the century turn has made certain achievements in the formulation of curriculum standards and the compilation of textbooks. The research in this paper also confirms this point. For the further reform and development, effort and involvement are required for every education dedicated personnel and the vast majority of teachers and students. We must constantly explore the law, accumulate experience, and overcome its weaknesses, to make progress, to achieve better results, and to meet the needs of the development of the times and the requirements of the basic education in China.

References

[1] Cao, Feiyu and Li, Runquan (1989). Reform of the primary mathematics textbook for general education over the last 40 years [in Chinese]. *Curriculum, Teaching Material and Method*, 9(10): 1–4.

[2] Kong, Qiping and Yang, Li (2012). A comparative study on Chinese and Japanese primary school pupils' academic achievements [in Chinese]. *Global Education*, 41(1), 77–81.

[3] MOE of PRC (Ministry of Education, People's Republic of China) (2001). *Mathematics Curriculum Standards for Compulsory Education (Experimental)* [in Chinese]. Beijing: Beijing Normal University Press.

[4] MOE of PRC (Ministry of Education, People's Republic of China) (2012). *Mathematics Curriculum Standards for Compulsory Education (2011 edition)* [in Chinese]. Beijing: Beijing Normal University Press.

[5] NIEPR (National Institute for Educational Policy Research) (2007a). *Heisei the 19th nationwide academic achievement and study situation survey: The arithmetic test papers A and B for Grade 6* [in Japanese].

[6] NIEPR (National Institute for Educational Policy Research) (2007b). *The results of Heisei the 19th investigation on the curriculum and subject* [in Japanese].

[7] NIEPR (National Institute for Educational Policy Research) (2007c). *Heisei the 19th nationwide academic achievement and study situation survey: Primary school study report* [in Japanese].

[8] NIEPR (National Institute for Educational Policy Research) (2010a). *Heisei the 22nd nationwide academic achievement and study situation survey: The arithmetic test papers A and B for Grade 6* [in Japanese].

[9] NIEPR (National Institute for Educational Policy Research) (2010b). *Heisei the 22nd nationwide academic achievement and study situation survey: Primary school study report* [in Japanese].

[10] NIEPR (National Institute for Educational Policy Research) (2010c). *The results of investigation on the curriculum and subject* [in Japanese].

https://doi.org/10.1142/9789814713955_0012

Chapter 12

A Study of the Writing Features of Primary School Mathematics Textbooks

Naiqing Song*, Yunming Song, Qiu Jiang, Lizhen Luo and Qingyou Cai

Faculty of Education, Southwest University
Beibei, Chongqing 400715, China
**songnq@swnu.edu.cn*

Textbooks are the materialized carriers of curriculum. As an important media for teaching and learning, they play an irreplaceable role in classroom instruction. They do not only provide an authoritative version of didactics in a particular knowledge domain, but also present each generation of students an official approved authoritative version of human knowledge and culture (Monica, 2006). The writing quality of the textbooks, to a large extent, determines the quality of teaching. In recent years, textbook studies, particularly those on textbook writing, have received more and more attentions in mathematics education research. The first large-scale study focusing on textbooks is known as the TIMSS study, which involved thousands of textbooks from approximate 50 countries (Schmidt *et al.*, 1997). During the 10th International Congress on Mathematical Education (ICME-10) held in Denmark in 2014, textbook was proposed as the theme of one discussion group for the first time in the history of ICME. In addition, there have been many overseas scholars probing into the representation and task design of primary school mathematics contents (e.g., fraction) in their mathematics textbook studies. The domestic studies on primary school mathematics textbooks mainly focused on the contents, structure, representation, usage, mathematical culture, history, comparison as well as writing features, etc. Nevertheless, the existing studies, both domestic and overseas, on the writing features of primary school textbooks remain far from adequate, hardly able to satisfy the needs of the development of primary school mathematics textbooks. Therefore, this paper intends to make a relevant investigation in order to provide appropriate references to the development and characteristic improvement of primary school mathematics textbooks in China.

1. Recognition of Primary School Mathematics Textbooks and Their Writing Features

The study on the writing features of primary school mathematics textbooks shall base on the existing research and establish a scientific recognition of two concepts, namely, the primary school mathematics textbooks and their writing features.

1.1 Primary school mathematics textbooks

Regarding textbooks, scholars have different understandings.

The most popular viewpoint is to regard textbooks as the most important tools and resources for instruction. The International Encyclopedia (Husen *et al.*, 1985) defines textbooks (in the entry "textbooks: an overview" by I. Westbury) as "the books which surround and support teaching of all kinds at all levels of instruction are the central tools and objects of attention in all schooling. The information carried within books defines, for many, the tasks of education; books are the most important resources which teachers have as they do their work of educating". The same entry of the Encyclopedia D. Gowin (1981) is referred. However, Gowin was talking about "educative materials" (the term "teaching materials" is more popular in China), a little broader concept, instead of textbooks. D. Gowin claimed that educative (teaching) materials have at least five roles in education, including (1) as vehicles of criteria of excellence, (2) as records of prior events used to make new events happen, (3) as the authority of the record, (4) as conceptual organizers, and (5) as multipliers of meaning.

The Dictionary of Education (Mingyuan Gu, 1998) regards "textbooks" as "books compiled according to the syllabus (or curriculum standards) of each subject, the main body of the educating materials. They serve as the main media for teachers and students in teaching and learning, the main basis for assessing learning achievements, and the important foundation for students to expand their knowledge field after class … Textbook adoption or accreditation system includes customizing system, approving system, and free system". Note that "teaching material" is another entry (though its English translation is "subject matter" there) in The Dictionary of Education. Educating materials are explained there as "the materials used by teachers and students to carry out teaching activities, the main media of teaching... They include textbooks, lecture notes, teaching outlines, pictures and charts, and related reference books, as well as audiovisuals".

Qiquan Zhong (2003) noted that textbooks are the materials applied in schools or any study groups when learning knowledge of a certain field, and books compiled to facilitate teaching. Ou Shi and Yu Shi (2012) deemed that textbooks, as the epitomized embodiment of state will, national culture, social progress and scientific development, are the most direct carrier to realize educational objectives. Textbooks are the text materials with most readers, being most special and trustworthy, even most reliable, energy and time consuming and influential. Tianshan Zeng (1997) defines textbooks as the educating materials which illustrate subject content systematically in accordance to syllabi (curriculum standards).

The relation between textbooks and educational policy (including national or local syllabi or standards, for example) is mentioned by many Chinese authors. However, the clearest statement of this relation is given by Valverde *et al.* (2002) in the preface of a TIMSS volume: "Textbooks are commonly charged precisely with the role of translating policy into pedagogy. They represent an interpretation of policy in terms of concrete actions of teaching and learning."

As mentioned above, there are two concepts (terms) related to our theme — textbook and teaching material. They are sometimes distinguished clearly (e.g., Mingyuan Gu, 1998), but in many circumstances they are identical (e.g., Husen *et al.*, 1985). This paper regard textbooks and teaching materials as synonyms. In our view, primary school mathematics textbooks refer to teaching materials written in accordance with mathematics curriculum standards and other relevant documents for compulsory education to reflect primary school mathematical contents, basic clues followed by teachers in school classroom instructions to organize primary school students' learning, and one kind of paper material. In the actual teaching, textbooks, as the basic clues for mathematics classroom instruction, play a leading role in what teachers to teach, how to teach, what students to learn, and how to learn.

Many studies have shown that teachers have strong dependence on textbooks, particularly primary school mathematics teachers. For instance, the TIMSS research found that when selecting teaching methods, most mathematics teachers used textbooks as the main resource (Beaton *et al.*, 1996). Another example is from "National Training Plan" primary school mathematics teacher questionnaire survey across 16 provincial regions indicated that in the recent ten years, the primary school mathematics two-basic teaching content in China uses textbooks as the main source and its teaching methods attach great importance to studying textbooks (Kongxiu Huang *et al.*, 2013). As a matter of fact, if a theme is represented in a textbook, teachers are more than likely to explain it in classrooms; otherwise, teachers are inclined not to introduce it. In mathematics textbooks, one may find that the same mathematic theme can be represented and organized in different ways, the relations among themes can be described in different ways, and the implicit teaching methods may influence, to a large extent, teachers' teaching strategies. In brief, mathematics, to some degree, means what is written in the textbooks for teachers and students.

1.2 Writing features of primary school mathematics textbooks

As for the features of textbooks, many scholars have provided their explanations. Juntao Shi (2000) held that according to the basic principles of Marxist philosophy, the diversity of textbooks reflects the particularity of teaching materials and the special nature of textbooks; it is such particularity that forms the diversity of textbooks. Thus, the diversity of textbooks is simply the features of textbooks. Only textbooks with different features can form the diversity of textbooks. Da Lü (2001) argued that the diversification did not equal to quantification, so that having large number, many versions, and great variety does not mean the realization of diversification; the nature of diversification lies in its features, levels, quality, option as well as benefits. In order to facilitate the elaboration, this paper will not differentiate between characteristic and feature. The so-called features of primary school mathematics textbooks refer to the originality or the epitomized embodiment of advantages of the primary school mathematics textbooks. The features of a textbook consist of the embedded writing features and the overt binding features. In general, the writing features of primary school mathematics textbooks refer to the originality in writing philosophy, the selection and elaboration of mathematical knowledge as well as ideas and methods, the design, material selection and context setup of teaching activities and questions, the representation of pictures and texts, the stylistic rules, the section layout and its difficulty levels, etc.; they are also an advantage in facilitating teaching and learning. Whether the writing features of textbooks are obvious is an important indicator to measure the textbooks' quality and level, and by large, such features determine how helpful the textbooks are in facilitating teaching and learning.

To study the writing features of primary school mathematics textbooks followed by development may promote the development of the textbook diversification so as to improve the writing quality; it may help to arouse pupils' interest and boost their development with high personalities; it may help to improve the quality of primary mathematics teaching; it may help to satisfy the diversified needs for different regions, different primary school pupils and different primary school teachers, thus promoting a balanced development in education. The diversification and specialization of textbooks has become one important aim in the development of school textbooks in China, in which the Section Five "The Development and Management of Teaching Materials" of the Outline of Basic Education Curriculum Reform (temporary) has stipulated that the high quality and diversification of textbooks shall be realized (see, for example, Qiquan Zhong, 2001). In 2011, the Ministry of Education held a working conference on the revision of the experimental teaching materials for compulsory education curriculum standards, during which a then vice-minister, Limin Liu, regarded "highlighting features and providing diversified options" as one of the important requirements in textbook revision (see Xiaoya Zhao, 2011).

2. Research Methods

2.1 The analytical framework

Gao Lingbiao (2002) held that textbooks have the features of both overt aspect and internal nature. The overt aspect includes specific applicable users, different learning materials, original framework, distinctive activity forms, creative educational philosophy and perspectives, etc., while the internal nature mainly refers to the originality in the arrangement of framework, design of teaching activities as well as the selection and elaboration of knowledge contents. Some scholars (see, for example, Ding Zhaopeng, 2001) classified the evaluation dimension of textbooks into three aspects; the first is the objectives of the textbooks; the second is the features of textbook contents, including the choice and organization of knowledge and skills, competences, emotion, attitude and values; and the last one is the teaching characteristics of textbooks, including the design of functional modules, text, graphs as well the layout, etc. Yet, some other scholars (see, for example, Ying Wang, 2011) believe that textbook features refer to the differences in one specific book or a specific version of textbook reflected in the writing style, knowledge organization and arrangement, content selection and representation, value and function orientation as well as the targeted audience, etc. from other textbooks. While there is a variety of manifestation modes for primary school mathematics textbooks, they in general can be classified into two categories: One is the arrangement features as revealed in writing style and knowledge organization; the other is the content features demonstrated in content context and theme selection.

Based on the above introduction, together with the interviews of the textbook editors, writers and front-line teachers, the authors believe that the writing features of primary school mathematics textbooks can be studied from the dimensions of writing philosophy, content selection, content organization, content representation and section setting. To be specific, the writing philosophy is the "soul" guiding the writing of primary school mathematics textbooks, which reflects the writers' unique understanding of the teaching regularities in primary school mathematics education, and their interpretations of the subject objectives of primary school mathematics curriculum as well as the specific needs of teachers and pupils. Content selection refers to the writers, under the guidance of course objectives and educational philosophy adopted in the mathematics curriculum standards and syllabus, selects the content of primary school mathematical knowledge and thinking methods, and the corresponding depth and width according to different regions and different requirements of teachers and pupils at varied levels. Content organization means the order arrangement of the selected content as is to be presented on the textbook in accordance with the logic system and learners' mental development rules. The representation of content refers to the writers' design on the elaboration method of mathematical content or theme, mathematical activity and questions, material selection, context setting, text representation as well as writing stylistic rules with a purpose to facilitate teaching and learning. Section setting, by contrast, refers to the setup of some specific sections to carry on a specific educational philosophy or write some specific mathematical contents.

2.2 Objects of the study

2.2.1 Selection of overseas primary school mathematics textbooks

In ancient times, good mathematics works were usually chosen as mathematics textbooks. In this sense, the overseas primary school mathematics textbooks could be dated back to the era of Babylon, ancient Egypt and ancient Greece. By approx. 2000 BC, the Babylon kingdom has offered mathematics subjects in its scribe schools. By approx. 1650 BC, the Egyptian monk Ahmose transcribed on papyrus a mathematics work entitled *Rhind Mathematical Papyrus*. This book had been an earlier mathematics textbook in ancient Egypt (Wikipedia), and even deemed as the first mathematics textbook in the world (Underwood, 2002). By approx. 387 BC, Plato set a Platonic Academy, whose teaching subjects included arithmetic. However, to the common understanding, Euclid's *Elements* is the first recognized mathematics textbook in the world. The teaching content of mathematics in many schools is from this book, and for about

2000 years, this book has been taking the dominant place in geometry teaching.

In 1202, the Italian mathematician Leonardo Pisano Bigollo wrote the monumental work on arithmetic and algebra, entitled *Liber Abaci,* which turned out to be the model textbook for the later 300 years after it had been published (Dingling Ma *et al.*, 2008). In 1522, the Latin arithmetic textbook written by Tonstall was published. In 1543, *The Ground of Arts* written by Robert Record was published, which boasted both academic value and great influence; Record's second mathematics work, *The Pathway to Knowledge*, was published in 1551, which was hailed as the first work on geometry written in English in history, whose content was mainly drawn from the first four books of *The Elements*; Record's fifth mathematics work, *The Whetstone of Witte*, was published in 1557, which was considered as the first algebra textbook in the U.K. (Xiaoqin Wang, 2008). The three textbooks constituted the foundations of mathematics in the U.K. for many centuries, and even for most western countries (Naiqing Song *et al.*, 2009). In the absence of curriculum standards, the U.K. used a "National Test" to control the subject and standardize the textbooks, thus the textbook system was exceptionally free. The writing and publication of textbooks was totally free among civilians so that the versions of school textbooks are of high variety. In 1988, the U.K. established a national curriculum. Since then, textbooks should be entrusted by renowned publishers to be written by teachers or not-long retired teachers, and experts in various fields would be organized to check whether there were problems academically; when the first draft of a textbook was finished, it would be sent to an appointed school for trial use followed by revision according to the teachers' views and suggestions, and adding pictures and explications to finalize the book. Currently, there are 15 educational publishers in the U.K. with an output of textbooks approx. 3,500 types (Yueru Chen *et al.*, 2009). Under the influence of the development of primary school mathematics textbooks in the European and American countries, the writing of mathematics textbooks in Japan were mainly featured by imitation initially; after World War II, Japan adopted an examination and approval textbook system, and the writing of textbooks took on a feature of diversification, mostly by the progressive democratic scholars. No matter in the writing or examining for approval textbooks, the fundamental principles of democracy and peace must be followed. Currently, there are 54 publishers specialized in textbook publishing in Japan (Yueru Chen *et al.*, 2009).

For the overseas textbooks, not only the writing are open, whose writers can be school teachers, superintendents of school districts, college teachers or even national professional textbook writers; the publication is also open, the writing and publication of primary school mathematics textbooks, in many countries, are completely carried on in the public with great flexibility, and the textbooks are written with distinctive features in a huge variety; as mentioned above, there were approximately 3500 types of textbooks used in schools in 2009 in the U.K., which is indeed a flourishing pattern in textbook writing; moreover, the selection system for textbooks are also diversified, such as the local authorities of the U.K. determine which textbooks to use; the USA adopts both an open and state selection system; Germany follows a state selection system; and the Malaysia central government appoints the textbooks, etc.

In order to investigate whether the heavy burden of Chinese primary school pupils was mainly due to the high difficulties of textbooks, the mathematics education team from Southwest University has conducted a research project entitled "An International Comparative Study on School Science Textbooks (Primary Mathematics)" since 2012, in which 13 sets of mathematics textbooks were selected for comparison, involving 10 countries including China, the U.K., the USA, Singapore, France, Germany, Australia, Russia, Japan, and South Korea. In this paper, 6 sets of mathematics textbooks were chosen for the analysis, which are of distinctive features and have greater effect on their respective countries. They are *Arithmetic* (Shimizu, 2011) from Japan, *My Pals are Here! Maths* (2nd Edition) (Fong, 2007) from Singapore, *Primary Mathematics* (Clarke, 2007), *Mathematics Connects* (Altieri, 2009) and *Investigations (common core edition)* (Pearson, 2008) from the USA, as well as *Nelson Mathematics for Victoria* (Doyle, 2007) from Australia. The applied regions of the above-mentioned textbooks involve four continents — Asia, Europe, North America and Oceania, and mathematics education in the concerned countries is of important influence all over the world, and thus these textbooks will serve as a good reference to the improvement of primary school mathematics textbooks of China.

The USA Curriculum Reform has experienced the "New Mathematics Movement" in the late 1950s and the "Curriculum Standards Movement" in the 1990s, both of which aroused reconsideration among the developed countries over their own mathematics education, whose influences affected the whole world. Among numerous primary school mathematics textbooks in the USA, *Mathematics Connects* is deemed as a highly influential textbook for elite education around the world with high requirement and difficulty level. When this textbook was first used in the experimental schools of California and other states, it evoked strong repercussion; by comparison, *Investigations (common core edition)* is less demanding and less difficult, and is adopted by many primary and middle schools in the USA, whose role plays in the USA is as that of the textbooks from the People's Education Press in China, and thus the two sets of textbooks were selected in this paper for the analysis.

The problem that Japan's primary mathematics education encounters is partially similar to that of China's, such as "after-school learning time, learning desire, and the development of learning and living habits" (Yiming Cao, 2012). The countermeasures of these problems are all reflected in textbooks; it may possibly provide some references to the writing of China's textbooks via a meticulous study of its textbook. Among the 54 Japanese textbook publishers, this study selected the Shinko Shuppansha Keirinkan Co., Ltd who published *Arithmetic*. In 2011, this series of textbooks was ranked No.2 in Japan, taking a share of 34.3%.

The education quality of Australia is acknowledged as world-class, which requires all the school education to be set up according to individual needs, ability and interest so as to bring the best out of every pupil from various aspects and then to apply in all aspects of life. From the perspective of content representation, the Australian textbooks are closer to daily life, more vivid and of great comprehensiveness to lift the boundary between mathematics and life or other subjects (Jiajian Zhang, 2005). The study of its textbooks may shed some inspirations for the integration of the mathematics curriculum in China. *Nelson Mathematics for Victoria* is not only widely used in Australian schools, but widely used in many Asian countries like South Korea, Singapore and countries in North America such as the USA and Canada.

Singapore shares a similar oriental traditional educational philosophy with China; in addition to introduce the western instructional theories, Singapore also makes every effort to improve and renew its mathematics curriculum, textbooks and teaching methods. Due to its outstanding performance in TIMSS, the Singaporean textbooks are drawing more and more international attentions. Many countries want to introduce the Singaporean mathematics textbooks to help improve their pupils' mathematics academic achievements. *My Pals are Here! Mathematics* (2nd Edition) is the most frequently used primary school mathematics textbook in Singapore which emphasizes the cooperation of home education and school education and also is considered an ideal textbook for average pupils.

Mathematics education of the U.K. plays an important role in the global mathematics education with a unique position and the development trend of mathematics education in the world could be roughly seen from the reform process in the U.K. Meanwhile, the U.K. emphasizes differentiated mathematics education at the primary level, which can provide some references to the development of the diversification of China's textbooks. Among the 3,500 school textbooks published annually in the U.K., the *Primary Mathematics* is one of the most popular primary school mathematics textbooks used in the schools of the U.K.

2.2.2 Selection of primary school mathematics textbooks in China

Compared with foreign countries, there is no modern primary and middle schools in China until 1901, and therefore the development of domestic primary school mathematics textbooks began much later. In a modern sense, the first series of domestic primary school mathematics textbooks was born under the motivation of the "Renyin-Kuimao Educational System"[a]. Since then, China started the progress from simple transplantation to creation, from rough innovation to huge leaps. In

[a]Renyin and Kuimao refer to the year of 1902 and 1904 respectively, according to the Chinese year reckoning system.

retrospection, the development history of China's primary school mathematics textbooks can be roughly divided into the following stages:

The late Qing Dynasty and the Republic of China period (1901–1949) started from translating the Western textbooks and then the Japan ones, then learnt from the European and American countries, and finally progressed to write on our own. Under the examination and approval system, civilians were allowed to write textbooks freely, therefore during this period the writing and publication of textbooks were extremely active with a great contribution in variety and there even appeared many individual-made textbooks. The content arrangement and representation method of the majority of textbooks gave a refreshing sense to the readers. During the Republic of China period, there were three major views over textbook writing and the dominant one was that textbook writers should take the children's age characteristics and living environment into consideration, and children's real-life experiences should serve as the main line for textbook writing. That is also what we focus on in textbook writing nowadays.

During the initial period of People's Republic of China (1950–1957), the founding of the People's Education Press (PEP) broke the previous diversification pattern. Since then, all textbooks should be written by the PEP in a unified manner. The textbooks during this period were mainly modeled on those of the Soviet Union, emphasizing "three calculations" (i.e., mental calculation, written calculation, and abacus calculation).

During the period of socialist construction (1958–1965), the primary arithmetic textbooks began to cast off the bondage of Soviet experience due to the deterioration in the Sino-Soviet relations, and textbooks were written based on the research summaries over a series of educational reforms and practical experiences.

During the Cultural Revolution period (1966–1976), the education undertakings developed for 17 years since the founding of New China were totally repudiated; primary school mathematics textbooks were severely damaged and the educational quality drastically dropped to the lowest level in New China as well; the gap of mathematics education between China and the developed countries were further widened.

During the reform and opening-up period (1977–2000), textbook development had become the most important task in education rectification, yet textbooks were nonetheless written on a unified basis. In 1986, the National Compilation Committee for School Textbooks was established, terminating the situation in which a whole nation used one series of textbooks, and the primary school mathematics textbooks began to boom and flourish again (Jia Wei, 2009).

Since the new curriculum reform (2001–), achievements have been made in diversification development of textbooks in China. There are eight series of mathematics textbooks for primary schools approved by the educational authorities, which were written in accordance with the *Mathematics Curriculum Standards for Full-time Compulsory Education (Experimental)* (MOE of PRC, 2001; hereinafter referred to as "the Standards"), including the ones published by the People's Education Press, Beijing Normal University Press, Southwest China Normal University Press, Jiangsu Education Press, Hebei Education Press, Qingdao Publishing House, and Zhejiang Education Press, respectively. Hereinafter referred to as the PEP edition (Jiang Lu *et al.*, 2002), Jingshan edition (Jingrong Chen, 2006), BNUP edition (Jian Liu *et al.*, 2006), SCNUP edition (Naiqing Song, 2007), JEP edition (Ligu Sun *et al.*, 2002), HEP edition (Xingmei Zhao, 2008), QPH edition (Tao Zhan *et al.*, 2005), and ZEP edition (Tianxiao Zhang, 2010). These eight series of primary school mathematics textbooks are used in wide regions across China. In this research, the eight textbook series were chosen for the analysis of the writing features of domestic primary school mathematics textbooks.

2.3 The process of the study

The research methods used in this study included literature review, comparative study and survey study. With a view to enhance the reliability of the findings, the study process was divided into three phases:

First, specifying the analytical framework and the subjects of the study. As mentioned above, the textbook analytical framework and subject of the study were first determined via literature review and discussion; then in order to ensure the rationality and scientificity of the analytical framework, the authors further interviewed the writers of the primary school mathematics textbooks and front-line primary school mathematics teachers so as to make amendments to the framework. As for the writers, they were mainly inquired about what considerations they took in writing textbooks. As for the school teachers, they were mainly inquired about what perspectives they would focus on and analyze when using a textbook.

Second, analyzing and comparing the writing features of primary school mathematics textbooks. The six sets of overseas textbooks were mainly analyzed and compared by the senior members of the project "International Comparative Study on School Science Textbooks (Primary School Mathematics)", including mathematics education researchers and primary school mathematics teachers, followed by several discussions until reaching an eventual consensus. As for the eight domestic primary school mathematics textbooks, the comparison and analysis were accomplished by three steps: the first is to analyze the existing papers, books or other documents concerning the features of these eight textbooks and then summarize their respective features, similarities and differences; the second is to read over these textbooks and verify the previous conclusions, modify and supplement such conclusions; the final step is to ask the writers and users of the textbooks to revise and supplement the arrived conclusions.

Third, proposing suggestions for the development of the writing features of the domestic primary school mathematics textbooks. Upon the previous findings together with the findings in this study as well as the suggestions from the interviews of the experts and teachers, the researchers made relevant recommendations on how to improve the writing features of the domestic primary school mathematics textbooks.

3. A Brief Analysis of the Writing Features of the Chosen Overseas Textbooks

As the old Chinese saying goes "Stones from other hills may serve to polish the jade of this one". The analysis of the good experience of overseas primary school textbooks, especially their writing features, may provide valuable references to improve the writing features of the domestic primary school mathematics textbooks.

3.1 The writing features of the six overseas primary school mathematics textbooks

3.1.1 "Mathematics Connects" from the USA

Editors in chief: Mary Behr Altieri, Don S. Balka *et al.*
Publishers: McGraw-Hill Company Inc.
Publication year: 2009

Mathematics Connects is a series of school textbooks written by many experts and scholars under the US 2000 NCTM *Principles and Standards for School Mathematics*, which has passed the assessment of well-known mathematicians and educationists in the USA; the primary school part consists of Books 1–6. The writing features of this series are as follows:

(1) The design of the textbooks is mainly constructivism-oriented and supplemented with centralization. By planning appropriate themes and activities, this series is intended to arouse pupils' learning initiatives, let them carry out exploration activities, and then construct their own knowledge structure.

(2) This series focuses on an early penetration of algebraic thinking, geometric intuition, ideas in statistics and probability, as well as the early infiltration of modern technology. The content of the whole series consists of three parts, including Number and Algebra, Graphics and Geometry, as well as Probability and Statistics. Moreover, algebraic model, classification, statistics (scoring matrix), bar graph as well as the usage of calculator are taught since Grade 1.

(3) The content is required to reflect algorithms, problem solving and deduction. For instance, one task on reviewing "multiplying a two-digit number" in the Review section of Chapter 7 at Grade 4 is, to calculate 178×34, in which the result is first calculated in arrangement, then represented in the form of diagram by a geometric partition method; obviously, the algorithm is given additional attention.

(4) Estimation is emphasized. California does not set a high standard for accurate computation but more attention is given to verbal and written estimation in practice.

(5) A "small spiral type" content representation is adopted to ensure a stepwise increment in knowledge acquisition. Many knowledge points are distributed among multiple chapters so as to break down the difficulty levels; by this way, the same point can be represented from different perspectives, taking into account the intellectual development pattern of pupils. In some chapters, the textbook colligates the contents of different branches, thus strengthen the integrity in mathematics.

(6) Emphasis has been laid on the connection to life and science when selecting materials. Diversified teaching materials are adopted in this textbook series and a host of cases in real life and science cases are applied. In the examples of each section, the interrelation between mathematics and real life is fully demonstrated; with regard to the arrangement of exercise background, a large number of science cases are selected besides those in real life.

(7) The series pays attention to problem solving and open-ended questions. A "Problem Solving" column is designed in every unit with an attempt to foster pupils' problem solving ability. There are always many open-ended questions set in the Hot Topics column of each section with an attempt to develop the pupils' divergent thinking and innovation.

(8) There are rich assessment contents, sound evaluation structure and clear evaluation criteria. From the perspective of the textbooks' macroscopic evaluation structure, each book is designed with a "Let's Review!" module at the beginning, which is intended to help pupils recall what they have learned in the previous book. At the end of each textbook, a "Practice by Standard" module is designed to test pupils how much they have mastered, which may serve as a summative evaluation of one entire academic year. Two modules "Let's Review!" and "Practice by Standard" construct an overall evaluation framework for each textbook which connects the contents of the adjacent books more closely, and also helps pupils organize their learning into a coherent whole (Yujia Huo *et al.*, 2011).

3.1.2 "Investigations (common core edition)" from the USA

Editor: Pearson Scott Foresman
Publishers: Pearson Education Companies
Publication year: 2008

The common core edition of primary school mathematics textbooks contains a total of 5 books with one book for each academic year. There are several main features of the textbooks:

(1) Attaching importance to mathematical communication and emphasizing the usage of technology.

(2) Adopting a "spiral type" to the content representation so that various knowledges can permeate mutually. Such a spiral arrangement in textbook representation makes circulative-rise of knowledge become true. Moreover, knowledge points, more often than not, are intersected, some even interweave among the whole book so much as that they can hardly be named by section titles. Taking "multiplication of integers and single digit numbers" represented in unit one of Book IV as an example, it is designed throughout all the units of Book IV with completely different representation forms.

(3) Materials are selected from real life and other disciplines, focusing on the application of mathematics. This series advocates pupils to learn mathematical knowledge from real life and science-technology activities. It also boasts of a host of real-life situations and a large number of real photos to help pupils understand knowledge points. The point "multiplication of integers and single digit numbers", for instance, is used all in real-life cases with attached photos to facilitate the application of this point.

(4) The number of examples is small and pupils are encouraged to find the steps for solving problems or to imitate the examples to do exercises. For instance, in the "Addition" section, pupils are expected to further understand the addition through data splitting and combination, yet there is only one example in the first column in which the starting number 243 is clearly given, requiring a computation with $+10$, -20 and $+100$, followed by the conclusion of $243 + 10 - 20 + 100 = 333$, since which

pupils are given free to play so that they can finish this part by imitating the above operation.

(5) Less but well-chosen homework, which requires pupils to collect information extensively. There are very few extra-curricular assignments in design, most of which are open-ended questions with keys that can hardly be found from books, and thus pupils are required to study a "project" with what they have learned in the books very flexibly. In addition, pupils must learn how to look for information from a library or internet, and homework has to be finished by using both hands and brains through independent thinking.

(6) Reducing pupils' learning pressure, textbooks are getting thinner and more concise. The content of textbook, exercises and workbook are designed into one piece, and daily homework is finished on book with fewer extra-curricular assignments. When homework is finished, the pupils can tear the homework sheet off and hand it in, thus the textbook is getting thinner as time goes on.

3.1.3 "Arithmetic" from Japan

Editors in chief: Shizumi Shimizu and Syunsuke Funakoshi
Publishers: Shinko Shuppansha Keirinkan Co., Ltd
Publication year: 2011

There are altogether 11 books in the series *Arithmetic* (Keirinkan edition): one book for Grade 1 and two books for each grade from Grade 2 to Grade 6. The editorial board for this series of textbooks consists of in total 50 members, including college professors, associate professors, and teachers of middle schools, primary schools and kindergartens. The main features of this series of textbooks are as follows:

(1) Focusing on the penetration of mathematical thinking into various fields. Function thinking and statistical ideology are involved in the learning of quantitative relation; limit thinking is taken in the graphics and geometry sections; mathematics thoughts can be felt all through the mathematics activities. When questions on quantity and graphs are discussed, the textbooks give additional attention to the relationship between changes of two quantities to reflect function thinking; the cylinder

knowledge is learned through one prism of a regular prism whose sides are getting manifold gradually, and ultimately get close to a cylinder on an indefinite basis, which also implies a thinking of "Limit"; the pupils are enabled to feel the mathematical thinking in operation, and "Abstraction" is introduced and discussed when adequate understanding is prepared at appropriate time.

(2) Estimation being emphasized. Estimation is paid more attention to in the fields of number, algebra, length and measurement, such as estimating the length or area. Pupils are not only required to estimate the size of something, but to learn how to measure with relevant tools.

(3) Adopting a "spiral-up type" content representation. By this way, pupils are more liable to grasp the basic and fundamental knowledge and skills. Taking fraction for example, basic fraction is introduced at Grade 2; yet the concept of fraction, numerator and denominator are not taught until one year later in Grade 3, and then fraction is used to indicate the concentration of liquid, what the meaning of "one tenth" is, how to reckon a fraction as a number, such as five-fifths equals to one, how to indicate fraction on a straight line, the magnitude of fractions with a common denominator, as well as addition and subtraction of fractions with a common denominator; in Grade 4, pupils are taught the representation approach of fraction greater than 1, addition and subtraction of fractions (including mixed fractions and improper fractions), the meaning of identical fractions etc.; in Grade 5, pupils learn reduction to common denominator, reduction of a fraction, addition and subtraction of fractions with different denominators, how fraction is multiplied or divided by an integer as well as the relationship between fractions and decimals or integers; by Grade 6, multiplication and division of two fractions are taught. Obviously, the learning of fractions goes all the way from Grade 2 through Grade 6. The spiral-up structure of knowledge points reduces the difficulty in fraction learning.

(4) Both knowledge content and exercises are hierarchical in design. Each textbook is divided into two parts: "Main Body" and "Window of Arithmetic". Main Body includes the content to be studied by all the pupils, while Window of Arithmetic serves as a further expansion and complement of Main Body which aims to encourage pupils to explore by

themselves and summarize, generalize as well as apply what they have learned in Body Section. Meanwhile, the exercises of each unit are divided into two difficult levels, namely, Basic Exercises and Extension.

(5) Endowed with an integral structural arrangement, the textbook lays emphasis on the links among knowledge points. The content is made up of three columns, in addition to what to be learned in the textbook, it also lists the relevant content that has been learned and the main content of the next textbook. Each textbook has also specified the learning process, method, learning path and contents; among which the learning path mainly includes five phases, i.e., clarification of problems, independent thinking, communication and discussion, consolidation by exercises as well as review.

(6) Giving various activities of mathematics, enlightening thinking through mathematical communication. Mathematics Activities was a newly added section under the 1998 Japanese Course of Study for primary mathematics, including physical manipulation, computing exploration, graphing and investigation etc. The Mathematics Activities can help pupils, in relaxing atmosphere, to proficiently master the basic and fundamental knowledge and skills in quantitative relation, graphics and other fields through various mathematical activities. During the communication process in activities, the pupils may express what they think through numbers, formula, graphs and tables etc., and exchange their thinking process in problem solving, which can not only enhance their language competence but also develop pupils' earnest attitude to learning from others as well.

3.1.4 "Nelson Mathematics for Victoria" from Australia

Editor in chief: Tony Doyle
Publishers: Thomson Nelson
Publication year: 2007

As a federal country, Australia always has a lack of unified curriculum standards. However, the federal government has issued syllabus and guidance document. The textbook series *Nelson Mathematics for Victoria* was written according to the syllabus and the guidance documents, and it was an assignment book designed for pupils' independent study and self-evaluation. Besides, Australian schools have the concept of "Stage", which is made up by two grades. The main features of this textbook series are as follows:

(1) Highlighting the training of key competencies which are "necessary basic skills in mathematics beneficial to the participation in adult life (including further education and employment) on an active and effective basis". Here the key competencies mainly include the competence of collecting, analyzing and organizing information, the ability to exchange ideas and information, the ability to design and organize activities, the ability in cooperation and teamwork, the ability to apply the mathematical thinking and techniques, the ability to solve problems and the ability to use technology. In these key competencies, mathematical thinking and method, skill acquisition as well as teamwork are given additional attention with clear and specific objectives, compared to the practical abilities in China's well-rounded education objectives, which appears more operable.

(2) Focusing on the development rule of pupils' thinking. Mathematical knowledge is divided into many themes and arranged in line with the structure of "recognition — experience — thinking — application", thus leading pupils to transit from recognition to thinking during the activities, and to help them develop an adequate thinking method since childhood. Even the "problem solving" contents are not designed as per some specific mathematical knowledge; instead, they are designed in accordance with the development rule of pupils' thinking and intended to enable pupils to grasp the general skills and the integrated capability to solve problems.

(3) Paying attention to the relationship between mathematical knowledge and real life as well as the other subjects. The mathematical contents are organized in close relationship with multi-disciplines and many social activities, and the development of problem solving capability in real life is given additional emphasis. The textbooks provide open mathematics activities associated with real life so as to facilitate pupils to monitor and evaluate his/her own study.

(4) Attaching importance to the training of pupils' ability to measure or estimate. In order to show the importance of measurement and estimation, they are set as one independent teaching unit.

(5) The knowledge looks "wide and shallow" in design, yet the teaching requirements are not low. By knowledge representation, the primary school mathematics textbooks in Australia look "wide and shallow", but as a whole, the difficulty level in teaching requirements is by no means low. For example, in the textbooks, numbers smaller than 20 are first designed for pupils to understand and know before learning how to read and write them, then followed with the four arithmetic operations of such numbers, including estimation; finally, the pupils are required to find out the modeling relationships in quantity with these numbers.

(6) Keeping in compliance with the psychological development rule of pupils in content arrangement, paying attention to the interest, activities and flexibility of the textbooks. The contents of the textbooks consist of six branches, namely, Space, Number, Measurement, Probability and Data, Algebra, and Reasoning and Strategies; the knowledge of these six branches is represented using a repeated and spiral-up method; knowledge points are taught on a gradual basis and in compliance with pupils' cognitive features. For instance, the contents of figures and geometry in the book for Grade 4 pupils are illustrated in seven units (Units 6, 9, 17, 18, 30, 34 and 38) respectively, among which the knowledge of "lines" in daily life, as represented in unit 6, and the various relationships of "lines" are discussed in the background of plane and space so as to trigger pupils' full sense to understand the relationships among each element of the planar and spatial patterns, thus laying a foundation for the pupils to have an accurate understanding of the features and nature of planar and spatial patterns.

(7) Boasting an integral structural arrangement. Besides, each unit is designed with two or three activity sheets which correspond to teachers' resource guides; there are also many test and self-assessment sheets to facilitate pupils at different levels to conduct self-monitoring and evaluation.

(8) Focusing on life experiences and the representation of socio-cultural background. The textbook design is usually carried out in line with the pattern "posing questions directly from real life — introducing models for problem solving, learning the basic concepts — using the knowledge to solve problems in real life". The whole textbook is intended to construct a personal experience context, the application instances are designed in close relation to the real life in Australia, and the given context is also very closely related to problem solving. By large, it is a textbook with a strong flavor of life, a book valuing the origin and final destination of mathematical knowledge. For instance, in the Number Travel column, part one of "Four Arithmetic Operations" section, the writer selected the well-known cities in Australia as the learning materials to train pupils how to use the basic addition and subtraction knowledge, while knowledge of estimation and comparative thoughts are involved; then in Homework column, corresponding exercises are given to develop pupils' practical problem solving ability.

(9) More "problems" than ordinary "exercises" in the exercise part, intending to develop pupils' ability to solve problems, to accumulate basic mathematical activity experiences, and to personally participate and engage in such activities. Many problems are represented in the form of games and operation. There are fewer exercises requiring pure imitation, with relatively more exercises of transformation-type or inquiry-type.

3.1.5 "My Pals are Here! Maths (2nd Edition)" from Singapore

Editors in chief: Fong Ho Kheong, Chelvi Ramakrishnan and Gan Kee Soon
Publishers: Marshall Cavendish Education Press
Publication year: 2007

My Pals are Here! Maths (2nd Edition) was written in line with the syllabus prepared by the Singaporean government in 2007. Under an activity-centered philosophy, this textbook series was written based on an extensive investigation and research as well as the feedbacks from both teachers and pupils, and were reckoned as ideal textbooks for middle-level pupils; as to the pupils with higher learning ability, they may choose to exercise on the Intense Training book, thus to embody the flexibility and diversity of pupils' learning. The main features of this series are as follows:

(1) Focusing on mathematical activities. This textbook series was written under an activity-centered philosophy, so it is comparatively easier for pupils to understand. A large proportion of teaching contents in this textbook series is represented in the form of activities, and many activity columns are set, through which pupils are guided to find patterns or abstract concepts. Among such activity columns, "Let's Explore!" requires pupils to carry out investigation activity; "Carry out activities" and "Play games!" need pupils' collaboration to complete; "Home Maths" includes mathematics activities to be carried on by collaboration between parents and children; "Try IT!" provides chances to carry on mathematics activities with the use of information technology tools. Each chapter is designed with many individual and group activities thus creating opportunities for pupils to participate in mathematics activities. [41]

(2) Attaching importance to mathematical problem solving skills and the use of modeling methods. This series of textbooks objectively reflect the core concept of "attaching importance to pupils' mathematical problem solving skills", as required in the Singapore curriculum, which is mainly manifested in the following areas: diversified types of problems — not only the general and closed questions are offered, a certain amount of non-traditional or challenging questions are also represented in textbooks with additional attention to pupils' mathematical problem solving abilities; problem solving skills are trained in combination with "word problems" and there is a special section entitled "Solving Word Problems" which explains how to solve word problems on a systematic basis, thus paving the way for pupils to solve mathematics problems in real life situations. Model method refers to the means via which a model or a diagram is drawn to represent the problem yet to be solved, which is an exploratory approach included in mathematical processes. A rectangular block model is one of the most widely used and most influential models in mathematics, which is a set of straight line segments assigned with fixed width, and mainly used to represent quantity. This model has attracted the attention of the international mathematics education world, and is called the "bar model" (Wenyu Zhang *et al.*, 2011a).

(3) Focusing on the integration of information technology and mathematics. The use of information technology in this series of textbooks is mainly reflected in the following three aspects: First, the use of calculators. The use of calculators starts in the Grade 5 and Grade 6. Calculators are not only used in mathematics activities to add the diversity of problem solving methods, but also in the study of estimation so as to evaluate the reasonableness of the estimated results. These methods are intended to reach a good balance between the computation skills and the problem solving skills. Second, the use of computer-assisted drawing tools; the use of drawing tools fastens on the geometry and fraction parts to deepen pupils' understanding as well as help them grasp the knowledge. Third, the use of internet; the pupils may use the internet to search for and study related information they have learned (Wenyu Zhang *et al.*, 2011b).

(4) Adopting a "spiral-up type" content representation with good cohesion and consistency. In content arrangement, each book is divided into seven to ten units, including the modules of number and algebra, shapes and geometry as well as probability and statistics; the general topics are the same with knowledge of these three categories interspersed throughout each textbook. The relevance and cohesion among the knowledge points are well designed. With the increment of grades, the contents involved are gradually deepened with higher level of difficulty, which go in compliance with the pupils' age characteristics and cognitive development levels.

(5) A "Home Maths" column is set, emphasizing the combination of home and school education. In the foreword of the textbooks, the functions of each column are introduced in details and both words and graphs are used to elaborate on the design idea of each column; the detailed requirements on teachers' usage of textbook, pupil activities as well as parents' tasks are also expounded. The Singaporean textbooks break through the conventional limitations in textbook itself, and assign tasks for parents, which not only helps pupils consolidate what they have learned in school, but also is beneficial to enhance the emotional communication between parents and children, and thus this series of textbooks are considered as a typical model to realize an effective combination of family and school education.

3.1.6 "Primary Maths" from the U.K.

Editor in chief: Peter Clarke
Publishers: Harper Collins Education Co.
Publication year: 2007

Primary Maths was first written in accordance with the national curriculum requirements stipulated in *The 1998 Education Reform Act* and *The National Curriculum Standards for Mathematics* in the U.K. This series of textbooks consist of six books. The main features are as follows:

(1) Focusing on differentiated mathematics education. The textbooks fully embody the teaching principle of "providing effective learning opportunities for all students". Textbooks are organized according to pupils' academic level, which reflects the tradition that the differences of pupils are always considered in the U.K. Therefore, it becomes inevitable for British pupils to use different textbooks in the same classroom. There is a great variety of textbooks, apart from the usual textbooks and guiding books for teachers, and there are also evaluation textbooks as well as textbooks applicable to pupils with different learning abilities.

(2) Focusing on the "fun" of textbooks, textbooks are designed in multiple bright colors with a great visual perception. The U.K. mathematics textbooks are designed with many activities. The acquisition of knowledge and the development of skills are typically incorporated in such activities that are organized according to the experience of life, which will arouse pupils' enthusiasm in study, and make them amused in learning.

(3) The contents cover a wide range, and are frequently replaced. This textbook series pays little attention to the development and consolidation of knowledge; instead, it focuses on the coverage degree of the National Curriculum and how to represent mathematics as a discipline of fun, even consideration of teaching quality in the textbook has to give way to "fun". For example, the concept of multiplication is illustrated by $m \times n$ arrangements, and the pupils are required to display stickers; this design is perhaps from the consideration of "real life", yet the excessive details in arrangement and the likelihood to distract pupils' attention in using these colorful stickers are not considered by the writers.

(4) Contents of each section are integral and consistent, and the arrangement of contents takes a web form in the sense of connection. Some basic principles, concepts or knowledge points that are required to be mastered by the pupils are often represented repeatedly and extended gradually. For instance, when "divided by 2, 3, 4, 5 and 10" is to be learned, a lengthy review of "table of multiplying by 2, 3, 4, 5 and 10" is designed, which shows the textbooks put emphasis on acquiring new knowledge on the original foundation. In addition, the application of mathematics is set throughout the whole textbooks.

(5) The teaching contents are often embedded in "Activities" and "Processes". For instance, when learning "measurement", pupils are required "to estimate the size of objects in appropriate words, and sort them by order via a direct comparison; to compare and measure objects by the same non-standard units, and then by standard units in length (centimeter or meter) ... To know the standard units for length, mass and volume, and use them to give reasonable estimation in daily life..." As can be seen from "measurement" in this textbook series, its teaching contents are mainly represented in the form of "Activities", and pupils are enabled to experience mathematical knowledge via the process of such activities. Therefore, it is not very hard to find that the U.K. textbooks lay more emphasis on the acquisition of knowledge. Meanwhile, in the introduction of knowledge points, the creation of problem situations is paid additional attention. The pupils are first guided to recognize and identify the models from context, and then such models are described in mathematical language and signs before introducing the knowledge point. Specific examples are the major targets to be discussed at the beginning, pupils are guided to learn how to conclude the general rules, and the ability to obtain data from real life is also emphasized.

(6) Focusing on the use of open-ended questions. There are relatively more open questions for pupils to explore the methods of problem solving or to summarize the governing rules, thus they are required to use the knowledge points more flexibly and reasonably; by using such open questions, they are further required to use these knowledge points into other fields and to make summaries on their own or to give suggestions to others.

On the whole, *Primary Maths* stresses more on the discipline level of mathematics rather than a simple theory pursue or systematic integrity, thus it may be more appropriate to address their primary and middle school mathematics education as “mathematical culture”. In classroom, instead of pure theories, more emphasis is put on mathematical consciousness and activities as well as the application of mathematical knowledge.

3.2 A comparison of the writing features in the six series of foreign primary school mathematics textbooks

Based on their requirements in education, the overseas primary school mathematics textbooks hereof introduced are fairly diversified and different, and each is endowed with its own features. Still, they have quite a number of features common to share.

Due to the different situations in each country, the cultural and educational reality in particular, there are inevitably differences among the textbooks from different countries, mainly manifested as follows:

(1) In terms of content, the three series from the USA and Australia have more difficult exercises while the rest are relatively easier. The California series emphasizes the consolidation of basic knowledge and skills; the Japanese series attach importance not only to the consolidation of basic knowledge and skills, but also to the extension of knowledge.

(2) In content arrangement, each textbook series has its own priorities. The Japanese textbooks attach importance to “Mathematics Activities”, while the Australian counterpart lays more stress on “Reasoning and Policy”; meanwhile, the different editions of textbooks within the same country may have differences as well.

(3) In content representation, the California textbooks lay more emphasis on the method of abstraction, while the Australia, Singapore and the U.K. editions attach more importance to the description of the induction methods and processes.

There are also big differences between *Mathematics Connects* and *Investigations (common core edition)* from the USA: (i) Difference in content selection; for instance, when statistics and probability are introduced for Grade 4 pupils, the California edition introduces the bar and pie charts, while the core edition represents broken line graphs; (ii) Difference in content representation; though both textbooks represent contents in an intuitive manner, the California edition also attaches importance to the means of abstraction, while the core edition gives more consideration to the methods of induction. (iii) Difference in material selection. There are differences in the selection of exercise materials between the two series of textbooks as well; the exercises of the California textbooks are largely represented in the real-life context, while those of the core edition are largely not set in such a context. (iv) Difference in exercise setting. The California textbooks put emphasis on the mastery of basic knowledge, thus Mid-Chapter Check, Chapter Check and Test Practice sections are also set in the textbook; by comparison, the core edition is designed with fewer exercises. (v) Differences in content requirements and difficulty. It is worth mentioning that the USA adopts a decentralization system in education where each state and school district is required to develop their own syllabus (Yueru Chen *et al.*, 2009); therefore, there is a sharp contrast between these two series of textbooks in both difficulty level and content requirements. *Mathematics Connects* sets higher requirements in theory study and its contents are more challenging, which is deemed as an influential textbook for elite education. Of all the 13 selected textbooks, it ranks No. 1 in difficulty level; by comparison, *Investigations (common core edition)* is of lower difficulty level, ranking No. 12 among all these textbooks.

Yet such differences, more often than not, reflect the implicit features of these textbooks. They are indeed the unique advantages of their own. For instance, the Singaporean textbooks value the involvement of parents, thus it not only specifies what the parents should do in the foreword, but also sets a unique "Home Maths" column in the textbooks. Under the writing philosophy of "providing effective learning opportunities for all students", the U.K. textbook writers produce different types of textbooks in line with different learning abilities of children; therefore it is not uncommon to see children in the same class using different textbooks. The Japanese textbooks put special emphasis on mathematics activities to help pupils, in relaxing atmosphere, to proficiently master the basic and fundamental knowledge and skills in quantity, shapes and other fields through various mathematical activities. Besides, they also attach great importance to the development of pupils' communication and cooperation abilities in these activities. The US textbooks give higher priorities to the early penetration, including the knowledge of algebraic thinking, geometric intuition, statistics and probability as well as modern technology. In content arrangement, the US textbooks adopt a "minor spiral type", i.e., many knowledge points are dispersed in multiple chapters and represented from different perspectives. The contents of the six branches in the Australian textbooks are "wide and shallow", taking a spiral-up pattern; they are applicable for pupils at different levels to carry on self-check and self-evaluation. Meanwhile, the pupils' personality development, participation and experience are put additional emphasis, and the development of pupils' problem solving ability in real life is also stressed. Yet the Australian textbooks do not over-emphasize a purely theoretical, strict knowledge structure, but represent pupils with the life experiences of human being and the socio-cultural background, they emphasize practical application of mathematical knowledge with multi-level and life-like features.

Commonality is absolute, while individuality is relative and conditional. Though there are some differences among the textbooks due to the cultural diversity and other factors in these countries, they share the following similarities: (i) Writing philosophy: paying attention to inquiry learning and open-ended questions, emphasizing the abilities of problem solving and reasoning; attaching importance to mathematics activities to offer a relaxing study environment for the pupils. (ii) Content selection: although the titles or names of such contents are different, they all include three major parts, i.e., number and algebra, shapes and geometry as well as probability and statistics; estimation is also much valued in these textbooks. (iii) Content organization: most of the textbooks adopt a "spiral" style in writing. (iv) Content representation: they all lay emphasis on an intuitive representation of contents and the relationship between mathematics and real life, thus the material selection is closely related to the pupils' life; they also highlight the linkage of mathematics with other disciplines.

4. A Brief Analysis of the Writing Features of the Domestic Primary School Mathematics Textbooks

The currently used series of primary school mathematics textbooks in China have common advantages in facilitating teaching and learning, yet they have their own features as well. Under the guidance of *The Standards*, these textbooks are all intended to facilitate teaching and learning and contributing their own efforts for the diversified development of the Chinese textbooks. In December 2011, the *Mathematics Curriculum Standards for Compulsory Education* (2011 version) (MOE of PRC, 2012) (hereinafter referred to as the 2011 Standards) was officially issued, and the PEP edition, BNUP edition, SCNUP edition, JEP edition, HEP edition and the QPH edition were correspondingly revised in line with the 2011 Standards, whose features are thus made more distinct. The followings are a brief introduction of the main writing features of the eight selected domestic primary school mathematics textbooks.

4.1 Writing features of eight series of domestic primary school mathematics textbooks

4.1.1 The People's Education Press (PEP) edition

Editors in Chief: Jiang Lu and Gang Yang
Publishers: People's Education Press

The using scope and quantity of this textbook series are both large. Its main writing features are as follows:

(1) Fully demonstrating the ideas of curriculum reform. For example, the teaching content is to show the knowledge formation process in a logical order as far as possible. In order to strengthen the cultivation of pupils' statistical sense and improve their comprehensive practical ability, the textbook adds the "Statistics and Probability" content on one hand, and sets the practice activities involving statistical knowledge on the other hand.

(2) Reflecting on the learning approach of self-exploration, cooperation and exchange. It encourages pupils to be more proactive so

that pupils can understand and master mathematical knowledge through observation, speculation, operation, thinking, communication and other activities, which contributes to developing and improving pupils' awareness of innovation and their practical ability.

(3) Selecting appropriate learning materials to design activities based on children's living background. For example, during studying the length unit of millimeter and decimeter, it selects the mathematics book that is known to pupils, asking them first to estimate its length, width, and thickness, then measure it with a ruler. The question "When a length is not an integer by centimeter, what should we do?" triggers pupils to think, and thus the concept of millimeter is introduced.

(4) Scientific and logical material structure. First, each section of the textbooks uses the pattern of "example-practice-supplementary exercise", so that pupils can understand teacher's instruction in class, and they can also timely review and consolidate the knowledge after school by using what they have learned. Second, the final chapter of each textbook is "General Review", which summarizes and reviews what they have learned throughout the semester.

(5) Presenting a wealth of new forms. The style of the entire textbook series is lively and bright. The examples and exercises are illustrated in various forms, which can stimulate pupils' interest in learning, and attract them to read.

(6) Adding a unique "Mathematics Wide-angle" column which centers on exercising mathematical thinking, enhancing problem-solving skills and revealing mathematical thinking approaches. One example of "Mathematics Wide-angle" on the Grade 3 mathematics book is "how many different dressing ways are there" has enlightened pupils to learn the classification knowledge through a tree diagram. Another example, as represented on the Grade 5 mathematics book, explains the encoding function of numbers.

Prominent features are represented after the textbook's revision in 2011: (i) reflecting more logical teaching order and pace according to the rules about pupils' mathematics learning, thus facilitating pupils' understanding of mathematical knowledge and helping them to develop mathematical ability; (ii) strengthening the penetration of mathematical thinking and paying more attention to make pupils be involved in mathematics activities; (iii) strengthening the cultivation of "four abilities" (ability to identify problems, ask questions, analyze problems, and solve problems), and providing clear clues and steps for pupils to solve problems; (iv) well-designed cases and steps for study and training pupils to build a reasonable mathematical knowledge structure; (v) setting a process evaluation section for pupils to make self-reflection and self-evaluation, so that pupils can get a good experience of learning mathematics, forming good study habits (Jiang Lu, 2012).

4.1.2 The Beijing Normal University Press (BNUP) edition

Editors in Chief: Jian Liu and Qiping Kong
Publishers: Beijing Normal University Press

The using scope of this textbook series is relatively large. Its main writing features are as follows:

(1) A close connection between mathematics and reality based on pupils' life experience. It starts from pupils' prior knowledge and experience, takes the real and funny matters that pupils are likely to experience and understand as teaching material, so that pupils can learn and understand mathematical knowledge in a familiar context.

(2) Demonstrating the process of mathematical knowledge generation, formation and application, forming the narrative mode of "problem situation — model building — interpretation and application". Guiding pupils to participate in the entire process of problem solving not only train and exercise pupils' mathematical thinking, but also develop and improve pupils' awareness of innovation and application.

(3) Advocating diverse learning styles and cultivating pupils' innovation sense. Most of the problem solving processes present different solution strategies so as to make pupils understand that there is more than one solution to the same problem. For example, when studying a one-digit number multiplying a one-digit number, in order to get the number of

pandas in the picture, two schemes of counting vertically and horizontally are provided, and pupils are encouraged to express their views.

(4) Step-by-step learning and spiral enhancement. To comply with the children's cognitive rules and optimization of learning effect, the textbooks take the escalating manner appropriate for children's receptivity rather than the linear layout of content.

(5) Encouraging pupils to participate, explore, and interact voluntarily via mathematics activities. All the exercises offer many situations which require pupils to carry on hands-on practice, cooperation and communication. For example, when teaching two-digit numbers multiplying another number, a picture of long queue to buy tickets in the amusement park for taking spacecraft is shown, which arouses pupils' desire to operate by themselves, and thus they actively look for a way to the total amount of money needed to pay.

(6) Concerning with pupils' emotional experience by creating a relaxing and harmonious atmosphere through a variety of ways. For example, the introduction of the cartoon characters "Naughty" and "Wise old man" loved by pupils aims to add to the fun of learning content. Other small columns like "Mathematics Story", "Mathematics Kaleidoscope" and "Do you know?" let pupils learn mathematics knowledge through storytelling and understand the relation between mathematics and other disciplines.

The textbooks revised in 2011 have added the following features besides the original ones: on the base of the "situation + serial questions" representation mode, trying to realize the integration of four processes (i.e., the development of course content, pupil learning, teaching process and teaching goals), thus promoting pupils to experience the thinking process from the beginning to the completion, obtaining the necessary basic knowledge, skills, ideas and activity experience appropriate to their age, as well as the ability to discover problems, ask questions, analyze and solve problems (Jiang Lu *et al.*, 2012).

4.1.3 The Southwest China Normal University Press (SCNUP) edition

Editor in chief: Naiqing Song
Publishers: Southwest China Normal University Press

Primary school mathematics textbooks published by the Southwest China Normal University Press have been widely used in Chongqing, Sichuan, Guizhou, Henan and other provinces. Its main writing features are as follows:

(1) Every lesson is designed with "Classroom Activity" which bridges teaching and learning. The textbooks use the structure of "Example — Classroom Activity — Exercises", rather than the common stylistic rule of "Example — Classroom Exercises — Exercises" used by most textbooks, thus forming the pattern of "learning new knowledge and accumulating experiences in activities via problem posing and solving". Examples, classroom activities and exercises are forming the trinity which interrelates with each other, and consolidates and enhances pupils' mathematics understanding. Meanwhile, it also provides resources and activity clues for the mathematics course, which benefits both learning and teaching.

(2) Selecting abundant and interesting materials that are close to pupils' real life, the forms of which are various with excellent pictures, thus promoting pupils' awareness and understanding of mathematics. For example, this series of textbooks uses things that pupils are familiar with and love as the background of knowledge learning, such as flag-raising (they can learn numbers by watching the flag-raising), vehicle riding, animal games, tree planting and fairy tales. It also presents the learning materials intuitively with the help of pictures, cartoons, games, tables as well as vivid word description. The textbooks use the pattern of "Life Situation — Asking Questions — Establishing Pattern — Explaining Application", which can stimulate pupils' learning interests and also pay attention to their awareness and understanding of mathematics.

(3) Paying attention to the penetration of mathematics culture. First, this series of textbooks sets the column "Do you know" for mathematics culture learning. Every book has three parts, except the first book with only two parts, whose contents involve mathematics thinking, the origin

and development of mathematics knowledge, mathematics applications, classical mathematics books, stories of mathematicians and many others. Second, the textbooks include exercises implying with mathematical culture. In particular, from Grade 1, the textbooks have included the content of mathematical culture in the forms of comics, which are easy to understand, interesting and informative.

(4) Paying attention to themes of the rural and the West areas of China and the Three Gorges. This series of textbooks has selected many materials about the rural, the West and the Three Gorges as examples and exercises. Particularly, it chooses the west landscape as the back cover, which not only has the ideological education function, enriches the mathematics resources, but also meets the learning needs of pupils in different areas.

(5) Highlighting the dominant role of pupils, focusing on their self-exploration, cooperation, communication as well as practical innovation. The columns in the textbook, such as "Look and Say", "Problem Solving", "Comprehensive Application", "Do You Know?" and after-class problems have given pupils freedom to think independently as well as raised their awareness of innovation.

(6) Reflecting on the learning approaches, beneficial to both learning and teaching. Its mathematical content system is built step by step and spirally enhanced in line with children's cognitive development. For example, in arranging "Numbers and Calculations", the textbook starts from "understanding numbers 0–10" to "understanding numbers 0–100" and then to "understanding numbers 0–10,000", from the "addition and subtraction with numbers 1–10" to "addition and subtraction with numbers 1–100" and then to "one-digit number multiplying two-digit and three-digit numbers", "mixed arithmetic calculations" and so on. Such an arrangement increases the difficulty of learning contents gradually along with the improvement of pupils' thinking level and is helpful in practical teaching.

(7) This series of textbooks selected novel materials for the columns "Integration and Practices", focusing on the balance of urban and rural themes with distinctively-featured activities. Each volume of the textbook contains three such columns, with contents based on the knowledge

presented in the text and closely linked the related knowledge in other disciplines, considering balance of contents and diversity of themes.

The textbooks as revised in 2011 have added the following two features besides the original writing features: (i) using the pattern of "Examples — Classroom Activities — Exercises", thus to guide pupils to do exercises in the forms of game, practice, communication and exploration, etc., and to promote pupils' basic knowledge, basic skills, basic ideas and basic activity experiences (hereinafter referred to as the "Four Basics" development); (ii) practical activities are closely linked to the learning contents; mathematics knowledge are integrated into an organic system, thus providing a selection of workable activities concerning urban and rural themes which can be represented in series (Naiqing Song, 2012).

4.1.4 The Jiangsu Education Press (JEP) edition

Editors in Chief: Ligu Sun and Lin Wang
Publishers: Jiangsu Education Press

The textbooks published by the Jiangsu Education Press are used by the primary schools not only in Jiangsu, but also in Shanxi, Anhui, Shaanxi, Guangdong and other provinces. Its main writing features are as follows:

(1) This series of textbooks selects questions, which highlight the process of knowledge generation and guide pupils to explore by themselves. The columns in the textbooks, such as "Have A Try" and "Think and Hands on", provide space for pupils' self-directed learning, let them acquire knowledge, train ability and accumulate learning approaches through observation, operation, experiment and reasoning during the process of active thinking, cooperation and communication. For example, one question is proposed in the textbook to let pupils try to use a rectangular piece of paper to roll out cylinder paper tubes of different shapes, which guides pupils to find that different rolling methods result in paper cylindrical tube of different lengths and sizes.

(2) Focusing on various strategies of problem solving. For example, in explaining question "there are 8 bugles in one box and 7 bugles in another, how many bugles are there all together?", the textbook gives two solutions: to add 8 up to 10 or to add 7 up to 10. Another question in the textbook

asks pupils to classify red and green fruits and vegetables in line with different criteria (i.e., color, type).

(3) Horizontal connections across interdisciplinary knowledge of the three major categories. Firstly, set appropriate small sections, arranging the contents of number and algebra, space and graph, statistics and probability in a cross manner so as to combine the contents of different categories organically. Secondly, combine the contents of different categories together for teaching. For example, "enlargement and reduction" and "proportion" come from the categories "Space and Graph" and "Number and Algebra" respectively. However, "enlargement and reduction" is that each part of a graph changes proportionally, and such a change is intuitive and vivid which can make the pupils sense the connotation of proportion. Thirdly, every textbook has arranged certain amount of the column "Integration and Practices", which further links up the cross communication of contents in different areas and emphasizes the mastery and digestion of knowledge so as to improve pupils' awareness of application and experience the close connection between mathematics and life.

(4) Arranging teaching contents in a vertical spiral-up form. For example, in arranging the contents of space and shapes, this series of textbooks complies with pupils' psychological development rules and follows pupils' mathematics thinking level, it creatively starts from the three-dimensional solids that are direct and easy to perceive, and let pupils make an initial understanding of two-dimensional figures with the help of solids, then understand the one-dimensional line segments, straight lines and rays. With the improvement of pupils' thinking level, when helping pupils further understand shapes, the textbooks arrange the contents oppositely to that of initial understanding of the shapes, i.e., arranging from one-dimensional lines to two-dimensional figures, and then to three-dimensional solids.

(5) The textbooks have intuitive and vivid materials, which are illustrated with pictures and words, lively and interesting. The learning materials are represented with vivid and intuitive pictures, which can not only stimulate pupils' curiosity, but also improve their enthusiasm and interest, as well as meeting pupils' diverse learning needs.

(6) The contents of the textbooks are closely connected with real life, and realistic situations are frequently used. For example, in learning one-digit numbers, the textbooks illustrate with examples that pupils are interested in and familiar with, such as skateboarding, swing, rotation, and butterfly.

(7) Rich contents of mathematical culture. The textbooks skillfully infiltrate with mathematical culture through the column "Do you know?" There are more than 60 pieces of reading materials in this column, which are related to various kinds of mathematical culture background, including its formation, development, application and so on.

While keeping the main content reasonably intact, the textbooks revised in 2011 explore the arrangement ways that are more suitable for pupils to learn; in addition, the teaching processes can be further enriched by this revision so as to compensate for the lack of experimental materials. The textbooks for Grade 1 pupils have been put into use, which mainly take on the following features: highlighting the basic role of mathematical ideas in knowledge formation and application, guiding pupils to sense problem solving processes with strategies, focusing on providing mathematical activity clues operable in class, and putting emphasis on the pertinence, hierarchy and selectivity of exercises (Lin Wang, 2012).

4.1.5 The Hebei Education Press (HEP) edition

Editor in chief: Zhao Xingmei
Publishers: Hebei Education Press

The main writing features of this series are as follows:

(1) Learning in the exploration of problems and promoting the development of mathematical thinking. The modules of "Let's Try", "Let's Talk", "Discussion", "My Own World" and the columns like "Adventure Land" and "Problem Solving" enable pupils to learn mathematical knowledge in exploration, and to learn how to observe things from different perspectives, so that they can experience such learning processes as "finding problems — posing problems — communication and discussion — forming their own ideas".

(2) Selecting materials from real life and laying emphasis on mathematical learning in life. The selected contexts are in line with pupils' ages, and mathematics activities are rich, interesting and full of novelty. For example, when learning measurement, pupils are required to measure the length of their most familiar objects — pencils. As another example, the notions "up" and "down" are introduced by storytelling with pictures and a melodious song called "The Crescent Moon", so that pupils can understand the mathematical notions suggested by the pictures and the song in a pleasant and relaxing atmosphere.

(3) Represented with vivid pictures and texts. The textbooks adopt lively dialogues and multi-colored pictures to represent the contents, thus to stimulate pupils' interests and arouse their initiatives in learning.

After the revision in 2012, Grade 1 textbook shows the following features: designed in line with class hours to facilitate teaching and learning activities; the content arrangement is in compliance with pupils' learning processes; it respects pupils' age characteristics and highlights the feature to "learn mathematics in real life"; activities in "Integration and Practices" are meticulously designed to help pupils accumulate knowledge from activities in the hands-on operation process and experience the methods of mathematical thinking (Haijiang Cui, 2012).

4.1.6 The Qingdao Publishing House (QPH) edition

Editors in chief: Tao Zhan, Pinyi Liu and Yunhong Xu
Publishers: Qingdao Publishing House

The QPH edition has the following features:

(1) Highlighting the development of problem awareness. This textbook series gives priority to guiding pupils to analyze mathematical context, find out and pose questions. Usually in the integrated exercises, for example, two types of questions are asked: one is the explicit and leading questions, and others are implicit questions that can only be posed after careful observation and analysis by pupils, which aim to guide pupils to pose questions to their concern, and to improve their ability in data selection and organization from a mathematical perspective.

(2) Encouraging independent exploration and interaction between pupils and training them to organize and reflect the knowledge points on one's own initiative. The exploratory research spirit in mathematics study is fully manifested in writing philosophy, content representation and column design of this textbook series. For example, each "Comprehensive Information Map" is followed by a "Cooperative Exploration" module, which aims to, on the one hand, give pupils some time and space for exploration, and on the other hand, serve as a guide to encourage them to carry out exploratory activities.

(3) Paying more attention to training pupils' awareness of statistics, estimation and pattern searching. The awareness of statistical estimation and pattern searching is playing a more and more important role in the modern world, thus has been reckoned as an indispensable part in pupils' mathematical literacy. Therefore, such contents are prepared in textbooks since Grade 1 and run through various aspects in number and algebra, space and shapes as well as statistics and probability. Besides the specific chapters or sections in textbooks, these contents are, more importantly, penetrated into the teaching of each unit and section to increase their appearance frequency in textbooks. For example, in the learning of "volume and volume units" section, pupils are encouraged to estimate the volume of a shoebox in cubic decimeter, a just learned volume unit, then teachers could use wood blocks to fill in the shoebox and check pupils' estimations; when the unit "cubic meter" is taught, pupils are asked to estimate how many pupils one cubic meter can hold. In both cases, teachers pay attention to the penetration of estimation awareness.

(4) The unique plate structure. The teaching contents are represented with different plates. For example, each subsection consists of three basic plates, "Comprehensive Information Map — Cooperative Exploration — Practices". The questioning plate is named "Smart Cabin", and there are other plates like "Have I got it?", "Harvest Garden" as well as "Knowledge Corridor".

(5) Materials are selected according to pupils' life experience and cognitive characteristics. In the textbooks, the well-known objects from family, school, society and the nature are selected to enlighten pupils that mathematics is derived from life with great practicability and fun. For

example, there are knowledge points represented in the form of "I'm growing new teeth", "Visit my grandma" and "The underwater world", etc.

(6) Materials are selected meticulously to form consistent series of circumstances and questions, thus organizing the whole unit into an organic system. With the exception of a few small units, almost all the units in this series of textbooks take on the "question series" features. For example, in Unit One of Book 1, "Happy Schoolyard — to Know the Numbers 1–10" section is constituted by a "circumstances series" of "Tech Group Activities", "Hawk Catches Chicken", "Tug of War" and "Happy Class", so as to facilitate the pupils' learning of numbers 1–10 in a series of challenging questions derived directly from the real life. In addition, the awareness and ability to solve basic problems are fostered.

After being revised under the 2011 curriculum standards, there are a few changes, and the main features in the QPH edition mathematics textbook (for Grade 1) are as follows: the exploratory process is made more detailed for pupils to learn how knowledge is obtained through observation, experiments, guesswork, reasoning, idea exchange as well as reflection, by which both the processes and results are emphasized; attaching importance to the teaching targets on "basic ideas" and "experience of the fundamental activities", and carrying out the "Four Basics" targets in practice; highlighting the procedural and operational property of "Integration and Practices" section, and paying more attention to the accumulation of knowledge from mathematics activities; concerning about the integrity of knowledge, the basic structure of word problems is penetrated into pupils' knowledge beforehand, thus guiding them to perceive the quantitative relation (Guihua Pan, 2012).

4.1.7 The Zhejiang Education Press (ZEP) edition

Editor in chief: Tianxiao Zhang
Publishers: Zhejiang Education Press

This series of textbooks are used in schools of Hangzhou City as well as a number of famous schools from 20 provincial regions. Its main writing features are as follows:

(1) Innovative “two-basics” contents. This series of textbooks pays more attention to the two-basics education and has its original views on two-basics contents and the methods in implementation. For example, when “Numbers 0–9” is taught, the stereotype conventional exercises are no longer represented, only a few numbers are selected for practice rather than listing all the numbers for repeated computation. The “principle of composition and decomposition of numbers” is only implied in some specific changeable context, which may guide the pupils to solve real problems with this principle.

(2) Unique structure. Breaking the traditional arrangement system, it reconstructs and adjusts the internal structure of mathematical knowledge. Taking Book 1 for Grade 1 as an example, this book first introduces the basic geometric shapes before actually starting to teach the concept of number. Such a design may find its grounds as follows: for one thing, pupils live in a three-dimensional space and they have formed some intuitive, basic knowledge on shapes; for another, this part is mainly learned through hands-on activities, thus helpful to stimulate pupils' interest and to form a positive emotional experience from the very beginning; moreover, compared with the concept of number, this teaching target is easier to be achieved (Min Chen, 2012).

(3) Content arrangement is designed in series by a systematic order. For instance, the pure mathematical contents in “Number and Algebra” are designed into eight series, namely, number array, equivalent substitution, graph and equation, number puzzle, equation set, generalization and reasoning of quantitative relation, numeral-figure combination as well as integer division. There are ten series for training pupils' space concept in Graphic and Geometry, i.e., sense of shapes, the composition and decomposition of graphs, finding the hidden figures, counting the numbers of cubes, generalization of figures, figure reasoning, judgment on graphic combination, folding and unfolding of graphics, identification of the graphs, and comparisons of area and space. It also preliminarily penetrates the ten dialectical relations into contents: the one and the many, separation and combination, the equal and the unequal, composition and decomposition, change and unchange, accuracy and approximation, the

concrete and the abstract, the discrete and the continuous, the finite and the infinite, as well as the constant and the variable (Min Chen, 2012).

(4) The column "Paradise of Wisdom" is set to strengthen mental training. This series of textbooks pays more attention to fostering pupils' abilities in generalization, deduction and creativity. Taking the "Paradise of Wisdom I", as set in Book 1 for Grade 1 pupils, as an example, the question offers four glasses with different volumes of water, and pupils are asked water from which glass would be the sweetest if the same amount of sugar is added.

4.1.8 The Jingshan edition

Editor in chief: Jingrong Chen
Publishers: People's Education Press

This series of experimental primary school mathematics textbooks was written by Beijing Jingshan School for its pupils. It has four main writing features:

(1) Breaking through the "6 years + 3 years" convention for compulsory education textbook writing. The Jingshan School is a nine-year compulsory education school with five years for primary school study and four years for middle school level study. Correspondingly, the Jingshan series of primary textbooks is divided into five grades with ten volumes, and therefore by nature, it is endowed with the features of "breaking away from convention" and "experimental attempt for educational reform".

(2) Closely related to classroom teaching. Since most writers of the textbooks are classroom teachers, they can note down the problems they have ever noticed in class; then such problems will be reappeared in the textbooks they write, thus making them very close to classroom teaching. The Jingshan edition textbooks are produced, as it were, by repeated trials: written by teachers — tested in teaching practice — revised by teachers; the process repeats until the books are well accepted.

(3) Training pupils' problem solving skills and practical abilities. A specific "problem solving" chapter is designed to guide pupils to pay

attention to life and the world, teach pupils how to think, and train their problem solving skills and practical abilities.

(4) Circumstantial inquiry, unique teaching approaches and high adaptability. The circumstantial inquiry learning model is rooted in western nations' advanced educational concept and models, which can further generate a variety of new learning models to adapt to pupils with different personalities and learning habits; this model attaches more importance to pupils' ability in independent thinking. The Jingshan edition uses this model as reference to make the content representation of these textbooks more vivid and intuitive, and the difficulty level is set on a gradual basis.

4.2 A comparison of the writing features in the eight series of domestic primary school mathematics textbooks

The above analysis over the writing features of the eight series of textbooks clearly shows that these textbooks not only share many commonalities, such as educational philosophy, content selection, organization, representation, the setting of special columns as well as their emphasis on mathematical culture, but also take on respective characteristics of their own. China adopts a censorship system for textbooks, and their writing shall be based on the Curriculum Standards. Therefore, there are convergence phenomena in many ways among all these textbooks mentioned in this paper; different from the overseas primary school textbooks, the Chinese counterparts vary within a small range in difficulty. Their shared points as follows:

Writing philosophy: All the domestic textbooks here mentioned are written in line with the Curriculum Standards, give priority to the target of "Four Basics" and "Four Abilities", emphasize the changeover of learning approaches featuring independent exploration, cooperation and exchange, as well as pay attention to train pupils' innovative awareness and practical ability, so as to improve their comprehensive quality in mathematics and lay a solid foundation for their lifelong sustainable development.

Content selection: Some complicated and difficult parts are deleted as appropriate, thus to reduce the difficulty level of the textbooks.

Content organization: They all adopt a "spiral-up" type; the four major fields ("Number and Algebra", "Shapes and Geometry", "Statistics and Probability", and "Integration and Practices") are interpenetrated in contents.

Content representation: (1) They go hand in hand with children's real life of children and offer a selection of vivid and funny materials; pupils' existing experience is used as the basis for the generation and development of new knowledge thus to stimulate pupils' interests. (2) Contents are represented with vivid pictures and texts, which can not only stimulate pupils' interests, but also show the content in front of children in an intuitive manner so as to facilitate their understanding and memorizing such contents. (3) Mathematical models are built directly from problem situations thus to help pupils feel how close mathematics is to their real life and how practical mathematical knowledge is, so that their thinking ability and emotional attitude can be improved and developed further. (4) Displaying the knowledge acquisition process by designing various cooperative and exploratory activities. For example, there are a number of experiments and activities in the textbooks so that pupils can accumulate knowledge and acquire skills by experience and inquisition. Such a design is beneficial to the cultivation of pupils' comprehension quality. (5) The contents of probability and statistics, problem solving as well as comprehensive practice are highlighted. (6) The characteristics of the times and humanism are valued; for instance, mathematical knowledge, as suggested in the textbooks, can be used to solve problems, such as environment pollution, scarcity of resources and food safety.

Mathematical culture: Mathematics is, as it were, one kind of human culture, the content, thinking, methods and language of which are an indispensable part of modern civilization (MOE of PRC, 2001). The PEP, BNUP, SCNUP, JEP and QPH primary school mathematics textbooks all attach great importance to the infiltration of mathematical culture; contents like the masterpieces in mathematics, mathematical thinking and methods, the origin and development of mathematical knowledge as well as mathematical applications are all included in the textbooks. They are not only included in the examples and exercises, but also introduced by specific modules in some textbooks, like the "Do you know" module of the SCNUP and JEP editions and the "Knowledge Corridor" module of the QPH edition.

Meanwhile, each series of textbooks is designed with unique features of their own. For instance, "Mathematics Wide-angle" in the PEP edition, the narrative mode of "Problem Situation — Model Building — Interpretation and Application" in the BNUP edition, "Adventure Land" of the HEP edition, the combination of "Circumstantial Series" and "Question Series" in the QPH edition, the series-based and order-based content arrangement in the JEP edition, the unique "Five-year and Ten-book" design in the Jingshan edition, the unique "Classroom Activity" section in the SCNUP edition paying much attention to issues of the rural, the West and the three Gorges, and the JEP edition highlighting the diversified methods in algorithm, to name a few.

5. Thoughts on the Development of Writing Features in the Chinese Primary School Mathematics Textbooks

Over the years, our featured development in textbooks has achieved remarkable results. Lingbiao Gao (2002) used to summarize that a successful textbook had to take on features in the aspects of discipline system, teaching contents, abidance by pupils' psychological rules, functions of textbooks, combination with modern information technology as well as the integration of emotion, attitude and values as objectives reflected in the curriculum. The writing features of primary school mathematics textbooks have also reached unprecedented levels. The common characteristics and their respective features of the eight series of domestic primary school mathematics textbooks, as elaborated in the above, are the embodiment of such progresses. The researchers of this paper hold that the PEP, BNUP, SCNUP, JEP, QPH and HEP textbooks share the following features: they comply with the rules and characteristics of primary school pupils' mathematics learning, and keep at the unification of science with humanity and disciplinary characteristic with life nature in mathematics which are conducive to pupils' mastery of basic knowledge, basic skills, and ideas in mathematics as well as their accumulation of basic mathematical experience from activities, thus to raise pupils' overall quality (EOBET, 2008).

However, some limitations and deficiencies in the development of textbooks are also exposed. For instance, as revealed in the General Report on Social Sciences Foundation of China's Project "Comparative Study of Textbook Characteristics after the New Curriculum Reform" (Hongyu Zhang, 2012): There is still some discrepancy from the overall diversity requirements despite the large selection of various textbooks available, particularly related to the requirements about how to understand and represent the features of textbooks. For example, when interpreting their own characteristics, each edition of textbooks is inclined to apply some teaching or educational theories or some "advanced" teaching philosophies mechanically, or simply put, "difference" is considered as "feature". And rural teachers generally hold that the materials selected in textbooks are, to a certain degree, of some urbanism tendency. As to the writing of primary school mathematics textbooks, there are still rooms for improvement. For instance, a number of textbooks take a trend of convergence in multiple aspects such as content selection, stylistic rules in writing, content representation and difficulty levels; and the materials selected in some individual editions are not extensive enough, and part of the practical activities are not very operable, etc. The main reasons for the convergence tendency in the writing features of primary school mathematics textbooks may be as follows: First, the influence from the inertia in textbook writing and teaching experience for the long term "One Syllabus, One Textbook" educational system; Second, all the editions were written in line with the mathematics curriculum standards and designed in reference to the "Suggestions of Coursebook Writing" in them; Third, due to the strict textbook censorship in China, the writers are afraid that too personalized textbooks would not pass the national examination.

China is a big country with vast territory and a huge population, striking regional differences, so the requirements of primary school mathematics textbooks are also manifold; the diversification of textbooks, by nature, means distinctive design, considering the cultural and traditional differences and social and economic development situation, under the "One Standard", which seems more than necessary to accelerate the characteristic development of China's primary school mathematics textbooks. As a matter of fact, the corresponding requirements on the writing of China's primary school mathematics textbooks have been stipulated already, as the "Suggestions of Coursebook Writing" in *The Standards* as specified, "...textbooks shall take pupils' differences into consideration and highlight their own features when the basic requirements can be ensured, and also should be given a certain degree of flexibility" (MOE of PRC, 2001), while the "Suggestions of Coursebook Writing" in the *2011 Standards* stated that "...textbook writers shall make an effort to highlight features of their works and probe into the diversity of textbooks on a proactive basis" (MOE of PRC, 2012). In view of the above analysis together with the development status of primary school mathematics textbooks at home and abroad, the authors herein put forth the following development suggestions on the writing features of the primary school mathematics textbooks:

(1) The primary school mathematics textbooks shall be diversified in the design of content selection, organization as well as difficulty level.

The most important task in primary school mathematics textbook writing is to select appropriate contents; the curriculum standards shall relent the relevant requirements as appropriate; the writers may also make corresponding tradeoffs according to the essential requirements of the curriculum standards and the characteristics of different pupils, such as the "mixed fraction" or the simple equation "$ax \pm bx = c$", thus to realize diversification of content selection in textbook writing.

Making specialized arrangement over the contents based on the researches on both theory and practice, thus to ensure they comply with pupils' cognitive features. As stated in "Suggestions of Coursebook Writing" in the *2011 Standards*: "On the basis of not violating the logic relation of mathematics knowledge, the textbook can reasonably arrange the learning content according to the pupils' cognitive rules of mathematics learning, their knowledge background and their activity experience, form its own arrangement system and reflect its own style and characteristics" (MOE of PRC, 2012). For instance, the order-based content arrangement in JEP edition is a useful attempt.

Intuitively, the difference among each edition of the textbooks is not very striking. Therefore, writers may try to make diversified difficulty levels from the aspects of content requirements, examples and exercises, etc. Yet in view of the political, cultural and educational situation in China, especially the nationally unified curriculum standards and the strict censorship system for school textbooks, there would be a long way to go before diversified content selection and difficulty level could actually come true.

(2) Writing materials related to pupils' real life and making activity-oriented textbooks.

These are also the common trends in school mathematics textbook writing. By "keeping at the approach to writing materials related to pupils' real life", the writers shall make the best use of pupils' life experience to help them understand the abstract mathematical concepts and knowledge, and meanwhile overcome the all-time malpractice in which mathematics education seriously deviates reality. By saying "activity-oriented", here it aims to highlight the active exploration and hands-on practices (Yuxin Zheng, 2003). In textbook writing, the writers shall comply with children's age characteristics and cognitive rules. It is advisable to start from pupils' life experience and select the teaching materials which can be sensed directly and intuitively by pupils, so as to help them understand and accept what has been taught; moreover, pupils shall be given more chances to operate, think and exchange ideas; for instance, at the 1st key stage, lectures may give ways to games in design, which may help to stimulate pupils' interests in mathematics study, accumulate their activity experience and improve their mathematical literacy. Each edition of the above-mentioned primary school mathematics textbooks has made positive efforts for such purpose, the "Classroom Activity", as designed in the SCNUP primary school mathematics textbooks, is an effective attempt in this sense.

(3) Paying attention to the local features of the primary school mathematics textbooks, especially the mathematics learning requirements of pupils from rural areas or the areas inhabited by the ethnic minorities.

China is a big agricultural country as well as a multiethnic country; most of the pupils are from rural areas. Therefore, the writers shall try their best to provide balanced opportunities for mathematics learning. Promoting equity in mathematics education is an unavoidable issue that has to be faced in textbook writing. In addition, since there is a wealth of valuable mathematical culture elements in rural areas and the areas inhabited by the minority nationalities, it might be feasible to write primary school mathematics textbooks desirable in these areas; or rather, the rural-themed teaching materials or those closely related to the ethnic minorities shall share an appropriate proportion in textbooks. In this respect, the SCNUP primary school mathematics textbooks have made a proactive exploration.

(4) Learning the best practices of writing primary school mathematics textbooks from developed countries.

This requires a meticulous study on the writing features of primary school mathematics textbooks from developed countries as well as the motivations behind them. Many developed countries adopt a "Multiple Standards and Multiple Textbooks" model. Compared with the "One Standard and Several Textbooks" countries, there boasts a wealth of primary school mathematics textbooks with more prominent features; the Californian and core edition textbooks in the USA, for example, take on very different styles in content representation and difficulty level. The writing features of primary school mathematics textbooks from developed countries, as described in the previous paragraphs, have provided us with great references and revelation in our own practice, such as parents' involvement in the Singaporean edition, learning hierarchy emphasis in the U.K., and the "wide and shallow" content representation to show the life experiences of human being and the socio-cultural background in the Australian edition.

(5) Conducting an in-depth exploration over the cut-in points of textbook writing features, and carrying on theory and experiment research.

To write featured primary school mathematics textbooks, it is an indispensable premise to probe into the cut-in points of textbook writing features based on the pupil development-oriented education concept. The "situation + serial questions" representation mode presented in the BNUP edition makes an effort to realize the integration of four processes, and the design of the combination of "Circumstance Series" and "Question Series" are all good cut-in points of textbook writing features. Textbook writing is a highly specialized work; if a textbook has to be written on an advanced education concept and unique content representation structure, or has to be dedicated to a special group of pupils, the writers must conduct a systematic theoretical research as well as meticulous and thorough investigation and experiment combined with the teaching practices, thus realize the purpose to facilitate teaching and learning in a real sense. As to the theoretical investigation and experimental research level, currently there is still a long way to go before the gap between China and the developed countries could be bridged, which requires the joint efforts of all the textbook researchers, writers, users as well as the administrators.

(6) The education administrative authorities shall duly loosen the restrictions on textbooks censorship, esp. on the formats of textbooks.

According to the Interim Regulations on the Certification and Management of School Textbook Writing, as required by the Ministry of Education, a textbook has to go through two stages before it could be actually used, i.e., "Applying and Approval" and "Review of Textbook". It is recommended that in the stage of "Review of Textbook", the textbooks, which are proved to be with unique styles and features in educational philosophy, content arrangement or themes, shall be encouraged and supported, as for the concerns on format, pages as well as other trivial restrictions shall be given less regulation.

References

[1] Altieri, M. B. (2009). *Mathematics Connects*. New York: McGraw-Hill Companies.

[2] Beaton, A. E., Mullis, I. V. S., Martin, M. O., Gonzalez, E. J., Kelly, D. L., and Smith, T. A. (1996). *Mathematics Achievement in the Middle School Years: IEA's Third International Mathematics and Science Study.* Chestnut Hill: TIMSS International Study Center, Boston College.

[3] Cao, Yiming (2012). *Review the Mathematics Curriculum Standards of Thirteen Countries* [In Chinese]. Beijing: Beijing Normal University Press.

[4] Chen, Jingrong (2006). *Mathematics* [In Chinese]. Beijing: People's Education Press.

[5] Chen, Yueru and Liu, Xin (2009). *A Comparative Study on School Textbook Policies at Home and Abroad* [In Chinese]. Jinan: Shandong Friendship Publishing House.

[6] Chen, Min (2012). Attentions to the possibility of children's thinking development — A brief introduction of the features of the primary textbook "New Thinking Mathematics" by Zhejiang Education Press (the first volume for Grade 1), [In Chinese]. *Primary Education (Mathematics)*, (Z1): 88–93.

[7] Clarke, P. (2007). *Primary Maths*. Scotland: Collins Education.

[8] Cui, Haijiang (2012). Make mathematics learning become children's happy life — revision note of primary mathematics textbooks of Hebei Education Press edition and introduction of the contents of Grade 1 *textbook* [In Chinese]. *Primary Education (Mathematics)*, (07): 91–94.

[9] Ding, Zhaopeng (2001). Analysis of textbook structure and evaluation of content quality [In Chinese]. *Theory and Practice of Education*, (8): 61–64.

[10] Doyle, T. (2007). *Nelson Mathematics for Victoria*. Melbourne: Thomson Nelson.

[11] EOBET (Examination Office of Basic Education Textbooks, The Ministry of Education, People's Republic of China) (2008). *Comments on Experimental Textbooks of Compulsory Education* [In Chinese]. Beijing: People's Education Press.

[12] Fong, H. K. (2007). *My Pals Are Here! Maths* (2nd edn). Singapore: Marshall Cavendish Education Publishers.

[13] Gao, Lingbiao (2002). Some reflections of the new curriculum textbooks [In Chinese]. *Curriculum, Teaching Material and Method*, 22(9): 17–19.

[14] Gowin, D. B. (1981). *Educating*. London: Cornell University Press.

[15] Gu, Mingyuan (chief ed.) (1998). *The Dictionary of Education (Supplementary One-Volume Edition)* [In Chinese]. Shanghai: Shanghai Education Press.

[16] Huo, Yujia and Huang, Xiang (2011). Exploration of the assessment function in primary school mathematics textbooks — Take California primary school mathematics textbooks for example [In Chinese]. *Journal of Mathematics Education*, 20(3): 76–79.

[17] Husen, T. and Neville Postlethwaite, T. (chief eds) (1985). *The International Encyclopedia of Education: Research and Studies*. SERBIULA (Sistema Librum 2.0).

[18] Kuang, Kongxiu and Song, Naiqing (2013). Situations and reflections about the "two basics" mathematics teaching in primary schools in the context of new curriculum reform — based on the survey in seminars for primary school mathematics teachers in "The National Training Plan" [In Chinese]. *Curriculum, Teaching Material and Method*, 33(2): 66.

[19] Liu, Jian and Kong, Qiping (2006). *Mathematics* [In Chinese]. Beijing: Beijing Normal University Press.

[20] Liu, Jian, Kong, Qiping and Zhang, Dan (2012). Establishing a mathematics home which aims to promote flourish for children — The introduction of the features of the fourth edition of the new century primary textbook *Mathematics* (Beijing Normal University Press Edition) [In Chinese]. *Primary Education (Mathematics)*, (7): 57.

[21] Lu, Jiang (2012). A new textbook of innovative and enterprising, rigorous and realistic, reflecting the spirit of the times — Introduction of the revised PEP's compulsory education textbook *Mathematics* (Grades 1 to 6) [In Chinese]. *Primary Education* (Mathematics), (07): 50–51.

[22] Lu, Jiang and Yang, Gang (2002). *Mathematics* [In Chinese]. Beijing: People's Education Press.

[23] Lü, Da (2001). Deepening the theoretical basis of the curricular reform [In Chinese]. *China Education Daily*, 10/20/2001 (page 4).

[24] Ma, Dingling and Ji, Zhigang (2008). On Fibonacci historical significance and research value in the exchange of mathematics between China and the West [In Chinese]. *Journal of Shanghai Jiaotong University (Philosophy and Social Sciences)*, (2): 54–58.

[25] MOE of PRC (Ministry of Education, People's Republic of China) (2001). *Mathematics Curriculum Standards for Compulsory Education (Experimental)* [In Chinese]. Beijing: Beijing Normal University Press.

[26] MOE of PRC (Ministry of Education, People's Republic of China) (2012). *Mathematics Curriculum Standards for Compulsory Education (2011 Version)* [In Chinese]. Beijing: Beijing Normal University Press.

[27] Monica, J. (2006). Teaching mathematics with textbooks — A classroom and curricular perspective. Luleå University of Technology, Department of Mathematics, Sweden.

[28] Pan, Guihua (2012). New standards, new ideas and new textbooks — revision note and instructional advices of Qingdao edition of compulsory mathematics textbooks [In Chinese]. *Primary Education (Mathematics)*, (7): 81–85.

[29] Pearson (2008). *Investigations in Number, Data, and Space (Common Core Edition)*. New Jersey: Pearson Education.

[30] Schmidt, W. H., McKnight, C. C., Valverde, G. A., Houang, R. T. and Wiley, D. E. (1997). *Many Visions, Many Aims: A Cross-national Investigation of Curricular Intentions in School Mathematics*. Boston: Kluwer Academic Press.

[31] Shi, Juntao (2000). On diversification of teaching materials [In Chinese]. *Curriculum, teaching material and methods*, (9): 31.

[32] Shimizu, S. (2011). *Arithmetic* [In Japanese]. Osaka: Shinko Shuppansha Keirinkan Co., Ltd.

[33] Shi, Ou and Shi, Yu (2012). On the basic characteristics of textbooks [In Chinese]. *Educational Research*, (4): 92.

[34] Song, Naiqing (2007). *Mathematics* [In Chinese]. Chongqing: Southwest Normal University Press.

[35] Song, Naiqing (2012). *Mathematics* (the Second Volume of Grade 1) [In Chinese]. Chongqing: Southwest Normal University Press.

[36] Song, Naiqing and Xu, Binyan (2009). *An Introduction of Mathematics Curriculum* [In Chinese]. Beijing: Beijing Normal University Press.

[37] Sun, Ligu and Wang, Lin (2002). *Mathematics* [In Chinese]. Nanjing: Jiangsu Education Press.

[38] Underwood, D. (2002). The world's first mathematics textbook. *Mathematics Horizons*, (4): 8–11.

[39] Valverde, G. A., Bianchi, L. J., Wolfe, R. G., Schmidt, W. H. and Hoang, R. T. (2002). *According to the Book: Using TIMSS to Investigate the Translation of Policy into Practice through the World of Textbooks*. Dordrecht: Kluwer Academic Publishers.

[40] Wang, Ling (2010). Optimizing the textbook structure and enriching education course — revision note of primary mathematics textbooks of Jiangsu Education Press edition [In Chinese]. *Primary Education (Mathematics)*, (7): 74.

[41] Wang, Xiaoqin (2008). Robert Recorde: the first British mathematics educator [In Chinese]. *Journal of Dialectics of Nature*, (5): 92–101.

[42] Wang, Ying (2011). Using content analysis method inspecting characteristics of the primary school mathematics textbooks [In Chinese]. *Theory and Practice of Education*, (8): 26.

[43] Wei, Jia (2009). Evolution and development about the content of the primary mathematics textbooks in China during the 20th century [In Chinese]. Doctoral dissertation, Chongqing: Southwest University.

[44] Wikipedia. Mathematics education, http://en.wikipedia.org/wiki/Mathematics_education.

[45] Zeng, Tianshan (1997). *On Educating Materials* [In Chinese]. Nanchang: Jiangxi Education Press.

[46] Zhan, Tao, Liu, Pinyi and Xu, Yunhong (2005). *Mathematics* [In Chinese]. Qingdao: Qingdao Publishing House.

[47] Zhang, Jiajian (2005). The six impressions on primary school mathematics textbooks in Australia [In Chinese]. *Young Teacher in Primary Education*, (2): 47–48.

[48] Zhang, Hongyu (2012). The inadequacy of Compiling Characteristics of school textbooks should not be ignored [In Chinese]. *China Press and Publishing Journal*, 5/28/2012 (Page 6).

[49] Zhang, Tianxiao (2010). *Mathematics* [in Chinese]. Hangzhou: Zhejiang Education Press.

[50] Zhang, Wenyu and Fu, Hailun (2011a). Characteristics of the primary school mathematics textbooks in Singapore and its inspiration [In Chinese]. *Journal of Mathematics Education*, 20(4): 63–66.

[51] Zhang, Wenyu and Fu, Hailun (2011b). A Comparative study on primary school mathematics textbooks between Singapore and China [In Chinese]. *Studies in Foreign Education*, (7): 36–39; 82.

[52] Zhao, Xingmei (2008). *Mathematics* [In Chinese]. Shijiazhuang: Hebei Education Press.

[53] Zhao, Xiaoya (2011). Revising experimental textbooks of compulsory education according to curriculum standards on full swing [In Chinese]. *China Education Daily*, 07/12/2011 (Page 1).

[54] Zheng, Yuxin (2003). On the trends of activitization, individualization and life orientation in mathematics curriculum reform [In Chinese]. *Educational Research*, (6): 90–93.

[55] Zhong, Qiquan (2001). *For the Rejuvenation of the Chinese Nation, for Each Student's Development — The Interpretation of the National Program for Primary Education Curriculum Reform (Trial)* [In Chinese]. Shanghai: East China Normal University Press.

[56] Zhong, Qiquan (2003). *Modern Theory of Curriculum* [In Chinese]. Shanghai: Shanghai Education Press.

https://doi.org/10.1142/9789814713955_0013

Chapter 13

On the Representations of Basic Mathematical Thoughts in Mathematics Textbooks

Fanzhe Kong[1] and Jiali Yan[2]

[1]*Faculty of Education, Northeast Normal University*
Changchun, Jilin 130024, China
fzkong@nenu.edu.cn

[2]*Faculty of Education and Public Administration*
Suzhou University of Science and Technology
Suzhou, Jiangzu 215000, China
yanjiali86@126.com

In this chapter, the representations of basic mathematical thoughts in mathematics textbooks are chosen as the object of study; using the methods of content analysis and speculation with typical topics in school mathematics textbooks as examples, we analyze the likelihood and forms to represent the basic mathematical thoughts in mathematics textbooks. Research shows that as one of the important components in mathematics curriculum, basic thoughts should be regarded as an important element in mathematics textbooks and must be well represented in textbooks; meanwhile, the representations should be subject to the criteria whether they could facilitate students to understand such thoughts so as to gain further development in mathematics. There are mainly three basic channels to represent the basic mathematical thoughts in textbooks: first, specific chapters could be set to explicitly and systematically elaborate the thoughts; second, they could be represented by integrating with the specific content of mathematical concepts, formulas and algorithms, by which the knowledge, skills and thoughts could be represented clearly; third, a narrative interposed or sparse infiltration could also be adopted in the explicit representation of knowledge and skills. Whether the basic mathematical thoughts could be consciously represented in textbooks depends basically on the value orientation of editors over curriculum content, or simply put, whether the textbook writers take a knowledge-based orientation or thoughts-and-methods-based orientation, or lays emphasis on both. Currently, an all-purpose textbook representation mode for all basic mathematical thoughts has not yet been discovered.

The National *Mathematics Curriculum Standards for Compulsory Education (2011 Edition)* (MOE of PRC, 2012) of China stresses in its rationale that "curriculum content ... should include not only the mathematical results, but also the process how the mathematical results are obtained as well as the implied mathematical thoughts and methods", and for the first time, the term "basic mathematical thoughts" and "basic activity experiences" were listed in the overall objectives of the mathematics curriculum.

As a materialized form of curriculum, textbook is not only the carrier of basic knowledge and skills, but also the important carrier and necessary materialized form of basic mathematical thoughts.

Taking typical cases in the school mathematics textbooks as examples, this paper analyzes the representations of basic mathematical thoughts in mathematics textbooks.

1. Clarification of Basic Notions

1.1 The meaning of basic mathematical thoughts

The so-called *thought* normally refers to some objective existence reflected in one's consciousness and the generated results of thinking; it is the foundation of all human behaviors. Simply put, *thought* is the "thinking of mind."

The so-called *basic mathematical thoughts* refer to the thoughts upon which the mathematical science was generated and developed, and which could offer a lifetime benefit to students once grasped.

1.2 Mathematical thoughts in the three core phases of mathematization

Hans Freudenthal, a renowned Dutch mathematician and mathematics educator, once said, instead of saying "mathematics should not be learned as a closed system but rather as an activity of mathematizing reality and if possible even that of mathematizing mathematics." (Freudenthal, 1991) In this sense, the core objective of school mathematics instruction is to "learn mathematization". By the detailed representation process in the school mathematical curriculum, "mathematization" can be subdivided into three core phases, namely "mathematization of real problems", "internal regularization of mathematics" and "contextualization of mathematical content". Each phase corresponds to one kind of basic mathematical thoughts (Fanzhe Kong, 2012).

1.2.1 Mathematization of real problems — abstraction

"Mathematization of real problems" refers to the process of appropriate abstraction of real problems and then turning such problems into mathematical problems, among which mathematical abstraction plays the core role. Though abstraction is by no means unique in mathematics, mathematical abstraction targets at the abstraction of quantitative relation and spatial form, so it is a special kind of abstraction. The nature of mathematics is to study abstract objects, and the abstraction is based on the real world and the objects for mathematics are the real things being abstracted. The mathematical abstraction can be the quantitative relation and the spatial form in the real world, or those merely in our mathematical thinking. Genuine knowledge is rooted in perceptual experience and obtained by intuition and abstraction, and such abstraction cannot be independent of human thinking. Abstraction is the basis of thinking because only by possessing a certain degree of abstraction could human gain substantive characteristics of things (events or material objects) from their perceptual knowledge, thus lifting to a level of rational knowledge.

1.2.2 Internal regularization of mathematics — reasoning

The internal development of mathematics is benefited by mathematical reasoning which consists of deductive reasoning and inductive reasoning, in a general sense; while the building of mathematical structure is based on axiomatization, i.e., the process to sort out mathematical knowledge into an internally simple and rational, self-contained and complete system. The fundamental work for the internal structure building of mathematics requires the assistance of axiomatization whereas the internal development of mathematics requires the help of reasoning. Though there is a little bit axiomatization thought penetrated in primary school mathematics, there is a clear expression, however, of axiomatization thought in the secondary school mathematics textbooks, containing an explicit and relatively complete axiom system for geometry. Therefore, the basic thoughts of mathematical reasoning should include (the general) induction, logical reasoning, and axiomatization.

1.2.3 Contextualization of mathematical content — modeling

One important task for mathematics (also an important impetus for the development of mathematics) is the "contextualization of mathematical content", or simply, the process to find, on a proactive basis, the real prototype of mathematical content, to understand problems in the real world mathematically, and then to formulate suitable mathematical problems, to analyze and solve the mathematical problems; among which mathematical models serve as the core.

To sum up, in school mathematics, the most important mathematical thoughts include mathematical abstraction, mathematical reasoning, and mathematical modeling. They will benefit the lifetime sustainable development of students in mathematics.

2. Literature Review

As a core constituent in mathematical thinking, the basic mathematical thoughts and their research context are developed along with the advancement of mathematics. Though the basic mathematical thoughts have been born when mathematics came into being, the systematic study on them has not begun until the early 20th century.

"As the primary carrier of mathematics curriculum, mathematics textbooks are the major channels for the implementation of mathematics education ... The reform of mathematics textbooks is one of the core tasks of mathematics curriculum reform, and it is also one of the focuses in each mathematics education reform" (Fanzhe Kong, 2008b). However, the systematic study of textbooks in Mainland China has started quite late, even later than Taiwan, where the relatively systematic study of textbooks by Yung-Sheng Ou (see Meizhi Zhuang, 2003), and other researchers has also cast profound effects on Mainland China. As mathematics textbooks, though both the *Elements* in the West and the *Nine Chapters on the Mathematical Art* in China are successful and sold well (Fanzhe Kong, 2008b), a systematic research method for textbooks, including the study of textbook representation, has not started until a few decades ago (see, for example, Shuen De Lan, 2004 and Fanzhe Kong, 2008a). As David L. Elliott *et al.* (1990) pointed out, "Until recent years research in the field of textbooks has been modest." As a matter of fact, the research on textbook analysis and evaluation are relatively weak and the case is true to both China and other countries such as the USA.

2.1 Researches on mathematical thoughts

2.1.1 Overview

With the emergence of major thinking methods in mathematics since the early 20th century, researchers have paid attention to the internal connection among various branches in mathematics and started to focus on exploring the generation and development patterns of mathematical thoughts themselves. During that period, a large number of theoretical works on mathematical thoughts were published, and the focuses of these researches can be roughly divided into six perspectives (Enze Xie and Benshun Xu, 1995) as follows.

(1) From the perspective of content and application of mathematical thoughts themselves

David Hilbert, a famous German mathematician, used to brilliantly expound the features of major mathematical problems and the role they played in mathematics advancement in his speech *Mathematical Problems* (Hilbert, 1900) as was addressed in the 1900 International Congress of Mathematicians in Paris, in which he also listed 23 thought-provoking major mathematical problems, later these problems were named "23 problems of Hilbert". Henri Poincaré, a French mathematician, discussed the issue of mathematical methodology in specific chapters and sections in his works *Science and Hypothesis*, *The Value of Science*, and *Science and Method* (Poincaré, 1902–1908). Otto Ludwig Hölder, a German mathematician, published *Methodology of Mathematics* (Hölder, 1924), in which he discussed deductive, inductive, axiomatic and hypothetical methods on a systematic basis.

Lizhi Xu, a Chinese mathematician, also attached great importance to the study of mathematical methodology. In the works he published successively, such as *Introduction to Mathematical Methodology* (Lizhi Xu, 1980), *Concept of Mathematical Abstractness and Its Analytical Method* (Lizhi Xu, 1985) and *Topics on Mathematical Methodology* (Lizhi Xu, 2000). He put forward many insightful views on mathematical thoughts. The *Vertical and Horizontal Analysis of Mathematical Thinking and Methods* (Enze Xie and Shuzhi Zhao, 1987) offers a systematic analysis on the formation and development of mathematical thinking by a vertical and horizontal comparison. Ningzhong Shi holds that "in nature, mathematical thoughts consist of abstraction, reasoning and modeling"; he made a detailed discussion over each type of thoughts and summarized his idea into three booklets, respectively, entitled *Introduction to Mathematical Thoughts* (Ningzhong Shi, 2008–2012).

(2) From the perspective of historical development

The *Mathematical Thought from Ancient to Modern Times* (Kline, 1972) by Morris Kline, an American mathematician, is the most worth mentioning book. "This book treats the major mathematical creations and developments from the ancient times to the first few decades of the twentieth century. It aims to present the central ideas, with particular emphasis on those currents of activity that have loomed largest in the main periods of the life of mathematics and have been influential in promoting and shaping subsequent mathematical activities." (Kline, 1972) Its core features lie in that it illustrates the infiltration and inheritance for the formation and development of important mathematical thought with a focus on the development of mathematics. Just as Di Li, a Chinese mathematics historian, used to point out that, "The so-called mathematical thought is understood differently. Some think that it refers to the methods of thinking for mathematics itself; some believe that it is related to issues of philosophy in mathematics and the others regard the mathematical thought as a combination of both ideas above.... While the thought described by Kline is basically of the first kind." (Di Li, 2001). Di Li also pointed out that Kline's book has limitations that it "excludes the mathematics and relevant thoughts of China, Japan and Maya" and "its century-based segmentation does not reflect the actual development of mathematical thoughts" (Di Li, 2001).

Viewed from a global perspective, *Mathematics: Its Essence, Methods and Role* (Aleksandrov *et al.*, 1956) by A. D. Aleksandrov and his colleagues from the former Soviet Union is also one of the books with major influence over mathematical thoughts. This book on the one hand summarizes the historical evolution of mathematics as a whole, and on the other hand expounds each branch of modern mathematics with a focus on its history, essence, methods and role, highlighting the analysis and evaluation of mathematical thoughts. Comparatively, the book by Kline emphasizes more on the historical development of mathematics and mathematical thoughts.

(3) From the perspective of mathematics education and mathematical ability cultivation

G. Pólya, a famous Hungarian American mathematics educator, published a series of books including *Mathematics and Plausible Reasoning* (Pólya, 1954). Through his practice in education, Pólya realized that mathematical finding was often derived from estimation and guess; however, only rigorously demonstrated conclusions instead of the process for drawing such conclusions were given in mathematics textbooks. Pólya's books list a large number of cases in the history of mathematics, analyzing the origin for ideas of mathematical results, exploring the internal interaction between deductive and plausible reasoning, which illustrate the needs of emphasizing both deductive and plausible reasoning.

Kunihide Yoneyama, a Japanese mathematical educator, published *The Spirit, Thoughts and Methods of Mathematics* (Yoneyama, 1968) and mentioned ten mathematical ideas in the book: The idea of completely free thinking as the essence of mathematics, the idea of limit, the idea of forming a "set of undefined mathematical terms" and a "set unproved statements," the idea of sets and groups, the idea of regarding limited length as unlimited one, the idea of regarding curves as straight lines, the idea of enabling the appearance of different geometries, mathematics and operations, the idea of 2-dimensional, 4-dimensional and higher dimensional spaces, the idea of transfinite numbers, as well as the idea of mystery and beauty of mathematics.

In *Lectures on Modern Mathematical Thoughts* (Dianzhou Zhang and Chengjie Zhu, 1991) Zhang Dianzhou pointed out some original and important mathematical thoughts in the development of modern mathematics. In particular. he listed 15 mathematical thoughts which may be infiltrated into school mathematics teaching in his last lecture.

(4) From the perspective of mathematical philosophy

As a philosophy work, *Mathematical Manuscripts of Karl Marx* (Marx, 1881) mainly illustrates the application of dialectical thinking methods in mathematics and the historical evolution of mathematical thoughts. It is a book for studying mathematical thoughts. *Proofs and Refutations* (Lakatos, 1976) by Imre Lakatos, a British mathematical philosopher and *Mathematics, Matter and Methods* (Putman, 1975) by Hilary Putnam, an American philosopher, systematically discuss the so-called quasi-empirical method from the perspective of philosophy.

A Brief Analysis of Some Dialectical Contents in Mathematics (Fengpu Liu, 1980) makes a relatively systematic discussion of objective foundations of mathematics, dialectical nature of mathematical concepts, and the unity of opposites for mathematical operations. It also mentions some achievements in mathematics and their characteristics since the 19th century. *Philosophical Explorations on the Frontier of Science* (Shixia Fu, 1983) makes an analysis of philosophy on the new branches of modern mathematics including fuzzy mathematics, catastrophe theory and nonstandard analysis, etc.

(5) From the perspective of existence forms for mathematics

In 1979, some Chinese researchers pioneered a new area of study called "potential science". It discusses about the law that new scientific ideas are derived from the perspective of potential scientific forms and is quite popular among the academic field. Among the publications concerning potential science, *A Collection of Important Debates in Science History* (Xinmin Zhu, 1986), *A Collection of Scientific Disasters* (Enze Xie, 1998),

Puzzles and Prospects in the Frontier of Science (Dingguo Hong, 1988), *Greatest Musikalität in Ideological Sphere: Case Study of Scientific Finding* (Xingmin Li *et al.*, 1988), as well as *Introduction to Potential Science* (Enze Xie, 1987) all discuss about latent forms of mathematics.

(6) From the perspective of critical biography of figures

In addition to biographies of mathematicians that introduce some ideas and methods in a decentralized manner, such as *Men of Mathematics* (Bell, 1986), *Biographies of Women Mathematicians* (Pinfang Xu, 2005), *Zu Chongzhi* (Yihuan Tan, 1976), *Hilbert* (Reid 1970), and *Evariste Galois, Révolutionaire et Géométre* (Dalmas, 1956), there are still some other publications that discuss the ideas of mathematicians in a centralized way, such as *Ways of Thought of Great Mathematicians* (Meschkowski, 1961) analyzes and introduces the characters of some famous mathematicians including Pythagoras, Archimedes, Pascal, Leibniz, Gauss and Galois, etc.

It is easy to find that the above researches mainly combine specific development history of mathematics and philosophical views, etc. to illustrate the historical development of mathematical thoughts.

2.1.2 Various views of mathematical thoughts

Shisun Ding (1956) considers that mathematical thoughts are people's views towards mathematics, which include role of mathematics in the knowledge system of human beings, relationship between mathematics and production practice, relationship between mathematics and other disciplines, rules of mathematics development, features of mathematical research methods, etc.

Yuxin Zheng (1993) believes that the mathematical thoughts should have two meanings. First, it is applicable in the meanings opposite to specific mathematical knowledge contents (especially the ultimate and rigorous forms of expression); this means that clear distinction is made between thinking activities in mathematical research and final product of the thinking activities. Second, separated from specific mathematical contents, it is the thinking mode or principle with wider universal meanings.

Lixue Qu (1994) believes that, mathematical thoughts are people's deep understandings of the essence and rules of mathematical science research. The subjects of such understandings are the previous, present and future well-known and nameless mathematicians in the history of human beings, and the topics of the understandings include objects and characteristics of mathematical science, characters of ways and methods for mathematical research, cultural and spiritual value of research achievements and their actual impact on the physical world, interaction and support among various results or conclusions inside, etc.

Dianzhou Zhang and Boxiang Guo (1996) believe that mathematical thought has not yet become a specific terminology. People often use it to refer to mathematical results and findings with great significance, rich content and complete system. The same mathematical achievement is called method when it is used to solve other problems and thoughts when its value and significance is evaluated in the system of mathematics.

Shanghe Cai (1997) thinks that the so-called mathematical thoughts are the results produced when the spatial forms and quantitative relations in the real world are reflected in the people's consciousness and subject to thinking, and they represent people's understandings towards the essence of mathematical facts and theories. The mathematical thoughts are part of scientific thoughts and a kind of scientific thoughts may be converted into the mathematical thoughts only when they are subject to mathematization.

The reason for the differences in the understanding towards mathematical thoughts is researcher's different views towards them. Some of them observe mathematical thoughts inside mathematics; some view them from the philosophical epistemology, and still others give interpretations from the perspective of mathematics education. Despite of the different views, it can be discovered that the history of mathematical thoughts is a history of formation and development for basic concepts and important theories of mathematics as well as the history of the development of mathematicians' and philosophers' views on mathematics.

2.1.3 What are basic mathematical thoughts

Just as Di Li, a Chinese mathematical historian, used to point out that, "the so-called mathematical thoughts are understood differently. Some think that they refer to the methods of thinking for mathematics itself; some believe that they are issues of philosophy in mathematics; and the others regard the mathematical thought as a combination of both...," "while the 'thought' described by M. Kline in his book is basically of the former." (Di Li, 2001)

In China's *Mathematics Curriculum Standards for Compulsory Education (2011 Edition)* (MOE of PRC, 2012), "Four Basics" are explicitly proposed for the first time, and basic (mathematical) thoughts and basic (mathematical) activity experiences are added to the traditional "Two Basics", i.e., basic (mathematical) knowledge and basic (mathematical) skills.

Shanghe Cai (1997) believes that the ideas with fundamental and summative thinking results that are or should be reflected in core mathematics are called basic mathematical thoughts, which include the ideas of symbols and variable representation, the ideas of sets and correspondences, the ideas of axiomatization and structures, the ideas of combination of numbers and figures, the ideas of reductions, the ideas of unity of opposites, the ideas of integrity, the ideas of functions and equations, the ideas of sampling and statistics, the ideas of limits, etc.

Xiaotian Sun (2012) believes that basic mathematical thoughts mainly have two aspects of meanings. One is the thinking activities and results in promoting and supporting the formation and development of mathematical science. The reflection in mathematics curriculums as an educational requirement is the meaning of "basic thoughts" in curriculum standards. The other means the mathematical ideas that can be acquired or "rediscovered" by students during the process of mathematics learning.

Ningzhong Shi, head of revision group of the compulsory education mathematics curriculum and professor of the Northeast Normal University, proposed two criteria for the identification of basic mathematical thoughts (Ningzhong Shi, 2011). First, the thoughts are the ones that the formation and development of mathematics must rely on, and, second, they are the features of thinking acquired by those who have learned mathematics.

Ningzhong Shi believes that basic mathematical thoughts essentially include abstraction, reasoning and modeling. By abstraction, the external world and matters in relation to mathematics may be abstracted into mathematics and form objects of mathematical rcscarch; by reasoning, the propositions and computation methods may be obtained so as to facilitate the development of mathematics; by modeling, expressive mathematics language may be created to bridge mathematics and the external world.

Pei Gu (2012) supplements the ideas of Ninzhong Shi and believes that the basic mathematical thoughts also include mathematical aesthetic ability. Through mathematical aesthetics, we could "see the essence through the appearance", observe the element of beauty in "the harmonious and unified matters", feel the joys presented by "controlling complexity by using simplicity" in mathematics and the "flawless nature" of mathematics, and discover and create new mathematics from the perspective of beauty.

In a word, mathematical thoughts represent the essential understandings upon summarization of mathematical facts and theories, while basic mathematical thoughts refer to the fundamental, summative and most extensive mathematical thoughts that are reflected or should be reflected in core mathematics. It involves the essence of traditional mathematical thoughts and basic characters of modern mathematical thoughts and evolves with times. The core of basic mathematical thoughts is characterized by "the most important and central ideas necessary for development of mathematical science" (Ningzhong Shi and Fanzhe Kong, 2012).

2.1.4 Mathematical thoughts in school mathematics

Jingzhong Zhang, an academician of the Chinese Academy of Sciences, used to point out that, "the ideas of functions, number-figure combination and reasoning in computation are belong to good mathematics" (Jingzhong Zhang, 2007). These ideas could infiltrate in the early phase of education, introduced through real-life problems without further elaborations.

Weiliang Huang (2009) believes that the mathematical thoughts and methods focused in primary school textbooks should be measured in three dimensions: first, some important ideas and methods embodied in the mathematics knowledge of primary schools; second, some common ideas and methods that could produce positive influences over the successive learning of students; and third, some specific methods and skills that directly act on the knowledge acquisition and problem solving of students and could be easily accepted by them.

Xiangbin Chen (2010) considers that, according to the characters of mathematics contents in primary schools and cognitive level of pupils, the following mathematical ideas and methods should be infiltrated in the mathematics teaching of primary schools: classification, reduction, number-figure combination, induction, sets, equations, symbolizations, functions and correspondences, limits, etc.

Yanwen Wang and Guangming Wang (1996) summarized the basic characteristics of mathematical thoughts in secondary schools: Mathematical thoughts in secondary schools should be the most common and fundamental part in mathematical thoughts, should have fundamental and summative characters, and should include two main components, i.e., one aiming to creating mathematics and the other aiming to learning mathematics. Different from specific skills and methods, the thoughts are high-level cognitive strategies and effective modes for problem solving and are continuously enriched and developed. Mathematical thoughts in secondary schools mainly include: ideas of sets and correspondence, reduction, mathematization, structure and dialectics.

Guangming Wang and Wengui Zhang (1997) believe that, the mathematical thoughts that should be paid attention to in mathematics for secondary schools mainly include ideas of sets, reduction and correspondence. The ideas of symbolization, axiomatization and limits are reflected to some extent in the mathematics for this period and should be penetrated in teaching according to specific situations.

It should be noted that, before the issue of the *Mathematics Curriculum Standards for Compulsory Education (2011 Edition)* (MOE of PRC, 2012), the basic mathematical thoughts for school mathematics discussed in China were mixed at different levels and grades. Some are the most central mathematical thoughts and some are quite microcosmic ideas (subordinate content, even approaching specific mathematical problem solving methods). Thus, the school mathematics in China habitually adopted the expression of "mathematical thoughts and methods". Just as pointed out in Ningzhong Shi (2011) and Fanzhc Kong and Ningzhong Shi (2012), basic (mathematical) thoughts mentioned in the Standards refer to the most central ideas that play an important role for the sustainable development of students in their whole life. Once they are acquired, they will benefit the learners in their whole life.

That is to say, the basic thoughts are the core ideas relied on by science and essential for the discipline development of human beings. "Various mathematical ideas could be evolved, derived and developed from them" (Pei Gu, 2012). Therefore, the basic mathematical thoughts are actually the "big ideas" in mathematics (Di Li, 2001).

2.1.5 Comments

In conclusion, it could be easy to find certain problems from the existing research on mathematical thoughts, basic mathematical thoughts or basic mathematical thoughts in school mathematics, both in China and abroad. They are mainly reflected in the following aspects:

On the one hand, the research on mathematical thoughts is hybrid and extensive, mixing the mathematical ideas and methods, and mixing the problems of all levels together. Also, most of the research illustrates the thoughts in a macro manner based on the development of a specific mathematics branch, does not have the character of general applicability of mathematics science and fails to study mathematical thoughts from the perspective of influence over the development of mathematics science, regardless of the influential literatures like *Mathematical Thought from Ancient to Modern Time*s (Kline, 1972) and *Mathematics: Its Content, Methods and Meaning* (Aleksandrov, 1956), or researches focusing on the local mathematical ideas embodied in the illustration of specific mathematical content or the mathematical ideas illustrating a particular development period of mathematics, such as "the creation of non-Euclidean geometry was the most consequential and revolutionary step in mathematics since Greek times" or "each geometry can be characterized by a group of transformations" (Di Li, 2001).

On the other hand, as for the basic thoughts in school mathematics, the research mixes the superior mathematical ideas (core content involving mathematical philosophy), mathematical ideas and methods as medium view (such as relation, mapping, inversion, etc.) as well as mathematical methods as specific solutions (such as factoring method of grouping and method of substitution in solving equations); meanwhile, the current research only studies from the perspective of the role of those mathematical ideas in relevant mathematics content but little of it involves students' development in mathematics and their demands for sustainable development during their life.

2.2 Review of textbook research

2.2.1 Perspectives and methods for textbook research

By analyzing relevant documentations, Pepin and Haggarty (2001) proposed a framework of textbook research including four major dimensions: mathematical principle, pedagogical principle, social background, and cultural tradition of the textbook.

Jiansheng Bao (2002) proposed in his doctoral dissertation a composite difficulty model used to measure and compare the difficulty levels of mathematics curriculum in both China and UK. Textbooks are also regarded as an intended curriculum. The model includes five factors, namely content coverage, cognitive demand, context, reasoning and computation.

Ningzhong Shi *et al.* (2005) constructed a model describing the difficulty level of curriculum through three factors, namely breadth, duration and depth. Breadth means the scope and the extensive degree of the field concerned in the content of curriculum; duration means the time necessary for completion of curriculum, which could be quantified by using the commonly said "teaching hours"; depth means the overall depth of thinking for completion of curriculum.

Fanzhe Kong (2008a) believes that the internal and external quality of textbooks should be analyzed from two dimensions including internal static text and external dynamic implementation. The influential factors for internal quality are mainly reflected in the following aspects: content, organizational structure, printing feature, and compiling characteristics; the influential factors for external quality mainly include curriculum implementation ability of teachers, quality guarantee mechanism, richness and diversity of curriculum resources as well as guidance of value orientation evaluation.

R. Y. Kim (2009) developed a framework for evaluation of non-text elements (e.g., pictures and charts) in mathematics textbooks through the interview with 21 mathematics teachers and 7 textbook editors. This framework is composed of five elements, namely accuracy, continuity, contextuality, simplicity and sense of beauty.

C. Y. Charalambous *et al.* (2010) classified textbook analysis into three categories: horizontal analysis, vertical analysis and context analysis. For horizontal analysis, textbooks are regarded as a unit of research and their universality (such as physical attributes and organization of content) are focused on; vertical analysis mainly concentrates on the arrangement and processing of a specific mathematics concept or theme in textbooks; context analysis pays attention to the way of textbook use and belongs to curriculum implementation field.

Some researchers in Taiwan made a reflective research on textbook research, with Yung Sheng Ou and Pei Yi Zhou, etc. as representatives. Pei Yi Zhou (2002, 2005) used to point out that, "textbook research mostly remains teachers' interpretation of the text in textbooks" and "content analysis is its main research method"; while Shuen De Lan (2004) indicates, "Textbook evaluation research should become one of the hot topics"; Pei Yi Zhou and Mingchang Zheng (2008) pointed out that according to their research, "local textbook research has gained huge increment and the focus on textbook research is shifted from increment in quantity to improvement of quality. It is advised that researchers should not only understand various examples and methods for textbook researches but also understand their applicability and limitations as well as their mutual relationship. Also, they should choose proper research methods according to the nature and characteristics of textbooks and jointly develop the textbook research methodology featuring example integration, method integration and interdisciplinary integration". Fanzhe Kong (2007) made a profound analysis in his doctoral dissertation to the quantitative method in textbook research, including constructing a textbook difficulty model and adopting a horizontal quantitative model on teachers' use of textbooks. The observation was also echoed by overseas

Chinese researchers. For example, Lianghuo Fan *et al.* (2013) claimed that, "it is important to examine textbooks not only in terms of their content and structure, but also their use in real classrooms".

2.2.2 Research of textbook representation

From an overview of the current research on "school textbook representation" both in China and abroad one can discover the following characters of the research:

(1) The disciplines involved: The research of textbook representations involves diversified disciplines, including mathematics, Chinese, English, geography, history, politics, physics, chemistry, computer and others.

(2) The time of the research: In China the research on textbook representation started after 2000, while it started earlier in some other countries. For example, W. R. McConnell (1920) did a representation research in geography textbooks, and a representation research of fairy tales in the school literature textbooks was done by P. C. Marion (1969). The related researches in China have gained increasing trend since 2006 and most of them were done by graduate students in their doctoral and master theses.

(3) The research content: Some researches concentrate on the patterns of representation modes. For example, Li Jun (2004) indicates that representations of chemistry experiment textbooks include four models, namely functional-modular model, task-driven model, chart model and scene model Zongfang Zhao and Junming Wu (2005) believe that the tasks and activities of chemistry experiment textbooks are represented in three ways: direct statement, statement plus activity list, and task promptings; there are also three forms of expression of content, including text-only statement, text statement plus columns, and charts (pictures and/or tables). Yongxiang Fang (2012) considers that good ideological textbook representation should help to facilitate teaching method reform, strengthen columns, focus on cultivation of beauty, and pay attention to the integration of humanity and sciences. Jiaoguo Wang and Xianyan Chen (2013) made attempts to construct a comparative model for representation of high school physics textbook content in macro and micro levels.

Some researches focus on the representation features of specific discipline knowledge arenas. For example, in the mathematics field, there are research on representation of plausible reasoning in the mathematics textbooks (Chunmin Gao, 2009; Wenwen Wu, 2009), research on representation of geometric figures and transformations in the mathematics textbooks (Shukun Lu, 2009), research on representation of mathematical concepts in textbooks (Shengli Zhang, 2013), research on representation of mathematics history in mathematics textbooks (Liu Yun and Weizong Zhu, 2012; Nainan Xu, 2012), as well as analysis on the representation of “fraction” and “division” in the US school mathematics textbooks (Winberg, 1998), etc. As for the chemistry field, one can find research on representation of chemistry history in the high school chemistry textbooks (Qing Sun, 2009) and research on representation of illustrations in textbooks (Bo Pei *et al.*, 2013). For the history field, there are research on representation of female figures (Colbert-Lewis, 2005) and research on representation of gender images (Nowakowska, 1997), etc.

In addition, from the perspective of abstraction degree of research content, there is a division of “concrete” and “abstract”. For example, the research on representation of mathematical culture in mathematics textbooks (Lele Song, 2012), research on representation of cultural values in English textbooks (Xiaohong Zheng, 2009) and research on representation of scientific literacy in geography textbooks (Garcia, 1985) can be regarded as researches in a relatively abstract way.

In conclusion, textbook representation research covers an extensive view of angle. It has not only the macro and micro angles, but also comparative ones. However, there are not many researches on the representation of basic mathematical thoughts and methods in textbooks, and the research on mathematical thoughts in mathematics textbooks for primary schools is later than that for secondary schools. In this direction, we have, as examples, the following literatures:

Shanghe Cai (1997) points out that secondary school mathematics textbooks shoulder the responsibility of teaching basic mathematical thoughts and ideas, and the way to do that may be divided into “infiltration”, “introduction” and “highlight” types depending on the degrees of emphasis.

Infiltration means integration of some abstract mathematical thoughts or ideas into real and concrete mathematical knowledge thus enabling students to obtain initial intuition or feeling of the thoughts and ideas rather than know them in a rational manner. The ideas required to be permeated include the idea of sets and correspondence, the idea of axiomatization and structure, the idea of sampling and statistics, and the idea of limits, etc.

Introduction means explicit statement of some mathematical thoughts and ideas together with mathematics knowledge in proper time, enabling students to obtain an initial image of the thoughts and ideas, which is the first step towards a rational understanding. The ideas to be introduced include the idea of symbol and variable representation, the idea of number-figure combination, the idea of reduction, the idea of functions and equations, the idea of sampling and statistics, the idea of limits, etc.

Highlight means regular emphasis of some mathematical thoughts and ideas and combination with a large number of comprehensive drills, thus achieving the goal of flexible application. Emphasis is made based on introduction, which aims to play the functions of these mathematical ideas to their maximum. The ideas to be emphasized include the idea of number-figure combination, the idea of reduction, the idea of functions and equations, etc.

Xiaoying Man (1999) classified the representations of mathematical thoughts and methods from two dimensions "expression means" and "obviousness degrees" into four types: first, frank statement, with terminology introduced; second, expressing an idea with a long "bedding", which could be understood only in case of coherent contexts and also requires call and response; third, reminding, enlightening and comprehensive expressions; fourth, obscure and hinting wording, with implied meanings.

Lin Wang (2010) summarized the characters for representation of mathematical ideas and methods in the mathematics textbooks for primary schools published by Jiangsu Education Publishing House. On the one hand, the textbooks emphasize the integration of the most fundamental and most adaptive mathematical ideas and methods into the process for formation and development of knowledge according to the actual cognitive level of pupils and though proper explicit knowledge carriers, striving to make students receive edification of mathematical ideas and methods while obtaining explicit mathematics knowledge and have some initial intuition and awareness of mathematical ideas and methods. On the other hand, the books, by selecting some problems with rich mathematical connotations and potential mobility, enable students to gradually enrich their own experiences in mathematical ideas and methods and accumulate initial knowledge of the ideas and methods during the process of applying the knowledge acquired to analyze and solve these problems.

Zhiming Lu (2011) believes that the following mathematical ideas shall be involved in the mathematics textbooks for primary schools, including transformation, statistics, correspondence, symbolization, number-figure combination, classification and sets. Lu also summarized the infiltration means of mathematical ideas in textbooks, including thematic illustration and situational illustration, mathematical problems, simple language and wide angle of mathematics.

Hua Xing (2011) proposed his own idea on the representation of mathematical ideas in the wide angle of mathematics in textbooks. He believes that representations in textbooks should reflect the process of actual operation and exploration. On the one hand, the actual operation should be presented; on the other hand, the process of students' exploration should also be presented. Also, the supporting excise design should also reflect the application of mathematical ideas.

2.2.3 Comments

According to the research on the basic mathematical thoughts both in China and abroad, basic mathematical thoughts in mathematics curriculum and textbooks in primary and secondary schools, as well as relevant textbook studies, it is not hard to find that, despite some obvious progress and development in the current researches, there remain some problems unsolved, which are mainly reflected in the following aspects:

First, the current research on mathematical thoughts is hybrid and extensive and mixes the mathematical ideas and methods of all levels together. Although known as "mathematical ideas and methods", they actually have not reflected the beauty and value of methodology of mathematical ideas and methods but only mention the application of such ideas and methods, with part of the content limited to influencing students' future mathematics study.

Second, the current research is only limited to the content of mathematics textbooks but little of it involves students' sustainable development.

Third, the current textbook research mainly regards the textbook representation of mathematics knowledge and skills as object of study, including both customary linear arrangement and spiral arrangement (Fanzhe Kong, 2007).

Finally, the current research on textbooks does not cover the textbook representation with basic mathematical thoughts as the main line, without specific successful cases or relevant theoretical analysis.

In conclusion, textbook representation covers a relatively extensive view of angle for research, including macro, micro and comparative ones. However, there are not many researches on the representation of basic mathematical ideas and methods in textbooks. Zhiming Lu (2011) finds through research that, the mathematical ideas and methods in the textbooks for primary schools are represented by means of topic map (scene map), mathematical problems, language guide and wide angle of mathematics, etc.; the model thought is usually fully permeated; the induction method is usually used to find rules and deduction method normally used for rule application; and the infiltration of mathematical ideas and methods is usually characterized by sequencing and systematicness. The research made by Lu provides a good foundation for the study of the representations of mathematical thoughts in textbooks; however, it is only limited to mathematics for primary schools (reflecting the style of representation featuring strong infiltration), which is neither profound nor systematic.

Thus, this paper tries to make a further discussion on the concrete representations and value orientation of basic mathematical thought in textbooks.

3. Several Representations of Basic Mathematical Thoughts in Textbooks and Analysis of Their Typical Cases

Mathematical thoughts are an important content in mathematics. As the core, basic thoughts play an indispensable role in illustrating the mathematical thinking mode and characteristics of mathematical learning, which proves to be true for mathematics science itself and school mathematics. Thus, as an important component of mathematics curriculum, basic thoughts should be reflected in the curriculum. But their detailed representations are restricted by the basic regulations of curriculum, particularly those related to curriculum organization. According to the regulations as well as the characters of textbook compilation, basic thoughts can be presented in textbooks in an explicit way or by means of infiltration, as it is the original way that mathematical thoughts were implied in the content of mathematics.

3.1 Explicit type

The chapters and sections that mainly focus on mathematical thoughts are specially set up and inserted into mathematical knowledge and presented.

In China's *Mathematics Curriculum Standards for High Schools (Experiment)* (MOE of PRC, 2012), the theme "reasoning and proofs" is contained in the optional course 2-2. Accordingly, a special chapter in the high school textbooks of different versions are provided for the theme, including an introduction of the idea of (generalized) induction in a systematic manner.

For example:

Example 1: Inductive reasoning

In Optional 2-2 of the series of high school mathematics textbooks (version B) published by People's Education Press as per the Standards, Chapter 2 is an exposition of "Reasoning and proofs" in which Section 2.1

is devoted to an introduction to plausible reasoning and deductive reasoning. The following capture from that section is a typical case presenting the basic thought "inductive reasoning" (Fig. 13-1).

这种根据一类事物的部分对象具有某种性质，推出这类事物的所有对象都具有这种性质的推理，叫做归纳推理（简称归纳）. 归纳是从特殊到一般的过程.

下面，我们通过一个例子来得出归纳推理的一般步骤.

例如，当你看到这样的几个关系式

$$10=3+7,\ 20=3+17,\ 30=13+17$$

时，你会发现：3，7，13 和 17 这些数字都是奇质数，偶数 10，20 和 30 都可以表示为两个奇质数的和. 其他的偶数又怎么样呢？它们也有类似的性质吗？显然，第一个等于两个奇质数之和的偶数是

$$6=3+3,$$

接下去，还有

$$8=3+5,$$
$$10=3+7=5+5,$$
$$12=5+7,$$
$$14=3+11=7+7,$$
$$16=3+13=5+11.$$

这样下去总是对的吗？无论如何，所观察到的个别情况，可以启发我们提出一个一般性的命题：任何一个大于 4 的偶数都是两个质数之和❶.

归纳推理的一般步骤：

1. 通过观察个别情况发现某些相同性质；

2. 从已知的相同性质中推出一个明确表述的一般性命题（猜想）.

一般地，如果归纳的个别情况越多，越具有代表性，那么推广的一般性命题就越可能为真.

例 1　用推理的形式表示等差数列 1，3，5，…，$(2n-1)$，…的前 n 项和 S_n 的归纳过程.

解：对等差数列 1，3，5，…，$(2n-1)$，…的前 1，2，3，4，5，6 项和分别进行计算：

注

❶ 这个命题叫做哥德巴赫猜想，是由数学家哥德巴赫首先提出的. 简称"一加一"，简记为"1+1". 这个猜想至今没有得到证明. 我国数学家陈景润对证明此猜想作出了重大的阶段性成果，证明了"1+2"，即大偶数 N 都可表示为 $N=p_1+p_2$ 或 $N=p_1+p_2p_3$，其中 p_1，p_2，p_3 都是质数.

Fig. 13-1 "Inductive reasoning" in a Chinese textbook

Translation of Fig. 13-1

The way in which a property held by all the objects in a category is deduced from the property held by some objects in the category is called *inductive reasoning* (abbreviated, *induction*). Induction is a process from particularity to generality.

We use an example to illustrate the general steps in inductive reasoning.

Suppose that, for instance, you are given equalities

$$10 = 3 + 7,\ \ 20 = 3 + 17,\ \ 30 = 13 + 17.$$

You will find that 3, 7, 13, and 17 are all odd prime numbers, and even numbers 10, 20, and 30 can be expressed as sums of two odd primes. You may wonder how about other even numbers. Do they also have the similar property? It is easy to see that the smallest even number which equals the sum of two odd primes is

$$6 = 3 + 3.$$

Next, we have

$$\begin{aligned} 8 &= 3 + 5, \\ 10 &= 3 + 7 = 5 + 5, \\ 12 &= 5 + 7, \\ 14 &= 3 + 11 = 7 + 7, \\ 16 &= 3 + 13 = 5 + 11. \end{aligned}$$

Can this process always continue? Nevertheless, the observation of individual cases suggests a general statement: *Any even number bigger than 4 is a sum of two odd prime numbers.*

Remark: This statement is called Goldbach Conjecture, since it was first proposed by mathematician C. Goldbach. People often call it "one plus one conjecture", or simply noted it as "1+1". So far this conjecture has not been proved yet. A Chinese mathematician named Chen Jingrun contributed an important step to the conjecture by proving "1+2", i.e., proving any big even number N can be expressed as $N = p_1 + p_2p_3$, where p_1, p_2, and p_3 are prime numbers.

General steps of inductive reasoning are:

1. To discover some consistent properties through observing individual cases;

2. To infer a general explicit statement from the known consistent properties (conjecture).

Generally speaking, more individual cases are observed, better representativeness may be expected, and thus more possibly the general statement is true.

Example 1. Deduce the sum formula of the first n-terms for the arithmetic sequence $1, 3, 5, \ldots, (2n - 1), \ldots$ by inductive reasoning.

Solution: Calculate the sums of the first $1, 2, 3, 4, 5$, and 6 terms of the sequence $1, 3, 5, \ldots, (2n - 1)$:

This is an example of typical explicit type.

In addition, a reading material with title *Calculus and Idea of Limit* on pages 49 to 50 (Fig. 13-2A & Fig. 13-2B) in the Chapter 1 of the same volume of the textbook is also a typical explicit representation of basic thoughts.

Example 2: Calculus and idea of limit

数 学 文 化 阅读与欣赏

微积分与极限思想

微积分思想雏形出现得很早，公元前两百多年，希腊科学泰斗阿基米德就用圆的内接正多边形无限逼近圆周，求得圆周率 π 的近似值. 阿基米德还用原始的积分观念得到球体积公式 $V=\frac{4}{3}\pi r^3$. 他是用球体“薄片”的叠加，并运用杠杆原理得到球的“薄片”相当于球的外切圆柱挖去一个正圆锥相应的“薄片”，从而得出了球体积公式(图1-25).

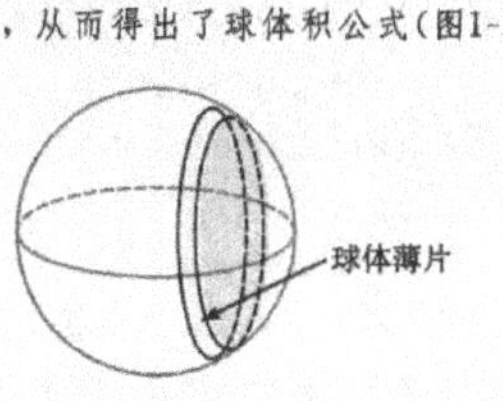

图 1-25

这是积分的雏形. 公元 5 世纪，中国数学家祖冲之、祖暅父子，得出了

“缘幂势既同，则积不容异”

的原理，现在我们称它为祖暅原理. 祖暅原理是说：

如果两物体在等高处的截面积都相等，则两者体积相同.

或者

如果两个平面的封闭图形在等高处截线段都相等，则两者面积相等（图 1-26）.

这也是一种朴素的积分思想.

微积分思想中的微分思想产生比较晚. 公元 16 世纪，伽利略发现了自由落体运动的规律：

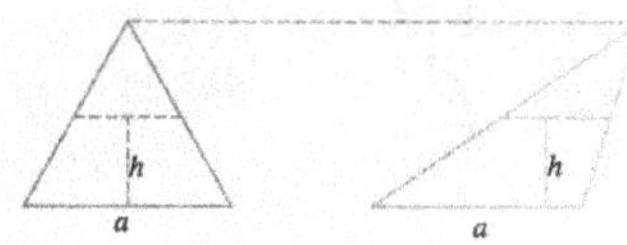

图 1-26

$$s=\frac{1}{2}gt^2.$$

这个变速运动的瞬时速度是 s 对 t 的平均变化率

$$\frac{s(t_0+\Delta t)-s(t_0)}{\Delta t}=\frac{1}{2}g\frac{(t_0+\Delta t)^2-t_0^2}{\Delta t}=gt_0+\frac{1}{2}g\Delta t.$$

当 $\Delta t\to 0$ 时，平均变化率趋向常数 gt_0. 这是导数概念的启蒙.

10 世纪初，笛卡儿建立了坐标系，使几何图形能够用函数表示. 函数也可以用几何图形直观地表示. 人们在探求曲线 $y=f(x)$ 的切线的时候，发现切线可由割线逼近而来. 当 $\Delta x\to 0$ 时，割线的斜率

$$\frac{\Delta y}{\Delta x}=\frac{f(x+\Delta x)-f(x)}{\Delta x}$$

所趋近的常数就是切线的斜率.

求变速运动的瞬时速度和求曲线切线的斜率是产生导数概念的直接动因.

下面，我们应用祖暅原理求球体的体积公式.

例 求球体积公式.

分析：设球体的半径为 r. 求解分两步：第一步由刘徽完成. 他创造了一个新立体图形，称为“牟合方盖”，并得到了球与“牟合

Fig. 13-2A “Calculus and the idea of limit” in a Chinese textbook

方盖”的体积比为$\frac{\pi}{4}$．如图1-27，在一个立方体内作两个互相垂直的内切圆柱，这两个圆柱体相交的部分，就是“牟合方盖”．“牟合方盖”恰好把立方体的内切球包含在内并且同它相切．第二步由祖暅完成．他求出了“牟合方盖”的体积，也就求出了相应球的体积．

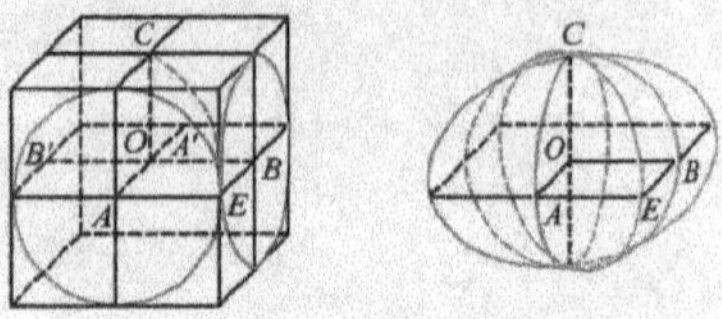

图 1-27

解：用一个水平截面去截“牟合方盖”及内切球，就得到一个圆（球的截面）和它的外切正方形（“牟合方盖”的截面）．设圆的半径为 a，则圆的面积等于 πa^2，外切正方形的面积等于 $4a^2$，于是得到在每一高度的水平截面截出的圆面与其外切正方形的面积比等于$\frac{\pi}{4}$．根据祖暅原理，球与“牟合方盖”的体积比也为$\frac{\pi}{4}$．

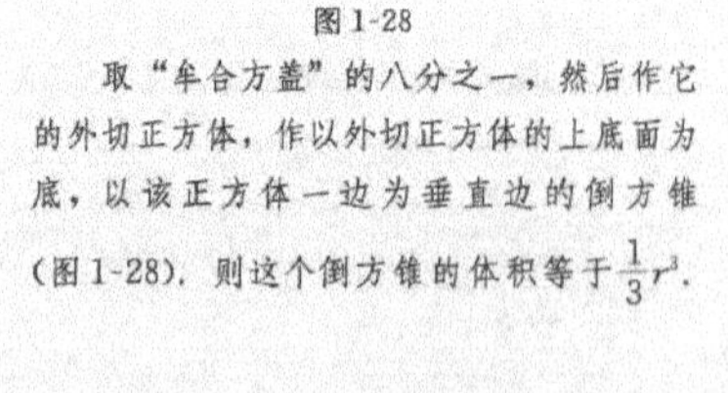

图 1-28

取“牟合方盖”的八分之一，然后作它的外切正方体，作以外切正方体的上底面为底，以该正方体一边为垂直边的倒方锥（图1-28)．则这个倒方锥的体积等于$\frac{1}{3}r^3$．把这个倒方锥与外切正方体放在同一平面上，用高度为 h 的水平截面，截倒方锥的截面面积为 h^2．截外切正方体的截面面积为 r^2，截“牟合方盖”的截面为正方形 $PQRD$，设 $AS = PQ = x$，则多边形 $ABCRQP$ 的面积为

$$S_{ABCD} - S_{PQRD} = r^2 - x^2 = h^2,$$

于是在任一相同的高处，外切正方体中挖去“牟合方盖”的部分的截面面积都与倒方锥的截面面积相等．根据祖暅原理，外切正方体中挖去“牟合方盖”的部分的体积与倒方锥的体积相等．因此八分之一“牟合方盖”的体积等于外切正方体的体积 r^3 减去倒方锥的体积$\frac{1}{3}r^3$．所以

$$V_{球} = \frac{\pi}{4}V_{牟合方盖} = \frac{\pi}{4}\times 8\times\left(r^3 - \frac{1}{3}r^3\right) = \frac{4}{3}\pi r^3.$$

在16世纪与17世纪之交，微积分观念得到了一定的发展与广泛的应用．在此基础上，两位科学巨匠牛顿和莱布尼茨分别独立地创造了微积分学．他们不仅提出了完整的微分与积分概念，建立了比较完整的学科体系，而且把微积分运算广泛应用于物理学、力学、几何学与函数分析，获得了巨大的成功．牛顿和莱布尼茨对微积分学最突出的贡献是，他们发现了微分与积分的内在联系，建立了微积分基本定理：如果 $F'(x) = f(x)$，$f(x)$是可积的，则

$$\int_a^b f(x)\mathrm{d}x = F(b) - F(a).$$

这样，困难的求积分问题就转化为求被积函数 $f(x)$的原函数$F(x)$的问题．这是微积分学历史上最重大的事件．

有了微积分基本定理，微积分学从繁琐的计算中解脱出来，并逐步被人们所应用、所接受．

Fig. 13-2B “Calculus and the idea of limit” in a Chinese textbook (cont.)

Translation of Fig. 13-2A & Fig. 13-2B

Mathematical Culture

Reading and Appreciation

Calculus and the Idea of Limit

The primitive idea of calculus can be traced back to ancient time. Around 200 B.C., a great Greece scientist Archimedes obtained approximate value of the circle-diameter ratio π by using regular polygons inscribed to a circle to infinitely approach the circle. Archimedes also deduced the volume formula $V = \frac{4}{3}\pi r^3$ for a ball through the raw idea of integral. He overlaid "thin slices" of a ball, identifying, by using the lever principle, a "thin slice" of the ball with a suitable "thin slice" of the circumscribed cylinder of the ball with a regular cone cut. The volume formula was obtained in this way (Fig. 1-25, omitted). This is the embryo of the idea of integral. In the 5th century AD, Chinese mathematicians Zu Chongzhi and his son Zu Geng arrived at a rule, called Zu Geng Principle nowadays. It stated as

As mi (sections) at equal shi (height) are always equal,
the ji (volume or area) cannot be different.

The principle implies that

If the areas of the sections of two solids at the same height
are always equal, their volumes are equal

and that (Fig. 1-26, omitted)

If the segments of two plane closed figures at the same height
are always equal, their areas are the same.

This is also a naïve idea of integral.

In calculus, the idea of differential appeared comparatively later. In the 16th century AD, G. Galileo found the law of freely-falling bodies

$$s = \frac{1}{2}gt^2.$$

The instantaneous velocity of this variable motion is the average rate of the change of s relative to the change of t:

$$\frac{s(t_0 + \Delta t) - s(t_0)}{\Delta t} = \frac{1}{2}g\frac{(t_0 + \Delta t)^2 - t_0^2}{\Delta t} = gt_0 + \frac{1}{2}g\Delta t.$$

When $\Delta t \to 0$, the average rate of change tends to the constant gt_0. This is the enlightenment of the notion of derivative.

In the early 10th century AD, R. Descartes invented the coordinate system. The geometric figures can be expressed as functions, and functions can be presented visually as geometric figures. When exploring tangent lines of a curve $y = f(x)$, people found they can be approximated by secant lines. When $\Delta x \to 0$, the slope

$$\frac{\Delta y}{\Delta x} = \frac{f(x + \Delta x) - f(x)}{\Delta x}$$

of a secant line tends to the slope of the tangent line.

The attempt to find the instantaneous velocity and that to find the slope of a tangent line directly lead to the notion of derivative.

Now we will derive the volume formula of a ball by using Zu Geng Principle.

Example. *Derive the volume formula for a ball.*

Analysis. Let the radius of the ball be r. The solution consists of two steps: Step 1 is done by Liu Hui, who created a new solid figure, called "mou he fang gai" ("two square umbrella", MHFG, for short), and obtained the volume ratio of the ball to the MHFG, which is $\frac{\pi}{4}$. As shown in Fig. 1-27 (omitted), inside a cubic box (with edge length $2r$) construct two inscribed cylinders perpendicular to each other. The intersection of the two cylinders is the MHFG. The MHFG has the ball inscribed. Step 2 is done by Zu Geng, who found the volume of the MHFG, and thus the volume of the ball.

Solution. Using a horizontal plane to cut the MHFG and the ball inside, one will obtain a circle (the section of the ball) and its circumscribed square (the section of the MHFG). If the radius of the circle is a, the area of the circle is πa^2, while the area of the square is $4a^2$. Thus, the ratio of the area of the section of the ball to the area of the section of the MHFG at any level is $\frac{\pi}{4}$. It follows from Zu Geng Principle that the volume ratio of the ball to the MHFG is $\frac{\pi}{4}$.

Cut down an octant of the MHFG. Construct its circumscribed cube, and construct an up-side-down pyramid with the up face of the cube as its bottom and one of the vertical edges of the cube as its edge (Fig. 1-28, omitted). The volume of the pyramid is $\frac{1}{3}r^3$. Put the up-side-down pyramid and the cube at the same level, and use a horizontal plane at height h to cut both of them. The section of the pyramid has area h^2, the section of the cube has area r^2, while the octant of the MHFG inside the cube has square $PQRD$ as its section. Suppose $AS = PQ = x$. Then the polygon $ABCRQH$ has area

$$S_{ABCD} - S_{PQRD} = r^2 - x^2 = h^2.$$

This means that at any height, the cube with the MHFG cut has the same section area as the up-side-down pyramid. Thanks to Zu Geng Principle, the cube with the MHFG cut has the same volume as the pyramid. Thus, one eighth of the volume of the MHFG equals r^3, the volume of the cube, minus $\frac{1}{3}r^3$, the volume of the pyramid. Therefore,

$$V_{\text{ball}} = \frac{\pi}{4}V_{\text{MHFG}} = \frac{\pi}{4} \times 8 \times \left(r^3 - \frac{1}{3}r^3\right) = \frac{4}{3}\pi r^3.$$

In the turn of the 16th century and the 17th century, the idea of differential and integral had been developed to a certain degree, and had been widely applied. With this basis, two great scientists Newton and Leibniz independently create calculus. They not only clearly proposed the notions differential and integral, built the scientific system in a good shape, but also successfully applied calculus in physics, mechanics, geometry, and functional analysis. The most outstanding contribution of I. Newton and G. Leibniz is the finding of the intrinsic relation of differential and integral, establishing the Fundamental Theorem of the Calculus: If $F'(x) = f(x)$, and if $f(x)$ is integrable, then

$$\int_a^b f(x)\mathrm{d}x = F(b) - F(a).$$

In this way, the difficult process of evaluating an integral is changed to the problem of finding the primitive function of the integrand. This is the most important event in the history of calculus.

With the Fundamental Theorem of the Calculus, calculus has been freed from complicated and cumbersome calculation. People gradually accept it and apply it.

Remarks: 1) Zu Geng Principle is known in the west as *Cavalieri Principle*, named after Italian mathematician B. F. Cavalieri by his book *Geometria indivisibilibus Continuorum Nova Quadam Ratione Promoat* published in 1635, about 1100 year after Zu Geng's discovery.

2) "Mou he fang gai" can also be called *bicylinder* by the way it generates. It is known in the west as Steinmetz solid, though the solid was known, even in the west, long before American mathematician C. P. Steinmetz (1865–1923) studied it.

Comments: Just as we have analyzed above, basic thoughts exist objectively and their representation can be in an explicit way but more in an implicit way.

Compared to explicit mathematics knowledge which could be observed and felt by students, the content of "plausible reasoning" in the high school mathematics curriculum seems difficult for some students. Just as some students commented, "Such content is method-based, without clear objects. We have no way to study it so as not be able to understand it. Once no proper example is given, we all will be at sea."

Indeed, as a component of mathematics curriculum, how can basic thoughts be properly presented in textbooks and benefit students' study more. There hasn't been a definite conclusion so far, even the idea has just aroused people's attention and the research remains in a primary stage. However, a basic principle lies in:

All representations shall bring more benefit to students' learning, acquisition and internalization.

In fact, in Example 1, the textbook writers present the idea of "inductive reasoning" in the text according to their own understandings of the curriculum standards by using a common representation method for explicit contents, such as "triangle" and "number sequence", which is a good attempt. Its advantage lies in that the explicitization of "inductive reasoning" is of some help for enhancing the role of the content in daily teaching.

While Example 2 adopts the method of "reading window", highlighting the ideas of calculus and limit through expanding horizons in order to refine the implied calculus thought featuring "replacing curves with straight lines, approaching accuracy from approximation" in sorting out the content of calculus and limit. The beauty of thinking may be felt by some students with strong sense of appreciation.

What ways should we adopt to explicitly present mathematical ideas, "knowledge interpretation type" as mentioned in Example 1 above (regarding an idea as a conclusive knowledge or skill and making systematic illustration according to its logic clue) or any other means? Is there any other means? It is worthy of consideration.

3.2 Semi-explicit type: integration into knowledge clues

According to the connotation of ideas, mathematical ideas are implied in the mathematics knowledge rather than exist in an explicit way.

Example 3: An infiltration of induction

In the content of "two-digit number multiplied by two-digit number", according to the normal process, applications follow the introduction of the algorithm. At this time, if the following is designed, we can not only achieve the goal of applying such algorithms but also present the induction thought better (Fanzhe Kong and Yingmei Cui, 2011).

Questions: Calculate, think and guess:

(1) Vertical calculation: $12 \times 11, 14 \times 11, 15 \times 11, 17 \times 11$.

(2) Through a careful observation of the factors and product of each arithmetic expression, what can you find?

From calculation

$$12 \times 11 = 132,$$
$$14 \times 11 = 154,$$
$$15 \times 11 = 165,$$
$$17 \times 11 = 187$$

Students may find that the product is a three-digit number, with the hundreds digit being 1 and the tens digit seemingly related to the two-digit numbers (possibly sum of the ones digits of the two two-digit numbers). The ones digit is the same with that of the first two-digit number.

For students, that is the result by observing the four mathematics problems above followed by inductive guessing, which is the first "**finding**".

However, is it right? It still needs verification. Based on the above, the textbook raises another question to students:

(3) Guess the results of "11×24" and "11×45". List an arithmetic expression to explain your guess.

According to the above "finding", students may have the following guess: the result of "24×11" should be "164" and that of "45×11" should be "195".

Then, students may find the following through vertical calculation:

$$24 \times 11 = 264,$$

$$45 \times 11 = 495.$$

They will find that the hundreds digit is not "1" anymore. So the guess should be modified. That is the second "**finding**" inducted upon first "finding" with calculation.

But how to correct it?

Based on the second finding, the textbook guides students to consider the fourth question:

(4) Observe the six equalities. What common characters do they have? What can you find? Verify your findings.

Students will soon find that "all of them are characterized by a two-digit number multiplied by 11" and "the hundreds number of the product is the same with the tens number of the two-digit number". Then, some students may get the following conclusion: "the product is to separate the two digits of the factor that is multiplied by 11 and putting the sum of the two digits in the middle" and further induce as follows: "pulling two digits towards both sides and adding their sum in the middle". That is the third "**finding**" upon observation.

In order to verify the "finding", the textbook gives the fifth question.

(5) Propose a two-digit number, say 36. Guess the product of 11×36 by using your above finding and verify your guess through vertical calculation.

Indeed! Our guess is correct.

Then, the textbook will give the sixth question when the conclusion for the third "finding" proves to be correct.

(6) Guess the result of "11×57" based on your "finding". Then, verify it using vertical calculation.

Students may find that, when the sum of "$5 + 7$" is larger than 10, the previous "finding" needs to be modified.

According to the cognitive conflict, students may be guided to induce as follows: The rule of "pulling two digits towards both sides and adding their sum in the middle" is applicable only when the sum of the tens digit and ones digit of the two-digit number is smaller than ten.

But, if the sum is ten or larger, how the "ten" shall be treated? Is it be expressed as "5127" or "627"?

Analyze the vertical calculation again. Students may find that the "12" in the tens digit should carry 1 towards the hundreds digit and get "627".

For the modified "finding", students may have a try by calculating "11×59" and "11×67", etc. The result is actually "$11 \times 59 = 649$" and "$11 \times 67 = 737$".

Finally, students modify the previous "finding" to "pulling two digits towards both sides, adding their sum in the middle, carry one towards the hundreds digit if the middle one is ten or larger".

That is the fourth "finding" obtained by students by observing several special cases, induction, guess, verification and modification.

Is the above process wordy? In fact, such wordy process is never unnecessary (Yingmei Cui and Fanzhe Kong, 2011). By making students experience such a process, its true intention is as follows: Let students experience the thinking process of induction and guess for several times during the teaching process for enhancing the basic knowledge and skills of "two-digit number multiplied by two-digit number", obtain the direct experience of

> *Case 1, Case 2, ..., Case n → induce a common rule and finding → guess → verify the finding and modify when necessary → get a general conclusion*",

experience a mathematician-like thinking process, appreciate the process of wisdom formation and enjoy the happiness of innovation.

Of course, for Question (6) above, where the sum of the two digits is 10 or larger, it seems more appropriate for some students to study after class and may demand higher than common requirements for all classmates in the whole class.

Example 4: Design ideas for multiple infiltrations of abstractions

Different levels of textbooks series may be designed according to different levels of abstractions. Thus, we should coordinate the mathematical abstraction and teaching of different levels in a reasonable manner and acquire the principle of persistence, synchronism, discipline and hierarchy.

First, the daily textbook-based mathematics teaching requires long-term persistence in the infiltration of mathematical abstraction. Students' mathematical abstraction ability will never be formed through a few simple abstraction processes but requires long-term persistence and step-by-step infiltration in classroom teaching. Never act with undue haste.

Second, the mathematical abstraction process needs to be permeated in a repeated manner in the classroom teaching of the same area and to keep the synchronism between different areas. For example, for "recognition of numbers" and "calculation of multiple-digit numbers" in the area of "Number and Algebra", small sticks or any other "counter" can be used to help students experience the mathematical abstraction process. "Recognition of numbers" is a mathematical abstraction process in a static level while "calculation of multiple-digit numbers" is in a dynamic level. Also, the repeated experiencing of mathematical abstraction process in the learning of mathematics knowledge of the same area is also helpful for students to understand the abstraction at a higher level.

Third, in the different areas of mathematics contents, the mathematical abstraction process will be realized via choosing proper methods in accordance to the characters of the respective areas so as to reflect their disciplinary attributes.

For example, while learning "understanding plane figures", we may use printed solid figures, edge-tracing ways, or projection to help students to experience the abstraction from solid figures to plane figures. Although this mathematical abstraction process differs from that in the study of calculation, the concept of "abstract objects are derived from the real world and are subject to human being's abstraction" remains the same.

Fourth, the mathematical abstraction process in classroom teaching should reflect the principle of hierarchy and needs to be improved step by step (Shengli Zhang and Fanzhe Kong, 2012). Each mathematics lesson should help students to experience a relatively complete mathematical abstraction process, which should not simply stay at one level but proceed step by step with all linked to each other. The mathematical abstraction process of different levels should be both connected and differentiated, which could be helpful for improving students' abstraction ability. At the same time, such a treatment provides a good opportunity for stratification teaching and also the possibility for abstraction thought in mathematics textbooks to be reflected in different layers and improved step by step.

Comments: Just as Louis Sullivan, a pioneer in the school of architecture of Chicago, said, "Form follows function". It is the same case with mathematics textbooks. No matter what representations the textbook adopts, its nature depends on its function and what goals are expected to be achieved through the learning of relevant content.

In Examples 3 and 4, in order to help students to understand the ideas of abstraction and induction, induction should be adopted in the process of basic knowledge and skill formation with students' current rich activity experience as a base. Students are invited to get involved actively in exploration. Driven by curiosity, student will discover new knowledge they expected naturally. By this, students can experience actively the process of induction and abstraction. So it is natural for them to accumulate direct experience and to sense the basic mathematical thoughts.

3.3 Deliberately implicit type (intentional infiltration)

A deliberately implicit method is often adopted to penetrate basic mathematical thought. Take the introduction of the concept "perimeter of figures" as an example (Fig. 13-3 − Fig. 13-5). The following methods are adopted from the mathematics textbook volume 1 for Grade 3 published by Beijing Normal University Press, page 44.

(1) Scene introduction

First, the textbook introduces the concept of perimeter through the scene of "an ant climbing on a leaf", set up an example familiar to students in their real life, and enables them to have a visual understanding of perimeter (Fig. 13-3).

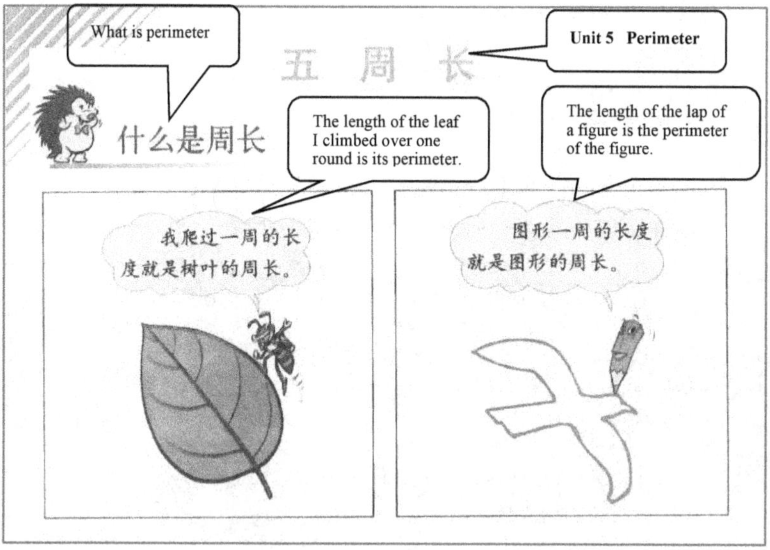

Fig. 13-3 Scene introduction of perimeter in a Chinese textbook

(2) Hands-on manipulation

By depicting the sidelines of leaf, desk top and mathematics book cover, students may have a profound personal feeling and initial experience of perimeter, gain a further understanding of the presentation of the concept and obtain an initial awareness of its significance (Fig. 13-4).

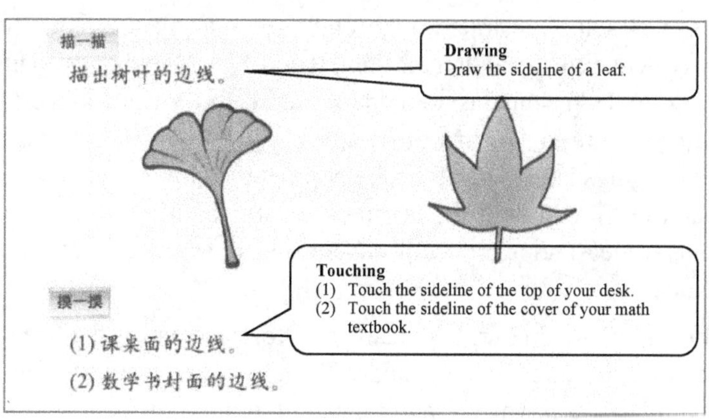

Fig. 13-4 Hands-on manipulation on perimeter

(3) Practical activities

实践活动

(1) 量一量你的腰围和头围，并与同伴说一说。

(2) 量一量一片树叶的周长，并与同伴说说你的方法。

Fig. 13-5 Practical activities on perimeter

Translation of Fig. 13-5

Practical Activities:

(1) Measure your waistline and the circumference of your head, then tell your partners.

(2) Measure the perimeter of a leaf, and tell your partners about your method.

Practical activities could help students to experience the meaning of perimeter again. Most important of all, through the study methods characterized by independence, cooperation and exploration, students could choose their favorite measuring methods and obtain a further awareness and understanding of perimeter. The activities mainly aim to help students to touch the concept by themselves and feel the great practical significance.

In a global view, the textbook version presents the perimeter in mathematics in an abstract way through irregular figures in our life, which is a typical representation of "induction-type" textbooks. This action expects to help students to understand the following aspects through learning: There are plenty of irregular figures in our life and not only regular figures have perimeter; the perimeter of figures exists everywhere in our life. In addition, the textbook creates multiple natural scenes for perimeter measurement to help students feel the diversity of the measuring methods.

In our view, this arrangement tries to systematize the scattered life experience accumulated in students' daily life (even cartoon games), which relates to perimeter, and then get the induced and abstract concept of the perimeter for general figures. Furthermore, the concept of "perimeter" is not instilled by the editor but is abstracted from the existing life experience and mathematics activity experience of students with the aid of the textbook (Fanzhe Kong, 2009). It is a typical idea and expression method of constructionism.

3.4 Unintentional implicit type (infiltration type)

In the preparation of traditional mathematics textbooks for primary schools, the scheme of "real situation(s) and/or example(s) — induction and summarization — concept formation or definition" is used for the introduction of almost every concept, which aims to introduce new concepts by means of induction and then put the induction aside.

For example, the primary school mathematics textbook volume 1 for Grade 3 published by People' Education Press presents the concept of "perimeter" on page 85 as follows (Fig. 13-6):

First, the textbook presents eight figures, including regular and irregular, figures in life and those abstracted from life. The concept of perimeter is introduced through the simple scene composed of eight figures, or through the sentence "the length of the lap of a closed figure is its perimeter" attached to the figures.

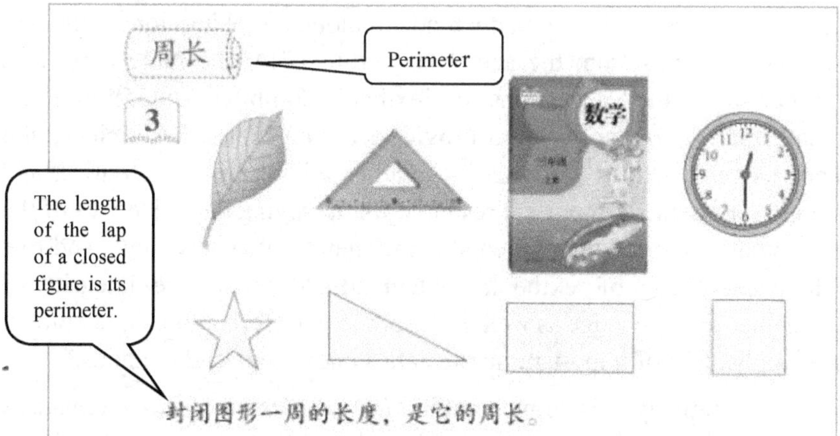

Fig. 13-6 Visual introduction of perimeter in a Chinese textbook

Comparatively, other versions of textbooks adopt the following method: perimeter is introduced according to the life scene and the exact concept of perimeter is not given; however, the concept of perimeter is implied on the scenes. It is proposed that the concept is introduced from normal irregular figures to emphasize the formation process of the perimeter. Especially, the concept of "perimeter" in the textbooks by Beijing Normal University Press emphasizes more on the process of induction while the treatment of the other versions is some kind of deduction-type concept presentation.

That is a typical "unintentional implicit type".

Comments: In *Mathematics Curriculum Standards for Compulsory Education (2011 Edition)* (MOE of PRC, 2012), it is explicitly specified in a macro way in the "Overall Objectives" (p. 8) that "through the mathematics study in the stage of compulsory education, students can … obtain basic knowledge, basic skills, basic thoughts and basic experiences necessary for adapting to social life and further development …", and the "curriculum design idea" (pp. 3–4) (naturally including the textbook compiling) is "the design of mathematics curriculum in the stage of compulsory education, …". "While presenting the mathematics result as knowledge and skills, it emphasizes the existing experience of students and enable students to experience the process of abstracting mathematical problems, constructing a mathematical model, seeking for result and solving problems from the actual background." At the same time, it is pointed out in the Suggestions for Textbook Compiling (pp. 59–60) that "mathematics textbook should provide learning topics, basic clues, and knowledge structures for students' mathematics learning activities, and should represent an important resource for achieving the objectives of the mathematics curriculum and conducting mathematics teaching", and that "the presentation of textbook content should reflect the integrity of mathematics knowledge as well as the process for formation, development and application of important mathematics knowledge and methods."

Comparatively, it is pointed out in *Mathematics Curriculum Standards for High Schools (Experiment)* (MOE of PRC, 2003) that based on basic thoughts, "mathematics education occupies a special place in school education. It can help students to understand basic knowledge, skills and ideas of mathematics, enables them to articulate and think clearly in an orderly manner, cultivate their practical attitude and persistent spirit and teach them to solve problems and learn about the world using the mathematical thinking mode." "High school mathematics curriculum should pay attention to the improvement of students' thinking ability in mathematics, which is one of the basic objectives of mathematics education. While studying mathematics and applying mathematics to solve problems, we continuously experience the thinking processes including intuitive perception, observation and discovery, induction and analogy, spatial imagination, abstraction and summarization, symbol representation, computation and evaluation, data processing, deductive proof, reflection and construction". These processes are concrete manifestation of mathematical thinking abilities, which can help students to make considerations and judgments on the mathematics models implied in the objective matters.

Thus, it can be inferred that the core ideas of curriculum representation pertaining to basic thoughts (textbook representation), *Mathematics Curriculum Standards for Compulsory Education (2011 Edition)* (MOE of PRC, 2012) and *High School Mathematics Curriculum Standards (Experiment)* (MOE of PRC, 2003) lie in:

First, the objective for basic thoughts should be achieved at the same time with basic knowledge and basic skills.

Second, the mathematical thoughts (including basic thought), methods and thinking abilities (even common mathematical thinking modes) are important process targets for the mathematics curriculum, which should be reflected in activities and processes more.

Third, the representation of important basic thoughts should reflect the characters of integrity and process, and the materials for presentation should approach the actual conditions of students.

The core of the above requirements is to help students better "master necessary basic knowledge and basic skills", "cultivate their abstract mathematical thinking and reasoning abilities", "cultivate their innovation awareness and practice ability" and "promote the development of students in terms of emotion, attitude and view of value" so as to lay a solid foundation for their future living, work and study.

It should be noted that the school mathematics curriculum standards in China raises macro but relatively clear requirements for the objectives of basic thought; however, for how to specifically present basic thought, only some macro requirements (suggestions, if more accurately) including synchronization of result and process are advised.

For basic mathematical thoughts, in what ways will textbooks present in order to fully reflect the core objectives of the mathematics curriculum — help students to think over problems from the perspective of mathematics? Regarding this the editors of current different versions of mathematics textbooks usually have their own understandings. Some only regard basic thought as "decoration" of mathematics textbooks; some are more than willing but lacking necessary textbook compiling skills; some are trying to make exploration. The focus and difficulties are characterized by the following:

(1) Will we make basic thought independent of other curriculum content or integrated into basic knowledge and basic skills for textbook representations? Both advantages and disadvantages are unclear.

(2) What is the relationship between basic thoughts and basic knowledge, basic skills and basic experience of mathematics in the organizational structure of mathematics curriculum? Acting as a supporting role for the leading role or in parallel with those factors?

As we have noted, in the cases above, Examples 1 and 2 are the methods adopted by the current mathematics textbooks for primary and secondary schools; Example 4 is the method adopted by some versions of the current mathematics textbooks for primary and secondary schools and Example 3 is the representation method tried in daily teaching, which proves to be quite successful. Also, the existing research (Shukun Lu, 2009) shows that the methods adopted in Examples 3 and 4 could be realized successfully in the teaching of mathematics concepts.

4. Analysis on Value Orientation of the Representation of Basic Thought in Textbooks

As an important content of mathematics science, basic thought could be represented in a concrete way in the mathematics curriculum for primary and secondary schools. In fact, different representations closely relate to the value orientation of curriculum designers and textbook writers, "In order to construct an effective mathematic textbook ... the first question we need to answer is what basic value is inherited in the mathematics textbook and what rationale is based for the determination" (Dingqiang Zhang, 2011).

4.1 Knowledge-based orientation

For long, deduction is mainly adopted in the content arrangement of primary school mathematics textbooks of China, i.e., from general to special. It is the same case with the classroom teaching for primary school mathematics. However, the special content is normally arranged in the phase of "extended learning" and only those students who study better and fast could benefit from such "dessert". Thus, it seems quite difficult for ordinary students to experience the process from special to general.

The new mathematics knowledge is derived from visual induction and mathematics development relies on reasoning, whose main methods are induction and deduction. Inductive reasoning represents an important thinking mode for human beings to acquire knowledge.

Therefore, in order to cultivate students' inductive thinking, we need to present some content by means of induction in the mathematics textbooks for primary schools. For example, as for "two-digit number multiplied by two-digit number", the focus of the teaching is "to understand the algorithm of two-digit number multiplied by two-digit number and understand its vertical calculation". The "easy calculation" mentioned in Example 1 above is in fact "two-digit number multiplied by two-digit number". If "a two-digit number is multiplied by 11", we take "12×11" as an example and "14×11", "15×11", "17×11", "24×11", "45×11", "36×11", "59×11" and "67×11", etc. as exercises for induction, then use a common arithmetic expression, such as "24×12" and "33×23", to test if students "understand the algorithm of two-digit number multiplied by two-digit number and understand its vertical calculation". Is it more appropriate?

At least, such design will enable students to repeatedly experience the inductive thinking in the operations including experience, observation, discovery, guess and verification.

It should be noted that, “easy calculation” is only a carrier for provision of inductive thinking process. Not all students are required to master such method. After all, its arithmetic expression is normally special and needs to be based on some prerequisites.

Analyzed from the perspective of value orientation for knowledge, deduction-type textbooks mainly adopt the knowledge-based value orientation, aim to systematically present explicit knowledge and skills, and reflect the function of textbooks in passing on the existing knowledge of human beings. Therefore, the systematic integrity of knowledge and the clarity and preciseness of content have become the goals pursued by the textbooks.

As one of the important carriers for students’ learning, the special function of cultivating talents and especially improving students’ mathematical literacy, is the core in curriculum design and textbook preparation. Guided by the knowledge-based textbook preparation concept, mathematics study aims to acquire systematic and complete mathematics knowledge. The pursuit of logic and rigorous knowledge system has become the central target for textbook design, while whether students are able to acquire become a secondary factor.

4.2 Thought-and-method-based orientations

The representation of the thought-and-method-based textbooks aims to highlight the thinking methods behind explicit content and display the beauty of ideas and power of methods.

The relevant text of “inductive reasoning” in the current high school mathematics textbook is a typical case with this value orientation.

The textbook designer holding this value orientation tends to believe that a textbook is not only a place for students to acquire mathematics knowledge but also “a place to enrich their mathematics wisdom and a proper stage for students to improve mathematical literacy, understand mathematics expressions, expand mathematical thinking and apply mathematics knowledge to analyze and solve problems" (Dingqiang Zhang, 2011). From the related publications in China, one can take the *Horizontal Infiltration of Mathematics Wisdom: Mathematical Thoughts and Methods* (Youliang Zha and Yiyu Li, 1990) as example. This book regards the extensive infiltration of mathematical thoughts and methods into natural and social science as well as important role of these thoughts and methods in the development of mathematics science as main lines and fully interpret the beauty of mathematical thoughts and methods using plenty of classical cases (for example, Isaac Newton established classical mechanics using the thought of axiomatization and Charles Darwin borrowed from axiomatization to create the theory of evolution). Naturally, thoughts and methods play the leading roles in these great innovations.

4.3 Simultaneous emphasis of knowledge, skills, thoughts and methods

Just as stated in *Mathematics Curriculum Standards for Compulsory Education (2011 Edition)* (MOE of PRC, 2012), "Curriculum content... should not only involve the result of mathematics but also the formation process of such result as well as mathematical thoughts and methods implied". With the aim of fully presenting basic knowledge, skills, thoughts and activity experiences, the mathematics textbook should encourage simultaneous emphasis of knowledge, skills, thoughts and methods.

From the perspective of historical development, *Mathematics Curriculum Standards for Compulsory Education (Experiment)* (MOE of PRC, 2001) propose to reproduce the process for formation and development of mathematics knowledge and skills, which hasn't promoted thoughts and methods to the same level with knowledge and skills. Comparatively, the Standards of 2011 Edition place the basic thoughts on the same level with basic knowledge and skills, and propose the simultaneous emphasis of basic knowledge, skills, thoughts and activity experiences.

Thereby, the preparation of mathematics textbooks must reflect basic thoughts and basic activity experiences effectively, and illustrate the simultaneous representation of basic knowledge and basic thoughts. Regarding how to realize it, shall it be implemented by an explicit or implicit means?

In our views, as for primary and secondary school mathematics, relatively ideal methods include semi-implicit type and intentional infiltration type, with the aid of an explicit type.

The core reason why we insist on the opinions above lies in that, as an important carrier of curriculum implementation, the fundamental purpose of mathematics textbooks is to help students to achieve the planned target of mathematics curriculum, realize personal development in mathematics and lay a solid foundation for further study, work and living. The minimum requirements are to learn how to approach problems from the perspective of mathematics, how to increase one's happiness index and expand one's living space through mathematics study so as to create more value to feed oneself, help others and serve the society. Thus, the fundamental criteria for textbook presentation of basic thoughts is how to help students build their understanding of mathematics in a more favorable way and grasp basic thoughts quicker and better.

At the same time, basic thoughts as the special mathematics contents hidden in specific contents, can be perceived well by learners if learners have reached a certain level or stage. There are two difficulties in learning:

First, it must be based on the prerequisite that learners acquired specific contents which basic thoughts are attached; and

Second, learners' understanding of the basic thoughts requires certain thinking level together with concrete scenes and materials.

In order to resolve the two difficulties above, in the preparation of textbooks, the curriculum designers are required to combine the actual situations of basic thoughts, choose proper materials at a proper time, regard the effective facilitation of students to perceive basic thoughts as target so as to represent basic thoughts in mathematics textbooks better.

In the view of this chapter, the absolutely correct, unified and standard textbook representations for basic mathematical thoughts haven't been found yet, while the judgment of "not existing" seems lacking evidences.

5. Conclusion

As one important part of mathematics textbooks, basic thoughts should be regarded as an important element and should be and must be represented in a better way.

In textbooks, the basic channels for representing basic thoughts include:

First, to set up specific chapters and/or sections to illustrate basic thoughts. The representation of "inductive reasoning" in the mathematics curriculum for high schools is such a case;

Second, to integrate basic thoughts with mathematical concepts, formulas and rules, demonstrating the abstraction of concepts, induction and discovery of formulas and rules in a detailed manner so as to reproduce mathematicians' original thinking processes. Such a way can not only represent new concepts, new formulas, new rules and new conclusions well, but also return the basic thoughts in knowledge generation process to the original, represent them in textbooks, and realize the development of the basic knowledge, basic skills, basic thoughts and basic activity experiences at the same time.

Third, to highlight the ideas of mathematical abstraction and modeling in the "mathematization" process, or more specifically, in the process of "mathematization of practical problems", and to highlight the ideas of mathematical reasoning (especially, the ideas of inductions and deductive proofs) in the process of "internal regularization of mathematics". In particular, axiomatization thought is represented in a systematic manner for some content (e.g., plane geometry) and the idea of mathematical modeling is highlighted in the "contextualization of mathematical content".

Fourth, to illustrate different mathematical ideas associated with different mathematical contents. For example, in the compulsory education curriculum, algebraic abstraction and algebraic reasoning (i.e., algebraic induction and algebraic logical proof) are emphasized with the domain "numbers and algebra"; geometric abstraction and geometric reasoning (i.e., geometric induction and geometric proofs) and axiomatization are emphasized in the domain "graphs and geometry"; the ideas of random awareness and statistical inference are focused in the domain "statistics and probabilities"; and the idea of mathematical modeling is highlighted in the domain "integration and practice".

Whether the basic thoughts could be represented consciously in mathematics textbooks basically depends on designers' the value orientation towards the mathematics curriculum content, either knowledge oriented or thoughts-and-methods oriented, or simultaneous oriented. There has no sufficient theories or practical evidences to make a convincing judgment on the existence of mathematics textbook representations universal to all mathematical basic thoughts.

References

[1] Aleksandrov, A. D. *et al.* (1956). *Mathematics, Its Essence, Methods and Role*. Moscow: USSR Academic of Sciences. Chinese version translated by Xiaoli Sun. Beijing: Science Press, 2001.

[2] Bao, Jiansheng (2002). A comparative study of the composite difficulties of Chinese and British middle school mathematics curriculum [In Chinese]. Doctoral dissertation. Shanghai: East China Normal University.

[3] Bell, E. T. (1986). *Men of Mathematics*. New York: Simon & Schuster.

[4] Cai, Shanghe (1997). Mathematical thinking and mathematical method [In Chinese]. *Secondary School Mathematics*, (9): 1–4.

[5] Charalambous, C. Y., Delaney, S., Hsu, H.-Y. and Mesa, V. (2010). A comparative analysis of the addition and subtraction of factions in textbooks from three countries. *Mathematical Thinking and Learning*, 12(2): 117–151.

[6] Chen, Xiangbin (2010). Mathematical thinking in mathematics teaching in primary schools [In Chinese]. *Curriculum, Teaching Material and Method*, 30(7): 37–42.

[7] Colbert-Lewis, S. C. D. (2005). *An Analysis of the Presentation of Women in Five High School World History Textbooks*. Charlottesville, VA: University of Virginia.

[8] Cui, Yingmei and Kong, Fanzhe. (2011). An investigation of "Redundant Links" in classroom teaching [In Chinese]. *Shanghai Research on Education*, (8): 63–66.

[9] Dalmas, A. (1956). *Evariste Galois, Révolutionaire et Géométre*. Paris: Fasquelle. Chinese version translated by Xundai Shao. Beijing: Commercial Press (1981).

[10] Ding, Shisun. (1956). Development of mathematical thoughts [In Chinese]. *Journal of Dialectics of Nature*, (6): 15–16.

[11] Elliot, D. L., Woodward, A. and Rehage, K. J. (1990). *Textbooks and Schooling in the United States*. Chicago: University of Chicago Press.

[12] Fan, Lianghuo, Zhu, Yan and Miao, Zhenzhen (2013). Textbook research in mathematics education: development status and directions. *ZDM: The International Journal on Mathematical Education*, 45(5): 633–646.

[13] Fang, Yongxiang (2012). Study on the representation of ideological and moral education textbooks [In Chinese]. *Teaching Reference of Middle School Politics*, (9): 26–28.

[14] Freudenthal, H. (1991). *Revisiting Mathematics Education: China Lectures*. The Nederland: Kluwer Academic Publishers. Chinese version translated by Changping Chen and Ruifen Tang. Shanghai: Shanghai Education Press, 1995.

[15] Fu, Shixia (1983). *A Philosophical Exploration of the Frontier of Science* [In Chinese]. Shenyang: Liaoning People's Press.

[16] Gao, Chunmin (2009). A comparative study of the representation on plausible reasoning in middle school textbooks: Cases of PEP edition, BNU edition, ECNU edition, and SEP edition [In Chinese]. Master thesis. Changchun: Northeast Normal University.

[17] Garcia, T. D. (1985). An analysis of earth science textbooks for presentation of aspects of scientific literacy. Ed. D. Dissertation. Houston: University of Houston.

[18] Gu, Pei (2012). Development of "Four Basics" from "Two Basics" in basic mathematics education [In Chinese]. *Journal of Mathematics Education*, 21(1): 14–16.

[19] Hilbert, D. (1900). Mathematical problems. *Göttinger Nachrichten*, pp. 253–297. English version translated by M. W. Newson, *Bulletin of the American Mathematical Society*, 1902. pp. 437–479. Reprinted in *Proceedings of Symposia in Pure Mathematics*, vol. 28. Providence R. I.: American Mathematical Society, 1976, pp. 1–34.

[20] Hölder, O. L. (1924). *Die Mathematische Methode*. Berlin: Springer-Verlag.

[21] Hong, Dingguo (ed.) (1988). *Puzzles and Prospects on the Frontier of Scienc*e [In Chinese]. Changsha: Human Science and Technology Press.

[22] Huang, Weiliang (2009). Mathematical thoughts and methods: Teaching material processing and enlightenment of teaching [In Chinese]. *Research and Review on Education (Primary School Edition)*, (2): 43–45.

[23] Kim, R. Y. (2009). Text + book = textbook? Development of a conceptual framework for non-textual elements in middle school mathematics textbooks. Doctoral dissertation. East Lansing, MI: Michigan State University.

[24] Kline, M. (1972). *Mathematical Thought from Ancient to Modern Times* (Vol. I, II, III, IV). UK: Oxford University. Chinese version translated by Lijing Zhang, Jinyan Zhang and Yihan Jiang. Shanghai: Shanghai Scientific & Technical Publishers, 1979.

[25] Kong, Fanzhe (2007). Research on the spiral ascending design of curriculum and arrangement of teaching materials on the basis of new curriculum of basic education [In Chinese]. *Educational Research,* (5): 62–68.

[26] Kong, Fanzhe (2008a). *Exploration of the Research Methods for Textbook Quality — A Case Study of Textbooks of Mathematics as per Curriculum Standards for Compulsory Education (Experimental)* [In Chinese]. Beijing: People's Education Press, 19–21.

[27] Kong, Fanzhe (2008b). Preliminary exploration of the tradition and development of Chinese mathematics education: A perspective of textbooks [In Chinese]. *Shuxue Tongbao* (*Bulletin des Sciences Mathematics*), 47(04): 37–42.

[28] Kong, Fanzhe (2009). A comparison of textbooks of different versions and inspiration to the implementation of curriculum [In Chinese]. *Research and Review on Education (Primary School Education and Teaching)*, (4): 39–43.

[29] Kong, Fanzhe (2012). Implication, function and infiltration of basic thoughts [In Chinese]. *Fujian Education*, (10): 44–46.

[30] Kong, Fanzhe and Cui, Yingmei (2011). The subject sense behind the wit calculation — A new look at the teaching of knowledge and skills [In Chinese]. *People's Education (Semimonthly)*, (11): 44–46.

[31] Kong, Fanzhe and Shi, Ningzhong (2009). Process analysis and level measurement of textbook use by teacher. *Frontiers of Education in China*, 4(2): 268–285.

[32] Kong, Fanzhe and Shi, Ningzhong (2012). Interpretation of mathematics curriculum standards for compulsory education (version 2011) [In Chinese]. *Fujian Education*, (6): 30–33.

[33] Lakatos, I. (1976). *Proofs and Refutations*. Cambridge: Cambridge University Press.

[34] Lan, Shuen De (2004). Analysis of local masters and doctors theses on textbooks in the last 20 years [In Chinese]. *Journal of the National Institute for Compilation and Translation*, 32(4): 2–25.

[35] Li, Di (2001). Comments on Kline's *Mathematical Thought from Ancient to Modern Times* [In Chinese]. *Studies in College Mathematics*, 4(3): 44–48.

[36] Li, Jun (2004). Analysis of the representation mode of the new-curriculum-standards-based chemistry experiment textbooks (People's Education Press edition) [In Chinese]. *Teaching Reference of Middle School Chemistry*, (7): 29–31.

[37] Li, Xingmin *et al.* (eds.) (1988). *Greatest Musikalität in Ideological Sphere: Case Study of Scientific Finding* [In Chinese]. Changsha: Human Science and Technology Press.

[38] Liu, Fengpu (1980). *An Analysis of Mathematics Dialectical Contents* [In Chinese]. Beijing: People's Educational Press.

[39] Liu, Yun & Zhu, Weizong (2012). An analysis of the representation of mathematics history in high school compulsory mathematics textbooks [In Chinese]. *Journal of Mathematics Education*, 21(02): 86–89.

[40] Lu, Shukun (2006). A Study of the representation of figures and transformations in experimental textbooks (Grade 7 – Grade 9) as per the curriculum standards for compulsory education [In Chinese]. Master thesis. Changchun: Northeast Normal University.

[41] Lu, Zhiming (2011). A study of permeation points of mathematical thoughts in new primary school mathematics textbooks [In Chinese]. Master thesis. Guangzhou: Guangzhou University.

[42] Man, Xiaoying (1999). Discussion and exploration of mathematical thoughts and teaching in middle schools [In Chinese]. *Journal of Teaching and Management*, (5): 46–47.

[43] Marion, P. C. (1969). The presentation of mythology in literature textbook series for the primary and secondary grades. Doctoral dissertation. Chicago: University of Chicago.

[44] Marx, K. (1881). Yanovskaya Sofya, ed. (1983). *Mathematical Manuscripts of Karl Marx*. London: New Park Publications Ltd.

[45] McConnell, W. R. (1920). A criticism of the presentation of an industry in the present type of geography textbook, with special reference to the cotton industry. *The Journal of geography*, 19(7), 260–266.

[46] Meschkowski, H. (1961). *Denkweisen Großer Mathematiker — Ein Weg zur Geschichte der Mathematik.* Berlin: Friedr Vieweg and Sohn Braunschweig. English version: *Ways of Thought of Great mathematicians — An Approach to the History of Mathematics.* Translated by J. Dyer-Bennett, published by Holden-Day, San Francisco, CA, 1964.

[47] MOE of PRC (Ministry of Education of People's Republic of China) (2001). *Mathematics Curriculum Standards for Compulsory Education (Experimental)* [In Chinese]. Beijing: Beijing Normal University Press.

[48] MOE of PRC (Ministry of Education of People's Republic of China) (2003). *Mathematics Curriculum Standards for High Schools (Experimental)* [In Chinese]. Beijing: People's Education Press.

[49] MOE of PRC (Ministry of Education of People's Republic of China) (2012). *Mathematics Curriculum Standards for Compulsory Education (2011 edition)* [In Chinese]. Beijing: Beijing Normal University Press.

[50] Nowakowska, H. Z. (1997). The presentation of gender images in reading textbooks for early elementary school children. Doctoral dissertation. Stockton, CA: University of the Pacific.

[51] Pei, Bo, Hu, Yanyan and Ji, Chunyang (2013). A study on the representation of illustrations in chemistry textbooks from a perspective of psychology [In Chinese]. *Journal of Chemical Education*, (3): 12–15.

[52] Pepin, B. and Haggarty, L. (2001). Mathematics textbooks and their use in English, French and German classrooms: A way to understand teaching and learning cultures. *Zentralblatt für Didaktik der Mathematik*, 33(5): 158–175.

[53] Poincaré, H. (1902–1908). *The Foundations of Science*. New York: Science Press; reprinted in 1921. This book includes the English translations of *Science and Hypothesis* (1902), *The Value of Science* (1905), and *Science and Method* (1908).

[54] Pólya, G. (1954). *Mathematics and Plausible Reasoning* (Vol. I & II). Princeton: Princeton University Press. Chinese version translated by Xincan Li. Beijing: Science Press, 1984.

[55] Putnam, H. (1975). *Mathematics, Matter and Method* (2nd ed.). *Philosophical Papers, vol. 1*. Cambridge: Cambridge University Press.

[56] Qu, Lixue (1994). Discussion on mathematical thinking [In Chinese]. *Journal of the Mathematics Education*, 3(1): 32–35.

[57] Reid, C. (1970). *Hilbert*. New York: Springer-Verlag.

[58] Shi, Ningzhong (2008–2012). *Introduction to mathematical thoughts* (Vol. I – IV) [In Chinese]. Changchun: Northeast Normal University Press.

[59] Shi, Ningzhong (2011). On basic thoughts of mathematics [In Chinese]. *Journal of Mathematics Education*, 20(4): 1.

[60] Shi, Ningzhong, Kong, Fanzhe and Li, Shuwen (2005). A model of curriculum difficulty: Comparisons of difficulty levels of geometry curriculum in China's compulsory education [In Chinese]. *Journal of Northeast Normal University (Philosophy and Social Sciences Edition)*, (6): 151–155.

[61] Song, Lele (2012). An analysis of the representation of mathematics culture in primary school mathematics textbooks: A case of BNU Edition [In Chinese]. Master thesis. Changchun: Northeast Normal University.

[62] Sun, Qing. (2009). An analysis of the representation of chemistry history in Chinese high school chemistry textbooks from a perspective of the nature of science [In Chinese]. Master thesis. Changchun: Northeast Normal University.

[63] Sun, Xiaotian (2012). Some understanding and thinking on the basic mathematical thoughts. *Jiangsu Education*, 2012(12): 6–8.

[64] Tan, Yihuan (1976). *Zu Chongzhi* [In Chinese]. Shanghai: Shanghai People's Press.

[65] Wang, Guangming and Zhang, Wengui (1997). Study on secondary school mathematical thoughts and methods and their teaching [In Chinese]. *Research of Mathematic Teaching and Learning*, (1): 6–8.

[66] Wang, Jiaoguo and Chen, Xianyan (2013). A comparison of content representations of Chinese and American high school physics textbooks — A Case of "Electrics" [In Chinese]. *Journal of Physics Teaching*, 31(2): 1–5.

[67] Wang, Lin (2010). Practice and thinking on permeation of mathematical thoughts in primary schools [In Chinese]. *Curriculum, Teaching Material and Method*, 30(9): 53–58.

[68] Wang, Yanwen and Wang Guangming (1996). Characteristics and significance of mathematical thinking in secondary schools [In Chinese]. *Journal of Taian Teachers College*, (6): 41–43.

[69] Weinberg, S. L. (1998). Fractions and division: Researcher conceptualizations, textbook presentations, and student performances. Doctoral dissertation. Chicago: The University of Chicago.

[70] Wu, Wenwen (2009). A comparative study on the presentation of plausible reasoning in high school textbooks: Cases of PEP-A edition, PEP-B edition, BNU edition, and SEP edition [In Chinese]. Master thesis. Changchun: Northeast Normal University.

[71] Xie, Enze (ed.). (1987). *An Introduction to Potential Science* [In Chinese]. Beijing: Guangming Daily Press.

[72] Xie, Enze (1998). *A Collection of Science Disasters* [In Chinese]. Changsha: Human Science and Technology Press.

[73] Xie, Enze and Xu, Benshun (1995). (Eds). *Mathematical Methods* [In Chinese]. Jinan: Shandong Education Press, pp.13–20.

[74] Xie, Enze and Zhao, Shuzhi (1987). *The Vertical and Horizontal Analysis of Mathematical Thinking and Methods* [In Chinese]. Beijing: Science Press.

[75] Xiong, Hua (2011). Strengthening the permeation of mathematical thoughts and developing mathematical thinking ability [In Chinese]. *Curriculum, Teaching Material and Method*, 31(9): 61–66.

[76] Xu, Lizhi (1980). *Introduction to Mathematical Methodology* [In Chinese]. Shenyang: Liaoning People's Publishing House.

[77] Xu, Lizhi (1985). The concept of mathematical abstractness and its analytical method. *Journal of Mathematical Research and Exposition*, (2): 133–140.

[78] Xu, Lizhi (2000). *Topics in Mathematical Methodology* [In Chinese]. Hubei: Huazhong College of Technology Press.

[79] Xu, Nainan (2012). Research on mathematics history in Chinese and Russian high school mathematics textbooks [In Chinese]. Doctoral dissertation. Changchun: Northeast Normal University.

[80] Xu, Pingfang (ed.). (2005). *Biographies of Women Mathematicians* [In Chinese]. Beijing: Science Press.

[81] Yoneyama, K. (1968). *The Spirit of Mathematics, Ideas and Methods*. Tokyo: Tokai University Press. Chinese version translated by Zhengzhong Mao and Suhua Wu, published by Sichuan Education Press, Chengdu, 1986.

[82] Zha, Youliang and Li, Yiyu (1990). *Transverse Permeation of Mathematical Wisdom, Mathematical Thoughts and Methods* [In Chinese]. Chengdu: Sichuan Education Press.

[83] Zhang, Dianzhou and Guo, Boxiang (1996). *Manuscript of the Mathematical Methodology* [In Chinese]. Shanghai: Shanghai Education Press.

[84] Zhang, Dianzhou and Zhu, Chengjie (1991). *Lectures on Modern Mathematical Thoughts* [In Chinese]. Nanjing: Jiangsu Education Press.

[85] Zhang, Dingqiang (2011). On the value of mathematics textbooks [In Chinese]. *Shuxue Tongbao* (*Bulletin des Sciences Mathematics*), 50(8): 5–10.

[86] Zhang, Jingzhong (2007). Appreciation of the power of mathematical thoughts in primary schools [In Chinese]. *People's Education*, (18), 31–34.

[87] Zhang, Shengli (2013). *Research on the Representation of Mathematical Concepts in Textbooks: A Case of Middle School Mathematics* [In Chinese]. Changchun: Northeast Normal University Press, 2013.

[88] Zhang, Shengli and Kong, Fanzhe (2012). Application of mathematical abstraction in mathematics teaching [In Chinese]. *Education Exploration*, 2012(1): 68–69.

[89] Zhao, Zongfang and Wu, Junming (2005). On the representation of new-curriculum-standards-based chemistry textbooks [In Chinese]. *Curriculum, Teaching Material and Method*, 25(7): 70–74.

[90] Zheng Xiaohong (2009). On the presentation of cultural value orientations in college English teaching materials [In Chinese]. *Foreign Language World*, (2): 86–91.

[91] Zheng, Yuxin (1993). Mathematical thought, its methods and methodology of mathematics [In Chinese]. *Science, Technology and Dialectics*, 10(5): 1−7.

[92] Zhou, Pei Yi (2002). Elementary school teachers' textbook decoding modes [In Chinese]. *Journal of National Taipei Teachers College*, (15): 115−138.

[93] Zhou, Pei Yi (2005). Analysis of textbook studies in Taiwan: 1979-2004 [In Chinese]. *Curriculum and Instruction Quarterly*, 8(4), 91−116.

[94] Zhou, Pei Yi and Zheng, Minchang (2008). Exploration of textbook research methods [In Chinese]. *Curriculum and Instruction Quarterly*, 11(1): 193−222.

[95] Zhuang, Meizhi (2003). *An Overview of Professor Yung Sheng Ou's Textbook Research.* Taipei: Association for Research & Development of Teaching Materials.

[96] Zhu, Xinmin (ed.) (1986). *A Collection of Important Debates in Science History.* Changsha: Human Science and Technology Press.

https://doi.org/10.1142/9789814713955_0014

Chapter 14

History of Mathematics in American Mathematical Textbooks of the Late 19th and Early 20th Centuries

Xiaoqin Wang

Department of Mathematics, East China Normal University
Shanghai 200241, China
xqwang@math.ecnu.edu.cn

This paper investigated the usage of history of mathematics (HM) in 27 mathematical textbooks published between the late 19th century and the 1920s in the United States. The result shows that there are considerable HM materials in these textbooks, and their presentation forms can be divided into two categories: explicit and implicit. The explicit presentation includes famous quotations, mathematical characters, notes on specific topics, historical sketches of mathematical disciplines, as well as mathematical terminology, literature, problems, thoughts and methods in proving theorems; while the implicit presentation are involved in compiling mathematical exercises and introducing basic mathematical concepts. Among them, mathematical problems and historical notes on specific topics are most-used presentation forms. Then, five methods of approaching the HM materials in the textbooks were classified as: ornamentation, supplementation, duplication, accommodation, and reconstruction. The characteristics of the usage were also analyzed, and it is found that: the HM usage in a mathematical textbook is closely related to the HM knowledge at the time when the textbook published, the HM literacy of the editor of the textbook, as well as the value of HM and the scientific humanism ideological trend at that time. Compared with the textbooks one century ago, today's Prentice Hall edition textbooks have made little significant progress in the HM usage.

1. Introduction

With the development of the study on History and Pedagogy of Mathematics (HPM), the usage of history of mathematics (HM) in textbooks has been receiving more and more attentions. At the end of the last century, J. Fauvel *et al.* organized scholars from 16 countries (including China) to examine the role the HM played in mathematics curriculum in each country (Fauvel and van Mannen, 2000); however, they mentioned relatively little about textbooks. In the 6th European Summer University (ESU-6) on the "History of Mathematics and Epistemology in Mathematics Education" held in Vienna in 2010, scholars from France, Italy, Poland, Greece and other countries had a panel to communicate the HM usage in mathematics textbooks in their or related countries; they also made a rough discussion on the goal, method and assessment criteria of integrating the HM into textbooks; meanwhile, some ideas objecting such a kind of integrations were presented at the panel (Boyé, 2010).

In recent years, relevant researches are increasing in China, which can be divided into four categories: (1) theoretical investigation about how to integrate the HM into textbooks (Mingzhen Li and Kun Pang, 2006; Shuping Pu and Xiaoqin Wang, 2012; Zhenhui Wang and Xiaoqin Wang, 2003); (2) statistical analysis on the contents, presentation, etc. of HM in existing textbooks (Xinbing Luo *et al.*, 2012; Xiaoqin Wang, 2012; Yuhui Yang *et al.*, 2007); (3) comparative study on the HM contents in different textbooks (e.g. Kaihui Li, 2006); and (4) research on the views of teachers and students on the HM in textbooks and current applications (e.g. Mingyue Zhang, 2012).

Until now, however, the researches of HM usage at home and abroad have mainly focused on existing textbooks, paying little attention to early textbooks.

In order to enhance the study of HPM, it is necessary to look into mathematical textbooks published in early time, about which we may ask some interesting questions like: which parts of the HM are mostly used in these textbooks, what are the ways of using HM, and what is the relation between the HM and them?

This paper tried to answer above questions by investigating 27 American high school mathematics textbooks published during 1880 and 1923, among which 13 are geometry textbooks, 10 algebra ones, 2 trigonometry ones, 1 analytical geometry one, and 1 comprehensive one, with 4 being published at the late 19th century and the other 23 published at the early 20th century.

At the end of the 19th century, scholars like D. E. Smith (1860–1944) and F. Cajori (1859–1930) were dedicated to the research and education of the HM. In 1891, Smith opened a course on the HM at Michigan State Normal College; in the early 20th century, the HM has become one of the most important Ph. D. course for mathematics education at the Teachers College, Columbia University. The 1910s and 1920s saw the rise of scientific humanism movement in the United States.

Smith and Cajori had made very fruitful researches in the field of HM: they accumulated a wealth of resource in the HM, had deep understandings of the educational value of the HM, and both considered the HM as the source of thought of mathematics education. Cajori was one of the ten members of the "sub-committee on mathematics" under the ten-number committee appointed by the National Education Association for the formulation of middle school mathematics curriculum standard (1892). Both Cajori and Smith were members of the "fifteen-member committee" appointed jointly by the American Math and Science Teachers Union and National Education Association for the development of the geometry curriculum (1910–1913). Therefore, the two HPM pioneers played important roles in the reform of mathematics curriculum in the United States at that time.

At that time, scientific humanism movement also paved the way for the integration of HM into mathematics textbooks; furthermore, rich HM materials were available due to fruitful HM researches made by Smith, Cajori, as well as European scholars.

2. Content Analysis over History of Mathematics in Textbooks

In mathematics teaching, there are explicit and implicit ways of using the HM (Fauvel and van Mannen, 2000). Similarly, the presentation of HM in mathematics textbooks can also be divided into explicit and implicit forms. Those textbook contents that can be obviously identified into the HM belong to the explicit presentation (hereinafter referred to "explicit history of mathematics"), and those contents that cannot be identified directly into, but are derived from the HM belong to the implicit presentation (hereinafter referred to "implicit history of mathematics").

2.1 Explicit history of mathematics

2.1.1 Famous quotations

At the frontispiece of *Elementary Algebra* (Slaught and Lennes, 1915), Roger Bacon's (1214–1294) sayings were cited to emphasize the value of mathematics.

- "All sciences rest ultimately on the mathematics."
- "Mathematics should be regarded as the alphabet of all philosophy."
- "Divine mathematics, which alone can purge the intellect and fit the student for the acquirement of all knowledge."

There are few textbooks that cite quotations.

2.1.2 Mathematical characters

After a brief introduction on the history of geometry, the appendix of the *New Plane & Solid Geometry* (Beman and Smith,1899) included a biographical table containing a total of 34 mathematicians from the ancient Egyptian Ahmes (Ca. 1650 BC) to Austrian G. Vega (1756–1802) that appeared in the book. Similarly, the appendix of the *Elements of Algebra* (Beman and Smith, 1900) included a biographical table containing a total of 43 mathematicians appeared in the book.

Durell (1911) in his *Plane and Solid Geometry* used the portraits of Thales, Plato, Pythagoras and Euclid, calling them "the discoverer and founder of geometry". Durell (1912) in his *Introductory Algebra* used the portraits of F. Viète, R. Descartes, I. Newton and C.F. Gauss on the frontispiece.

Cajori and Odell (1915, 1916) used the portraits of F. Viète, R. Descartes, J. Wallis, I. Newton, L. Euler, A. De Morgan with a brief description (Fig. 14-1).

Slaught and Lennes (1915) used the portraits of J. Widmann, F. Viète, W. R. Hamilton, Newton, Pythagoras, Wallis, Descartes, Gauss, and Pascal, with a brief account of their biographies.

Hallett and Anderson (1917) used the portraits of Pythagoras, Euclid, Viète, Pascal, Descartes, Newton, and Gauss, with a brief account of their biographies.

Slaught and Lennes (1919) used the portraits of Thales, Archimedes, B. Cavalieri, and A. M. Legendre.

Fig. 14-1 Portraits of the mathematicians in Cajori and Odell's *Elementary Algebra*

Rider and Davis (1923) used the portraits of Euler, Napier, Viète, and Descartes, with a brief account of their biography.

In addition to portraits of mathematicians, some textbooks introduced the stories and anecdotes of mathematicians in the historical notes of relevant topics. Tab. 14-1 shows parts of the historical notes in (Hart and Feldman, 1912).

Tab. 14-1 Introduction of mathematicians in the historical notes in (Hart and Feldman, 1912)

Column	Content	Historical notes
Exercises	Prove with the graph of Euclid: base angles of isosceles triangle are equal.	Euclid's biography
Exercises	Given two mutually perpendicular axes, find out the locations of points that meet the conditions.	Descartes and analytical geometry
Text	The property of the angle of circumference	biography of Thales
Text	The Pythagorean theorem	biography of Pythagoras
Text	Regular polygons	biography of Gauss
Text	Area and circumference of a circle	biography of Archimedes
Text	Volume of cuboids	biography of Plato

Rider and Davis (1923) introduced the life of A. de Moivre (1667–1754) following the De Moivre's theorem.

2.1.3 Literature of HM

Literature of HM refers to the prints of mathematics books (including manuscripts) in history, measurement tools used by astronomers and mathematicians in history, paintings reflecting the mathematics topics, and so on.

Wentworth, Smith and Brown (1917) gave a lot of pictures of various measurement tools and methods in history in the geometry section of the textbook *Junior High School Mathematic*, for instance, the quadrant (Fig. 14-2), level, ancient astrolabe and armillary sphere in 17th century's mathematics books, distance measurement method in 16th century (Thales' Method). In addition, prints of the earliest English edition of the *Elements of Geometry*, prints of a page of Recorde's *The Whetstone of Witte* where the sign of "equal" appeared for the first time (Fig. 14-2), signs of the plus and minus in mathematics books in the 16th century, the expression of power in mathematics books in the 16th century, prints of the Latin version of Al-khowarizmi's *Algebra*, and so on. They fully reflect David Eugene Smith's style: He attached great importance to the pictures of mathematicians and mathematical literatures both in teaching and writing.

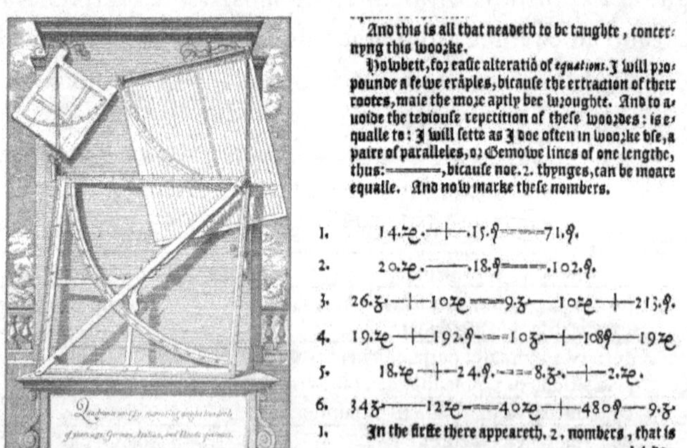

And this is all that neadeth to be taughte, concernyng this woorke.

Howbeit, for easie alteratiō of *equations*. I will propounde a fewe exāples, bicause the extraction of their rootes, maie the more aptly bee wroughte. And to avoide the tediouse repetition of these woordes: is equalle to: I will sette as I doe often in woorke use, a paire of paralleles, or Gemowe lines of one lengthe, thus: ═══════, bicause noe. 2. thynges, can be moare equalle. And now marke these nombers.

Fig. 14-2 Illustration of the *Junior Middle School Mathematics*: the left is the ancient quadrant and the right is a page of *The Whetstone of Witte*

Betz and Webb (1916) also used a number of illustrations, such as the woodcut in Durer's *Painter's Manual* (1525) that reflects the perspective drawing method as is shown in Fig. 14-3.

Fig. 14-3 Perspective drawing of artists in drawing the vase

2.1.4 Mathematical terminology

Beman and Smith (1899) gave a table of etymologies of more than 130 terms used and defined in elementary algebra as an appendix. Similarly, Beman and Smith (1900) gave a table of etymologies 118 terms of geometry in the appendix.

While introducing the history of geometry at the opening, Wells and Hart (1916) firstly explained the origin of the word "geometry" — it is formed from the Greek words "land" and "measure".

Rider and Davis (1923) explained in the opening introduction of the textbook the origin of the word *Trigonometry* which was formed from the Greek words *trigonon* (triangle) and *metria* (measure). In the chapter on "logarithm", the origin of the word *logarithm* was introduced: it was formed from the Greek words *logos* (ratio) and *arithmos* (number). In the chapter on "Trigonometric Functions", the origins of the names of six trigonometric functions were introduced.

The word *sine* comes from the Indian word "jiva" (half chord); after it was passed into Arab, it was transliterated as "jiba", which was changed again by the Arabs into their more familiar "jaib" (gulf). Plato Tiburtinus translated "jaib" into Latin "sinus", and finally, *sinus* was translated into *sine* in English.

The Danish mathematician Thomas Finck (1561–1656) created the words *tangent* and *secant* in 1583. In the 17th century, the Englishman Edmund Gunter (1581–1626) made the earliest use of *cosine* and *cotangent*, meaning "sine of the complementary angle" (Latin as *complementi sinus*) and "tangent of the complementary angle".

The Dutch mathematician Albert Girard (1595–1632) made the earliest use of simplified symbols such as sin, tan, sec, which were widely used after Thomas Simpson and Euler.

The above materials can also be found in Durell (1910).

2.1.5 Mathematical problems

Tab. 14-2 Historical problems used by Beman and Smith (1900)

No.	Problems	Source	Time
1	If 9 porters drink 12 casks of wine in 8 days, how many will last 24 porters 30 days?	Tartaglia	The 16th century
2	Demochares lived 1/4 of his life as a boy, 1/5 as a young man, 1/3 as a man, and 13 years as an old man. How old was he then?	Metrodorus	The 4th century
3	Of 4 pipes, the first fills a cistern in 1 day, the second in 2 days, the third in 3 days, and the fourth in 4 days. How long will it take all running together to fill it?	Heron	The 1st century
4	In the center of a pond 10 ft square grew a reed 1 ft above the surface; but when the top was pulled to the bank it just reached the edge of the surface. How deep was the water?	*Nine Chapters on the Mathematical Art*	The 1st century [a]
5	A horse and a donkey, laden with corn, were walking together. The horse said to the donkey: "If you gave me one measure of corn, I should carry twice as much as you; but if I gave you one we should carry equal burdens." Tell me their burdens, O most learned master of geometry.	Euclid	The 3rd century BC
6	Heap, its whole, its seventh, it makes 19. (That is, what is the number which when increased by its seventh equals 19?)	Rhind Papyrus	The 17th century BC
7	Find the number, 1/3 of which and 1, multiplied by 1/4 of which and 2, equals the number plus 13.	Al-Khowarizmi	The 9th century
8	In a pond the top of a lotus bud reached 1/2 ft. above the surface, but blown by the wind it just reached the surface at a point 2 ft. from its upright position. How deep was the water?	Bhaskara	The 12th century
9	Two anchorites lived at the top of a perpendicular cliff of height h, whose base was mh distant from a certain town. One descended the cliff and walked to the town; the other flew up a height, x, and then flew directly to the town. The distance traversed by each was the same. Find x.	Brahmagupta	The 7th century
10	An ancient problem relates that Titus and Caius sat down to eat, Caius furnishing 7 portions and Titus 8, all of equal value. Before they began Sempronius entered and they all ate equally and finished the food. Sempronius then laid down 30 denarii (pence) and said: "Divide these equitably between you in payment for my meal." How much should each receive?	Unknown	Unknown

In the chapter on Equation of the textbook, Shoup (1880) chose two historical problems as exercises: the Euclid's "horse and mule" question and the epitaph question of Diophantus. In the chapter on Linear Equation

[a]Beman and Smith mistaken it for 2600 BC.

with One Unknown of the textbook, Beman and Smith (1900) selected 10 historical problems as exercises, as is shown in Tab. 14-2.

Betz and Webb (1912) introduced some historical problems, such as the reed problem in *The Nine Chapters on the Mathematical Art.* In the exercises, an approximate way of A. A. Kochansky's (1631–1700) construction of the circumference of a circle was given.

Young and Jackson (1909) introduced many mathematical problems raised by the famous mathematicians in the 17th and 18th centuries, such as Clairaut, Newton, Saunderson and Euler. Here are some examples:

— A workman can perform a units of work in b units of time; a second workman can perform c units of work in d units of time; a third can do e units of work in f units of time. In what time will these three workmen together perform g units of work? (Clairaut's Algebra, 1746)

— Two messengers, A and B, set out towards each other from two places 59 mi. apart, B starting 1 hr. after A. A goes 7 mi. in 2 hr., and B 8 mi. in 3 hr. How far will A have gone when he meets B? (Newton's Arithmetica Universalis, 1707)

— A Company at a tavern, when they came to pay, found that if the same bill were divided among three persons more, the amount would be one shilling less per person; and, if it were divided among two persons fewer, it would be one shilling more per person. Find the number of persons in the original company, and the amount of the bill. (Saunderson's Algebra, 1740)

— The price of a house is 100 dollars. A could pay for it if he had half of B's money in addition to his own; B could pay for it if he had one third of C's; and C could pay for it if he had one fourth of A's money. How much had each? (Euler's Algebra, 1770)

In the chapter "Linear Equation Group", Young and Jackson (1913) presented Sanderson's "debt problem" and Euler's "house purchase problem". Stone and Millis (1915) introduced the problem recorded on Rhind's papyrus.

In the appendix, Wentworth and Smith (1913) introduced, in the form of exercises, some geometrical fallacies prevailing in Europe in the 19th century:

- Any triangle is an isosceles one,
- the right angle equals to an obtuse angle,
- in a line segment, a section equals to the whole,
- an arbitrary point can divide a line segment equally,
- through an external point, two lines perpendicular to a line can be drawn, etc.

Rider and Davis (1923) included some historical triangular problems into the exercises, such as in the chapter on Oblique Triangles, there is a "Pothenot Problem": "given three Points *A*, *B*, *C*, and the known sides *AB*, *BC*, *CA*, find the distance from *D* to *A*, *B*, *C*; $\angle ADB$ and $\angle CDA$ being known." (Fig. 14-4) In 1730, L. Pothenot (1650–1732) solved this problem; in fact, Dutch mathematician W. Snell (1591–1626) had worked it out as early as in 1617.

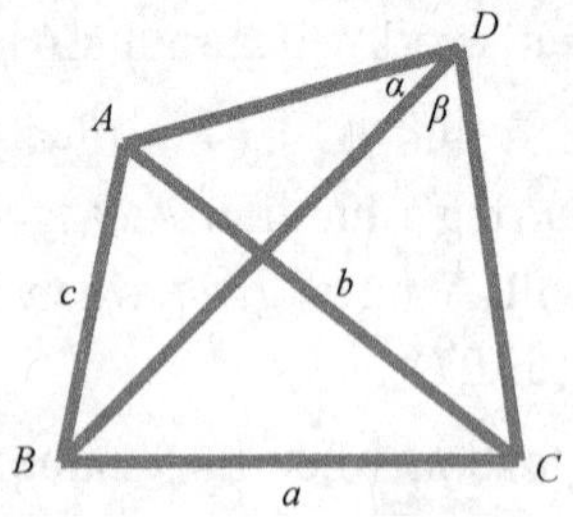

Fig. 14-4 The Pothenot Problem

2.1.6 Mathematical topics

Historical topics, mainly on the origin, the discoverers, or historical sketches of some formulas, theorems, usually appear in the form of notes in these textbooks.

Beman and Smith (1899) started the tradition of adding historical notes in a textbook. They pointed out in the preface that it had become a consensus that historical notes can enhance students' interest in a textbook. On the problem of "constructing an equilateral triangle on a given line segment", the historical note says it was "the first proposition of Euclid's *The Elements*". On the theorem of the sum of angles in a triangle, the note stated that it was discovered by the Pythagorean School, and the theorem of the sum of squares of four edges in a parallelogram was attributed to

Euler, etc. In addition, the textbook also introduced history of some other topics, such as the history of π, squaring the circle, construction of equilateral n-polygons inscribed in a circle, Platonic solids.

Thereafter, many textbooks followed the practice of Benman and Smith. Failor (1906) made historical notes of some topics, including the Pythagorean theorem, constructions with ruler and compasses of equilateral n-polygons inscribed in a circle, circular constant, etc. Hart and Feldman (1912) made numerous historical notes, among which, besides the aforementioned biographies of mathematicians, there are histories of such topics as the incommensurable magnitude discovered by the Pythagorean school, Zeno's paradoxes — Archilles chasing the tortoise, using a triangle with the side lengths 3, 4 and 5 to construct a right angle by ancient Egyptians, the origin of geometry in ancient Egypt, constructions of equilateral n-polygons inscribed in a circle with ruler and compasses, Plato's solids, the formula for the volume of a cylinder recorded on Rhind's papyrus.

Slaught and Lennes (1915) added historical notes at the end of each of the 18 topics in the textbook: the origin of Hindu-Arabic numerals, the origin of the operational signs, the origin of the parentheses (including curly brackets, square brackets and brackets), the origin of the distribution law, the origin of the word "algebra", using a letter to represent the unknown, the development of negative numbers, the laws of association and commutation of addition and multiplication, the exponent, the Pythagorean Theorem, the notations of fraction, ratio and proportion, the Indian formula for the solutions of an quadratic equation, equations represented by graphs, radicals, quadratic equations and imaginary roots, and the binomial theorem.

Betz and Webb (1916) made historical notes on some topics in solid geometry, such as Archimedes' work on the surface area and volume of a sphere, the tombstone of Archimedes, the association between Menelaus and the properties of spherical triangles.

Rider and Davis (1923) made historical notes on such topics as angular measurements, concept of logarithm, laws of sine, cosine and tangent.

Young and Jackson (1913) gave historical notes at the end of each chapter, as shown in Tab. 14-3.

Tab. 14-3 Historical notes made by Young and Jackson (1913)

Chapter	Topic	Content
Literal notation and its uses	History of literal notation	Diophantus' epitaph; Diophantus' use of literal notation.
Definitions of elementary terms	History of mathematical symbols	Diophantus's expressions of quadratic polynomials; Invention of the symbols of addition, subtraction, multiplication, division, square root, and equal.
Positive and negative numbers	History of negative numbers	Indians were the first to use negative numbers and interpreted them as "debts"; Work of Bhaskara, Descartes and Cardan, etc.
Equations	History of linear equations	Linear equation on the papyrus of Ahmes $x+\frac{1}{7}x=19$ and its solution.
Division	Origin of the term "algebra"	Al-Khowarizmi's Algebra.
Equations	History of factoring	Thomas Harriot's using of factoring to solve equations.
Graphs of linear equations	History of analytic geometry	Descartes and analytic geometry
Involution and evolution	History of the binomial theorem	Indian and Arabian mathematicians and related work of Viète, Pascaal and Newton.
Quadratic equations	History of quadratic equations in one unknown	Indian mathematicians and related work of al-Khowarizimi, Stifel and Stevin.
System of quadratic equations	History of systems of auadratic equations	Related work of Diophantus, Gerbert and Sylvester
Review and development	History of perational rule	Hamilton and law of association, commutation and distribution
Exponents and roots	History of Exponents	Viète's way of expressing powers; the life and work of Wallis; Wallis' use of negative integral Exponetns.
Logarithms	History of logarithms	John Napier's invention of logarithm; Henry Briggs' invention of common logarithms; The significance of logarithms.
Imaginary and complex numbers	History of complex numbers	Wessel and Gauss represented imaginary numbers geometrically; the biography of Gauss, "the prince of mathematics".
Quadratic equations	History of solution of equations	Tartglia and Cardan and the solution of cubic equations; Ferrari and the solution of biquadratic equations; Abel's statement
Series	History of Series	Various number sequences on the Babylonian clay tablets; Geometric Series on the textbook of 16th century; Bhaskara and the arithmetical and geometric series.

Unlike the other textbooks in which historical notes are usually attached as appendix to the end of a certain chapter or the entire textbook, Cajori and Odell (1915, 1916) wrote a whole section for each historical topic in the body of the textbook (but with smaller letters), such as the sections “the Beginning of Algebra”, “History of Quadratic Equations”, “History of Fractions”, “History of Operational Laws”, “the Invention of Logarithms”. For the topic of early history of algebra, they introduced the method of writing equations used by the ancient Egyptians, the notations of powers employed respectively by Diophantus (The 3rd century), F. Viète and P. Nunes (1502–1578), Robert Recorder’s invention of the equal sign “=”. For the operation law, they told the readers that despite the history of algebra could be traced back to 2000 BC, people acquiesced to the correctness of commutative law, associative law and distributive law, but no proof had been given; it was not until the 19th century that mathematicians from German and the United States began to make algebra complete in all respects; the French mathematician F. J. Servois (1768–1847) was the first to give the name of “commutative law” and “distributive law”, while the British mathematician W. R. Hamilton (1805–1865) was the first to give the name of “associative law”. For the history of logarithms, Cajori and Odell told a story about the Scottish mathematician J. Napier (1550–1617) before he invented logarithms: he had lived in a beautiful castle on the banks of the Endrick River for a long time; on the opposite side of the river was a lint mill, and its clack disturbed Napier so much that he had hoped that the factory miller could stop the mill so that the train of his ideas might not be interrupted.

2.1.7 Thoughts and Method

“Thoughts and methods” here refers to the methods used in textbooks to derive formulas or to prove theorems. Although the US geometry textbooks in the early 20th century have shaken off the shackles of *The Elements*, they were still deeply engraved with the mark of *The Elements*. A large number of geometrical propositions originate from *The Elements*; Euclid's proofs are still exemplary. For example, we take 7 sets of textbooks to investigate the proofs of the Pythagorean theorem: Bowser (1890), Beman and Smith(1899), Milne (1899), Failor (1906), Gore (1908), Durell (1911), Betz and Webb (1912), Hart and Feldman (1912), Wentworth and Smith (1913), Wells and Hart (1916), and Palmer and Taylor (1919). All proofs of the theorem appeared in these textbooks are listed in Tab. 14-4. We found that all of the textbooks introduced Euclid’s proof (no. 1 in Tab. 14-4) by area comparison as their primary choices. The proofs no. 2 (Zhao Shuang's proof, also known as Bhaskara's proof in the west), no. 3 (due to Pythagoras, in the speculation of mathematical historians), and the proof of Arabian mathematician Tābitibn Qorra (826–901) (no. 4 in the table) appeared in four textbooks, while proof of James A. Garfield (1831–1881) (no. 9) appeared in two textbooks. The other proofs were only included in one certain textbook.

Tab. 14-4 Proofs of the Pythagorean Theorem in 11 geometry textbooks

No.	Figure	Description	Proved by	Time
1	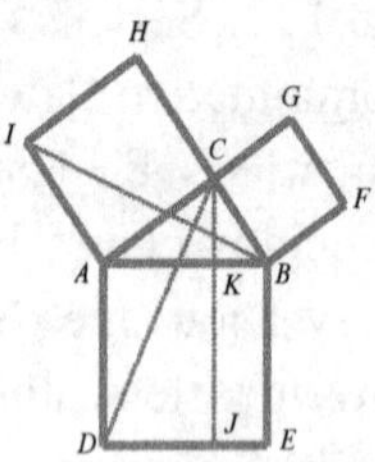	$\Delta ABI \cong \Delta ADC$, £ $AH = 2\Delta ABI$, ú$AJ = 2\Delta ADC$, ⇒ £ AH = úAJ Similarly, £ BG = úBJ	Euclid	The 3rd century BC
2	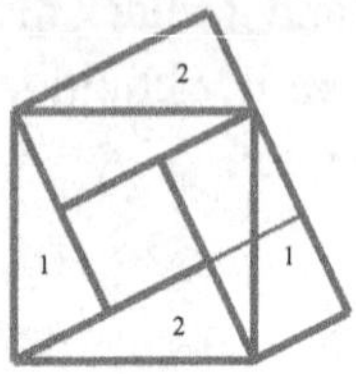	Put together four congruent right triangles and one small square in two different ways to get $a^2 + b^2$ and c^2 respectively.	Zhao Shuang Bhaskara	The 3rd century The 12th century
3	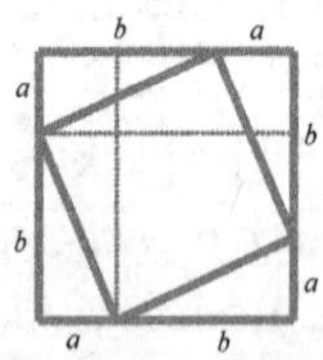	From the large square, remove four right triangles to obtain c^2; remove two rectangles to obtain $a^2 + b^2$.	Pythagoras	The 6th century BC
4	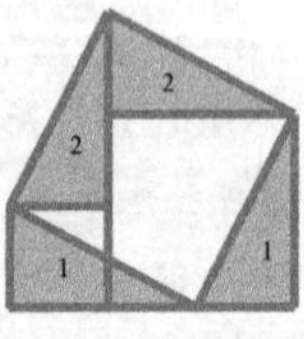	From the entire picture, remove the right triangles labelled 1 to obtain c^2; remove the right triangles labelled 2 to obtain a^2+b^2.	Ibn Qorra	The 9th century

<table>
<tr><td>5</td><td></td><td>Divide b^2 into four parts, and combine them with a^2 to form c^2.</td><td>H. Perigal</td><td>The 19th century</td></tr>
<tr><td>6</td><td>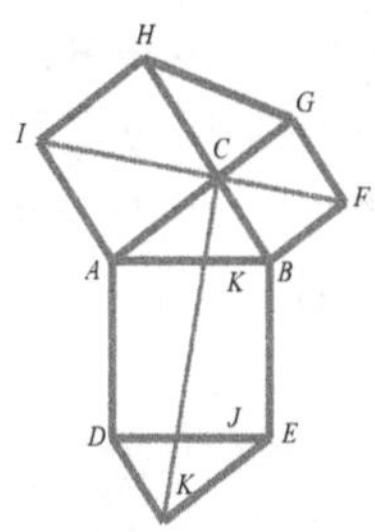</td><td>In the picture, quadrilaterals ACKD, EKCB, AIFB, HIFG are congruent pairwise.</td><td>Da Vinci</td><td>The 15th century</td></tr>
<tr><td>7</td><td>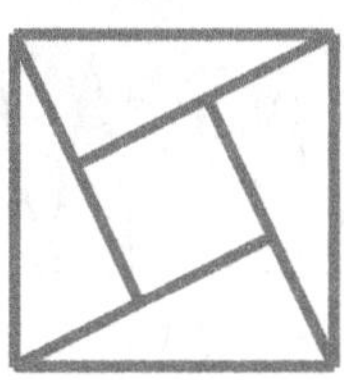</td><td>$c^2 = (b-a)^2 + 4 \times \frac{1}{2}ab$</td><td>Algebraic form of Zhao Shuang/ Bhaskara's proof</td><td></td></tr>
<tr><td>8</td><td>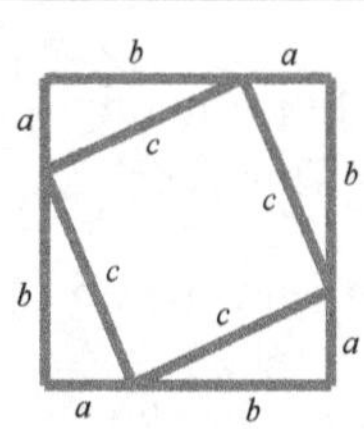</td><td>$c^2 = (a+b)^2 - 4 \times \frac{1}{2}ab$</td><td>Algebraic form of Pythagoras' proof</td><td></td></tr>
<tr><td>9</td><td>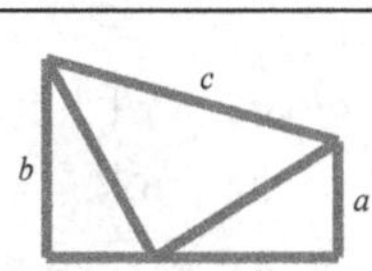</td><td>$2 \times \frac{1}{2}ab + \frac{1}{2}c^2 = \frac{1}{2}(a+b)^2$</td><td>Garfield</td><td>The 19th century</td></tr>
</table>

Of course, Euclidean methods were not always used in proving geometric theorem in the textbooks at this period; instead, some methods of later mathematicians were applied. For example, when proving the side-side-side theorem for congruent triangles, Wentworth and Smith (1913) adopted a method due to Philo, a mathematician in Byzantine period in the first century: Let one pair of the corresponding sides of the two triangles be coincided, so that the third vertices locate on the different sides. Connect the third vertices, then complete the proof by using the property of isosceles triangle and the side-angle-side theorem.

Wentworth, Smith and Brown (1917) derived the formula for the area of a circle by means of a method due to J. Kepler (1571–1630): Divide the circle into equal small "triangles", the height of each "triangle" being the radius of the circle, and the lengths of the bases summing to the circumference of the circle, thus the sum of the areas of these "triangles" is $\frac{1}{2}\times$ radius $\times$ circumference. However, the textbook points out that a strict proof will be given in subsequent geometry curriculum.

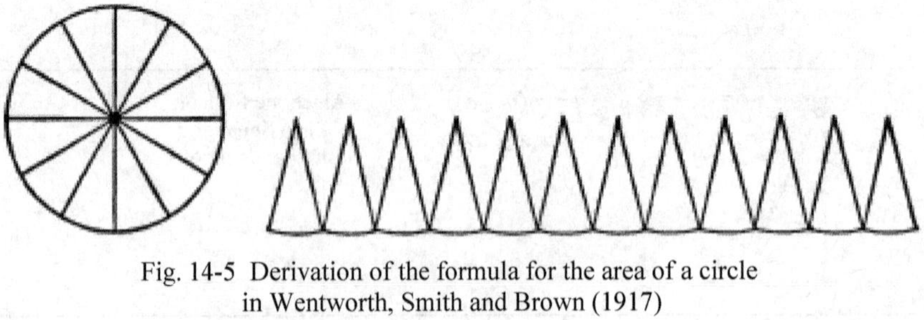

Fig. 14-5 Derivation of the formula for the area of a circle in Wentworth, Smith and Brown (1917)

Bowser (1890) derived the volume of a sphere by imitating Kepler's method (although no mention of Kepler was made). This is an example of duplicating application of mathematical thoughts.

2.1.8 Historical sketches of specific topics

Most of the textbooks under consideration gave brief historical accounts of some topics at the beginning or the end of the book.

Durell (1910) spent a whole chapter to introduce the history of trigonometry, including the historical stages of trigonometry, important mathematicians in the history of trigonometry (Hipparchus, Ptolemy, Regiomontanus, Euler), the history of trigonometric functions, the history of angular measure, the history of trigonometric tables, solution of triangles and the history of trigonometric formulas, etc.

In Durell (1911), a dedicated appendix was used to introduce the history of geometry, including the origin of geometry, historical stages of geometry, history of geometrical methods, history of geometric theorems (plane geometry and solid geometry), etc. In Durell (1912), the last chapter was devoted to the introduction of the history of elementary algebra, including the stages of algebra, history of algebraic symbols, history of algebraic quantities, history of solving equations, etc. Wells and Hart (1916) introduced the history of geometry at the beginning.

In the opening of their textbook, Betz and Webb (1912) introduced the origin of geometry in ancient Egypt, ancient Greek geometry (Fig. 14-6), the value of geometry, geometric methods. Siciloff, Wentworth and Smith (1922) introduced the history of analytical geometry in the appendix.

2. Greek Geometry. The early geometry, however valuable it may have been for practical purposes, was deficient in that it consisted simply of a set of rules obtained after centuries of experimenting. *How* to get a result seemed to be its important question. That this condition did not last indefinitely is due to the genius of the Greeks. Being a race of thinkers and poets, they wished to know *why* a certain result must follow under a

Fig. 14-6 A Page of Betz and Webb's *Plane Geometry*

Rider and Davis (1923) told of the history of trigonometry in the foreword of the book, involving in mathematicians such as Ahmes, Hipparchus, Theon, Ptolemy, Regiomontanus, John Bernoulli, De Moivre, Lambert, Fourier and Abel. The author expressly stated that Regiomontanus separated the trigonometry from astronomy and made it a part of the geometry; while Euler turned trigonometry into an independent subject.

2.2 Implicit history of mathematics

In these early textbooks, we can also find a wealth of implicit HM materials, including mathematics problems, mathematical concepts presentation, method of thinking, etc. Here we examine only the first two types of materials.

2.2.1 Mathematics problems as exercises

History of mathematics is a huge treasure which provides not only mathematical problems but also raw materials for the preparation of mathematics problems. A large number of exercises in the early textbooks are compiled based on the historical problems or materials. For example, Robbins (1907) designed an exercise by referring to Archimedes' propositions on the Shoemaker's Knife (Fig. 14-7):

- If from any point in a semi-circumference a line be drawn perpendicular to the diameter and semicircles be described on the two segments as diameters, the area of the surface bounded by these three semi-circumferences will equal that of a circle whose diameter is the perpendicular first drawn.

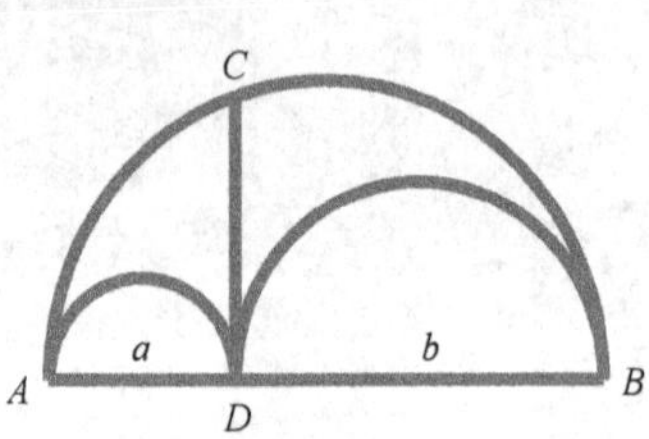

Fig. 14-7 Archimedes' "Shoemaker' Knife"

Gore (1908) designed an exercise by referring to the mathematics materials from ancient Egypt:

- The Egyptians said: "Construct a square the side of which is 8/9 of the diameter of a circle, and its area will equal that of the circle." From this compute their value of π.

Hart and Feldman (1912) prepared exercises by using history of mathematics, such as allowing students to solve geometric fallacy: 64 = 65, as is shown in Fig. 14-8.

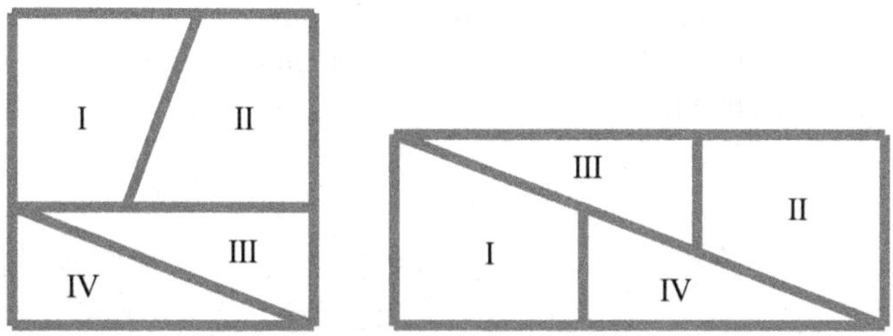

Fig. 14-8 Geometry fallacy: 64 = 65

Betz and Webb (1912) prepared a number of propositions or exercises according to HM. For example:

- Make a square having the same area with the given rectangular.

The author did not use Euclid's method in Proposition II. 14; instead he adopted Euclid's method in Proposition I. 47. As shown in Fig. 14-9, extend *AD* to *E* with *AE* = *AB*, then draw a semicircle with *AE* as the diameter, meeting *C*'s extension at *F*. The square with *AF* as a side is what we need.

- Construct with ruler and compass the picture of the arch and rose window of Gothic architecture as shown in Fig. 14-10, and explain the construction method.

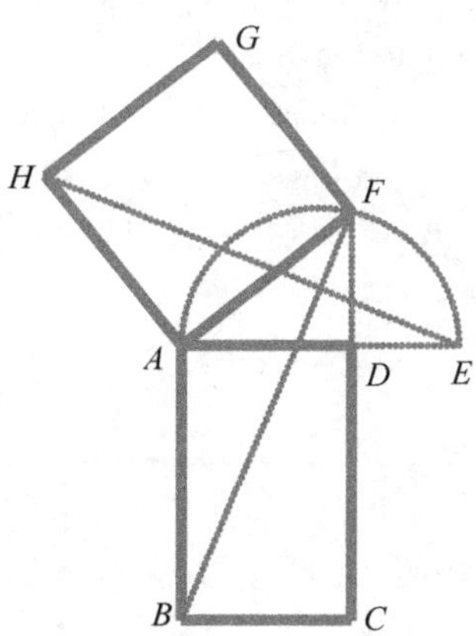

Fig. 14-9 Turn rectangle into square

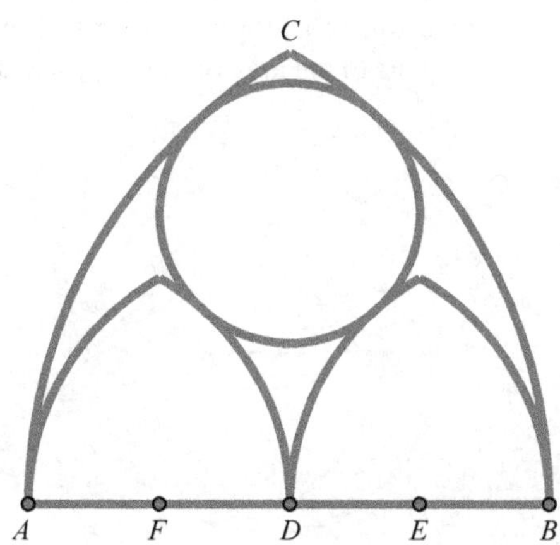

Fig. 14-10 Design of the Gothic window

- Explain Thales' method of measuring the distance between the ship and the shore, as shown in Fig. 14-11.

➢ The Greek historian Thucydides (The 5th century BC) estimated the area of the island of Sicily by the time spent on sailing around it. What's wrong with his method?

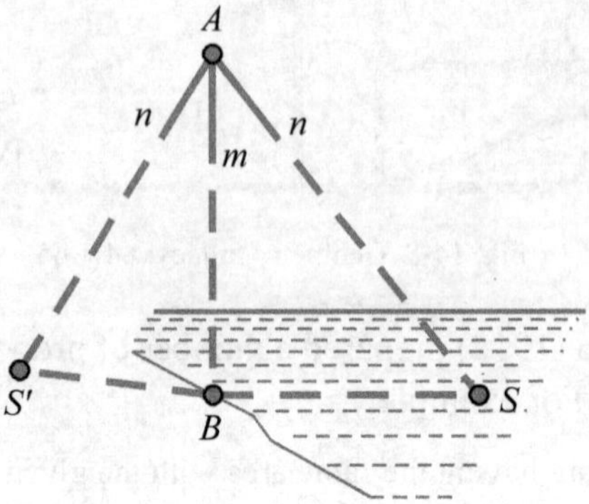

Fig. 14-11 Thales's method of distance measurement

➢ Archimedes gave the following proposition: The area of a circle is equal to that of a triangle whose base is equal to the circumference and whose height is equal to the radius. Construct this triangle by using Kochansky's method.

➢ The ancient Greek mathematician Hippocrates gave the method of turning the circle into a square: As is shown in Fig. 14-12, construct a square inscribed to a given circle, and construct four semicircles with the side of the square as their diameters. Then the sum of areas of the four lunars equals to the area of the inscribed square. Try to prove it.

Fig. 14-12 Hippocrates theorem

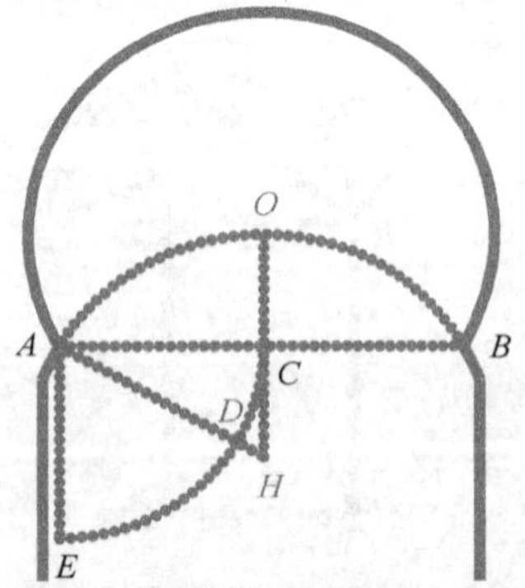

Fig. 14-13 The horseshoe arch

➢ Horseshoe arch is one of the important features of the Moorish architecture, and the constructing method is shown in Fig. 14-13: Draw a quarter-circle with A as the center and AC as the radius, connect A and the trisection point D of the quarter-circle, and extend AD to meet the perpendicular bisector of AB at H. With H as the center and HA as the radius, construct an arc to meeting the extension of HC at O. With O as the center and OA as the radius to get the horseshoe arch. If it is known that AB = 2a, find the length, height and area of horseshoe arch.

Some exercises designed by Betz and Webb (1916) are also based on historical materials, for instance:

- Verify Euler's polyhedron theorem with known examples.
- Pyramid of Khufu in Egypt is originally a right square pyramid with height 156.4 meters and a bottom-base of side 232.2 meters. In the later time its top collapsed and it was turned into a square frustum with height 147.8 meters. Regardless of the irregularity, calculate the area of the upper-base and the lateral area of the truncated pyramid.

To show the necessity of mathematical proof, Wentworth and Smith (1913) used the illusion graph of the German astrophysicist J. K. F. Zöllner (1834–1882) and German physicist J. C. Poggendorf (1796–1877) in the exercises of introductory part. For example, in Fig. 14-14, are both ends of the line segment *AB* and *CD* equidistant? Which of the three lines below the shaded portion is the extension of *AB* in Fig. 14-15?

Fig. 14-14 Zöllner's illusion graph

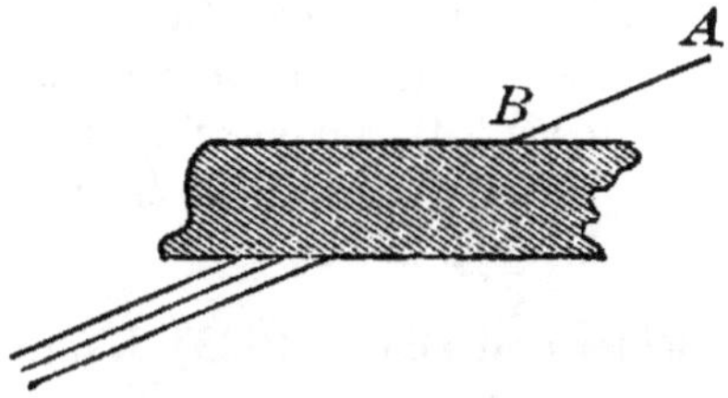

Fig. 14-15 Poggendorf's illusion

Some exercises of Wentworth, Smith and Brown (1917) are prepared based on historical materials, and the following two cases are more typical:

- Construct the crenellated window pattern of medieval churches in Italy (Fig. 14-16) with ruler and compass.
- Fig. 14-17 is an illustration to show Thales's measuring method in an Italian mathematics book in the 16th century. In the upper part of the figure, a person makes visual measurement of the other bank of the river through his brim of hat, then, by turning around, he makes visual measurement to the land side through his brim of hat. Question: how does this man get the width of the river?

Fig. 14-16 Crenellated window patterns of medieval churches in Italy

Fig. 14-17 Thales' method of measuring the width of a river

Stone and Millis (1915) also prepared mathematics problems by using HM, for instance:

- Surveyors, in order to mark off a right angle on the ground, surveyors, sometimes set three stakes so that when a tapeline is stretched around them so as to form a triangle, it is divided at the stakes into parts whose lengths are as 3, 4, and 5. The ancient Egyptians used this method in laying out their pyramids, etc. If a 100-foot tapeline is used, find the lengths of the parts into which it must be divided at the stakes. What is the answer if a 50-foot tapeline is used?
- A method used several hundred years ago of finding the distance from A to the inaccessible point B was to erect a vertical staff AC, place upon it an instrument resembling a carpenter's square, pointing one blade towards B, and note the place D on the ground to which the other blade pointed. AC and DA were measured. If AC = 52 in., and DA = 6 in., find AB (Fig. 14-18).

Rider and Davis (1923) designed the ancient "broken bamboo" and "ladder" problems into trigonometric ones:

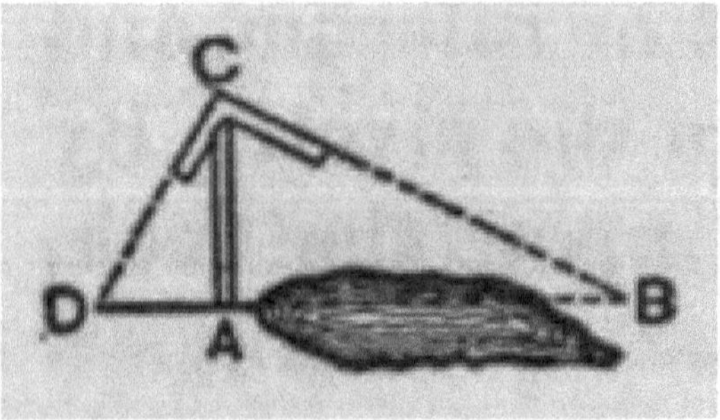

Fig. 14-18 Measurement

- ➢ A tree was broken by the wind, constituting a right triangle with the ground. If the broken part and the ground form an angle of 50°, and the treetop is 55 feet off the root, find the height of the original tree.
- ➢ A 36-feet ladder is tilted against the wall, the bottom part is 25 feet away from the wall. Find the angle formed by the ladder and the land.

2.2.2 The introduction of mathematical concepts

When introducing a series of basic mathematical concepts in a textbook, if differences exist between logical and historical sequences of the occurrence and development of knowledge, appropriate selection must be made at the time of the compilation of textbooks in accordance with students' pycho-genetic sequence. In early textbooks, examples of introduction of new concepts in historical sequence can also be found.

Historical and logical sequences of expansions of the number system are not the same. Historical sequence was employed by Beman and Smith (1900) to present such process, as shown in Fig. 14-19. Moreover, introduction of zero was being placed behind negative numbers, which was consistent with the history of mathematics.

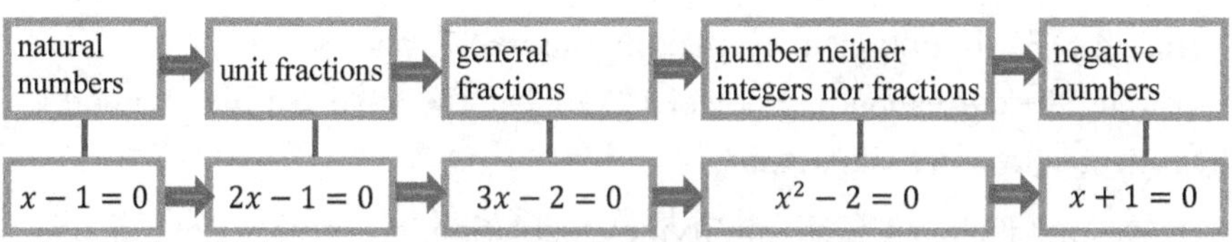

Fig. 14-19 Introduction of negative numbers

Another typical example in school mathematics showing differences in logical and historical sequences is the introduction of exponent and logarithm concepts. According to logical sequence, the concept of exponent comes before that of logarithm concept, and the latter was defined based on the former. This method is being widely used in today's textbooks. However, the historical sequence goes in the other way, which was employed by Rider and Davis (1923).

First, let students recall the exponential laws (Tab. 14-5).

Tab. 14-5 Exponential laws

Exponential Law	Example
$b^m \times b^n = b^{m+n}$	$b^6 \times b^3 = b^{6+3} = b^9$
$b^m \div b^n = b^{m-n}$	$b^6 \div b^3 = b^{6-3} = b^3$
$(b^m)^n = b^{mn}$	$(b^6)^3 = b^{6\times 3} = b^{18}$
$\sqrt[n]{b^m} = b^{\frac{m}{n}}$	$\sqrt[3]{b^6} = b^{\frac{6}{3}} = b^2$

The laws show that multiplication and division of powers of a fixed base are reduced to the addition and subtraction of exponents, while involution and evolution are reduced to multiplication and division of exponents. This gives a way to simplify calculation — if one can find the corresponding exponent of a number (on a certain base), the complex operations on numbers can be reduced to simpler operations on the corresponding exponents.

Take $b = 2$ as an example, one can list correspondence between exponents and powers as follows:

Exponent	...	1	2	3	4	5	6	7	8	9	...
Power	...	2	4	8	16	32	64	128	256	512	...

Then 4×32 is obtained by making sum of 2 and 5, which is 7, and locating the corresponding power 128 of 7 in the table. Similar method is applicable to other situations.

Despite its benefit in simplifying calculation, the table is not practical due to its limited entries. It will be useless except for positive integers' powers of 2. To expand the table, a careful observation of the laws of its row sequences is necessary. In the upper sequence, each number, beginning from the second one, is the arithmetic mean of the two adjacent numbers, while in the lower sequence, each number, beginning from the second one, is the geometric mean of the two adjacent numbers. Following the laws, the above two number sequences can be refined by inserting new numbers between two adjacent numbers, i.e., insert $\sqrt{16 \times 32} = 22.624$ between 16 and 32 in the lower sequence, and insert $\frac{4+5}{2} = 4.5$ between 4 and 5 in the upper number sequence. Then the table becomes

Exponent	...	4	4.5	5	...
Power	...	16	22.624	32	...

This approach to an applicable table is still a difficult task. On the one hand, the precise value of geometric mean is impossible to get; on the other hand, there is no way to obtain the corresponding "exponent" for a given power. In fact, the power of a non-integer exponent does not make sense at this point. From the logical view, power for any exponent should be defined in mathematics.

However, through the above the author actually guided readers to witness the complete process of how logarithm was invented.

3. Five Methods and Characteristics of Approaching HM Materials

3.1 Five methods of approaching the HM materials in textbooks

According to the roles of the HM played in textbooks, we have categorized five methods of approach to such materials in mathematics textbooks, namely, *ornamentation*, *supplementation*, *duplication*, *accommodation* and *reconstruction* approaches. Among which, the former three are considered as explicit forms, and the latter two as implicit forms.

The ornamentation approaches are basically used for the purpose of "embellishment," "beautification" and "humanization." Ornamentation-based materials include portraits of mathematicians, ancient mathematics books, mathematical symbols, pictures of measurement or drawing tools, artistic works, etc. In the mathematics textbooks of the late 19th and early 20th centuries, portraits of mathematicians are common place. Of course, the scope of this type of materials is not limited to the history of mathematics, and also covers buildings, constructional and decorative patterns, etc.

The ornamentation approaches are adopted not only for the sake of beautification but also compliments with the contents of the textbooks. For example, when introducing Viète and Descartes' expression of equations in the history of algebra, Cajori and Odell (1915, 1916) supplemented them with portraits of two mathematicians and their equations as footnotes. Generally speaking, only portraits of mathematicians who made outstanding contributions to relevant subjects or topics, were included in textbooks.

The supplementation approach refers to integration with aims of "tracing historical origins, supplementing historical knowledge and providing supplementary materials". Supplementation-based materials usually take the forms of appendices and notes, which can be separated from the main content. In early textbooks, the histories of geometry, algebra, trigonometry and coordinate geometry were included in appendices, or relevant historical notes were added to mathematical concepts, theorems, formulae, rules, terms contained in the main part and mathematical problems in the exercises. This type of materials includes "famous quotations", description on mathematicians and mathematical topics in explicit materials in the history of mathematics, as well as the history of a topic apart from the body text.

The duplication approach refers to the direct quotation of historical problems, solutions and proofs of theorems, etc., or histories of relevant topics in the main part. Duplication-based materials are an integral part of the textbooks, serving the functions of providing mathematical problems, evoking ancient wisdoms and promoting mathematics learning. This type of materials includes "mathematical problems", "thoughts and methods" in explicit materials in the history of mathematics, as well as "the history of a topic" in the main part.

The accommodation approach refers to the application of adapted historical materials or adapted historical problems to facilitate students' learning. Accommodation-based materials are also an integral part of the textbooks. Adaptation is necessary since today's students do not know much about the situations of the ancient mathematical problems; and integration of historical materials into textbooks can be beneficial to students in the sense of transcending time and space, communicating with the ancients, and making mathematical problems more meaningful. So, this type of materials serves the functions of providing mathematical problems, increasing exploratory opportunities and stimulating students' interest in learning. In early textbooks, the percentage for accommodation-based materials is generally bigger than that of duplication ones.

The reconstruction approach refers to the application of knowledge in the sequence of occurrence and development of the history of mathematics. Reconstruction does not mean repeat of history, but rather re-creating a topic naturally by referencing historical foundation, knowledge's logical sequence and students' psycho-genetic sequence, serving the functions of mastering cognitive foundation, stimulating learning motivation and promoting mathematical understanding. Reconstruction-based materials include "mathematical concepts" in implicit materials in the history of mathematics. However, such an approach was rarely used in early textbooks due to lack of fixed patterns and profound understanding of the history of mathematics by authors.

Tab. 14-6 and Fig. 14-20 summarize the five methods of approach and corresponding functions contained in early American textbooks.

Tab. 14-6 Approaches of using the history of mathematics in textbooks

Category	Contents	Functions
Ornamentation	Illustrations such as portraits of mathematicians, historical measurement tool and ancient mathematics book-related pictures, mathematics-themed drawings, etc.	Compliment the main body; embellishment, beautification and humanization.
Supplementation	Reading materials on the history of mathematics such as historical notes added to some topics, history of independent special topic, historical description or biography added to the texts and origins of mathematical terms, etc.	Trace historical origins, supplement historical knowledge, and provide supplementary materials.
Duplication	Historical mathematical problems, solutions and proofs of theorems, etc, or the history of a subject at the beginning of textbooks.	Provide mathematical problems, evoke ancient wisdoms and promote mathematics learning.
Accommodation	Exercises adapted from historical mathematical problems, or mathematical problems adapted from historical materials, or thoughts and methods originated but simplified from the history of mathematics.	Provide mathematical problems, increase exploratory opportunities and stimulate students' interest in learning.
Reconstruction	Mathematical concepts introduced by referencing or reconstructing occurrence of knowledge and history of development, or knowledge arranged in a way that suits modern students' cognition.	Understand cognitive foundation, stimulate learning motivation and promote mathematical understanding.

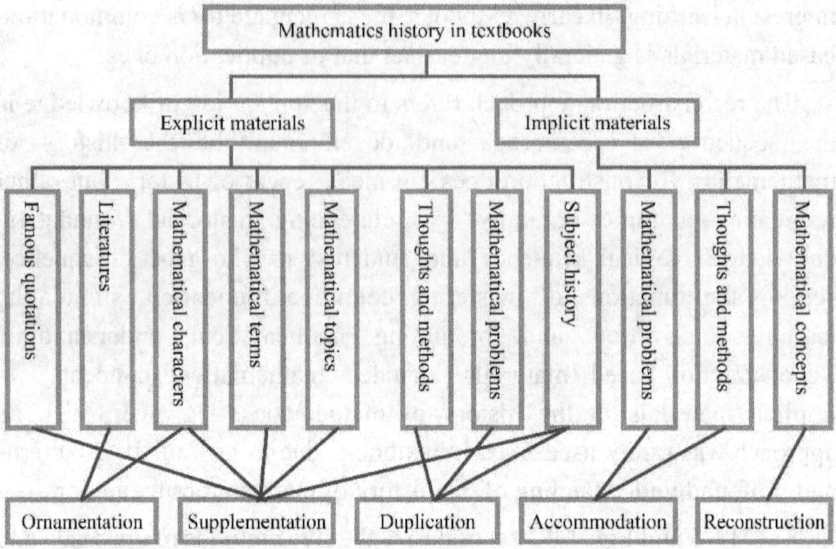

Fig. 14-20 Different approaches to integrate HM into textbooks

3.2. Characteristics shown in the uses of HM in textbooks

After examination and analysis over the contents of mathematics history used in early mathematics textbooks, several noticeable characteristics were found.

(1) Close relations exist in the use of the history of mathematics in mathematics textbooks and researches on the history of mathematics in the same period.

Before Smith and Cajori, only a few authors used the history of mathematics in their textbooks, and they had little access to materials in this regard, except little knowledge points on the Pythagorean Theorem, Pi, etc., used with the Ornamentation and Supplementation approaches. Thanks to their in-depth understanding of the history of mathematics and access to rich materials on this topic, Smith and Cajori were capable of integrating more historical elements into textbooks. However, their perceptions on the history of mathematics were rather narrow and limited by the times they lived in. For example, they never mentioned works of ancient Chinese mathematicians on negative numbers, the Pythagorean

Theorem, Pi, the binomial theorem, higher-order equations, arithmetic progressions, which testified western scholar's ignorance of mathematics achievements made by ancient Chinese mathematicians.

(2) The use of the history of mathematics varies in different textbooks of the same subject.

Researches on the 27 textbooks showed that historical materials were included more or less in them, but the quantities and approaches were by and large different and there was no uniform standard. Further analysis on more textbooks in the same period would show: no explicit historical materials in textbooks could be found. Evidently, the contents, quantities, approaches of using the history of mathematics in mathematics textbooks are subject to authors' understanding of the history of mathematics and recognition of its educational values. Those who integrated historical topics had become an exemplary of their times, including Beman and Smith, Young and Jackson, Betz and Webb, Cajori and Odell, Slaught and Lennes, Rider and Davis, and other mathematicians.

(3) Noticeable differences exist in the use of the history of mathematics in textbooks of different subjects.

Researches on the 27 textbooks showed that, even for the books written by the same author (such as Betz and Webb, Wentworth and Smith), the proportion of the history of mathematics in plane geometry textbooks was greater compared to that of solid geometry, the geometry textbooks greater than algebra ones, and geometry and algebra textbooks greater than other subjects such as trigonometry and analytic geometry. Though large amount of historical materials was integrated by Smith and his partners into geometry and algebra textbooks, strangely, there were no explicit historical materials in Wentworth and Smith (1915), which were far inferior compared to the same type of textbooks by Durer. From Smith (1925), it is safe to assume that he had intimate knowledge of the history of trigonometry (including Arabic mathematicians' work). It can be presumed that Smith thought the history of trigonometry was not suitable for textbook. In this sense, when historical materials cannot be used directly in textbooks, cropping, adaptation and conversion shall be carried out. This process is not easy and requires author's in-depth research.

(4) Significant influence of scientific humanism on mathematics textbooks.

In the 27 textbooks, except Beman and Smith (1899, 1900), materials for the history of mathematics were basically from publications in the 1910s, such as Betz and Webb (1912), Young and Jackson (1913), Wentworth and Smith (1913), Slaught and Lennes (1915), Cajori and Odell (1915). During this period, the scientific humanism movement was still in the ascendant. The movement, without a doubt, played an important role in the development of mathematics textbooks.

4. The History of Mathematics in Today's Prentice Hall Textbooks

Historical materials are found scattering in today's American mathematics textbooks. Take algebra and geometry textbooks published by the Prentice Hall for example: in *Algebra II*, when introducing the quadratic equation with one unknown, the textbook used a stamp commemorating the 1200th birth anniversary of al-Khowarizmi, an Arabic mathematician, together with a brief biography; when introducing the fundamental theorem of algebra, a stamp commemorating the 200th anniversary of Gauss's birth was given; when introducing the Pascal Triangle, information on the calculator invented by Pascal was given. Unit fraction in Rhind Papyrus was introduced in the chapter on "Operation of Rational Numbers" in *Algebra 1*; "the diagram of involution up to the 8th powers of an ancient method" in Zhu Shijie's book *Jade Mirror of the Four Unknowns* (1303) was introduced in the chapter on "factoring of polynomials"; Garfield's Proof of the Pythagorean Theorem was introduced in the chapter on "Area" in Geometry. Gauss' proof of fundamental theorem of algebra and history of Pascal Triangle were introduced in the main body of *Algebra 2*. The above materials involved the Ornamentation, Supplementation and Duplication approaches.

The Accommodation approach was used more often, for example:

- Given the first five rows of the Pascal Triangle, determine each number at the sixth row and their sum;
- Given the first 7 terms of the Fibonacci number sequence, determine the eighth and eleventh terms;

- Determine the removed and remaining areas at step n during the construction process of the Sierpinski triangle;
- Use the Gelosia Method (this method was originated in India, and was called Gelosia method by the Europeans and "Pudijin" by Cheng Dawei, a Chinese mathematician in Ming dynasty) to seek the product of two integers;
- The Problem of "Cattles and Chickens in a Pen", which was adapted from the Chickens and Rabbits: there are 40 heads and 100 feet, calculate the number of chickens and cattle;
- Determine the number sequence generated by Koch curve;
- Determine the number sequence generated by the triangular numbers;
- Use Eratosthenes' method for determining the size of the Earth to draw parallel lines: in Fig. 14-22, how did Eratosthenes know $\angle 1 = \angle 2$?

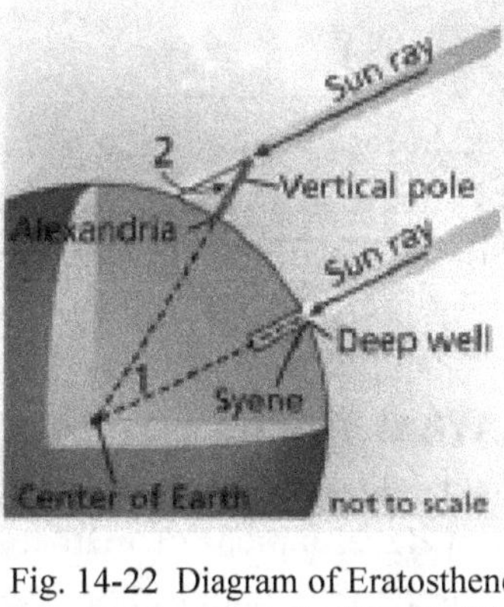

Fig. 14-22 Diagram of Eratosthenes's congruent measurement of the Earth's size

Fig. 14-23 Application of triangles

- Based on a story of measuring the width of a river by an officer from Napoleonic army: prove EF = EG in Fig. 14-23;
- Based on the diagram engraved on the tombstone of Archimedes, determine the ratio of the volumes of a sphere and a circumscribed cylinder.

Considering the length of textbooks, there is little increase in historical materials in the Prentice Hall textbooks compared to the textbooks of one century ago, and the applications are mainly Ornamentation, Supplementation and Accommodation approaches with little Reconstruction. It is safe to say that no obvious progress in integrating the history of mathematics into mathematics textbooks.

5. Conclusions

In mathematics textbooks of the late 19th and early 20th centuries, many historical materials were found which could be grouped into two categories — explicit and implicit materials. Explicit materials include famous quotations, mathematicians, mathematical terms, literatures, mathematical problems, thoughts and methods, history of special topics, history of subjects, etc.; while implicit materials include mathematical problems, mathematical concepts and textbooks, and thoughts and methods. Among which, mathematical problems and the history of special topics were used most frequently. There are five approaches to integrate the history of mathematics into textbooks, i.e., Ornamentation, Supplementation, Duplication, Accommodation and Reconstruction approaches.

During this period, the use of the history of mathematics in American mathematics textbooks was closely related to researches on the history of mathematics in the same period. Ancient Chinese mathematics was hardly known to the west world so the work of Chinese mathematicians was rarely mentioned in these mathematics textbooks. The use of the history of mathematics in different textbooks varies in different subjects and even the same subject, which is subject to author's understanding of the subject history and recognition of its educational values. Moreover, scientific humanism movement has played a role in development of mathematics textbooks. It is worth reflecting that, compared to textbooks of one century ago, no obvious progress has been made in integrating the history of mathematics into today's American textbooks.

It is a systemic project to integrate the history of mathematics into textbooks, which requires not only a deep understanding of the educational values of the history of mathematics but also rich relevant materials. Moreover, the textbook compilers also need to combine the logical sequence and students' cognition, and appropriately cut and adapt relevant historical materials. In spite of countless western mathematics textbooks, only a few of them have used historical materials. This fully testifies that it is far from easy to integrate the history of mathematics into textbooks. There are ample reasons to believe that it deserves long-term study.

References

[1] Beman, W. W. and Smith, D. E. (1899). *New Plane & Solid Geometry*. Boston: Ginn & Company.

[2] Beman, W. W. and Smith, D. E. (1900). *Elements of Algebra*. Boston: Ginn & Company.

[3] Betz, W. and Webb, H. E. (1912). *Plane Geometry*. Boston: Ginn & Company.

[4] Betz, W. and Webb, H. E. (1916). *Solid Geometry*. Boston: Ginn & Company.

[5] Bowser, E. A. (1890). *The Elements of Plane and Solid Geometry*. New York: D. van Nostrand Company.

[6] Boyé, A. *et al.* (2010). The History of Mathematics in School Textbooks: Panel Discussions of the 6th European Summer University on the History and Epistemology in Mathematics Education.

[7] Cajori, F. and Odell, L. R. (1915). *Elementary Algebra: First Year Course*. New York: The Macmillan Company.

[8] Cajori, F. and Odell, L. R. (1916). *Elementary Algebra: Second Year Course*. New York: The Macmillan Company.

[9] Durell, F. (1911). *Plane and Solid Geometry*. New York: Charles E. Merrills Co.

[10] Durell, F. (1912). *Introductory Algebra*. New York: Charles E. Merrills Co.

[11] Durell, F. (1910). *Plane and Spherical Trigonometry*. New York: Charles E. Merrills Co.

[12] Failor, I. N. (1906). *Plane and Solid Geometry*. New York: The Century Co.

[13] Fauvel, J. and van Maanen, J. (2000). *History in Mathematics Education*. Dordrecht: Kluwer Academic Publishers.

[14] Gore, J. H. (1908). *Plane and Solid Geometry*. New York: Longmans, Green, and Co.

[15] Hallett, G. H. and Anderson, R. F. (1917). *Elementary Algebra*. Boston: Silver, Burdett & Company.

[16] Hart, C. A. and Feldman, D. D. (1912). *Plane and Solid Geometry*. New York: American Book Co.

[17] Li, Kaihui (2006). Comparative analysis and revelation of mathematics history contents in mathematics textbooks of middle school in China, Singapore and the United States. *Teaching Reference of Middle School Mathematics*, (8): 53–55.

[18] Li, Mingzhen and Pang Kun (2006). Principles, methods and problems of mathematics history integrating into secondary school mathematics teaching materials. *Shuxue Tongbao* (*Bulletin des Sciences Mathematics*), 45(3): 23–25.

[19] Luo, Xinbing *et al.* (2012). On the characteristics and mode of distribution of mathematics history in high school mathematics textbooks — A case study of textbooks for compulsory course published by Beijing Normal University Press. *Journal of Mathematics Education*, 21(1): 30–33.

[20] Milne, W. J. (1899). *Plane and Solid Geometry*. New York: American Book Company, 1899.

[21] Palmer, C. I. and Taylor, D. P. (1919). *Plane and Solid Geometry*. Chicago: Scott, Foresman and Company.

[22] Prentice Hall Mathematics (2009). *Algebra 1*. Boston: Pearson Education.

[23] Prentice Hall Mathematics (2009). *Algebra 2.* Boston: Pearson Education.

[24] Pu, Shuping and Wang, Xiaoqin (2012). How to integrate history of mathematics into mathematics textbooks: Case study of secondary school textbooks in China and France. *Curriculum, Teaching Material and Method*, 32(8): 63–68.

[25] Rider, P. R. and Davis, A. (1923). *Plane Trigonometry*. New York: D. van Nostrand Co.

[26] Robbins, E. R. (1907). *Plane & Solid Geometry*. New York: American Book Company.

[27] Shoup, F. A. (1880). *The Elements of Algebra*. New York: E. J. Hale & Son.

[28] Slaught, H. E. and Lennes, N. J. (1915). *Elementary Algebra*. Boston: Allyn & Bacon.

[29] Slaught, H. E. and Lennes, N. J. (1919). *Solid Geometry with Problems and Applications*. Boston: Allyn & Bacon.

[30] Siciloff, L. P., Wentworth, G. and Smith, D. E. (1922). *Analytic Geometry*. Boston: Ginn & Company.

[31] Smith, D. E. (1925). *History of Mathematics*, Vol. II. Boston: Ginn & Co.

[32] Stone, J. C. and Millis, J. F. (1915). *Elementary Algebra*. Chicago: B. H. Sanborn & Company.

[33] Wang, Xiaoqin (2012). History of mathematics in French middle school textbook. *Shuxue Tongbao* (*Bulletin des Sciences Mathematics*), 51(3): 16.

[34] Wang, Zhenhui and Wang, Xiaoqin (2003). How the history of mathematics integrates into secondary school mathematics textbooks. *Shuxue Tongbao* (*Bulletin des Sciences Mathematics*), 42(9): 18–21

[35] Wells, W. and Hart, W. W. (1916). *Plane and Solid Geometry.* Boston: D. C. Heath & Company.

[36] Wentworth, G. A. and Smith, D. E. (1913). *Plane and Solid Geometry*. Boston: Ginn & Company.

[37] Wentworth, G. A. and Smith, D. E. (1915). *Plane and Spherical Trigonometry*. Boston: Ginn & Company.

[38] Wentworth, G. A. and Smith, D. E. & Brown, J. C. (1917). *Junior High School Mathematics*, Boston: Ginn and Co.

[39] Young, J. W. A. and Jackson, L. L. (1919). *Elementary Algebra*. New York: D. Appleton & Company.

[40] Young, J. W. A. and Jackson, L. L. (1913). *A High School Algebra*. New York: D. Appleton & Company.

[41] Yang, Yuhui *et al.* (2007). Exploration on content and its presentation of history of mathematics in primary mathematics textbooks. *Journal of Mathematics Education*, 16(4): 80–83.

[42] Zhang, Mingyue (2012). The research on teaching of topics related to history of mathematics in middle school mathematics based on the textbooks published by People's Education Press. Unpublished Master's Thesis, Hunan Normal University.

[43] Zhu, Shijie (1303). *Jade Mirror of the Four Unknowns.* A recension (edited by Zhaohua Li), Beijing: Science Press, 2007; A bilingual (modern Chinese and English) edition (Shuchun Guo, modern Chinese translator; Zaichun Chen, English translator), Shenyang: Liaoning Education Press, 2006.

https://doi.org/10.1142/9789814713955_0015

Chapter 15

Competition Mathematics as a Complement and Enhancement for In-Class Instruction

Bin Xiong* and Yijie He

Department of Mathematics, East China Normal University
Shanghai 200241, China
**bxiong@math.ecnu.edu.cn*

Competition mathematics was introduced to China in 1956, and was developed as a complement and enhancement for in-class instruction. This chapter first provides an introduction to the organization, the training system and learning materials, and then discusses the influence of competition mathematics on students' mathematical ability as well as the existing problems of competition-oriented education in China.

1. Introduction

According to Professor Luogeng Hua (1956a, 1956b) competition mathematics was introduced to China in 1956, and in the same year, mathematics competitions for high school students were held in Beijing, Tianjin, Shanghai and Wuhan respectively. Luogeng Hua (also known as Loo-keng Hua) served as the Chairman of the Beijing Mathematics Competition Committee and engaged in the preparation of test materials. Many other famous senior mathematicians, such as Zhongsun Fu, Jiangong Chen, Buqing Su, Xuefu Duan and Zehan Jiang, also gave lectures during the competitions (Ruiqing Sun, 1998).

The development of Chinese mathematics competition activities can be divided into the following stages (Chuanli Chen and Tongjun Zhang, 2013):

Stage I (1956–1964): Considered as the birth of China's competition mathematics, competitions were mainly advocated and personally directed by senior mathematicians and held in a few large cities of China.

Stage II (1978–1984): After the ten-year political turmoil came to an end, mathematics competitions were resumed. A working committee was set up by the Chinese Mathematical Society (CMS, hereafter) in order to standardize and institutionalize the development of mathematics competition activities in China.

Stage III (1985–present): This period sees a flourishing of mathematics competition activities. China began to participate in the International Mathematical Olympiad (IMO) and in 1990, the 31st IMO was successfully held in Beijing on an unprecedented scale. The level of mathematics competitions in China quickly caught up with the international standard and continued to maintain its leading position afterwards. Meanwhile, various kinds of competitions at all levels were launched and a wide and diverse range of learning materials was readily accessible. Competition-oriented training became the "second classroom" for a proportion of students, or even the "second school" for a few top students.

The following first three sections of this chapter provide an introduction to the organization, the training system and learning materials. The latter two sections discuss the influence of competition mathematics on students' mathematical ability as well as the existing problems of competition-oriented education.

2. Description of the Mathematics Competition Organization in China

Since the establishment of the Popularization Committee of Chinese Mathematical Society and after more than three decades' exploration and practice, an ever-enriching, relatively stable system has evolved in the Chinese high school mathematics Olympiad practice. The National Senior High School Mathematics Competition and National Junior High School Mathematics Competition are two important events in this framework. Over the selection of the IMO national team, CMS has also gradually set up a feasible working procedure.

2.1 The National High School Mathematics Competition

In 1980, with a view to popularizing mathematics on a national scale, CMS held its first national conference in Dalian, in which organizing mathematics competition was confirmed to be a regular task for CMS and its branches in each province, municipality and autonomous region. In addition, the National High School Mathematics Competition was agreed to be held in October annually (Huawei Zhu, 2009). Since 2014, the competition date has been adjusted to September.

The National High School Mathematics Competition is divided into two parts: Test 1 and Test 2 (also called the additional test). Test 1 mainly aims at educational spreading and popularization. Problems are mainly based on the high school mathematics syllabus or standards in use (see, for example, MOE of PRC, 2000, 2003), thus closely related to what students learn in class, but require some problem solving strategies as well as a flexibility in knowledge application. On the other hand, problems in Test 2 largely focus on capability enhancement, some of which are specially designed for testing mathematical ability and identifying mathematical talent. Contents are linked to the IMO, covering geometry, algebra, number theory and combinatorics. The difficulty of such test problems is slightly lower than that of the IMO problems.

Taking the *2013 National High School Mathematics Competition* as an example (see Coaching Staff of National Team, 2014). Test 1 problems were largely related to the main knowledge topics like functions, inequalities, sequences and analytic geometry, which effectively covered each chapter in the textbooks and were designed at an appropriate difficulty tier. Test 2 focused on the extensible content specified in the *Syllabus for High School Mathematics Competition* (PC of CMS). The following four examples are taken respectively from Problems no. 1, 3 and 8 of Test 1 as well as Problem no. 3 of Test 2 of the 2013 National High School Mathematics Competition.

Example 2.1: Let $A = \{2, 0, 1, 3\}$ and $B = \{x \mid -x \in A, 2 - x^2 \notin A\}$. The sum of all the elements in B is ______.

This problem only tests the basic knowledge of set representation. Students of Grade 11 may complete it successfully provided they know the relevant concepts.

Example 2.2: Let ABC be a triangle with $\sin A = 10 \sin B \sin C$ and $\cos A = 10 \cos B \cos C$. The value of $\tan A$ is ______.

The reference solution of this item is as follows: Note that

$$\begin{aligned}\sin A - \cos A &= 10(\sin B \sin C - \cos B \cos C) \\ &= -10 \cos(B + C) = 10 \cos A .\end{aligned}$$

Therefore, we have $\sin A = 11 \cos A$, i.e., $\tan A = 11$.

The formulas used to solve this problem were all basic contents of in-class instructions, but students were required to choose the right ones flexibly with awareness of "elimination" and all-rounded consideration.

Example 2.3: Let $a_1, a_2, \dots, a_9$ be a sequence satisfying $a_1 = a_9 = 1$ and $\frac{a_{i+1}}{a_i} \in \left\{2, 1, -\frac{1}{2}\right\}$ $(i = 1, 2, \dots, 8)$. The number of sequences with such property is ______.

This was a counting problem designed with a sequence background, which was obviously much more difficult than the exercises in textbooks, yet the knowledge and method involved were still within the teaching requirements of combination and permutation, and the amount of calculation needed was appropriately controlled, so it would not be too challenging for students with clear thinking and good logic judgment.

Example 2.4: There are m problems in a test and n students taking the test, where m and n are integers greater than 1. For each problem, the scoring rule is as follows: If a total of x students fail to get a correct solution to this problem, those presenting correct answers will get x points, and the ones with wrong answers will receive zero points. The grade of each student is the total of the points gained from m problems. If the grades of all the students are written as $p_1 \geq p_2 \geq \cdots \geq p_n$ from high to low, find the maximum possible value of $p_1 + p_n$.

This was a typical extremal value problem of discrete variables in competitions, but was unconventional for most students, since in-class instruction in senior high school was mainly concerned with continuous variables. As a Test 2 problem, the knowledge and approach to be applied were both extended to some degree. Generally, it required the students to translate the condition into quantity, correctly apply the properties of inequalities and the mean value inequality of n unknowns to complete the upper bound estimation, as well as to construct an optimal example.

The National High School Mathematics Competition is a public-oriented extracurricular activity with significant influences. In recent years, there are approximately 50,000 students participating in this competition annually (around 1 million students in total if the preliminary competitions organized by provinces or cities are included). Moreover, the National High School Mathematics Competition also has another function — to select the best contestants around China (currently 300–350 annually) to participate in the China Mathematics Olympiad (CMO).

2.2 The National Middle School Mathematics Competition

The National Middle School Mathematics Competition is another public-oriented competitive event organized by CMS to popularize mathematics. It aims to arouse students' interest in mathematical learning, develop their innovative awareness and capability as well as discover and cultivate mathematical talents.

Since 1985, the National Middle School Mathematics Competition has been held annually, normally in March or April (Zhu Huawei, 2009). The organization form is similar to that of the National High School Mathematics Competition.

The contents of the National Middle School Mathematics Competition cover numbers, algebraic expressions, equations and inequalities, functions, geometry, logic reasoning, etc., in which the contents listed in the national mathematics syllabus or mathematics curriculum standard are specified as the basic requirements of the competition (See MOE of PRC, 2012 for the latest national mathematics curriculum standard). High requirements are made with regard to comprehension level, flexibility in application as well as the level of proficiency in the grasp of methods and skills. The competition similarly consists of Test 1 and Test 2. The former focuses on basic knowledge and skills, while the latter focuses on the problem solving and analysis capabilities.

It is worth noting that though the contents of "Viete Theorem of quadratic equation", "criteria and properties of triangle similarity", "angle of circumference", "cyclic quadrilateral", and "tangent length of circle" have been diminished in the test requirements of the currently used curriculum standards, the mathematics competition for middle school still keeps these contents as a supplement for in-class instruction. In the *Syllabus for Middle School Mathematics Competition* (PC of CMS, 2006b), contents of "four concyclic points" and "circle-power theorem" were added in particular besides the above-mentioned knowledge. Take the problems no. 1 and no. 2 in Test 2 (Paper A) of the 2014 National Middle School Mathematics Competition as examples (Shenglin Xu, 2015):

Example 2.5: Let a, b be real numbers such that $a^2(b^2+1)+b(b+2a)=40$ and $a(b+1)+b=8$. Find the value of $\frac{1}{a^2}+\frac{1}{b^2}$.

Example 2.6: As shown in Fig. 15-1, in the parallelogram $ABCD$, point E is on diagonal BD, satisfying $\angle ECD = \angle ACB$. The circumcircle of ΔABD meets the extension line of AC at F. Prove that $\angle DFE = \angle AFB$.

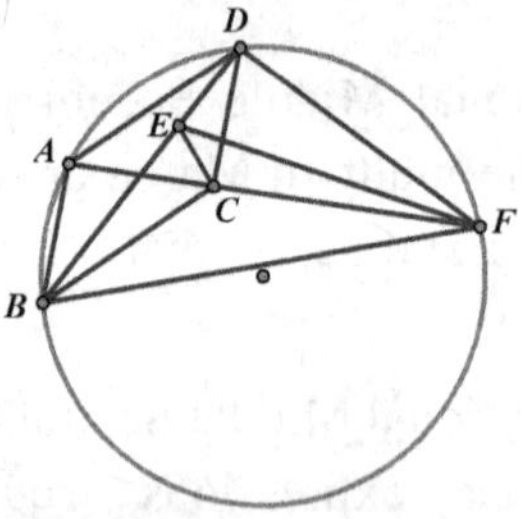

Fig. 15-1. Figure of Example 2.6

The problem in Example 2.5 could be simplified by the substitution $a+b=x,\ ab=y$. It was not difficult to see that the structural feature of algebraic expression in this problem was closely related to the Viete Theorem of quadratic equation; in Example 2.6, the knowledge of similarity of triangles and four concyclic points was tested.

If the curriculum standards are meant for the fundamental goal of compulsory education, then competition mathematics can be reckoned as an extension, in both breadth and depth, for students who have the capacity and desire for further study, to develop their thinking ability and avoid a large amount of low-level repetition in mathematical learning. The training in algebraic techniques can improve their ability in algebraic calculation; the knowledge of circle and similar triangles can provide abundant mathematical materials for in-class instruction and after-class study, and further improve the students' ability of geometrical reasoning. It should also be noted that, these are directly related to the analytic geometry and other contents to be studied in senior high school.

2.3 The selection procedures for IMO Chinese national team

Fig. 15-2 illustrates the selection procedures for IMO Chinese national team.

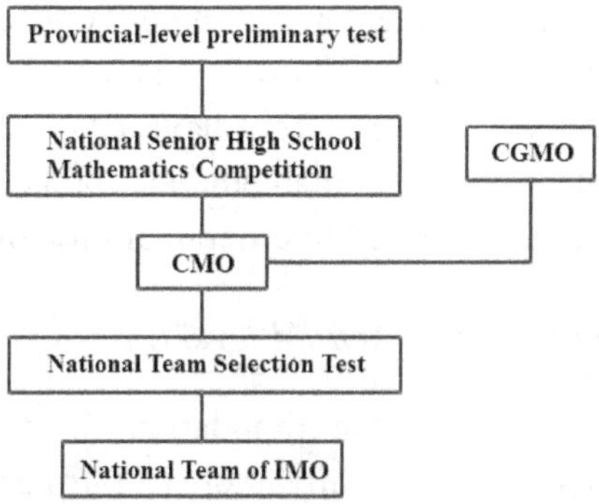

Fig. 15-2. The selection procedures for IMO Chinese national team

In Fig. 15-2, CMO refers to the "China Mathematics Olympiad", a mathematical competition organized by the Mathematical Olympic Committee of CMS aiming to select mathematical talents. It is also the top mathematics competition for secondary school students in China. CGMO refers to the "China Girls Mathematical Olympiad", a mathematics competition specifically for high school girl students designed by the Mathematical Olympic Committee of CMS aiming to provide a platform for girls to show mathematical talents, to increase their interest in mathematics, and to promote mutual learning and friendship of girls from various regions.

The participants in CMO consist of provincial winners of the National High School Mathematics Competition, the CGMO winners and the invited international teams. The top 60 participants from mainland China have qualification to enter the National Training Team and they also qualify for admission into any top university in China. The difficulty of the CMO test is roughly the same as the IMO.

As the final round of the entire selection process, the National Team Selection Test aims to select 6 members for the national team of IMO. Usually this round consists of several sets of extremely difficult problems.

3. Training System for Mathematics Competition in China

Along with various mathematics competitions, extensive educational activities related to these competitions have been organized all over China and a comprehensive training system have taken shape gradually. As a result, China's performances in mathematics competitions have improved a lot and have gradually formed international influence.

3.1 Organizational forms of school-level training

School-level training is the foundation for Chinese education on competition mathematics. It mainly adopts a form of "second classroom", such as interest groups and extension courses etc., which is accessible by each grade of secondary school. Such training activities mainly serve as a means of popularization for mathematics competitions and a supplement for classroom study. However, at some key schools, school-level training is highly specialized and these schools have become home bases for training high-level Mathematics Olympiad participants.

It is generally accepted that gifted education mainly consists of three basic forms: enrichment, differentiation and acceleration (Zhirong Sheng and Chao Zhou, 2010). Enrichment refers to enhanced study materials for gifted students, so as to extend their scope of knowledge. Differentiation refers to grouping students according to their capabilities with adapted courses to suit the needs of students at different level. Acceleration means intensive materials or more information for students, so as to accelerate the pace of teaching, and let gifted students acquire advanced knowledge system as soon as possible. The three forms of training coexist in school-level training for mathematics competitions.

For enrichment, to be specific, schools offer mathematics extension courses with different themes that include competition mathematics, mathematical problem solving, mathematical modeling, history and culture of mathematics. Schools offer a broad variety of choices to students with different interests and at different levels. These are beneficial to enhancing students' mathematical understanding, improving their mathematical literacy and broadening their horizons. Moreover, some schools also organize competition mathematics groups among gifted students by conducting extension lectures or training after normal school hours. Participation in this kind of activities is voluntary and mainly for mathematics competitions.

For differentiation, at some key schools, students with mathematical talent are assigned to the same class to receive separate small-size class teaching. With a tradition of specializing in competition mathematics, schools provide accelerated mathematics courses in each grade. The overall teaching plan for the entire semester is mapped out at the time of admission, and further differentiated teaching is adopted for students in the above accelerated programs. Shanghai High School, for example, has adopted differentiated teaching since 1990 and successfully trained the IMO gold medalists for consecutive years starting from 2008. Every year, the school selects 40 odd freshmen who are mathematically talented to form an experimental class, and designs specific courses for them. Since 1998, it began to further select about 10 most gifted students from the experimental class to teach them in small class. Finally, individual instruction is offered to 3–4 exceptionally gifted students who have shown great potential (Shengchang Tang and Zhigang Feng, 2011).

In many instances, differentiation is to accelerate the teaching process for gifted students, i.e., to let students quickly master textbook knowledge as well as the basic knowledge and skills required in mathematics competitions, so as to lay a better foundation. After this is completed, students are at liberty to further study for competition mathematics and improve their skills in solving mathematical problems.

Among the above three forms of gifted education on mathematics, enrichment mainly focuses on popularization of competition mathematics, while differentiation and acceleration serve the functions of both popularization and improvement. The latter two forms do not apply to most students so as to avoid excessive academic burden. However, they are suitable for students who can easily master mathematical knowledge and skills without adversely affecting the study of other subjects.

For secondary school mathematics competitions, many schools such as Shanghai High School adopt a "$1 + n$" tutor teaching model (Shengchang Tang and Zhigang Feng, 2011), which is to assign a long-term mathematics teacher for a certain group of gifted students. This model consolidates the collective wisdom of the school's mathematics teaching team and sometimes even a guest expert team. The core teacher must have intimate knowledge of every student to assess their potential, follow closely their performances in school and formulate individualized instructional plans. He/she must also set up a specific timetable for each milestone and invite professional experts to offer special guidance at appropriate times. This model has several advantages — putting equal emphasis on students' foundation and improvement, showing students how different teachers have different perspectives and thinking styles of mathematical problem solving.

School-level training normally starts from classroom teaching to the popularization and promotion of competition mathematics. This whole process can be roughly divided into four stages, i.e., basic training, thematic training, intensified training and pre-competition training (Department of Secondary School Mathematics of the People's Education Press, 2006). The emphasis for each stage is different.

The basic training stage aims to finish teaching the contents of high school mathematics, thus ensuring that students reach the level required for graduation. Forms and methods of teaching are similar to conventional ones, but contents are intensified and expanded to some degree, with an emphasis on basic thinking skills used in competition problems. Training at this stage enables students with a certain mathematical ability to perform well in public-oriented competitions.

The main task of the "thematic training" stage is to systematically impart knowledge and problem solving skills required in mathematics competitions. In the meantime, teachers need to encourage students to form good study habits, further improve their self-learning awareness and ability, as well as keep a positive learning attitude. After this stage, students can gain a deeper understanding of various topics in competition mathematics, be able to solve difficult problems such as those in Test 2 of National High School Mathematics Competition.

Intensified training is generally targeted at a few outstanding students (mainly CMO and higher-level participants). At this stage, challenging problems can be used for exercise, to keep top students moving forward by competing and cooperating with each other.

Pre-competition training is usually aimed at preparing students for a certain competition. Well-prepared simulation tests not only enable students to familiarize themselves with forms of competition, but also reveal students' lack of knowledge and skills so as to provide reference for them and their teachers. A week or two prior to some provincial or higher level mathematics competitions, a lot of schools will organize intensive training for participants during vacation time and after-class hours. This is also an integral part of school-level training.

Generally speaking, the first stage is closely related to the usual curriculum, while the second and the third mainly put emphasis on skill improvement, and the fourth mainly focus on competitive readiness.

3.2 Provincial and municipal-level competitions and training organizations

There are many high school mathematics competitions at provincial and municipal levels in China. Such competitions are usually hosted by the mathematical societies or associations in relevant provinces and cities. In Beijing and Shanghai, these competitions have become traditional events and test papers of previous competitions are systematically published. In some provinces, the provincial mathematics competition is reckoned as preliminary tests for the National Mathematics Competition.

In addition to hosting competitions, provincial and municipal mathematical societies also organize training programs for mathematics competitions.

Many provinces organize mathematics summer camps, and invite university professors, senior teachers and trainers to teach the participating students. In Zhejiang province, for example, the Department of Mathematics of Zhejiang University and Zhejiang Mathematical Society jointly hold mathematics summer camps every year, and organize training courses at levels from college entrance exam to the National High School Mathematics Competition, in order to improve students' performances in competitions and university entrance exams.

The various provincial and municipal mathematics summer camps are very popular among high school mathematics enthusiasts, and often attract students from other provinces and cities to participate. These camps have turned out thousands of talented mathematics students, among which there were IMO and CMO winners as well as top candidates in the National College Entrance Examinations.

A distinctive model can be found in Shanghai — the Secondary Extracurricular Mathematics School. Thanks to the training of the Secondary Extracurricular Mathematics School, students in Shanghai have got great achievements in domestic and international mathematics competitions in the recent two decades and more. The school is hosted by Shanghai Mathematical Society and guided by the Teaching and Research Office of the Shanghai Municipal Education Commission. Since its establishment in 1987, the school has been run for more than twenty years, and the late mathematician Chaohao Gu used to be the honorary principal of this school. Students come from the sixth to twelfth grades, with approximately 400 students from each grade, all of whom are Shanghai's students talented in mathematics. The school also invites professors from Fudan University, East China Normal University and the best local trainers for mathematics competitions. Every Sunday, the school offers 2 hours of extracurricular instruction on mathematics.

Courses offered by this school are roughly synchronized with general courses by high schools, conforming to the "enrichment" form of gifted education. Here, students not only have the opportunity to strengthen what they have already learnt in the classroom, but also to learn new knowledge, such as knowledge in elementary number theory and combinatorics, some famous theorems in Euclidean geometry, as well as mathematical problem-solving methods and skills. These students can quickly master them after teachers' instructions and group discussions.

Furthermore, entrusted by Shanghai Mathematical Society, the school has also been responsible for the proposition and organization of competitions and selected some exceptionally gifted students to receive special tutoring. Since 1987, almost all national team members coming from Shanghai have been trained in this school.

With a combination of provincial, municipal and school-level training, both mathematically gifted students and their tutoring team are offered the opportunity for further development.

3.3 Training system for mathematics competition

Generally, the structure of training system for China's mathematics competitions is of "school level — provincial & municipal level — inter-provincial level", as shown in Fig. 15-3, balancing popularization and improvement and covering students of different ages and learning levels. Inter-provincial mathematics competition activities consist of regional mathematics competitions and a variety of summer camps, which can generate many shared resources each year.

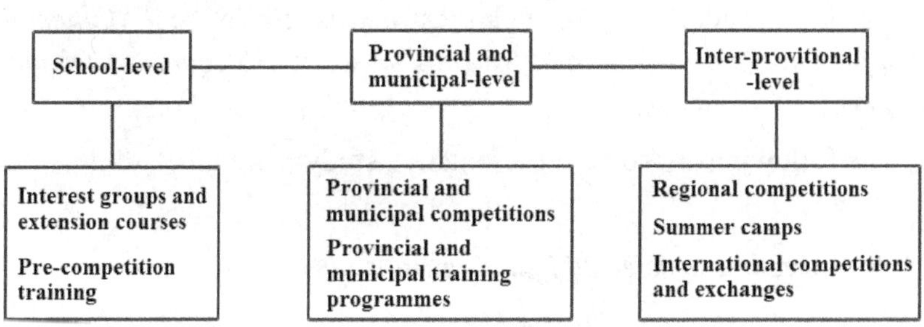

Fig. 15-3. The training system for China's mathematics competitions

Moreover, the training of tutors is also a part of the overall training system. The Popularization Committee of CMS established a "China Mathematics Olympiad Tutor Rating System" in 1988. CMS and its branches in each province organize training workshops for high school teachers who have little experiences in teaching competition mathematics, then evaluate their performances through their problem-solving skills and teaching plans. Tutors will receive some extra learning opportunities before workshops in addition to basic courses and teaching internship on mathematics education. Some normal universities also offer optional courses such as competition mathematics, problem-solving strategies and mathematical methodology to them for future development.

4. Competition Mathematics Learning Materials

Books are the best teachers.

Since China formally participated in IMO in 1985, research and reading materials on competition mathematics have flourished in China, covering a wide range of contents with a variety of styles. Take East China Normal University Press as an example, not only does it publish books for above-average students, such as *From Textbooks to Mathematics Olympiad*, *Mathematics for Excellent Students*, *Excellent Problem Solver*, but it has also published competition popularization books such as *Mathematics Olympiad Tutorial*, *Mathematics Olympiad Book Series*, providing problems and full solutions with helpful tips and tricks. Furthermore, it has published books that collect problems and solutions of domestic and international high school mathematics competitions at various levels and of various categories such as *High School Mathematics Olympiad Reference Manual*, *Towards IMO: A collection of Olympiad mathematics problems*.

The following are some representative study materials.

4.1 Introduction to selected learning tutorials

— *Mathematics Olympiad Tutorial* (published by East China Normal University Press):

This series of books was originally published in 2000, and the traditional Chinese version was subsequently published in Hong Kong and Taiwan. It is now in its sixth edition, covering various grades from primary school to high school. As a supplement, the supporting Study Manuals and Proficiency Tests were rolled out. It is one of the most influential series of competition-oriented mathematics books in China.

Most contents of this series are based on the requirements of the school syllabus and proceed at the same pace with the teaching process, while about a third of its contents are deemed as difficult, aiming to satisfy the needs of students for high-level mathematics competitions. Take the volume for Grade 10 (the sixth edition) (Bin Xiong and Zhigang Feng, 2014) as an example, the book is comprised of two parts with 30 lectures (Lectures 1–21 are designed for Foundation, Lectures 22–30 for Improvement), including sets, functions, sequences, trigonometry, vectors, etc., which are basically in line with high school teaching progress. Easy problems can be used for classroom examples. Problems of medium difficulty can be used as an extension for classroom teaching, which are of the Test 1 level of National High School Mathematics Competition. Difficult problems are of the Test 2 level of the National High School Mathematics Competition. Among the 254 problems in this book, the proportions of easy, medium-difficult and difficult problems in Foundation part is roughly 2:3:1, while that in Improvement parts is roughly 1:2:3.

— *Mathematics Olympiad Book Series* (published by East China Normal University Press):

This series of books (second edition) comprises of 8 and 14 books for middle and high school students respectively. Most of its authors are experts of competition mathematics. It covers a wide range and detailed topics, which not only include common methods in competition mathematics, but also detailed analysis and solutions, some of which are the author's own research results.

This series of books contains works of unique style. Take *Equations and system of equations* (Jun Ge, 2012) (for middle school) as an example. The book is comprised of 18 lectures, covering knowledge in linear equations, systems of linear equations, quadratic equations, higher-order equations, fractional equations, irrational equations, systems of quadratic equations with two variables. integral roots and related applications. The author arranges problems from simple to complex, putting emphasis on the methods of solving problems and making it easy for both teaching and self-learning. The author also utilizes mathematics as a "medium" to promote human wisdom and its educational values via introducing ancient poetry, quotations, historical puzzles, anecdotes about great men and interesting stories, etc.

— *National Middle School Mathematics Competition Tutorial* (published by Peking University Press):

This set of books comprises of 3 volumes (for Grades 7, 8 and 9) with rich materials ranging from curricular to extra-curricular, from theory to application. Each volume is further divided into three parts, namely, Foundation, Improvement and Application. Take the volume for Grade 9 (Ruiqing Sun, 1998) as an example. The foundation part consists of 14 lectures, specifically, "solving fractional equations", "solving irrational equations", "solving simple higher-order equations", "problems about systems of equations", "concepts and properties of functions", "quadratic functions", "maximum and minimum values of functions", "Viete Theorem and its applications", "discriminant and its application", "solving quadratic inequalities in one variable", "trigonometric functions", "solving right triangles", "basic properties of circles" and "positional relationship between two circles". The improvement part consists of 8 lectures, specifically, "concyclic points", "proportional segments in a circle", "valuation problems in plane geometry", "extremal value in plane geometry", "selected famous theorems in plane geometry", "coincidental points in a triangle", "proof by contradiction" and "topics on ideas and methods in problem solving". The application part consists of 3 lectures, i.e., "turning practical problems into mathematical ones (II) — an introduction to mathematical modeling", "mathematics in real-life (V) — light and shadow" and "mathematics in real-life (VI) — earth, sun and moon".

The foundation part basically covers contents stipulated in mathematics syllabus of middle school. By reading this part, students not only can review classroom knowledge, but also gain a deeper and systematic understanding of it. Students who are able to easily grasp this part can continue to read the improvement part. The application part focuses on nurturing students' ability to apply knowledge in textbooks to real life.

— *Hua Luogeng Mathematics School Olympiad Book Series* (published by Encyclopedia of China Publishing House):

Hua Luogeng Mathematics School is one section of the School Affiliated to Renmin University of China. It was jointly established by the University of Science and Technology of China, Hua Luogeng Laboratory of Mathematics of Chinese Academy and the School Affiliated to Renmin University of China, aiming to discover and train mathematically gifted students at an early stage. This school organized a team of experts to write the mathematics textbook series as well as the problem and solution series, which were published around 1995 and considered as representative study materials for competition mathematics back then.

The mathematics textbook series is a set of school-based textbooks of extreme difficulty focusing on gifted education. Take Grade 10 textbook (Liu Pengzhi, 1995) for example, the book includes 15 lectures: "sets and subsets", "functions and equations (I)", "functions and equations (II)", "divisibility of integers (I)", "divisibility of integers (II)", "the function [x]", "geometric transformation", "area", "straight-line geometry", "vectors", "solid geometry (I)", "solid geometry (II)", "graph theory (I)", "combinatorial problems (I) — game", and "combinatorial problems (II) — the Pigeonhole Principle and proof by contradiction". Almost all the lectures contain extension content and insightful points of view. There are 155 example problems, 90% of which are beyond the difficulty of the Test 1 of National High School Mathematics Competition.

The Problem and Solution series is easier compared to the textbook series. This series of books provides rich training materials for students at different levels.

4.2 Other study materials

Under the influence of thoughts on problem solving proposed by renowned mathematician and educator George Polya, some researches on problem-solving theory and practice in competition mathematics based on mathematical methodology have been carried out in China. These books mainly focus on methodologies and strategies, overlapping with classroom teaching, which are beneficial for deepening the understanding of knowledge and thoughts on problem solving. For example, *Study of Problem Solving* (Zun Shan, 2007) and some other books are excellent works that combine competition mathematics and problem solving strategies in mathematical problems, which are very inspirational for competition participants and tutors.

Due to the convenience of access and exchange of information, Chinese experts working in competition mathematics education collected a variety of international materials and gradually published problems and solutions of IMO Shortlist and Mathematics Olympiads in other countries. These collected materials have broadened students' horizons and become a window for Mathematics Olympiad enthusiasts to know about mathematics competitions around the world.

In the tide of research and popularization of competition mathematics, mathematicians, educational researchers and high school teachers (even their students) have all participated in reporting their findings on various mathematics education journals. Some of them offer good reading materials for competition participants and tutors. For example, *Secondary School Mathematics* is a specialized monthly journal sponsored by Tianjin Normal University, Tianjin Mathematical Society and the Popularization Committee of CMS, whose contents are mainly about secondary school mathematics competitions and after-class activities. This journal consists of several columns, namely, Lectures on Mathematical Activity Courses, Problem Posing and Problem Solving, Students' Writing, Research on Elementary Mathematics, Window of Competitions, After-class Training, etc. Every year, it also issues two special supplements. One is *Domestic and International Mathematics Competitions*, the other is *A Collection of Simulation Exams of the National High School Mathematics Competition.*

5. Training Mathematical Ability through Competition Mathematics

Mathematical ability refers to a relatively stable psychological characteristic for the successful completion of mathematics activities. Mathematicians, educators and psychologists at home and abroad have discussed mathematical ability from different aspects.

Krutetskii (1976) has determined the 9 key elements of mathematical ability according to the basic features of mathematical thinking:

1. The ability to formalize mathematical materials, and operate in the formal structure;
2. The ability to summarize mathematical materials;
3. The ability to operate by using numbers and other symbols;
4. The ability to use continuous and rhythmic logical reasoning;
5. The ability to shorten reasoning process;
6. The ability to reverse psychology process (ability of transferring from positive thinking to reverse thinking);
7. Flexibility in thinking;
8. Mathematical memory; and
9. Spatial concept.

In the *Principles and Standards for School Mathematics* (NCTM, 2000) released by American National Council of Teachers of Mathematics, "problem-solving, reasoning and proof, communication, connection and representation" are defined as the five criteria of process ability in mathematics.

It was pointed out by the *Mathematics Curriculum Standard for High School (Experimental)* (MOE of PRC, 2003) that when people were learning mathematics and using mathematics to solve problems, they constantly underwent thinking process such as intuitive perception, observation and discovery, analogical induction, spatial visualization, abstraction, symbolic representation, operation and problem solving, data processing, deductive reasoning, reflection and construction. It also indicated that mathematical thinking played a unique role in the formation of rational thinking.

Competition mathematics education is beneficial to the development of gifted students' mathematical ability in various aspects.

5.1 Mathematical ability and problem solving

We may as well combine with the NCTM standards to give some explanations to the educational value of competition mathematics.

In recent years, the teaching of geometry has been weakened in the compulsory education in China. The lack of training in deductive reasoning hinders the development of students' reasoning skills, which arouses the concern of mathematicians and mathematics educators. Yet in mathematics competitions at all levels, the proportion of geometry problems remains stable, especially in IMO of recent years, where team leaders from all countries always select some extremely difficult geometric problems as official competition problems. It is somewhat beneficial to maintain the level of reasoning and proving of those mathematically gifted.

In addition, mathematics competition problems require students to recognize and use "connections" with flexible mathematical thinking at a more advanced level. At the same time, they require students to be good at selecting, applying and converting mathematical representation, which contributes to the enhancement of students' ability in mathematical connection and mathematical representation.

Furthermore, mathematics competition problems require students to present their mathematical ideas and problem-solving process clearly. Students should also evaluate others' mathematical thinking and problem-solving processes by communicating with teachers and classmates. These activities are beneficial for improving students' ability in mathematical communication.

More importantly, competition mathematics provides rich source for mathematical problem solving. It is pointed out by Luogeng Hua that "the nature of mathematics competition is different from that of an exam in school, and also not the same as the university entrance exam. What we require is that students taking part in the competition can not only apply formulae and theorems, but also show their flexibility in thinking, a good understanding of mathematical principles, and the ability to use these principles to solve problems. They should even be able to discover new methods and principles to solve unfamiliar problems. Such a requirement can exactly test and train the students' mathematical ability" (Luogeng Hua, 1956b).

Example 5.1: Ten numbers $1, 2, \ldots, 10$ are written by order on a circumference (with 10 adjacent to 1). Two operations are permitted to be performed on the numbers: (a) swapping the locations of two adjacent numbers, (b) adding an integer simultaneously to two adjacent numbers. Question: Is it possible to turn all numbers into 10 by finite steps of such operations?

This is a relatively easy competition problem. The authors of this chapter used to ask this problem to high school students who had not undergone competition mathematics training. Students made repeated attempts and found it was always so close to the "goal" but always failed. They hardly caught the key point of the problem. Some of them were confident that the conclusion was "impossible", but they struggled to carry out the mathematical reasoning. As a matter of fact, one might as well consider these 10 numbers on an overall basis: No matter operating through (a) or (b), the sum of the ten numbers always maintains the original parity. Teachers often guide students in problem-solving strategy, inspire them to be aware of all-rounded consideration and to find the quantitative relationship which remains unchanged in the operation.

Example 5.2: Given real numbers m, x, y with $x, y > 0$ and $x + y < \pi$. Prove that

$$(m^2 - m)\sin(x + y) + m(\sin x - \sin y) + \sin y > 0.$$

This problem is somewhat difficult. The challenge lies in complex structure of the expression (quadratic form mixed with trigonometric form), too many parameters (three in total), and not knowing how to use the condition $x, y > 0$ and $x + y < \pi$. Here is an illuminating solution:

Step 1: Construct ΔABC with $A = x, B = y, C = \pi - x - y$. Let $a = BC, b = CA, c = AB$.

Step 2: Applying the law of sines, one has

$$\frac{\sin x}{a} = \frac{\sin y}{b} = \frac{\sin(\pi - x - y)}{c} = \frac{\sin(x + y)}{c} > 0.$$

Therefore, the initial inequality can be transformed into

$$(m^2 - m)c + m(a - b) + b > 0.$$

Step 3: Since $c > 0$, it suffices to prove that the discriminant of the quadratic form $cm^2 + (a - b - c)m + b$ with respect to m is negative, i.e., $(a - b - c)^2 - 4bc < 0$, or, equivalently, to prove

$$a(a - b - c) + b(b - c - a) + c(c - a - b) < 0,$$

which is an easy consequence of $a < b + c$, $b < c + a$, $c < a + b$.

Throughout the above solution, the original problem has been translated and reduced step by step. The first step is an application of the idea of "construction", which allows the subsequent steps to be removed from a lot of complex calculation. The second step is due to the condition $x, y > 0$, $x + y < \pi$ and the homogeneity of the law of sines. The third step is due to the quadratic structure of the inequality with respect to m.

When explaining problems such as Example 5.2, teachers should guide students to grasp the structure of the algebraic expression, to make association and connection with familiar knowledge in order to simplify the original problem. A helpful tip is to apply the idea of "construction", which often enables problem solving process to be "simple and delicate".

Solving problems by constructive method requires comprehensive knowledge, divergent thinking and keen intuition. The following Example 5.3, a problem quite impressive in the authors' teaching experience, is also related to the constructive method.

Example 5.3: For any given positive integer m, prove that there exist $2m + 1$ positive integers a_i $(1 \le i \le 2m + 1)$, so that the integers form an increasing arithmetic progression and their product is a perfect square (Bin Xiong and Yijie He, 2012).

The solution can be done in a sentence: Let $a_i = ik$ $(1 \le i \le 2m + 1)$ with $k = (2m + 1)!$. Then $a_1 a_2 \cdots a_{2m+1} = (2m + 1)!\, k^{2m+1} = (k^{m+1})^2$.

However, this brief answer does not reflect the hidden thinking process. How can this problem be considered? In fact, one can start from an arbitrary increasing arithmetic progression with $2m + 1$ positive integer terms $b_1, b_2, \cdots, b_{2m+1}$. Note that for any positive integer k, the numbers $b_1k, b_2k, \cdots, b_{2m+1}k$ also make up an increasing arithmetic progression. Therefore, one can freely select the value of k to satisfy the condition "product is a perfect square". For instance, one can choose $k = b_1b_2 \cdots b_{2m+1}$, then $(b_1k)(b_2k)\cdots(b_{2m+1}k) = (k^{m+1})^2$.

In brief, we have adopted a frequently used strategy that "relax a condition and then try to re-impose it", which is typically helpful in solving a number of construction problems.

The problem is not as easy as it seems in solution. Once we gave this problem to a number of high school students with some experience in mathematics competition. We observed that most students considered this issue from the following two approaches: The first one was starting from the simple cases (for example, the case of 3 terms or 5 terms), then tried to make a generalization. However, it would seem difficult for these students to extract the general rules. The second approach was to set out the two basic parameters of arithmetic progression (such as the common difference d and the middle term $a = a_{m+1}$), expressed the product of the $2m + 1$ terms, and then tried to solve the equation with three unknowns. We noticed that some students gazed longingly at the equation and got stuck because of the complexity of the structure. Only a few of them completed the construction within 20 minutes.

Indeed, in solving such a problem, a great deal of thinking is needed. Students should not only strive to plan their solving processes, but should always monitor their thinking to determine the feasibility of the scheme, and avoid the interference of invalid plans. Such kind of experience is beneficial for students to improve both their problem solving skills and their metacognitive monitoring strategies.

For ease of comparison with Examples 5.2 and 5.3, the following three problems relating to the same knowledge point:

Example 5.4: Determine all possible real number k such that the inequality $kx^2 - (k-2)x + k > 0$ holds for any real number x.

Example 5.5: A triangle ABC with circumcircle radius $\sqrt{2}$ satisfies $2\sqrt{2}(\sin^2 A - \sin^2 C) = (a - b)\sin B$. Find the degree of angle C and the maximum value of the area of the triangle.

Example 5.6: Let $\{a_n\}$ be an infinite arithmetic progression. For any positive integer k, denote the sum of $a_1, a_2, \cdots, a_k$ by S_k.

(1) If $a_1 = \frac{3}{2}$, and the common difference $d = 1$, find all positive integers k satisfying $S_{k^2} = (S_k)^2$.

(2) Find all infinite arithmetic progression $\{a_n\}$ with the property that $S_{k^2} = (S_k)^2$ for all positive integer k.

The three problems above are selected from *Mathematics Review Guide for College Entrance Examination* (Department of Secondary School Mathematics of the People's Education Press, 2006). They roughly correspond to the difficult problems in classroom teaching. The knowledge and method involved in Example 5.4 and Example 5.5 are close to that of Example 5.2, but the direction in problem solving is somewhat clear. In Example 5.5, one can apply a routine method to find the answer of Subproblem (1). For Subproblem (2), one can infer possible pairs (a_1, d) from the consideration of cases $k = 1, 2$ before a complete verification, or can directly write an equation of degree 4 in unknown k and solve it by the principle of polynomial identity. Both approaches require strong computational and reasoning skills, however, a specific formula and method can be applied at each step. In contrast, although no extra knowledge is needed in Examples 5.2 and 5.3, the two problems are not conventional in classroom teaching and require much more thinking. Example 5.1 usually does not appear in classroom teaching, though it only relates to the knowledge of addition and subtraction. However, it is good for high school students in the sense of enriching their thinking patterns.

For those who have a certain understanding of competition mathematics, most feel that even if one is able to read the methods and principles for 100 problems, one could still feel helpless when facing the 101st problem. In a sense, many mathematics competition problems are special or even unique, thus simple imitation does not always work. In fact, students should be able to transfer the thinking method and problem-solving strategy in one case to other cases by imitating and practicing, in which way they will not be at a loss when solving new problems in mathematics or even other disciplines. But at the same time, they should also be able to generate new ideas and try different methods, which is in line with the basic spirit of problem solving.

5.2 Competition mathematics and teaching of open-ended mathematics problems

Teaching of open-ended problems is another major issue related to problem-solving and creativity. It was first introduced from Japan by Zaiping Dai who was a strong advocate of it. Since then, abundant theoretical achievements have been achieved in China (Zaiping Dai, 2002).

Roughly speaking, open-ended mathematics problems are those with non-exclusive answers. In most cases, there is only one correct answer to high school mathematics problem; even for "a problem with several solutions", there are only a few which would be easy to arrive at. However, it seems that problems posed in higher-level mathematics competitions are more likely to be open-ended in terms of problem-solving approach, with solutions often beyond the problem designers' anticipation.

For example, among the four problems in Test 2 of the 2013 National High School Mathematics Competition, the solutions of three problems were simplified afterwards. In 2014 National Team Selection Test, the examining committee noted some brilliant solutions after going over the answer sheets, and the given reference answers to several other problems were also simplified, which accounted for more than half of the total. In Southeast China Mathematics Olympiad in 2014, a much simpler solution, unexpected by all experts from the examining committee, was discovered by a few of participants to unlock the most difficult problem. Historically, in the 1980 joint competition of Finland, the United Kingdom, Hungary and Sweden, the solution of a certain problem was rather complicated where mathematical induction was used four times, the Chinese translation of which took around 4,000 characters. Later Chinese experts gave some simpler solutions, the lengths of which were just a dozen lines (Huawei Zhu, 2009). However, the relevant algebraic skills were deeply concealed and difficult to be discovered.

In all previous sessions of IMO, the examining committee would present a special award to those participants who give particularly brilliant solutions and non-trivial generalizations. To date, this special prize has been awarded to more than 40 participants.

Below is a case from the author's personal teaching experience:

Example 5.7: Let ABC be a triangle with area S and circumcircle radius R.

(1) Prove that $S = 2R^2 \sin A \sin B \sin C$.

(2) Prove that $S = \frac{1}{2}R^2(\sin 2A + \sin 2B + \sin 2C)$.

This problem was designed for a test just after the completion of teaching, which was intended to test students' flexible application of the formula of triangle area, the law of sines and the sum-to-product formulas. The anticipated approach was to use the formulas of triangle area $S = \frac{1}{2}ab\sin C\left(\text{or } S = \frac{abc}{4R}\right)$ and the law of sines $\frac{a}{\sin A} = \frac{b}{\sin B} = \frac{c}{\sin C} = 2R$ to finish Subproblem (1), and further to solve Subproblem (2) by deducting the identity $\sin 2A + \sin 2B + \sin 2C = 4\sin A \sin B \sin C$ in ΔABC.

One of the students came up with a different approach to solve Subproblem (2):

Denote the circumcenter of ΔABC by O. The student first assumed that ΔABC is an acute triangle. In this case,

$$S_{\Delta AOB} = \frac{1}{2}OA \cdot OB \sin\angle AOB = \frac{1}{2}R^2 \sin 2C.$$

Similarly, $S_{\Delta BOC} = \frac{1}{2}R^2 \sin 2A$, $S_{\Delta COA} = \frac{1}{2}R^2 \sin 2B$. Hence,

$$S = S_{\Delta AOB} + S_{\Delta BOC} + S_{\Delta COA} = \frac{1}{2}R^2(\sin 2A + \sin 2B + \sin 2C).$$

When ΔABC is a right triangle, this equality also holds though the area of one part is zero.

When ΔABC is an obtuse triangle, the student assumed $A > \frac{\pi}{2}$ without loss of generality so that $S = S_{\Delta AOB} - S_{\Delta BOC} + S_{\Delta COA}$.

Note that $\angle BOC = 2\pi - 2A$. Then

$$S_{\Delta BOC} = \frac{1}{2}R^2 \sin(2\pi - 2A) = -\frac{1}{2}R^2 \sin 2A.$$

Therefore, still the conclusion could be derived by imitation of the above approach.

How did the student discover such an ingenious proof? He said he was inspired by the proof of another area formula of triangle

$$S = S_{\Delta ABC} = \frac{1}{2}(a + b + c)r,$$

where a, b, c are the opposite sides of angles A, B, C, and r is the radius of inscribed circle of ΔABC. He thought of "calculating the total area by summing the parts". Considering that the conclusion was related to the circumcircle radius R, it is more than natural to divided ΔABC into ΔAOB, ΔBOC and ΔCOA.

He was also aware of discussing the situation of obtuse triangle. After completing his reasoning, he made an analogy of this case with the derivation of area formula $S = \frac{1}{2}(b + c - a)r_a$, where r_a is the radius of escribed circle of ΔABC with respect to A.

This method shows a kind of unexpected yet reasonable artistic charm.

Definitely, many competition problems that have been studied and discussed are still highly "open in approaches". Other approaches or non-trivial relevant problems may also be obtained under the "attack" of new solution seekers (which is of great educational value although less so in academic value).

Meanwhile, there are plenty of explorative problems in competition mathematics, of which the target of problem-solvers is uncertain. As there is no ready method to follow, the solving process often requires thinking and exploration from multiple perspectives.

Therefore, although competition mathematics problems are different in content from open-ended mathematics problems, it is possible to partially achieve the effect of open-ended mathematics problem teaching for the development of gifted students' mathematical ability and creativity by taking advantage of competition mathematics problems.

6. Disputes over China's Competition Mathematics Education

Although quite a few Chinese and foreign mathematicians support and promote mathematics competitions by asserting their educational value, they also hold a very prudent attitude towards these competitions. At the very beginning of the launch of mathematics competitions, Luogeng Hua (1956a) had such a concern: "Will this work (mathematics competition) affect our school education negatively? Will it affect the students' overall development? It might happen if the job is not done properly." Nowadays the involvement of the "general public" and very young children in mathematics competitions have been disputed continuously, which echo Professor Hua's concerns to a certain extent.

In respect of universalization of competition mathematics education, opinions are widely divided among Chinese scholars. Some argue that competition mathematics education should be oriented to the entire high school student population in order to inspire creative thinking and cultivate problem-solving ability. On the other hand, others argue that competition mathematics education should only apply to a small group of students (e.g., 5%) and should have the scale of their effects controlled. To use an analogy, sports for the general public are for keeping fit, whereas sports practiced by athletes aim to achieve excellency and breakthroughs. Likewise, in terms of the function of popularization, competition mathematics can be targeted at a large population of students. But in terms of talent identification and selection, competition mathematics is only suitable for a small group of students with a strong interest and aptitude for mathematics. Krutetskii (1976) studied three groups of students with different mathematical abilities. Analysis showed that students with mediocre mathematical ability needed more time and efforts than students with strong ability to make mathematical achievements; they tended to feel very troubled when solving new types of mathematical problems and needed assistance to understand general methods; only through rote practice is it possible for them to shorten the reasoning process. Students with strong mathematical ability could work intensively on mathematical activities for a long period of time, without exhibiting any tiredness. On the contrary, mathematically weak students were prone to feeling more tired after a short period of time in studying mathematics than in studying other subjects. Thus, it can be concluded that it is against the interests of education to require the involvement of too many students in high-level mathematical training.

With regard to the involvement of younger children in mathematics competitions, actually younger-grade award winners are by no means rare in the IMO of previous years. One example is the Fields Medal winner Terence Tao who obtained his IMO gold medal when he was just 12 years old. The launch of mathematics competitions is instrumental for discovering the mathematical gift of such child prodigies. As the eminent mathematician Kolmogorov wrote in his preface to the book *Moscow Mathematical Olympiad* (Гальперин and Толпыго, 1986): "At the very beginning, the Moscow Mathematical Olympiad was only for students of Grades 9 and 10; it started to invite students of Grades 7 and 8 from 1940. This change in age of the participants was because students of such grades had already started to show their interest and talent in mathematics." However, the involvement of overly young students has shortcomings and disadvantages. Kolmogorov further expressed that, "although it is possible to invite even younger competitors, we could not help but notice that most of these Grades 5 and 6 students who had solved competition problems later lost their problem-solving ability and interest in mathematics as they progressed to higher grades." It shows that Kolmogorov had his reservations about the involvement of younger participants. According to the research of mathematicians and psychologists, such as V. A. Krutetskii's tracking study on the 26 gifted children in mathematics (1976), M. A. Clements' case study on the child Terence Tao (1984), to name a few, talent in mathematics is formed in early childhood, and gifted children demonstrate a remarkable talent for mathematics and learning speed. So, how to make such students develop their mathematical ability while still maintaining their interest in mathematics is an issue deserve attention when conducting mathematics competitions.

In recent years, as more and younger children are attending competition mathematics lessons, a substantial improvement in problem solving abilities among younger participants has been observed. On the other hand, the difficulty and complexity of the problems have increased and are sometimes out of touch with the classroom teaching. Take the first item in the 2014 National Middle School Mathematics Competition, for example (Xu Shenglin, 2015):

Example 6.1: Let x, y be integers such that

$$\left(\frac{1}{x}+\frac{1}{y}\right)\left(\frac{1}{x^2}+\frac{1}{y^2}\right)=-\frac{2}{3}\left(\frac{1}{x^4}-\frac{1}{y^4}\right).$$

The possible values of $x+y$ are ( ).

(A) 1 (B) 2 (C) 3 (D) 4

The essence of the problem is to find out, by observation, that $\frac{1}{x}+\frac{1}{y}$ and $\frac{1}{x^2}+\frac{1}{y^2}$ at the left-hand side of the equation are two factors of $\frac{1}{x^4}-\frac{1}{y^4}$ at the right-hand side, thus the problem could be simplified; yet after that, a quadratic indeterminate equation should be solved, whether x, y are zero also be considered and the case $\frac{1}{x}+\frac{1}{y}=0$ cannot be neglected. There are too many details to be considered, and the validity of the option set is also questionable. As the first item in a test, would it not be more advisable to lower the difficulty and try to design problems with the purpose of promoting in-class instruction?

There is another point of view that "when the new content that appears in mathematics competitions are familiar to and within the grasp of high school students and teachers, the competitions would then have successfully fulfilled their Olympiad mission — to integrate into high school mathematics, or in other words, to have popularized and disseminated mathematical knowledge" (Chen Chuanli and Zhang Tongjun, 2013, p. 15). Actually, two challenges are behind the successful popularization. For one, more and more extracurricular contents will be integrated into in-class instruction, resulting in excessive information for students to learn, thus adding to their burden in schoolwork, thereby raising the problem of controlling this tendency. For another, as the contents of competition mathematics become increasingly familiar with the general public, the function of selecting talents may be lost as it becomes more difficult to differentiate between participants, so the cooperation of experts in various fields including mathematicians is required in order to prepare the test problems conscientiously, to ensure the novelty and validity in the problems given. In competition mathematics, we should by no means propose those complicated problems in such fields that are known to all or systematically studied.

In conclusion, China's competition mathematics education serves as a supplement and enhancement for in-class instruction. However, much practices and explorations in the long run are essential for kicking a balance between popularization and selection, as well as public education and gifted education.

References

[1] Chen, Chuanli and Zhang, Tongjun (2013). *Competition Mathematics Course* [in Chinese], 3rd edn. Beijing: Higher Education Press.

[2] Clements, M. A. (1984). Terence Tao. *Educational Studies in Mathematics*, 15(3), 213–238.

[3] Coaching Staff of National Team (2014). *Towards IMO: A Collection of Olympiad Mathematics Problems* [in Chinese]. Shanghai: East China Normal University Press.

[4] Dai, Zaiping (2002). *Open-Ended Problems: A New Pattern of Mathematics Teaching* [in Chinese]. Shanghai: Shanghai Education Press.

[5] Department of Secondary School Mathematics of the People's Education Press (2006). *Mathematics Review Guide for College Entrance Examination* [in Chinese]. Beijing: People's Education Press.

[6] Feng, Yaofeng (2006). *Theory and Practice of Olympiad Mathematics Education* [in Chinese]. Shanghai: Shanghai Education Press.

[7] Гальперин, Г. А. and Толпыго А. К. (1986). *Московские Математические Олимпиады* [in Russian]. Moscow: Просвещеиии.

[8] Ge, Jun. (2012). *Equation and System of Equations* [in Chinese]. Shanghai: East China Normal University Press.

[9] Hua, Luogeng (1956a). Mathematics competition to be held in our country [in Chinese]. *Shuxue Tongbao (Bulletin des Sciences Mathematics)*, (1), 1–2.

[10] Hua, Luogeng (1956b). Writing after the mathematics competition in 1956 [in Chinese]. *Shuxue Tongbao (Bulletin des Sciences Mathematics)*, (6), 1–2.

[11] Krutetskii, V. A. (1976). *The Psychology of Mathematical Abilities in School Children.* Chicago: University of Chicago Press.

[12] Liu, Pengzhi (1995). *Mathematics Textbook of Hua Luogeng Mathematics School (Grade 10)* [in Chinese]. Beijing: Encyclopedia of China Publishing House.

[13] MOE of PRC (Ministry of Education of the People's Republic of China) (2000). *Mathematics Syllabus for Full-time High Schools (Revised Experimental Edition)* [in Chinese]. Beijing: People's Education Press.

[14] MOE of PRC (Ministry of Education of the People's Republic of China) (2003). *National Mathematics Curriculum Standard for High Schools (Experimental)* [in Chinese]. Beijing: People's Education Press.

[15] MOE of PRC (Ministry of Education of the People's Republic of China) (2012). *Mathematics Curriculum Standard for Compulsory Education (2011 Edition)* [in Chinese]. Beijing: Beijing Normal University Press.

[16] NCTM (National Council of Teachers of Mathematics) (2000). *Principles and Standards for School Mathematics*. Reston: NCTM.

[17] PC of CMS (Popularization Committee of Chinese Mathematical Society) (2006a). Syllabus for high school mathematics competition [in Chinese]. *Secondary School Mathematics*, (12), 3–5.

[18] PC of CMS (Popularization Committee of Chinese Mathematical Society) (2006b). Syllabus for middle school mathematics competition [in Chinese]. *Secondary School Mathematics*, (12), 2–3.

[19] Shan, Zun (2007). *Study of Problem Solving* [in Chinese]. Shanghai: Shanghai Education Press.

[20] Sheng, Zhirong and Zhou, Chao (2010). Review on the researches of international mathematics gifted education [in Chinese]. *Journal of Zhejiang Education Institute*, (3), 9–15; 60.

[21] Sun, Ruiqing (1998). *National Middle School Mathematics Competition Tutorial (Grade 9)* [in Chinese]. Beijing: Peking University Press.

[22] Sun, Ruiqing and Hu, Datong (1994). *Introduction to Olympiad Mathematics Education* [in Chinese]. Beijing: Peking University Press.

[23] Tang, Shengchang and Feng, Zhigang (2011). Early identification and education of mathematics talents: A case study [in Chinese]. *Shuxue Tongbao (Bulletin des Sciences Mathematics)*, 50(3), 11–16.

[24] Xiong, Bin and Feng, Zhigang (2014). *Mathematics Olympaid Tutorial (Grade 10)* [in Chinese]. 6th edn. Shanghai: East China Normal University Press.

[25] Xiong, Bin and He, Yijie (2012). *Problem Solving Methods and Strategies in High School Competition Mathematics* [in Chinese]. Shanghai: East China Normal University Press.

[26] Xu, Shenglin (2015). National middle school mathematics competition in 2014 [in Chinese]. *Secondary School Mathematics*, (1), 21–25.

[27] Zhu, Huawei (2009). *From Mathematics Competition to Competition Mathematics* [in Chinese]. Beijing: Science Press.

East China Normal University Scientific Reports
Subseries on Data Science and Engineering

Published (continued from page ii)

Vol. 10 *Clustering and Outlier Detection for Trajectory Stream Data*
by Jiali Mao (East China Normal University, China),
Cheqing Jin (East China Normal University, China) and
Aoying Zhou (East China Normal University, China)

Vol. 9 *Concurrency Control and Recovery in OLTP Systems: High Scalability and Availability*
by Peng Cai (East China Normal University, China),
Jinwei Guo (East China Normal University, China) and
Aoying Zhou (East China Normal University, China)

Vol. 8 *Network Data Mining and Analysis*
by Ming Gao (East China Normal University, China),
Ee-Peng Lim (Singapore Management University, Singapore) and
David Lo (Singapore Management University, Singapore)

Vol. 7 *Time-Aware Conversion Prediction for E-Commerce*
by Wendi Ji (East China Normal University, China),
Xiaoling Wang (East China Normal University, China) and
Aoying Zhou (East China Normal University, China)

Vol. 6 *Discovery and Fusion of Uncertain Knowledge in Data*
by Kun Yue (Yunnan University, China), Weiyi Liu
(Yunnan University, China), Hao Wu (Yunnan University, China),
Dapeng Tao (Yunnan University, China) and
Ming Gao (East China Normal University, China)

Vol. 5 *Review Comment Analysis for E-commerce*
by Rong Zhang (East China Normal University, China),
Aoying Zhou (East China Normal University, China),
Wenzhe Yu (East China Normal University, China),
Yifan Gao (East China Normal University, China) and
Pingfu Chao (East China Normal University, China)

Vol. 4 *Opinion Analysis for Online Reviews*
by Yuming Lin (Guilin University of Electronic Technology, China),
Xiaoling Wang (East China Normal University, China) and
Aoying Zhou (East China Normal University, China)

East China Normal University Scientific Reports

CPSIA information can be obtained
at www.ICGtesting.com
Printed in the USA
JSHW020506260121
11221JS00001B/4

9 789814 713931